Network Theory and Agent-Based Modeling in Economics and Finance

Anindya S. Chakrabarti · Lukáš Pichl ·
Taisei Kaizoji
Editors

Network Theory and Agent-Based Modeling in Economics and Finance

Springer

Editors
Anindya S. Chakrabarti
Indian Institute of Management Ahmedabad
Ahmedabad, Gujarat, India

Lukáš Pichl
College of Liberal Arts
International Christian University
Tokyo, Japan

Taisei Kaizoji
Division of Social Sciences
International Christian University
Mitaka, Tokyo, Japan

ISBN 978-981-13-8318-2 ISBN 978-981-13-8319-9 (eBook)
https://doi.org/10.1007/978-981-13-8319-9

This Springer imprint is published by the registered company Springer Nature Singapore Pte Ltd.
The registered company address is: 152 Beach Road, #21-01/04 Gateway East, Singapore 189721, Singapore

Contents

Complexity and Emergence: A New Paradigm for Economic Modeling

Anindya S. Chakrabarti , Lukáš Pichl and Taisei Kaizoji

1 Introduction

Just a little more than one decade back, we have witnessed the biggest economic and financial crisis that is known as the *great recession*. This was a once in a lifetime event that led to an abrupt awakening to the academic, corporate, and the policy world alike. Alan Kirman captured the state of uncertainty very succinctly through the phrase "*The economic crisis is a crisis for economic theory*" (Kirman 2010). Reflecting the confusion during that time, then European Central Bank head Trichet famously made the following statement (Trichet 2010):

> *When the crisis came, the serious limitations of existing economic and financial models immediately became apparent. Arbitrage broke down in many market segments, as markets froze and market participants were gripped by panic. Macro models failed to predict the crisis and seemed incapable of explaining what was happening to the economy in a convincing manner. As a policy-maker during the crisis, I found the available models of limited help. In fact, I would go further: in the face of the crisis, we felt abandoned by conventional tools.*

Naturally, this episode led to a proliferation of different ideas and modeling techniques in economics and finance. After the dust settled and the US (and the world) economy came back close to the normal growth path, it is important to take a look back and document the new developments that are useful for understanding the economic dynamics. In this book, we have collected a series of articles which possess

A. S. Chakrabarti (✉)
Economics Area, Indian Institute of Management, Vastrapur, Ahmedabad, Gujarat 380015, India
e-mail: anindyac@iima.ac.in

L. Pichl · T. Kaizoji
Graduate School of Arts and Sciences, International Christian University, Tokyo, Japan
e-mail: lukas@icu.ac.jp

T. Kaizoji
e-mail: kaizoji@icu.ac.jp

A. S. Chakrabarti et al. (eds.), *Network Theory and Agent-Based Modeling in Economics and Finance*, https://doi.org/10.1007/978-981-13-8319-9_1

new qualitative as well as quantitative features and provide new directions of modeling and making connections between large volume of economic data. Our focus in this volume is exclusively on a new paradigm for economic modeling, which explicitly models complexity of economic interactions and analyzes the corresponding emergence of system-level features.

Before we get into the specifics, let us first describe the core of standard economic models. In particular, we start from macroeconomic models since this literature received the most attention during and after the crisis. The prevailing structure depends critically on two fundamental assumptions, rationality with unbounded cognitive resources to deploy rationality and lack of interactions among granular agents. An informative example would be a standard real business cycle model. A generic model would consider forward-looking households who performs utility maximization over an infinite stream of consumption, firms who maximize profit by optimally choosing labor and capital; and the model would be closed using the general equilibrium channel that demand equals supply. Although there were many other mechanisms added (like imperfect competition, frictions in nominal prices), this formed the core of the modeling exercise. As such, the methodology is fairly useful and under suitable calibration, can indeed replicate properties of aggregate time series. However, limitations were twofold. One, idiosyncratic effects of aggregate shock can not be understood from this kind of models and two, they depended too much on exogenous shocks to produce fluctuations in all variables. These models themselves had very little in-built shock propagation/generation mechanism. Thus, the possibility of building up of risk and then eventual collapse when the accrued risk goes beyond a tolerable threshold is completely absent. Notably, financial sectors were largely absent in such framework.[1] Hence, this whole class of models would not be even useful for understanding effects of financial shocks like the one that affected the world economy in 2007.

This inability of the macroeconomic models led to a fundamental question: under what kind of realistic scenarios, spillovers from financial market to the real economy would take place? The question is not about explaining a past event, but is about prediction. In order to do that we need quantification of dis-aggregated risk that leads to an aggregate shock to the economy. This necessitates analyses of the economic entities at a much more granular level than what is found in the aggregative models. Thus, *granularity* and *networks* became two important tools to analyze economic phenomena. The mainstream economics also embraced the ideas very quickly and adopted the tools and methodology fairly rapidly. Two important papers (Gabaix 2011; Acemoglu et al. 2012) made it very clear that granular entities and their linkages contribute substantially to the aggregate behavior. Later literature builds on the same idea. However, economists were not the only ones who were interested in this development. Researchers from different domains including physics, computer

[1]For the sake of completeness, we should mention that there was a literature on incorporating financial sectors in a DSGE set up (two important papers are Kiyotaki and Moore 1997 and Bernanke et al. 1999).

science, applied mathematics, and other technical domains analyzed problems of similar nature with very different kinds of tools.

1.1 Behavioral Rules Versus Optimization: Why Agent-Based Models?

As we have stated above, most of the standard economics models implicitly endow unbounded cognitive resources to the agents to carry out complicated optimization problems. For example, a basic real business cycle model would assume that households receive an infinite sequence of shocks and they are aware of the distributional features of the process (but cannot foresee the exact path). Then they would solve an infinite horizon optimization problem with infinite number of budget constraints (one or more for each period) and come up with the famous Euler equation (it is customary to impose strong assumptions like additive separability of the utility function).[2] A standard response to the critique that the required computational ability is immense, is that it simplifies the calculations and reduces ad-hocism in terms of lifespan of the household. One can justify this strong assumption based on the positivist argument (Friedman 1953) that the results matter, not the assumptions. Thus, for modeling business cycle dynamics, it is fine to use such an assumption as long as it replicates the targeted behavior of the macro variables.

However, there are two problems with this approach. One, replication is not the same as explanation. Even if we ignore that distinction, the second problem is that its usability outside that very narrow domain is questionable. Often, such models would construct value functions over a series of state variables to describe the equilibrium behavior of agents. However, optimization is costly. Even solving those models require significant computational power (for convergence of the value/policy function). Thus, it is difficult to see how common people can perform such a calculation across the economy. Not only the informational requirements are huge, so are the computational requirements. Economics literature noticed such deficiencies and there are some proposed solutions, notably in the form of *rational inattentions* (started by Christopher Sims (see e.g., Sims 2003), see the series of papers by Filip Matějka (e.g., Matějka and McKay 2015) and *bounded rationality* by Gabaix (e.g., Gabaix 2014)). In contrast, the agent-based models take irrationality as given and instead of explaining why there is irrationality, starts with agents endowed with behavioral rules and try to figure out what happens when a million of such agents interact with each other.

The key difference from the mainstream model is the mode of interactions across granular agents. Standard models use complicated rules of trading over population with zero or minimal interactions, often characterized by complete homogeneity among the agents. A complementary view is to consider simple interaction rules

[2]The *Euler equation* when tested separately in data, does not really hold true. See, for example, the paper by Carroll (2001).

with nontrivial individual interactions, which leads to emergent macro behavior. Such behavioral rules provide many degrees of freedom and reduces computational burdens drastically. However, it comes at a cost of lacking the feedback effect coming from optimization of the agents, i.e., such models are susceptible to the *Lucas critique*. The second distinctive feature of such models is that these models are not micro-founded as opposed to the standard economic models where macro behavior is understood from a clear microeconomic optimization exercise. Here, the focus is more on the emergence of macro-level features from simple and often mechanical, micro-level behavioral rules.

1.2 Networks as Building Blocks of Economic and Financial Complexity

The basic building block of emergent phenomena are microscopic interactions. This idea reflects the observation that "By modeling one ant, it is almost impossible to understand how they build an ant colony" (Kirman 2017). In order to model interactions between granular entities, we need to define the mode of interaction as well. Networks as mathematical objects provide an abstract framework to describe many different types of linkages across economic entities. We would like to point out that networks as a paradigm do not provide a new theory per se. It is more of a descriptive tool. Compared to most of the other new frameworks, it is less contentious and has gained popularity in the mainstream economics literature.

However, because of the close association of networks and graph theory, it also provides a nice mathematical tool to model the interactions across entities. In recent times, the relationship between economic quantities and different topological measures (like centrality) have attracted a lot of attention. Especially, in the financial markets' context, a significant amount of work has been done on analyzing how shock propagation occurs from the core to periphery of the financial networks. The asset market interlinkages, banking networks, corporate networks among others have been under study for finding how the linkages affect the economic outcomes.

2 Interactive Models with Boundedly Rational Agents

Emergence of coordination among a large number of agents is an explicit demonstration of complex behavior exhibited by economies and societies. The impact of coordination is context-dependent. Lack of coordination could bring down political, social, and economic systems. However, a coordinated move in the financial markets might create bankruptcy. Therefore, it has been an important question that what causes coordination (or lack thereof). Typically, economists would start from a rational agent framework to analyze coordination. Aymeric Vie (chapter "Information Selection Efficiency in Networks: A Neurocognitive-Founded Agent-Based Model") starts from the opposite end of the spectrum on rationality. He analyzes a class of neuro-cognitive models based on agents with bounded rationality to model emergence of coordination across agents interacting through a network. The objective of the agents is to estimate signals from noisy signals obtained through local connections of the agents. The results show that the stability of aggregative opinion in the society is dependent on the network structure. More importantly, the aggregate behavior of society is substantially different for subsets of agents separated in modules.

Iwanaga and Kubo (chapter "Diversification by Self-reinforcement of Preferences and Change of Interaction Type") introduce coordination games along with their counterparts, anti-coordination games in a single framework. In their framework, agents can play either of the two types of games. They analyze stability of the resulting collective behavior and show that in a pure world of agents playing only one type of the game, the aggregate statistics are stable with respect to small perturbation in terms of errors in strategy choice. Their paper broadly allows to study reinforcement of preferences due to repeated interactions of agents in a stylized network.

Ruiz-Buforn et al. (chapter "Price Distortions and Public Information: Theory, Experiments and Simulations") analyze the effect of public information in price distortions in an agent-based framework. They introduce two types of agents with differential learning skills: naive and sophisticated learners. They show that when the public information is correct, it helps to coordinate market-wide activities whereas uninformative information typically destabilizes the market.

Alessio Emanuele Biondo (chapter "Order Book on Financial Networks") analyzes the connections between a social interaction network of agents and networked financial market. He introduces an order book model of where agent interacts with each other through a network and due to limited learning based on information flow on the network, they might show a very strong herding behavior. Spillover of such behavior is seen on the financial variables.

Yagi et al. (chapter "Detection of Factors Influencing Market Liquidity Using an Agent-based Simulation") propose a simulation-based method to analyze the availability of liquidity in a financial market populated by boundedly rational investors. They analyze multiple factors like volume, tightness, resiliency, and depth and show that volume has an opposite effect with respect to the remaining three variables.

Ishikawa et al. (chapter "Macroscopic Properties in Economic System and their Relations") propose an explicit agent-based model of firm–firm interactions and analyzes the emergence of macroeconomic properties. They show that the emergent properties are fairly robust to micro-level details of the firms as well as the micro-level descriptions of the linkages across the firms. This indicates that macro-level properties may not be understood from studying agent behavior, which provides a counterexample to microfoundation of macro models.

Venkat Venkatasubramanian (chapter "How Much Income Inequality is Fair? Nash Bargaining Solution and its Connection to Entropy) introduces and analyzes a new model of economic inequality through interaction between granular agents. He introduces a novel approach through modeling interactions via Nash bargaining solution. He provides an interesting perspective on a massively important policy issue: How much inequality is fair? He provides an additional idea that the income distribution has lognormal characteristics, which is also empirically documented earlier in the literature on income distribution.

Chakrabarti and Lahkar (chapter "A Limit Cycle View of Inequality") analyze a model with myopic agents endowed with evolutionary dynamics. They consider a set of buyers and sellers and build on a search model where buyers are looking for technology of different qualities. Sellers have higher cost for producing higher quality technology. Under certain assumptions, they show that there would be a limit cycle in the aggregate variables. In particular, inequality oscillates as a function of business cycle.

Jiří Kukačka (chapter "Simulated Maximum Likelihood Estimation of Agent-Based Models in Economics and Finance") provides a methodology through simulated maximum likelihood estimator to find credible parameter estimates of agent-based models. The results indicate that the methodology is suitable for linear as well as nonlinear interactions. Simulations also show that the methodology robustly identifies true parameter values from synthetic multivariate time series.

Yen et al. (chapter "Emergent Urban Morphologies in an Agent-Based Model of Housing") study urban morphology based on a spatial agent-based model. They picked up a classic question in economic geography: what is the process of formation of cities and how do they evolve over time? They propose and analyze a model with utility maximizing households whose key variables under consideration are rent and transportation costs. They show through simulations that urban agglomerations emerge as a result of complex interactions between economic variables impacting the households as well as other amenities.

Akira Shimada (chapter "The Transferability of Human Capital, the Brain Drain and the Brain Gain") analyzes the role of human capital in the development process of a country. Brain drain and brain gain are major issues in the context of developing countries as they have major impacts on the countries' future growth path. The author shows that the net effects would depend on multiple factors comprising wage dispersion and transferability of skills.

Menon et al. (chapter "Is Life (or at Least Socioeconomic Aspects of It) Just Spin and Games?") provide an interesting parallel between strategic behavior that economic models use and rule-of-thumb behavior that agent-based models use. Generally, there is a clear bifurcation in terms of acceptability of these modeling paradigms in these two different strands of literature. The authors have an exciting proposal for building a new literature on statistical mechanics using game-theoretic structure.

3 Financial and Economic Linkages

In the second part of the book, we focus more on the descriptive part of the economic and financial linkages. Ladley and Tran (chapter "An Agent-Based Model of BCVA and Systemic Risk") introduce an agent-based model of systemic risk in an interconnected financial market. They show that a straightforward way to increase the stability of financial markets would be to price credit worthiness of counterparties in a proper fashion.

Kaneko and Hisakado (chapter "Additional Default Probability in Consideration of Firm's Network") analyze credit risk arising out of firm–firm linkages. They propose a methodology to characterize the default probability of firms based on their neighbors' characteristics. They also derive some implications from pricing perspective as well.

Sasidevan and Bertschinger (chapter "Systemic Risk: Fire-Walling Financial Systems Using Network-Based Approaches") provide a summary of quantitative characterization of systemic risk. In presence of modern financial assets which creates entanglements and makes the market interdependent, it is becoming more and more important to construct useful measures of risk, where the classical methods fail badly. They provide a perspective on complex networks of assets as well as financial institutions.

In the statistical finance literature, a whole new set of tools and techniques have been developed to analyze large volume of financial data with the focus of filtering large data sets to extract the underlying patterns. Bhadola and Deo (chapter "Spectral and Network Method in Financial Time Series Analysis: A Study on Stock and Currency Market") discuss a number of such tools mostly focusing on the techniques that depend on eigenspectra analysis and also apply them on financial data.

Hashimoto and Kurahashi (chapter "Research on Loss-Absorption of Financial Group") analyze loss absorption across financial institutions and how that relates to systemic risk. They introduce a model with balance sheet-driven crisis and study spillover of distress across financial institutions. Their simulation exercises suggest that injection of external funds into a market should be dependent of the network characteristics. In particular, highly central entities might contribute substantially to distress spillover.

Along with quantification of systemic risk based on network characteristics, another major development took place in the recent times in the form of applications of machine learning techniques in data analysis to predict trends. Upreti et al. (chapter "Knowledge-Driven Approaches for Financial News Analytics") describe a rich set of tools and techniques to conduct news analytics in the context of financial markets. They discuss and compare different approaches to textual analysis and sentiment analysis based on many different sources of financial texts.

Fernando Garcia Alvarado (chapter "Network-Based Policies Versus Tax Evasion") analyzes a social planner's problem of reducing tax evasion of economic agents placed on a given social network with homophilic linkages. He devises an optimal income tax scheme in the presence of agents who remember the previous audits and the corresponding frequencies, which leads to learning by agents by pooling in local information. In this framework, a random scheme seems to work as well as a search-based scheme which treats the agents as homogeneous. Interestingly, an income-based audit policy identifying key sectors and auditing them, has superior efficiency in terms of raising the perceived likelihood to be audited.

Borkotokey et al. (chapter "Network Games: The Cooperative Approach") survey games on networks. In particular, they provide an in-depth analysis of cooperative and bi-cooperative games. They discuss possible applications in social as well as technological networks.

4 Summary and Outlook

In this book, we have collected a number of articles on network theory and agent-based models with a focus on economic and financial phenomena. Fundamentally, the goal is to bring together scientists working on distinct fields to provide perspectives so that cross-fertilization of ideas take place which might lead to the birth of newer tools and techniques. While thinking of a large-scale, evolving complex system, it will not be very useful to model it with only a limited set of tools. Therefore, our idea is not to necessarily substitute standard economic and financial theory and econometrics, but more in the line of complementing them. We would like to refer to an article by Menon, Sasidevan, and Sinha in this book, which presents an interesting discussion on merging statistical mechanics and game theory. This idea resonates nicely with the theme of this book and we hope such amalgamation will actually take place in the near future.

References

Acemoglu D, Carvalho VM, Ozdaglar A, Tahbaz-Salehi A (2012) The network origins of aggregate fluctuations. Econometrica 80(5):1977–2016
Bernanke BS, Gertler M, Gilchrist S (1999) The financial accelerator in a quantitative business cycle framework. In: Handbook of macroeconomics, vol 1, pp 1341–1393
Carroll CD (2001) Death to the log-linearized consumption euler equation!(and very poor health to the second-order approximation). Adv Macroecon 1(1)
Friedman M (1953) The methodology of positive economics
Gabaix X (2011) The granular origins of aggregate fluctuations. Econometrica 79(3):733–772
Gabaix X (2014) A sparsity-based model of bounded rationality. Q J Econ 129(4):1661–1710
Kirman A (2010) The economic crisis is a crisis for economic theory. CESifo Econ Stud 56(4):498–535
Kirman A (2017) The economy as a complex system. In: Economic foundations for social complexity science, pp 1–16. Springer
Kiyotaki N, Moore J (1997) Credit cycles. J Polit Econ 105(2):211–248
Matějka F, McKay A (2015) Rational inattention to discrete choices: a new foundation for the multinomial logit model. Am Econ Rev 105(1):272–98
Sims CA (2003) Implications of rational inattention. J Monet Econ 50(3):665–690
Trichet J-C (2010). Reflections on the nature of monetary policy non-standard measures and finance theory. Opening address at the ECB Central Banking Conference, Frankfurt, 18 November 2010. Accessed 28 July 2018

Interactive Models with Boundedly Rational Agents

Information Selection Efficiency in Networks: A Neurocognitive-Founded Agent-Based Model

Aymeric Vié

Abstract Heterogeneity in beliefs in financial markets constitutes a major obstacle to coordination. Beliefs being heavily driven by information, efficient coordination and selection of information in financial markets are interconnected matters. The dialectic between different mechanisms at play, behavioral, cognitive, and interactive, has received little investigation. However, it appears fundamental in opening the black box of agents' decision-making in the context of information selection. We introduce an agent-based model of information signal selection. In line with the Agent Zero neurocognitive-founded model (Epstein in Agent zero: toward neurocognitive foundations for generative social science. Princeton University Press 2014), agents are endowed with deliberative, affective, and social modules, through which they acquire disposition to adopt a signal. Agents receive signal values and estimate their variance with memory, i.e., ability to remember past signal values; they emotionally react to shocks and transmit individual dispositions through social networks. Information selection efficiency and opinion volatility are studied activating different combinations of cognitive modules and considering various network topologies. Human cognition is significantly more complex. Yet, this simple model generates a rich set of behaviors, and complex general mechanisms of information selection emerge. Simulation results obtained with novel model exploration methodologies outline that the impact of network structure and of memory abilities on market coordination, information selection efficiency and opinion stability appear to vary in magnitude and in sign depending on the combinations of cognitive modules considered. The behavior of the aggregate cognition framework appears to be different

---

Electronic supplementary material The online version of this chapter (https://doi.org/10.1007/978-981-13-8319-9_2) contains supplementary material, which is available to authorized users.

A. Vié (✉)
Paris School of Economics, Paris, France
e-mail: aymericvie78@gmail.com

New England Complex Systems Institute, Cambridge, MA, USA

Sciences Po Saint-Germain-en-Laye, Saint-Germain-en-Laye, France

A. S. Chakrabarti et al. (eds.), *Network Theory and Agent-Based Modeling in Economics and Finance*, https://doi.org/10.1007/978-981-13-8319-9_2

from the aggregate behavior of its modules taken separately, and interactions between cognitive modules appear crucial to explain information selection at the micro-level.

Keywords Agent-based computational economics · Agent zero · Information selection · Network topology · Social contagion

1 Introduction

As dynamic economic systems can be considered as expectations feedback systems, any theory of expectation formation becomes a fundamental part of any economic model (Hommes 2001). The role of beliefs heterogeneity has been central in previous explanations of market fluctuations in the context of asset pricing. It has inspired novel approaches of heterogeneous agent models in finance using adaptive beliefs systems (Brock and Hommes 1998, 1999; Chiarella and He 2001; Gaunersdorfer 2000; Lux and Marchesi 1999) fueled with heterogeneous expectations strategies and heuristics (Arthur et al. 1996; Chiarella and He 2001; Branch and McGough 2009; Dosi et al. 2017) or higher order thinking (Allen et al. 2006; Mauersberger and Nagel 2018). These models reviewed by Hommes (2006, 2013) have achieved significant progress in reproducing real financial markets stylized facts, giving birth to endogenous bursts, bubbles, and crashes related to beliefs changes. Studying expectations formation, these models have pushed toward more realistic approaches of financial markets and human behavior, putting an emphasis in behavioral dimensions of heterogeneous agent models, as well as social physics (Michard and Bouchaud 2005).

However, likewise asset prices are driven by expectations, the result of the expectation formation processes depends on the information these processes rely on. The differences in opinions or expectations, may have different sources: heterogeneous information and homogeneous processing, homogeneous processing and heterogeneous processing, or both heterogeneous information and processing. Heterogeneity in beliefs may arise from heterogeneous expectations strategies that yield different anticipations with the same information (Arthur et al. 1996), but also from homogeneous expectations strategies fueled with different information (Allen et al. 2006), or even both at the same time (Vié 2018). Ensuring great performance in the market thus requires both adequate expectation strategies and reliable information. In situations of uncertainty, when information is short lived or when its value understood as "forecasting reliability" is not known, selecting the "right" information can be tricky for financial agents, especially when several signals of unknown value are simultaneously provided. Through which channels does information selection intervene?

1.1 Emotions and Economic Decision-Making

Emotions are recognized to play a fundamental role in opinion dynamics within economic theory, notably in the context of decision-making. Individual panic has often been indicated as a major cause of financial crisis burst and transmission. The role of collective learning and collective knowledge in order to achieve coordination between agents have been central in the literature, as Orléan (1995) noted. Emotions and the associated learning play a crucial role in the approach of individual rationality of Kirman (2010). See additionally Loewenstein (2000) for a review of the role and importance of emotions in economic theory and economic behavior. The role of emotions in financial decision-making is recognized enough to have been studied in another discipline, e.g., by psychologists (Bracha and Weber 2012). Faith, trust, uncertainty, desire for predictability and control, (over)confidence, are complex emotions that are yet to be fully understood. Yet, their role in major opinion shifts, financial panics in economics, is recognized. Reinhart and Rogoff (2009) wrote on this topic the famous *This time is different: Eight centuries of financial folly*, while Akerlof and Shiller (2010) wrote *Animal spirits: How human psychology drives the economy, and why it matters for global capitalism*. Emotional responses to shocks, events or experience hence constitutes a relevant channel for information selection.

1.2 Deliberative Heuristics

However, markets do not appear to be only governed by emotions. Affective responses to various phenomena, and more generally, all kinds of information source, contribute to the formation of expectations strategies, i.e., the way information is processed in economic decision-making in financial markets. Dosi et al. (2017) extensively review the origin of heuristics, and their formation through learning. Facing uncertain environments, agents may receive imperfect signals on its state and attempt to infer from these sparse pieces of information the actual features of the environment. A large portion of the literature in opinion dynamics leads back to Bayesian theory (Bernardo and Smith 2009). The inference process faces important limitations in economies, notably bounded rationality, sample selection bias, and error. In the contribution of Epstein (2014) introducing a neurocognitive- founded model of cognition, a deliberative inference mechanism proceeds by local sampling. Agents receive pieces of information, with which they can form probability estimates of variables of interest. The memory of the agents become crucial in creating depth of reasoning to improve their estimates of relevant variables, e.g., in our context of information selection, the variance of the provided information signals. Receiving over time values drew from a distribution with unknown variance, agents estimate the variance of each information channel. Identifying a more precise (resp. more volatile) information sources drives disposition to adopt (resp. to reject) the corresponding information source. The memory in this framework indicates the number of values the agent will remember when forming her estimate: a higher memory band will intuitively improve the accuracy of the heuristics, and favor information selection efficiency, i.e., the ability of the agents to adopt the most precise signal.

1.3 Social Transmission

Finally, opinion acquisition and more specifically information selection may occur through social transmission and imitation. In situations of uncertainty, due to time constraints and heuristics limitations, the formation of agents' heuristics may rely on herding and more generally imitation rather than differential fitness as a primary evolutionary mechanism (Frank 2003). The first imitation mechanism to study beliefs adaptation in uncertainty was introduced by Kirman (1991) to explain contagion of opinions and crowd effects in financial markets by herd behavior. Orléan combined rational and imitative behavior under the notion of "mimetic rationality", developing models of mimetic contagion of information among investors in stock markets relying on processes of opinion formation (Orléan 1995). Vié (2018) considered the interaction between heterogeneous imitation decision rules in the context of information selection. Informationally, efficient markets were studied by Kristoufek and Vosvrda (2018) in the context of herding and minority from a physics spin model approach. Regarding imitation as an important engine of opinion dynamics, Epstein (2014) introduces in the Agent Zero cognitive model a social interaction mechanism based on dispositional contagion, moving beyond the "monkey see, monkey do" paradigm. Indeed, in simultaneous games, imitating another player's behavior is impossible as this behavior is not defined. On the contrary, prior to action, different physical signs, or any pieces of information, may contribute to identify not their resulting behavior, but their underlying dispositions to act. The mechanism of social transmission strongly depends on the structure of the connections between the agents, outlining the interest of network science approaches to model local social influences in the context of decision-making (Friedkin and Johnsen 1990). The structure of the network, i.e., its topology, is crucial in evaluating the propagation or percolation of opinions in a system (Amblard and Deffuant 2004; Vié and Morales 2019).

These three possible channels extensively studied by the literature on opinion dynamics in financial markets have been often considered separately, with little overlap. Yet, intuitively, they are strongly interconnected and often respond one to each other. Emotion-driven dispositions to adopt a given opinion may be enforced or going against the deliberative evidence, and both may interact with peers through social transmission, a mechanism of many forms. We present in this paper a simple model of information selection between two alternatives, where agents are endowed with a simple cognition architecture assembling these dimensions. The Agent Zero neurocognitive-founded model of Epstein (2014) associates these three main dimensions, reproducing stylized research results in psychology and cognitive sciences. See Epstein (2014) for a complete presentation of the Agent Zero framework and its connections with research in these disciplines. Essentially, this approach gives us a simplified yet realistic enough basis to study agents' behavior trying to design human-like cognitive architecture. We here derive it in our Agent Zero model of Information Selection. We adopt an agent-based perspective on the issue of information selection, emphasizing individual mechanisms of signal adoption, and the interactions between agents at the origin of collective dynamics. These important micro-level heterogeneity and interactions are adequately captured by the use of agent-based models (ABM).

Section 2 presents the Agent Zero Information Model. The structure of the information selection framework is introduced, and the deliberative, affective, and social modules are presented. The simulation procedures to explore the model behavior are detailed in Sect. 3. Section 4 presents and discusses the achieved information selection efficiency, and variance of opinions in various cognitive settings, considering different combinations of the three modules introduced. Section 5 summarize our findings, and attempts to provide a unified view of information selection through the different channels considered.

2 The Agent Zero Information Selection Model

2.1 *Information Signals*

Suppose two sources of information, denoted Blue and Red, exist. They can be asset evaluations, and the agents in the model aim at adopting the most precise of the two, in order to yield better economic performance. Both information signals have an equal and constant mean component and a Gaussian White noise term. A high variance White noise term for a signal corresponds to a noisy signal, while a low variance in the Gaussian White noise characterizes a more precise information source.

$$B = \mu + \varepsilon_B^2 \text{ with } \varepsilon_B^2 \sim WN(0, \sigma_B^2) \tag{1}$$

$$R = \mu + \varepsilon_R^2 \text{ with } \varepsilon_R^2 \sim WN(0, \sigma_R^2) \tag{2}$$

We further express the variance of signal R, σ_R^2:

$$\sigma_R^2 = \sigma_B^2 + \sigma_S^2 \tag{3}$$

σ_B^2 is the "base" level of variance in the signals, while σ_S^2 is denoted the "spread", the difference in variance across signals. While agents are interested in identifying the respective total variances of the signals, accounting for both the base and spread levels is relevant for the analysis. The efficiency of information selection between two signals through adoption of the most precise one does not only depends on the difference in variance between both (spread), but also on the base variance level considered across signals. In a very noisy environment, having a larger spread may be necessary to clearly identify the most precise signal, given that both exhibit high fluctuations. Equivalently, in a stable environment, only small differences in variances between signals may be necessary to adopt the least noisy one.

We denote as "information selection efficiency" the ability of the agents to adopt the most precise signal. Noise parameters σ_B^2 and σ_R^2 are set at the model initialization. The agents are stochastically assigned to either initial state blue or red with a given probability ρ. We restrict our analysis in this model to the binary case: each investor either adopts signal Blue, either chooses signal Red. The noise parameters are unknown to the agents. This illustrates the classical situation of an investor facing several, possibly contradicting, signals on the value of an asset, while the value, i.e., informational relevance of each signal is uncertain. This investor will attempt to infer what signal is the most relevant to describe the market state, by different means.

2.2 *Agent Cognition: The Deliberative Component*

In our variation of the deliberative component of the Agent Zero cognition model (Epstein 2014), agents receive a signal from each information source (blue and red). Noise terms are independently and identically distributed in the population, generating heterogeneity in signal reception.

Agents are endowed with limited memory abilities. They remember a given number of signal emissions according to a memory length parameter λ. λ is interpreted in our model the upper bound on the number of signal values they can remember. Then, agents estimate the variance of each signal based on their own memory sample. The comparison of the two estimated variances by agent i, denoted $\hat{\sigma}_{iB}^2$ and $\hat{\sigma}_{iR}^2$ generates incentives in favor or against adopting signal blue or red.

We denoted D_{iB} and D_{iR} the respective dispositions of agent i to select signal blue and red, arising from this deliberative cognitive module.

$$D_{iB} = \begin{cases} 1 & \text{if} \quad \hat{\sigma}_{iB}^2 < \hat{\sigma}_{iR}^2 \\ 0 & \text{if} \quad \hat{\sigma}_{iB}^2 = \hat{\sigma}_{iR}^2 \\ -1 & \text{if} \quad \hat{\sigma}_{iB}^2 > \hat{\sigma}_{iR}^2 \end{cases} \tag{4}$$

$$D_{iR} = \begin{cases} 1 & \text{if} \quad \hat{\sigma}_{iR}^2 < \hat{\sigma}_{iB}^2 \\ 0 & \text{if} \quad \hat{\sigma}_{iR}^2 = \hat{\sigma}_{iB}^2 \\ -1 & \text{if} \quad \hat{\sigma}_{iR}^2 > \hat{\sigma}_{iB}^2 \end{cases} \tag{5}$$

Accounting for bounded cognitive abilities, we impose a limit on the precision of the estimates of the variances $\hat{\sigma}_{iB}^2$ and $\hat{\sigma}_{iR}^2$. We introduce bounded reasoning abilities and fundamental uncertainty à la Heisenberg by setting the maximum number of decimal places on the agents' estimates of the variance.

2.3 Agent Cognition: The Affective Component

In the spirit of Epstein (2014), individual disposition to select a given signal varies with the emotional response associated to each. Emotional responses can be activated by observing asset crashes, peer failures, and gossips.

In our model, we distinguish shocks linked to signal blue S_B from those linked with signal red S_R. A signal blue shock, denoted blue shock, is symptomatic of high noise in the blue signal for the agents.

The probability of each shock to occur is an increasing function of the variance of its Gaussian White noise term. We set:

$$\begin{cases} \mathbb{P}(S_B) = \sigma_B^2 \cdot \frac{1}{100} \\ \mathbb{P}(S_R) = \sigma_R^2 \cdot \frac{1}{100} \end{cases} \tag{6}$$

In the affective component, agents may observe shocks, corresponding to news, market crashes, linked to each signal. Shocks occur locally and according to a probability depending on the variance of the white noise of each signal. Agents i are randomly located in a grid torus space. Each squared element j of the grid is subjected to the shock probabilities above. For an agent to observe a shock, that shock must then occur on the squared grid element the agent is placed on, denoted by the dummy variable S_{ij}^B equal to 1 if the blue shock occurs on this element, for an agent i at location j. This can be interpreted as local shocks, possibly individual specific given the random locations of the agents at initialization. If an agent observes a blue signal shock, it will interpret it as bad news for signal blue (B), and will have more disposition to adopt R. We assume here that if in a given squared element of the grid, both blue and red shock occur, the resulting effect is negative for both dispositions.

Denoting A_{iB} and A_{iR} agent i's affective disposition to, respectively, adopt signal blue (B) or red (R), we obtain:

$$A_{iB} = \begin{cases} 1 \text{ if } S_{ij}^B = 0 \text{ and } S_{ij}^R = 1 \\ 0 \text{ if } S_{ij}^B = 0 \text{ and } S_{ij}^R = 0 \\ -1 \text{ if } S_{ij}^B = 1, \forall S_{ij}^R \end{cases} \tag{7}$$

$$A_{iR} = \begin{cases} 1 \text{ if } S_{ij}^R = 0 \text{ and } S_{ij}^B = 1 \\ 0 \text{ if } S_{ij}^R = 0 \text{ and } S_{ij}^B = 0 \\ -1 \text{ if } S_{ij}^R = 1, \forall S_{ij}^B \end{cases} \tag{8}$$

2.4 Individual, Social, and Total Disposition to Act

From the deliberative and affective components, player i's own disposition to choose the blue or red signal, respectively, denoted δ_{iB} and δ_{iR} is computed as the sum of their dispositions to act.

$$\begin{cases} \delta_{iB} = D_{iB} + A_{iB} \\ \delta_{iR} = D_{iR} + A_{iR} \end{cases} \tag{9}$$

The social component in the Agent Zero cognitive architecture adds to these own dispositions the disposition of other agents within range of interaction. We study here different network structures organized with respect to different centralization levels, i.e., the tendency for one or few agents to occupy central positions in the social interconnections. We intend to study how network structure, coupled with your specifications of the cognitive architecture of the agents, influences information selection efficiency.

The centralized network model used is derived from the Alpha-centralization model developed by Vié and Morales (2019), that generates graphs with different levels of centralization. It constitutes a generalization of the preferential attachment mechanism (Barabási and Albert 1999) with an exponent controlling for the emergence and importance of hubs. This specification is able to generate diversity of different webs, from no centralization (randomly and independently distributed edges) to perfectly centralized networks (in one or a few hubs). In between the two extreme cases, a large range of scale-free networks are obtained, in which different situations of importance and number of hubs are delivered. The network generation process consists of allocating edges between agents (nodes) as a function of the attachment probability. The probability of node i to create an edge with node j is as follows:

$$p_{ij} \propto k_j^{\alpha} \tag{10}$$

where k_j is the number of connections of node j and α is the exponent we use to control the influence of the preferential attachment mechanism. If $\alpha = 0$, the attachment probability becomes equal among all nodes and we obtain a random network with no central hub similar to the Erdos–Ranyi model. If $\alpha = 1$, we obtain the standard Barabási–Albert network with a few hubs. If $\alpha = 2$, we create a network with full centralization where all nodes are linked to a single central one. This extension of the preferential attachment mechanism magnifies the degree heterogeneity among nodes for $\alpha > 1$, and reduces such attractive force for any $\alpha < 1$. An illustration of the model variants is shown in Fig. 1.

Accounting for imitation, communications, and interactions between agents, the total disposition of each agent to adopt signal blue or red, denoted for agent i, respectively Δ_{iB} and Δ_{iR} are given by the following equations. Agents do infer the dispositions of the other agents they are linked to in the social network, and compute the average disposition over the N agents they are connected to.

$$\begin{cases} \Delta_{iB} = D_{iB} + A_{iB} + \frac{1}{N}\sum_k \delta_{kB} \\ \Delta_{iR} = D_{iR} + A_{iR} + \frac{1}{N}\sum_k \delta_{kR} \end{cases} \tag{11}$$

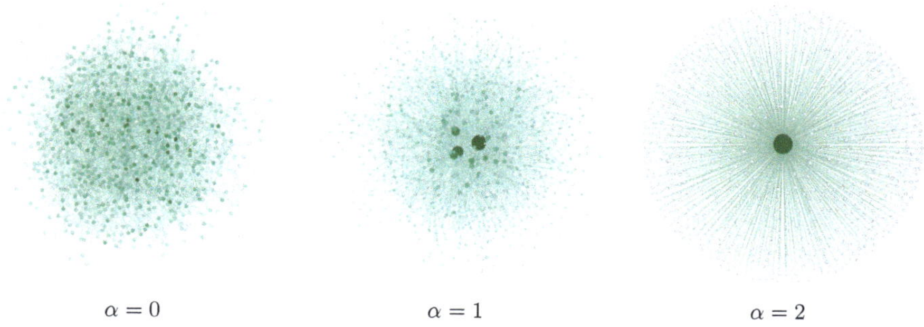

Fig. 1 Visualization of network topologies from the Alpha-centralization network process (Vié and Morales 2019)

This relative threshold-based decision rule for the agents proceeds as follows. If the total disposition (own deliberative, own affective and social) for a signal (e.g., blue) is higher than the total disposition for the other signal (e.g., red), then the signal with the higher disposition represents the signal with the most incentives to adoption. In such case, signal blue would be adopted. In case of equality in dispositions, we assume here that the agents conserves the same action as in the last period. Denoting action variable X with possible entries B (adopting signal blue) and R (adopting signal red), we note, with t the period considered:

$$X_t = \begin{cases} B \text{ if } \Delta_{itB} > \Delta_{itR} \\ R \text{ if } \Delta_{itB} < \Delta_{itR} \\ X_{t-1} \text{ if } \Delta_{itB} = \Delta_{itR} \end{cases} \quad (12)$$

In the social component, agents are linked by a given network structure (several can be considered). They evaluate the total disposition of their neighbors in the network and incorporate it to compute their final disposition, generating herding. Eventually, if the total disposition for adopting signal blue (red) exceeds the disposition for adopting the other signal, the signal blue(red) is adopted. Information selection efficiency can be measured as the share, evaluated in percentage, of agents adopting the most precise signal. The variance of the choices (volatility of opinions) is also measured as an indicator of the stability of information choices, confounded with opinions in our simplified model of information selection.

3 Implementation

The model is implemented in the agent-based environment Netlogo (Wilensky 1999). Agent-based modeling (AMB) offers a relevant tool to study complex systems, as they emphasize the role of individual interactions, that may be local, heterogeneous, and interdependent, and displays flexibility in network generation from the node perspective. As noted (Wildemeersch et al. 2016), the behavior of interconnected systems as network systems are too complex to be adequately modeled by traditional tools, suggesting quantitative and simulations methods as possible adequate modeling tools to study network resilience and fragility. Nevertheless, the complexity of the dynamics they describe has created a tough challenge to keep the model tractable and its results understandable. A major criticism addressed to multi-agent models and the use of simulation models, in general is indeed focused on the lack of tractability of the results and of the model dynamics. Dealing with several parameters, strong micro-heterogeneity justifying the use of ABMs, complex interactions, and issues of stochasticity and randomness, new tools in simulation exploration and analysis may be required to assess solid understanding of model behavior.

The OpenMole platform introduced by Reuillon et al. (2010, 2013) and more specifically its embedded Pattern Space Exploration (PSE) algorithm described in Chérel et al. (2015) provides a useful tool to improve model understanding by looking at unexpected patterns and exploring the space of outputs generated by the model. This method allows pushing the standards of model validation, commonly done by verifying that the model is able to reproduce the patterns to be explained, to test the validity of the model against unexpected input combinations. The Pattern Space Exploration (PSE) algorithm thus allows to identifying all different output patterns generated by a given range of parameters, which may be useful to understand causality in complex simulation models. It permits to identify regardless of observer's assumptions the "areas of interest" of the model, i.e., where variation in output occurs. It finally offers significant benefits in computation efficiency, as areas of little or no variation are not explored by the algorithm, in contrast with a classical sampling task. We thus obtain through the PSE method a more efficient and precise understanding of model areas of interest. Through the Calibration Profile Algorithms also embedded in the OpenMole platform, marginal effects of numerous parameters on explained variables can also be measured. These preliminary identification strategies of important model parameter settings to consider are adequately complemented by more traditional Monte Carlo methods, sampling and regression procedures.

4 Results and Discussion

In this section, we analyze the model behavior through information selection efficiency and opinion variance. We first explore the behavior of the model taking single cognitive modules into consideration, highlighting the mechanisms behind information selection in all cases. We first analyze the deliberative module in Sect. 4.1. The behavior of the emotional module is studied in Sect. 4.2. We then move to more complex settings in which cognitive modules are associated are then studied: deliberative and social modules in Sect. 4.3, affective and social in Sect. 4.4, and deliberative with affective in Sect. 4.5. Finally, the comprehensive model of Information Selection with the Agent Zero framework associating the three cognitive modules is analyzed in Sect. 4.6. We assume below for model runs a random initial distribution of opinions and iterate the model for 1,000 periods for each parameter combination. The model is static by nature, allowing these 1,000 independent experiments to return robust estimates of output values.

4.1 Deliberative Module Analysis

In the deliberative cognitive framework, agents receive blue and red numerical signals. They remember a given number of these emissions, this number being denoted "memory". The key parameters to consider in the analysis of the deliberative module is then the base and spread noise levels, as well as the memory. While it is intuitive that enhanced ability to remember past values is helpful in forming adequate estimates of the signal variances, the interaction of this ability with base and spread noise levels is a more complex phenomenon. Selecting an information signal based on relative variance is not a task of same difficulty when both signals are relatively stable, with respect to a very volatile environment.

Figure panel Fig. 2 graphically presents the impact of memory on information selection efficiency, i.e., the average percentage of agents in the population having adopted the most precise signal, and the variance of opinions, namely the variance of this percentage. It maps the system performance with respect to the parameters of interest. While Sect. 1.1 of the supplementary material presents this analysis from all memory levels from 1 to 10, we here illustrate the positive impact of memory in the deliberative framework on information selection efficiency and stability with two memory levels, respectively, 2 and 8. Through a larger sample for the agents to estimate the variance of both signals, a higher memory significantly contributes to a better identification and to a more efficient selection of information. At a constant base level of variance (y-axis), a higher variance spread enforces information selection efficiency. At a constant spread variance level (x-axis), a higher base variance level undermines information selection efficiency: both information sources show high variance, and a higher memory bandwidth is necessary in order to perfectly identify the most precise of the two. Spread variance level exhibit a positive effect on opinion stability; while the base variance levels negatively impact the stability of opinions. These insights are confirmed by Table 1: the impact of these parameters are of expected signs, and significant with p-value inferior to 0.01. The analysis allows to identify an unexpected positive impact of memory over system variance, contributing to higher volatility of opinions. As memory increases, the agents' estimates of the signal variances become more precise, triggering better information selection efficiency notably through more frequent adjustment of opinions, especially when the spread is low and makes identification of the most precise signal tricky. This explains the dual role of memory: improving selection efficiency and increasing opinion volatility.

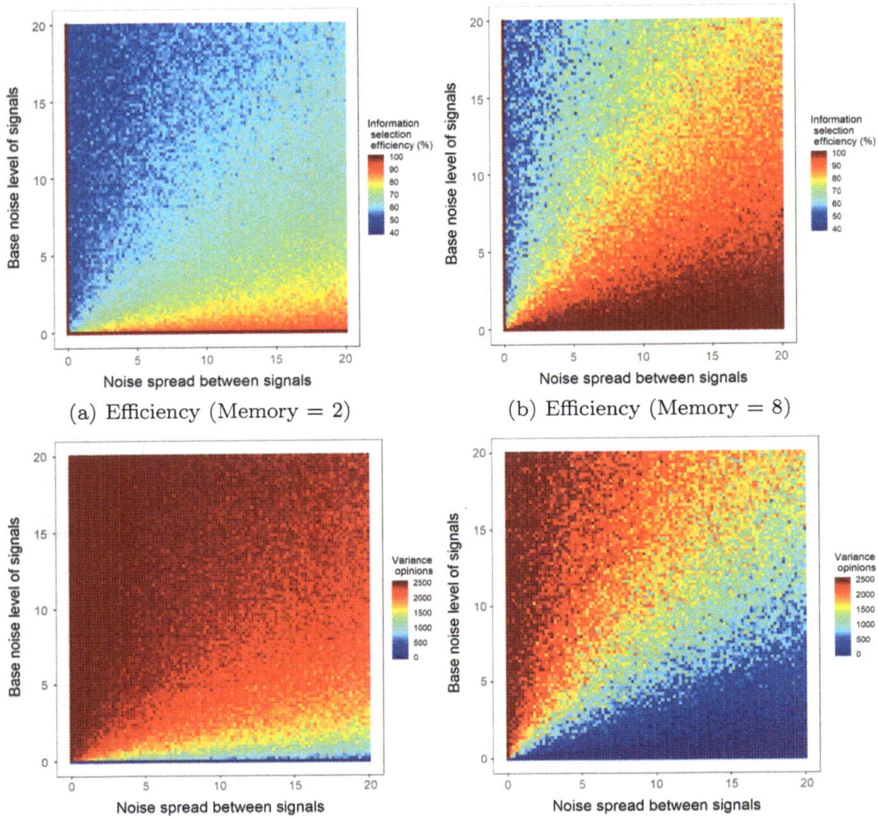

(a) Efficiency (Memory = 2) (b) Efficiency (Memory = 8)

(c) Variance (Memory = 2) (d) Variance (Memory = 8)

Fig. 2 Deliberative module analysis: memory, information selection efficiency, and opinion variance

4.2 Affective Module Analysis

The affective, or emotional module in the Agent Zero framework accounts in our model for the instinct reaction of investors to shock, such as bad news. The shock specification in our model accounts for these events and their reaction. The shock occurrence probability depends on the variance of their respective Gaussian noise term. We, therefore, study the information selection efficiency and the opinion variance across the space of possible base and spread noise levels.

Table 2 identifies positive effect of variance spread, and negative effect of base variance level, over information selection efficiency. As the former increases, relatively more shocks for the noisiest signal occur, triggering adverse disposition in the agents' cognition. However, as the latter is higher, both signals' shocks occur, making identification of the most precise signal harder, due to repeated conflicting disposi-

Table 1 Linear regression in the model (D)

	Dependent variable	
	Information selection efficiency (%)	Variance of opinions
Base	−1.295***	75.332***
	(0.005)	(0.396)
Spread	1.178***	−60.496***
	(0.005)	(0.396)
Memory	3.082***	25.493***
	(0.010)	(0.796)
Constant	56.828***	1,134.958***
	(0.099)	(7.604)
Observations	91,809	91,809
R^2	0.690	0.398
Adjusted R^2	0.690	0.398
Residual Std. Error	9.089 (df = 91805)	699.169 (df = 91805)
F Statistic	68,133.480*** (df = 3; 91805)	20,210.850*** (df = 3; 91805)

Note *p < 0.1; **p < 0.05; ***p < 0.01

Table 2 Linear regression in the model (A)

	Dependent variable	
	Information selection efficiency (%)	Variance of opinions
Base	−0.999***	0.108***
	(0.010)	(0.010)
Spread	1.194***	−0.627***
	(0.010)	(0.010)
Constant	71.336***	27.882***
	(0.154)	(0.153)
Observations	10,201	10,201
R^2	0.700	0.286
Adjusted R^2	0.700	0.285
Residual Std. Error	5.938 (df = 10198)	5.872 (df = 10198)
F Statistic	11,920.380*** (df = 2; 10198)	2,037.844*** (df = 2; 10198)

Note *p < 0.1; **p < 0.05; ***p < 0.01

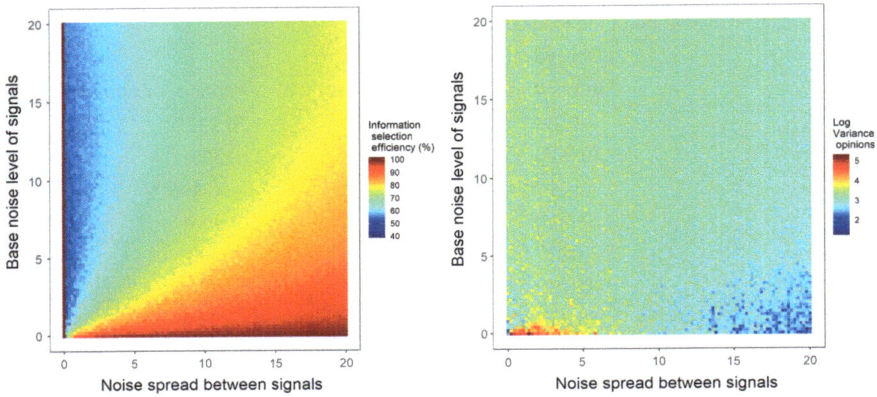

(a) Info. selection efficiency (b) Opinion variance

Fig. 3 Information selection efficiency and logarithm of opinion variance in the affective cognition framework

tion effects. The variance of opinions increases with respect to base, and decreases with respect to spread for the same reasons.

The variance of opinions exhibits is of significantly lower magnitude than in the deliberative framework presented in Table 1. Figure 3b plots the logarithm of the variance in the proportion of agents adopting the most precise signal. This level is overall quite low, due to the stochastic and localized nature of the shocks. As they occur identically and independently on each squared location in the grid constituting the model world, agents stimuli become quite rare. The preference for status quo and conservation of previous adoptions if no shocks of any kind occur, hence generates a stability of aggregate opinion dynamics. In low base—low spread in variance, this variance is higher, as it corresponds to a parameter region of numerous adjustments. In the inverse, the large gap between a quasi-stable information source and a very high one, corresponding to the dark blue region of Fig. 3b with low base and high spread, shocks almost exclusively occur from the noisiest process, guaranteeing through stimulation and preference for status quo a lower opinion variance as all agents settle on the most precise signal.

The ability of agents to adopt the most precise signal only based on the signals provided by the local stochastic shocks exhibits nonlinear behavior. As the difference in variance between both signals (spread, represented in the x-axis) increases, the increase in the percentage of agents adopting the most precise information source (y-axis) through conditioning by shocks varies differently according to the base variance level in the environment. In environments when the common noise level of the signals is low, shocks are rare. As the spread increases, the relative probability of shocks occurrence for the noisiest signal increases, and generates a strong disposition in favor of the alternative signal. However, this mechanism is less reliable for high common variance levels. The shocks do occur with much higher probabilities, possibly in identical locations. This results in confusion: agents' emotional disposition is less

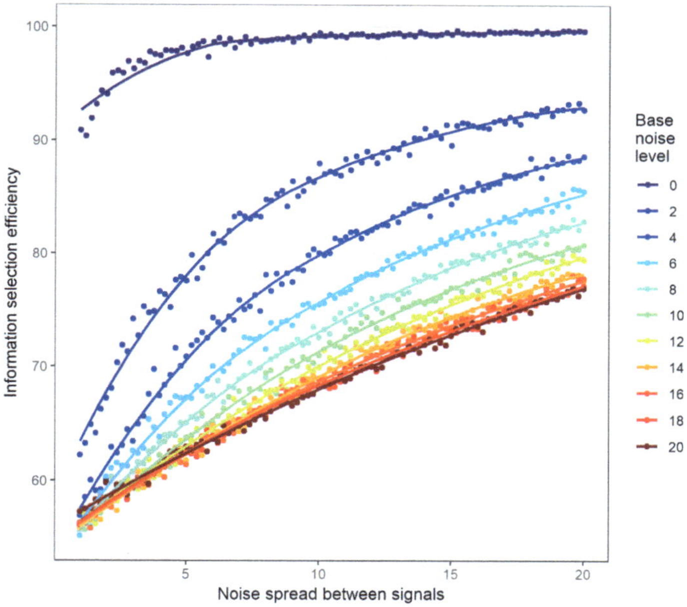

Fig. 4 Impact of noise spread over information selection efficiency, at given base levels

and less affected by differences in variance between signals when the overall noise is high. Figure 4 clearly outlines that for high levels of base noise level, the adoption curves tends to be confounded with respect to the noise spread. For low base levels (e.g., 1), the adoption curve plotted with respect to the spread is concave: at this given low base level, the increase in spread displays decreasing marginal returns to information selection efficiency.

We do not analyze the behavior of the social module alone. As it constitutes a vector for transmission and contagion of dispositions to act, the social cognition component does not display own dynamics when initializing the model with random opinions. However, the interaction of this module with the two previous one analyzed above deserve further treatment.

4.3 *The Deliberative and Social Components*

In Table 3, we perform linear regression analyses to evaluate in this simplified functional form, the impact of model parameters (base, spread, memory, and alpha, i.e., centralization) on information selection efficiency and opinion volatility. As before, base variance (resp. variance spread) level negatively (resp. positively) drives information selection efficiency, and positively (resp. negatively) impacts opinion volatility. However here, while memory keeps playing a positive role in enforcing

Table 3 Linear regression in the model (DS)

	Dependent variable	
	Information selection efficiency (%)	Variance of opinions
Base	−0.616***	27.152***
	(0.006)	(0.223)
Spread	1.090***	−41.426***
	(0.006)	(0.223)
Memory	0.343***	−12.572***
	(0.013)	(0.460)
Alpha	0.026	7.371***
	(0.046)	(1.595)
Constant	86.386***	553.301***
	(0.135)	(4.673)
Observations	153,015	153,015
R^2	0.200	0.246
Adjusted R^2	0.200	0.246
Residual Std. Error	14.725 (df = 153010)	509.297 (df = 153010)
F Statistic	9,571.959*** (df = 4; 153010)	12,493.430*** (df = 4; 153010)

Note *p < 0.1; **p < 0.05; ***p < 0.01*

information selection efficiency, it also plays as a stabilizing force, contrary to the model only initialized with the deliberative module. Different levels of centralization do not seem to have a significant impact on information selection efficiency, showing that in this simple framework, the structure of social networks do not significantly drives better or worse information selection. However, the impact of centralization over opinion variance is significant and negative. While more centralized networks do not matter for better information selection, over-centralized interconnection structure creates instability, making agents more vulnerable to the mistaken evaluation of a few, conveying erroneous disposition throughout the system. The overall variance level illustrated by the constant, is also much lower than in the only deliberative framework, illustrating social disposition effects to here play a stabilizing role over the opinion dynamics. These results are graphically illustrated in figure panels in Sect. 1.2 of the supplementary material.

4.4 The Affective and Social Interaction

We here compare the web-free case, the Erdös–Rényi random network, the Barabási scale-free network, and the perfect centralization network, as introduced in Fig. 1. We can first notice that the addition of the social component to the affective cognitive module enforces stabilization of the system toward the most precise signal as

Table 4 Linear regression in the model (AS)

	Dependent variable	
	Information selection efficiency (%)	Variance of opinions
Base	0.016***	−0.558***
	(0.004)	(0.157)
Spread	0.179***	−8.654***
	(0.004)	(0.157)
Alpha	−0.334***	32.920***
	(0.026)	(1.122)
Constant	97.415***	100.657***
	(0.062)	(2.652)
Observations	30,603	30,603
R^2	0.078	0.113
Adjusted R^2	0.078	0.113
Residual Std. Error	3.737 (df = 30599)	160.242 (df = 30599)
F Statistic	860.961*** (df = 3; 30599)	1,302.830*** (df = 3; 30599)

Note *p < 0.1; **p < 0.05; ***p < 0.01

the spread increases. But more surprisingly with respect to the other model specifications, Table 4 shows that the impact of the base variance level is positive and significantly different from zero. In the affective social module, higher background noise is slightly information selection improving, while the effect was clearly negative in the only affective framework. Social transmission here may act as a coordinating device around the correct signal. However, the impact of centralization on information selection is negative. Facing this surprising combination, we may conclude that social transmission between affective agents improves information selection efficiency, notably in random networks. However, as the interconnections structure change toward more centralized organizations, the total effect of social transmission becomes detrimental to information selection efficiency.

In the affective framework, agents receive shocks rarely, and status quo dominates. Adding the social disposition to their decision unit, the agents do receive permanent suggestions from their link neighbors. However, while we may expect that this mechanism enforces by link diversification a relevant signal adoption, Table 4 shows a significant negative although small effect of centralization over information selection efficiency. We may conclude that when affective processing is considered with our shocks, social transmission is detrimental to the adoption of the most precise signal. Excessive centralization may propagate throughout the system some bad luck of central nodes who noticed irrelevant shocks. Centralization also contributes to higher opinion volatility. For greater detail, in supplementary Sect. 1.3, figure panels display information selection efficiency and opinion variance (y-axis), i.e., the percentage of agents adopting the most precise information signal, and compares it between different network structure, with respect to base and spread variance levels.

4.5 *The Deliberative and Affective Interaction*

Figure 5 compares information selection efficiency and opinion variance (in color), with respect to memory, in the deliberative model as well as in the deliberative and affective model. The x-axis gives the spread (difference in variance between signals), while the y-axis indicates the base level of variance in both signals. The color, respectively, shows the percentage of the agents adopting the most precise signal (from 40 to 100%, denoted efficiency), and the opinion variance. Section 1.4 of the supplementary material provides additional analysis of the model under a larger set of memory values.

It graphically appears from Fig. 5 that the addition of the affective component to the deliberative component has a positive impact on information selection effi-

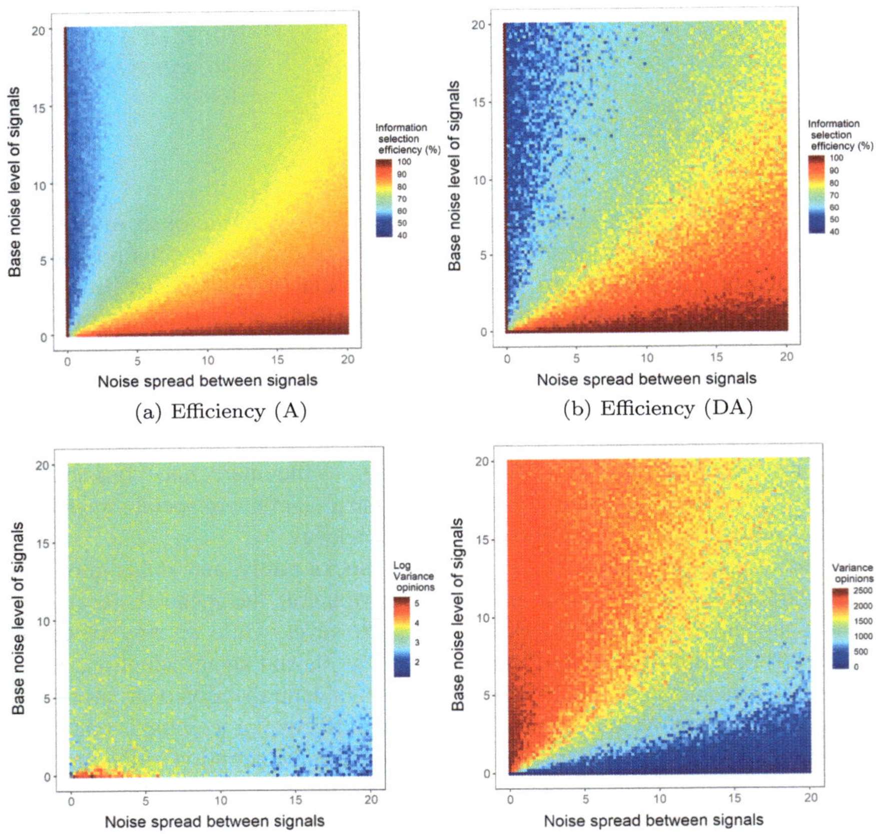

(a) Efficiency (A)

(b) Efficiency (DA)

(c) Log Variance (A)

(d) Variance (DA)

Fig. 5 Comparison between affective (A) and deliberative + affective (DA) model in selection efficiency and opinion variance, for memory = 4 ciency, particularly in high spread/intermediate base variance levels areas where the deliberative model was struggling in performance. The agents exploit more the information source through this dual cognition processing, which enforces their ability to identify and adopt the most precise information signal. The addition of the affective component to the deliberative setting, by allowing agents to rely on more informative objects such as shocks, has a very clear impact on the variance of opinions, notably in parameter regions with high base level of variance. In summary, adding the affective module to the deliberative model provides a complementary association, and enforces information selection ability. This improvement is notably located in parameter settings making information selection a tricky task, showing the richness of the combined framework and its utility to make financial agents systems operational in larger windows.

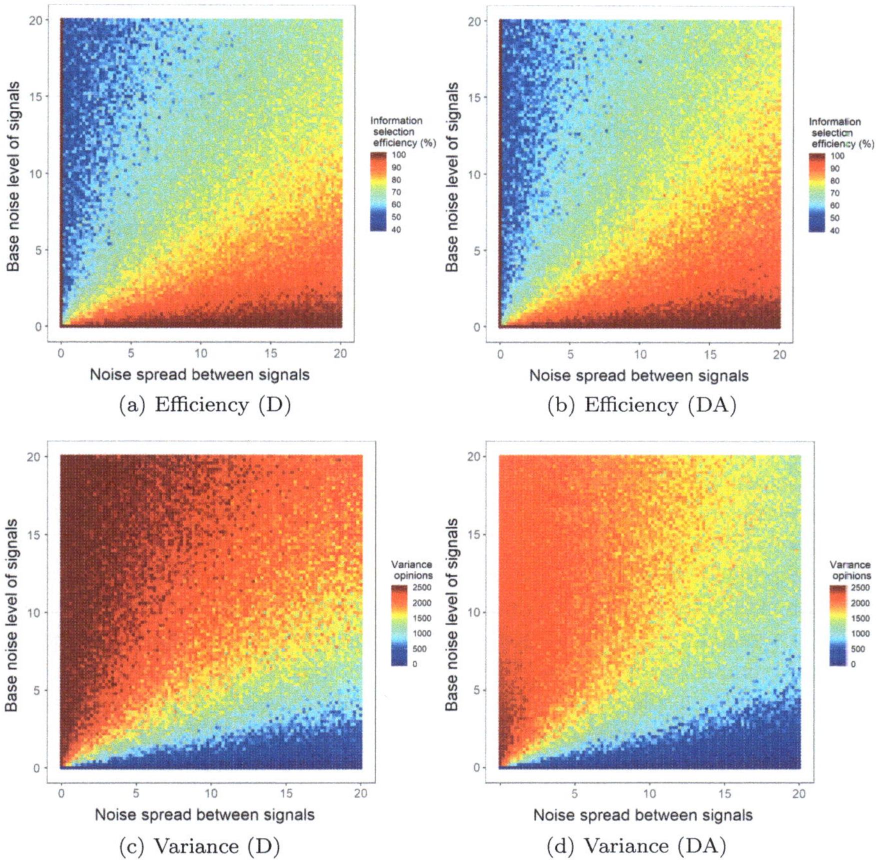

(a) Efficiency (D) (b) Efficiency (DA)

(c) Variance (D) (d) Variance (DA)

Fig. 6 Comparison between deliberative (D) and deliberative + affective (DA) model in selection efficiency and opinion variance, for memory = 4

Table 5 Linear regression in the model (DA)

	Dependent variable	
	Information selection efficiency (%)	Variance of opinions
Base	−1.314***	57.365***
	(0.006)	(0.276)
Spread	1.380***	−71.087***
	(0.006)	(0.276)
Memory	2.006***	−68.122***
	(0.012)	(0.569)
Constant	63.744***	1,927.758***
	(0.110)	(5.428)
Observations	51,005	51,005
R^2	0.741	0.708
Adjusted R^2	0.741	0.708
Residual Std. Error	7.385 (df = 51001)	363.342 (df = 51001)
F Statistic	48,520.340*** (df = 3; 51001)	41,316.840*** (df = 3; 51001)

Note *p < 0.1; **p < 0.05; ***p < 0.01

Considering the comparison the other way around, Fig. 6 compares the single affective (A) component to the combined deliberative and affective (DA) component, to study the impact of the addition of deliberative heuristics and memory functions to the model performance. The shocks considered by the affective module were quite rare events, with strong persistence of previous choices given our specification for status quo. This explains the relatively smooth dynamics observed. The deliberative heuristics appear to bring instability and volatility to these opinions, as they bring continuous and variant disposition effects to adopt each of the signals. As the deliberative heuristics information is delivered at each period, status quo disappears and make the system weight toward the deliberative framework messages. The variance of opinions considerably increases adding the deliberative framework, comparing Fig. 6c expressed in log and Fig. 6d. In spite of these differences, the overall trends in selection efficiency remain the same. Table 5 delivers coefficients of expected sign: variance spread and memory improve information selection efficiency and the stability of opinions, while the base variance level increases the opinion volatility and reduce information selection efficiency. The constant coefficient of opinion variance is quite high, from the addition of the deliberative module.

4.6 *Information Selection with Agent Zero Population*

We now consider the full extent of the Agent Zero framework, by activating all modules: deliberative, affective, and social. Table 6 presents the coefficients of a linear

Table 6 Linear regression in the model (ADS)

	Dependent variable	
	Information selection efficiency (%)	Variance of opinions
Base	−0.166***	9.348***
	(0.006)	(0.247)
Spread	0.544***	−29.587***
	(0.006)	(0.247)
Memory	0.031**	0.037
	(0.012)	(0.480)
Alpha	0.005	6.058***
	(0.045)	(1.765)
Constant	93.434***	351.619***
	(0.122)	(4.814)
Observations	61,206	61,206
R^2	0.118	0.205
Adjusted R^2	0.118	0.205
Residual Std. Error	9.053 (df = 61201)	356.539 (df = 61201)
F Statistic	2,052.309*** (df = 4; 61201)	3,943.132*** (df = 4; 61201)

Note *p < 0.1; **p < 0.05; ***p < 0.01

regression analysis. The base variance level in the model undermines information selection efficiency, and increases the volatility of opinions. The difference in variance level between signals contributes to greater information selection efficiency, and reduces the variance of opinions. The impact of memory is positive, but small (though significant) on the percentage of agents adopting the most precise signal. However, memory does not seem to have in this model any significant impact on the variance of opinions. Inversely, centralization (alpha) does not seem to have an effect on information selection efficiency significantly different from zero. Nevertheless, it has a significant positive impact on opinion volatility. Figure panels presented in Sect. 1.5 of the supplementary material graphically confirm these results. We compare the coefficient signs of marginal effects of model parameters over the output variables in the next Sect. 5.

5 Conclusion

We implement in this paper a neurocognitive-founded agent-based model of information selection. Agents endowed with affective, deliberative, and social cognitive mechanism acquire through these different channels representations of the world, adopting the signal they have the highest disposition in favor of. Environment parameters such as the base variance level and the difference in variance between signals

Table 7 Information selection efficiency and different model specifications

	Information selection efficiency					
	D	A	DS	DA	AS	ADS
Base	–	–	–	–	+	–
Spread	+	+	+	+	+	+
Memory	+	NA	+	+	NA	+
Centralization	NA	NA	0	NA	–	0

(Spread), and cognition parameters such as the degree of memory of the agents, or the centralization of their social networks, often strongly determine the market performance in achieving adoption of the most precise signal, a phenomenon denoted as information efficiency, as well as the volatility of opinions in this market.

However, the effects of these parameters happens to be sometimes counterfactual, shedding new light on the determinants of efficient information selection in this simple framework. The different behaviors of the model in some cognitive specifications shows the relevance of the Agent Zero neurocognitive-founded cognitive architecture for tackling this kind of social phenomena, and highlight the richness of the behaviors at stake.

Tables 7 and 8 plot the signs of the marginal effect coefficients in linear regression analysis in all the six cases considered. We notably see that while the spread parameter, often the only one considered in models of information selection, is reliable in the sense that its effect is always the same across cognitive architectures: favoring information selection by higher gap between signals, and improving the stability of opinions, as previously identified by Vié (2018). However, some associations such as the affective and social combination, delivers more surprising behavior, in which the overall base variance level improves signal selection and reduces the volatility of opinions. Deliberative parameters such as memory appear to have ambivalent impact on opinion variance—sometimes positive, sometimes negative, sometimes not significantly different from 0—and positive on information selection efficiency. When we include social transmission of dispositions to act, in some situations, this social mechanism has no impact on the system, while in others, it can be detrimental to system efficiency and stability. The Agent Zero information selection model exhibits surprising behavior, that is enriching to understand possible engines of instability and inefficiency in information selection in financial markets. The complexity of the behavior of the model, illustrated by the differences in coefficients signs depending on cognitive modules activation, outlines the relevance of simulation methods and of the Agent Zero framework to capture the emergent underlying dynamics of these phenomena.

Table 8 Variance of opinions and different model specifications

	Variance of opinions					
	D	A	DS	DA	AS	ADS
Base	+	+	+	+	–	+
Spread	–	–	–	–	–	–
Memory	+	NA	–	–	NA	0
Centralization	NA	NA	+	NA	+	+

References

Akerlof GA, Shiller RJ (2010) Animal spirits: How human psychology drives the economy, and why it matters for global capitalism. Princeton University Press

Allen F, Morris S, Shin HS (2006) Beauty contests and iterated expectations in asset markets. Oxford University Press, Oxford

Amblard F, Deffuant G (2004) The role of network topology on extremism propagation with the relative agreement opinion dynamics. Phys Stat Mech Appl 343:725–738

Arthur BW, Holland JH, Lebaron B, Palmer RG, Tayler P (1996) Asset pricing under endogenous expectations in an artificial stock market

Barabási AL, Albert R (1999) Emergence of scaling in random networks. Science 286(5439):509–512

Bernardo JM, Smith AF (2009) Bayesian theory, vol. 405. Wiley

Bracha A, Weber EU (2012) A psychological perspective of financial panic. FRB of Boston Public Policy Discussion Paper (12-7)

Branch WA, McGough B (2009) A New Keynesian model with heterogeneous expectations. J Econ Dyn Control 33(5):1036–1051

Brock WA, Hommes CH (1999) Rational animal spirits. In: Herings PJJ, van der Laan G, Talman AJJ (eds) The theory of markets, North-Holland, Amsterdam, pp 109–137

Brock WA, Hommes CH (1998) Heterogeneous beliefs and routes to chaos in a simple asset pricing model. J Econ Dyn Control 22(8–9):1235–1274

Chérel G, Cottineau C, Reuillon R (2015) Beyond corroboration: Strengthening model validation by looking for unexpected patterns. PLOS ONE 10(9)

Chiarella C, He X (2001) Asset price and wealth dynamics under heterogeneous expectations. Quant Financ 1(5):509–526

Dosi G, Napoletano M, Roventini A, Stiglitz J, Treibich T (2017) Rational Heuristics? Expectations and behaviors in evolving economies with heterogeneous interacting agents

Epstein JM (2014) Agent zero: Toward neurocognitive foundations for generative social science, vol 25. Princeton University Press

Erdos P, Rényi A (1960) On the evolution of random graphs. Publ Math Inst Hung Acad Sci 5(1):17–60

Frank H (2003) Natural selection, rational economic behavior and alternative outcomes of the evolutionary process. J Socio-Econ 32–6(12):601–622

Friedkin NE, Johnsen EC (1990) Social influence and opinions. J Math Sociol 15(3–4):193–206

Gaunersdorfer A (2000) Endogenous fluctuations in a simple asset pricing model with heterogeneous agents. J Econ Dyn Control 24:799–831

Hommes C (2001) Financial markets as nonlinear adaptive evolutionary systems

Hommes C (2006) Heterogeneous agent models in economics and finance. In: Tesfatsion L, Judd KL (eds) Handbook of computational economics, vol 2, Chap 23. Elsevier, pp 1109–1186

Hommes C (2013) Behavioral rationality and heterogeneous expectations in complex economic systems. Cambridge University Press
Kirman A (1991) Epidemics of opinion and speculative bubbles in financial markets. In: Taylor M (ed) Money and financial markets. Blackwell, Oxford, pp 354–368
Kirman A (2010) Complex economics: individual and collective rationality. Routledge
Kristoufek L, Vosvrda M (2018) Herding, minority game, market clearing and efficient markets in a simple spin model framework. Commun Nonlinear Sci Numer Simul
Loewenstein G (2000) Emotions in economic theory and economic behavior. Am Econ Rev 90(2):426–432
Lux T, Marchesi M (1999) Scaling and criticality in a stochastic multiagent model of a financial market. Nature 397:498–500
Mauersberger F, Nagel R (2018) Levels of reasoning in Keynesian beauty contests: a generative framework. In: Handbook of computational economics, vol 4, 10, pp 541–634
Michard Q, Bouchaud JP (2005) Theory of collective opinion shifts: from smooth trends to abrupt swings. Eur Phys J B-Condensed Matter Complex Syst 47(1):151–159
Orléan A (1995) Bayesian interactions and collective dynamics of opinion: Herd behavior and mimetic contagion. J Econ Behav Organ 28(2):257–274
Reinhart CM, Rogoff KS (2009) This time is different: Eight centuries of financial folly. Princeton University Press
Reuillon R, Chuffart F, Leclaire M, Faure T, Dumoulin N, Hill DRC (2010) Declarative task delegation in OpenMOLE. In: Proceedings of high performance computing and simulation (HPCS) international conference
Reuillon R, Leclaire M, Rey-Coyrehourcq M (2013) OpenMOLE, a workflow engine specifically tailored for the distributed exploration of simulation models. Future Gener Comput Syst
Vié, A (2018) Selecting information in financial markets herding and opinion swings in a heterogeneous mimetic rational agent-based model. In: International conference on complex systems. Springer, Cham, pp 120–128
Vié A, Morales AJ (2019) How connected is too connected? Impact of network connectivity on systemic risk and collapse of complex economic systems. In press
Wildemeersch M, Strelkovskii N, Poledna S, Leduc M (2016) Network resilience and systemic risk. In: European meetings on cybernetics and systems research, methodological approaches to address network resilience
Wilensky U (1999). NetLogo. Center for connected learning and computer-based modeling, Northwestern University, Evanston, IL. http://ccl.northwestern.edu/netlogo/

Diversification by Self-reinforcement of Preferences and Change of Interaction Type

Saori Iwanaga and Masao Kubo

Abstract There are situations where agents can benefit from coordinating their actions, i.e., agents gain payoffs by taking the same action as the others. Such situations can be modeled as coordination games. Other situations, i.e., agents gain payoffs by taking different actions, are modeled as complementary, asymmetric coordination, or minority games. In this study, we treat coordination and complementary games as types of interactions. Each agent can change the interaction type and faces binary choices. We examine the effects of decision errors on collective behavior and we show that a small decision error stabilizes collective behavior in a pure population of all agents who play either the coordination or complementary game. We also examine populations in which each agent can change the interaction type. We show that in a mixed population of agents who play either game, the behaviors and games become heterogeneous and the ratio of cooperative agents becomes 80% and the average utility is stable high.

1 Introduction

Studies on multi-robot and multi-agent simulations began in the late 1980s. Reynolds (1987) showed that collective behavior occurred when three rules were used simultaneously by Boids and that collective behavior was a complex system. The study by Sims (1994) of virtual block creatures and that by Terzopoulos et al. (1996) of artificial fish provide examples of emergent physical or functional characteristics. Kubo and Kakazu (1994) demonstrated the emergence of unique individuals in a competition for food among ant colonies. However, individual heterogeneity has often been ignored by most studies (Axelrod and Cohen 2001). For example, a recent study investigated the formation of multi-robot systems. Any heterogeneous robots in such

S. Iwanaga (✉)
Japan Coast Guard Acadamy, Kure, Hiroshima, Japan
e-mail: s-iwanaga@jcga.ac.jp

M. Kubo
Yokosuka, Kanagawa, Japan

A. S. Chakrabarti et al. (eds.), *Network Theory and Agent-Based Modeling in Economics and Finance*, https://doi.org/10.1007/978-981-13-8319-9_3

a system become removed (Rubenstein et al. 2014). There are limits to the adaptation performance of multi-robot systems with heterogeneous robots and computer simulations rarely reproduce emergent diversity in a population.

As for heterogeneity or diversity, Axelrod (1985), Axelrod and Cohen (2001) showed that the supersizing dynamics that occur in a complex system are often the consequences of diversity, and as individual agents experience more diverse contacts, the system can become less diverse. Therefore, the world is unable to keep their local ways. Sandholm (2001) presented a dynamic analysis of the evolution of preferences in a strategic environment. Dynamics simultaneously run at two speeds: while natural selection slowly reshapes the distribution of preferences, players quickly learn to behave as their preferences dictate.

There are situations where agents can benefit from coordinating their actions, i.e., agents gain payoffs by taking the same action as the others. These situations can be modeled as coordination games (Schelling 1978; Huberman and Glance 1993). Examples of situations where cooperation is important include trade alliances and choices about compatible technologies or conventions, such as choice of software or programming language. In other situations, such as complementary games, agents gain payoffs by taking different actions. Complementary games that have been receiving attention recently are asymmetric coordination games or minority games. Examples of situations where asymmetric cooperation is important are the division of labor, traffic congestion, or site selections for opening new shops. The Santa Fe bar problem and minority games are also modeled in the same manner. We treat coordination and complementary games as types of interactions.

Kirman (1997) shows that the performance of a collective system depends crucially on the type of interaction, as well as heterogeneity in the preference of the agents. In a former study (Iwanaga and Namatame 2017), we examined the coordination game. For the emergence of diversity, we demonstrated two different learning speeds: slow learning of preferences and quick learning of behavior. We used a pure population in which all agents played the coordination game and showed that isolated interactions are effective for the emergence of diversity.

Selten (7) showed that the introduction of slight mistakes may lead to rational strategies and the voluntary deviations influence the realization of equilibria. In this study, we examined various populations. First, we looked at populations in which all agents played the complementary game or only half of the agents played the complementary game and confirmed the emergence of diversity. Then, we focused on the effects of decision errors on collective behavior and showed that small decision errors stabilized collective behavior in a pure population in which all agents played the coordination or complementary game. Finally, we examined a population in which the agents could change their interaction types and showed that in a mixed population of agents who played both games, the behaviors and games become heterogeneous while the ratio of cooperative agents became 80% and the average utility was stable high.

2 Model

2.1 *Social Network*

We consider a lattice structure for a social network of agents and each agent interacts only with its neighbors. We use a von Neumann neighborhood model, as shown in Fig. 1, in which each agent interacts with its nearest four neighbor agents all at once. A lattice model is used to make the ecological characteristics of living beings clear (Axelrod 1985; Nowak 2006).

2.2 *Type of Interactions*

There are two behaviors, S1 and S2, for each agent. We adopted the payoff matrix shown in Table 1 (Iwanaga and Namatame 2002), which describes the outcomes of the interactions among agents. In this matrix, if two agents choose the same behavior, agent Ai can receive a positive payoff $1 - \theta i$ or θi, where $0 \leq \theta i \leq 1$, or 0. This is the coordination game. In contrast, if two agents choose different behaviors, agent Ai can receive a positive payoff $1 - \theta i$ or θi, where $0 \leq \theta i \leq 1$. This is the complementary game, whose payoff matrix is shown in Table 2 (Iwanaga and Namatame 2003).

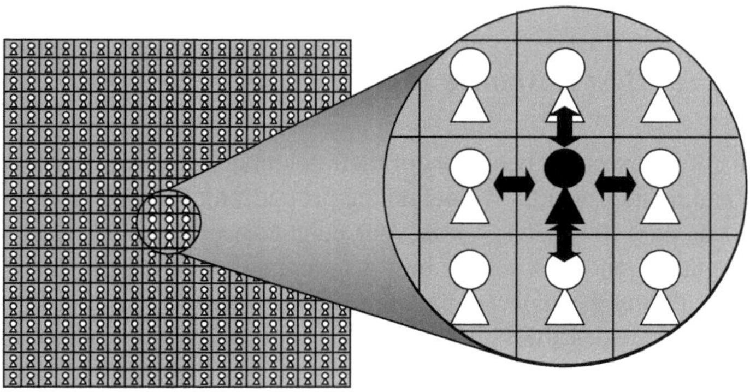

Fig. 1 von Neumann neighborhood model

Table 1 The payoff matrix of cooperative agent Ai

Coordination game		Behavior of agent	
		S1	S2
		pi(t)	1 − pi(t)
Behavior of agent Ai	S1	$1 - \theta i$	0
	S2	0	θi

Table 2 The payoff matrix of perverse agent Ai

Complementary game		Behavior of agent	
		S1	S2
		pi(t)	1 − pi(t)
Behavior of agent Ai	S1	0	θi
	S2	1 − θi	0

2.3 *Preferences and Heterogeneity of Agents*

Each agent is unique in the choice of behavior. In Tables 1 and 2, payoff parameter θi characterizes the preference of the agent. We examined the heterogeneity of agents. In a coordination game, each agent has a different payoff θi and 1 − θi, as shown in Table 1 (Iwanaga and Namatame 2002). If payoff 1 − θi is more than payoff θi, then agent Ai prefers S1 to S2. If payoff 1 − θi is less than payoff θi, then agent Ai prefers S2 to S1. For example, if payoff θi of agent Ai is less than 0.5, then the agent prefers S1. If payoff θi of agent Ai is more than 0.5, then the agent prefers S2. In contrast, in the complementary game, as shown in Table 2, if payoff θi is more than 1 − θi, then agent Ai prefers S1. If payoff θi is less than parameter 1 − θi, then agent Ai prefers S2.

3 Behaviors of Agents

3.1 *Choices: Best Response Rule*

Agents are not aware of the behaviors of their neighbor agents but know the choices taken by the neighbor agents. We define the proportion of agents who have chosen S1 at time t as pi(t) ($0 \leq$ pi(t) ≤ 1) in four neighbors of agent Ai. Each agent acts to maximize utility and chooses the best response against the behavior of neighbor agents by calculating the expected payoff. Rational agent Ai in a coordination game calculates the expected utilities by S1 and S2 at time t, which are expressed as follows:

$$\mathrm{Uc}\,(\mathrm{S1},\ \mathrm{t}) = \mathrm{pi(t)}(1-\theta \mathrm{i})$$

$$\mathrm{Uc}\,(\mathrm{S2},\ \mathrm{t}) = (1-\mathrm{pi(t)})\theta \mathrm{i} \tag{1}$$

Agent Ai chooses S1 if Uc (S1, t) > Uc (S2, t) or chooses S2 if Uc (S1, t) < Uc (S2, t) at time t + 1. The decision rule of agent Ai at time t + 1 is as follows (Iwanaga and Namatame 2002):

$$\text{Agent Ai chooses S1 if } \mathrm{pi}(t) > \theta \mathrm{i};$$

$$\text{Agent Ai chooses S2 if pi}(t) < \theta\text{i}. \tag{2}$$

The best response rule and the decision depend on the idiosyncratic parameter θ_{i}. When pi(t) = θi, we assume that the agent chooses the same behavior as that at time t. We define the parameter θi as the threshold of agent Ai. According to the threshold, the behaviors of the agents are differentiated from each other. Moreover, the proportion of agents having chosen S1 among the partners pi(t) of the agents is different from each other. Therefore, the decisions of the agents are different from each other, i.e., the threshold and social network of the agent affect the behavior. The agents form collective behavior, which is the behavior of the whole population. We define the collective behavior p(t) as the proportion of agents having chosen S1 in the population at time t. We regard the agent with the payoff matrix shown in Table 1 as a cooperative agent.

In contrast to the coordination game, Agent Ai in a complementary game calculates the expected utilities by S1 and S2 at time t as follows:

$$\text{Up (S1, t)} = (1 - \text{pi(t)})\theta\text{i}$$

$$\text{Up (S2, t)} = \text{pi(t) } (1 - \theta\text{i}) \tag{3}$$

The decision rule of agent Ai at time t + 1 is as follows (Iwanaga and Namatame 2003):

$$\text{Agent Ai chooses S1 if pi(t)} < \theta\text{i};$$

$$\text{Agent Ai chooses S2 if pi(t)} > \theta\text{i}. \tag{4}$$

When pi(t) = θi, we assume that the agent chooses the same behavior as that at time t. We regard the agent with the payoff matrix in Table 2 as a perverse agent.

3.2 *Movable for New Neighbors*

If an agent wants to receive a high payoff, an effective method is to move to another site and change the partners with which the agent interacts (Schelling 1978; Iwanaga and Namatame 2017). If the average payoff that agent Ai receives from interacting with four neighbor agents is above the threshold μ, then agent Ai stays at the current site or moves to another site. In cooperative and complementary games, the expected utilities by S1 and S2 at time t are given by (1) or (3). The rule for changing neighbor agents is as follows: where v = p (perverse agent), c (cooperative agent), and k = 1 (S1), 2(S2). Therefore, at time t + 1, agent Ai stays at the current site or moves to another site (Iwanaga and Namatame 2017).

$$\text{Agent Ai stays at the current site if } Uv(Sk,\ t) \geq \mu;$$

$$\text{Agent Ai moves to another site if } Uv(Sk, t) < \mu. \quad (5)$$

When some agents in the population want to move to another site, they form random pairs and exchange sites with each other. All agents who want to change can change their neighbor agents at the same time. At the new site, the agent chooses a behavior similar to that at the former site, as the agent is not yet familiar with the new site.

3.3 *Self-reinforcement of Preferences*

Each agent can update the payoff parameter through interaction. We term this as the self-reinforcement of preferences. For the coordination game in Table 1, if Agent Ai has chosen S1 for T times in succession, then the agent increases the payoff $1 - \theta i$ for S1 by Δ and decreases the payoff θi for S2 by Δ, as shown in Table 3a (Iwanaga and Namatame 2017). If agent Ai has chosen S2 for T times in succession, then the agent increases the payoff θi for S2 by Δ and decreases the payoff $1 - \theta i$ for S1 by Δ, as shown in Table 3b.

For the complementary game in Table 2, if Agent Ai has chosen S1 for T times in succession, then the agent increases the payoff θi for S1 by Δ and decreases the payoff $1 - \theta i$ for S2 by Δ, as shown in Table 4a. If agent Ai has chosen S2 for T times in succession, then the agent increases the payoff $1 - \theta i$ for S2 by Δ and decreases the payoff θi for S1 by Δ, as shown in Table 4b.

Table 3 The self-reinforced payoff matrix of cooperative agent Ai: **a** after choosing S1 T times in succession; **b** after choosing S2 T times in succession

a Coordination game		Behavior of agent	
		S1	S2
		pi(t)	1 − pi(t)
Behavior of agent Ai	S1	$1 - \theta i + \Delta$	0
	S2	0	$\theta i - \Delta$
b Coordination game		Behavior of agent	
		S1	S2
		pi(t)	1 − pi(t)
Behavior of agent Ai	S1	$1 - \theta i - \Delta$	0
	S2	0	$\theta i + \Delta$

Table 4 The self-reinforced payoff matrix of perverse agent Ai: **a** after choosing S1 T times in succession; **b** after choosing S2 T times in succession

a Complementary game		Behavior of agent	
		S1	S2
		pi(t)	1 − pi(t)
Behavior of agent Ai	S1	0	θi + Δ
	S2	1 − θi − Δ	0
b Complementary game		Behavior of agent	
		S1	S2
		pi(t)	1 − pi(t)
Behavior of agent Ai	S1	0	θi − Δ
	S2	1 − θi + Δ	0

3.4 *Decision Errors*

We set a very small probability of an agent's making a mistake. If an agent is not absolutely rational, then there is the possibility of mistakes, which we call the decision error. We define the probability as ε (Selten 1975). Each agent makes decisions according to (2) or (4) with probability 1 − ε. With probability ε, each agent makes a decision contrary to (2) or (4).

3.5 *Changes in Interaction Types*

When agent chooses S1 at time t

Agent Ai chooses the cooperative game if Uc (S1, t) > Ua (S1, t) or the complementary game if Uc (S1, t) < Ua (S1, t) at time t + 1. The decision rule of agent Ai at time t + 1 is as follows:

$$\text{Agent Ai chooses cooperative game if pi(t)} > \theta\text{i};$$

$$\text{Agent Ai chooses complementary game if pi(t)} < \theta\text{i}. \quad (6)$$

When pi(t) = θi, we assume that the agent chooses the same game as that at time t. Equation (6) is similar to (2).

When agent chooses S2 at time t

Agent Ai chooses the cooperative game if Uc (S2, t) > Ua (S2, t) or chooses the complementary game if Uc (S2, t) < Ua (S2, t) at time t + 1. The decision rule of agent Ai at time t + 1 is as follows:

$$\text{Agent Ai chooses cooperative game if pi(t)} < \theta\text{i};$$

$$\text{Agent Ai chooses complementary game or if pi(t)} > \theta\text{i.} \quad (7)$$

When pi(t) = θi, we assume that the agent chooses the same game as that at time t. Equation (7) is similar to (4).

4 Simulation Settings

We arranged agents for an area of 50 × 50 (2,500 agents) with four corners, no gaps, and the ends connected to the opposite side. First, we set the payoff parameter θi as 0.5 for all agents. We define the initial collective behavior p(0), which is the proportion of agents having chosen S1. Each agent makes its decision according to the best response rule (2) or (4) at each timestep, thereby changing the collective behavior p(t), which is the proportion of agents choosing S1.

We conducted three types of simulations, such as standard, decision error, and interaction type.

For the standard simulation, each agent decides the behavior S1 or S2 with self-reinforcement and mobility. The self-reinforced payoff matrix is given in Table 3 or 4. The rule of moving agents is given by (5). To detect the characteristic of the collective behavior, we set the initial collective behavior to 0.5, i.e., half of the agents choose S1 while the other half choose S1 at first timestep, when each agent chooses at random for three populations: all cooperative, all perverse, and mixed. We simulated 10,000 timesteps for a trial and simulate one trial for three populations. Then, we set initial collective behavior p(0) from 0.0 to 1.0 at intervals of 0.125, for which we assumed that each agent chose at random. We simulated 100 trials per initial collective behavior p(0) and investigated the final collective behavior p* for the three populations.

For the decision errors, we created three cases, in which agents make decision errors with 0.001 (0.1%), 0.01 (1%), and 0.05 (5%) for the three populations.

For the effects of the interaction types, each agent could select a game as expressed by (6) or (7). We set the initial ratios of the agents adopting the cooperative game to 100, 0, and 50%.

We set the following parameters on the basis of the preliminary experiments: the threshold to move μ to 0.3, the interval of self-reinforcement T to 10, increasing delta Δ to 0.01, and the interval of the value change R to 20. Increasing delta Δ must occur at a rate lower than the decision speed for the emergence of preference diversity. Similarly, we set the interval of value change R greater than the interval of self-reinforcement T for the emergence of interaction diversity.

5 Simulation Results

5.1 Self-reinforcement and Mobility

Figure 2 shows the results of the simulations of the population in which all cooperative agents have an initial collective behavior p(0) of 0.5, which means that half of the agents chose S1 at the first timestep. Figure 2a shows the last choices of the agents and about half of the agents having chosen S1. There are two big groups: S1 and S2. Each agent coordinates with its neighbors in the group where the four corners and the ends of an area are connected to the opposite side. Figure 2b shows the final preferences of the agents. The final preferences of half the agents become 0.0 and those of the other agents become 1.0 even though the initial preferences of all the agents were 0.5. Figure 3 shows a simulation result for the population of all perverse agents when the initial collective behavior p(0) is 0.5. Figure 3a shows that half of the agents chose S1. There are two groups: S1 and S2. Each agent complements its neighbors. Figure 3b shows that the final preferences of half the agents become 0.0 and those of the others become 1.0. Figure 4 shows a simulation result in the mixed population of 50% cooperative and 50% perverse agents when the initial collective behavior p(0) is 0.5. Figure 4a shows that about 72% of the agents chose S1. There are some groups of cooperative agents and perverse agents. Almost all cooperative agents coordinate with their neighbors by S1 and the perverse agents complement their neighbors. Figure 4b shows that the preferences of almost all cooperative agents become 1.0. For the perverse agents, the final preferences of half become 0.0 and those of the other half become 1.0.

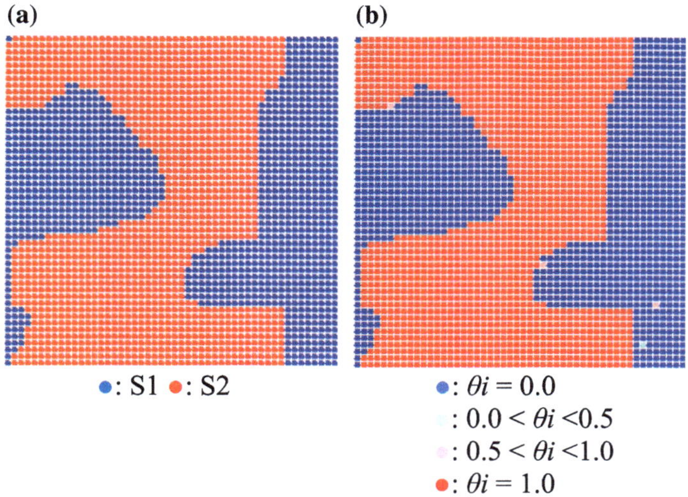

Fig. 2 A result of cooperative agents (p(0) = 0.5) **a** Behavior of agents, **b** Preferences of agents

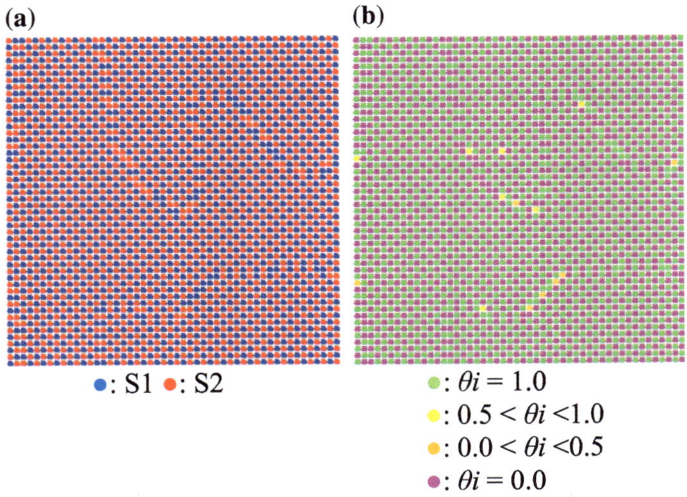

Fig. 3 A result of perverse agents (p(0) = 0.5) **a** Behavior of agents, **b** Preferences of agents

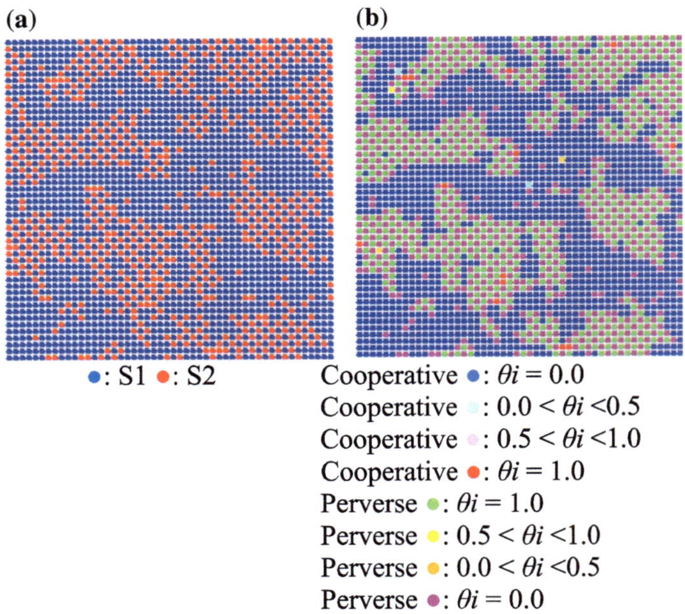

Fig. 4 A result of mixed agents (p(0) = 0.5) **a** Behavior of agents, **b** Preferences of agents

The final collective behavior for each initial collective behavior p(0) is shown in Fig. 5. The x-axis represents the initial collective behavior p(0) and the y-axis represents the average final collective behavior p*. Each plot shows the average of 100 trials and the error bar shows the standard deviation of the final collective behavior p*. Figure 5a shows the result of all cooperative agents. If the initial collective behavior p(0) is more than 0.625, then the collective behavior converges to 1.0 and almost all agents select S1 in any trial. If the initial collective behavior p(0) is less than 0.375, then the collective behavior becomes 0.0 and almost all agents select S2 in any trial. The majority behavior spreads but the minority behavior disappears from the population. However, if the initial collective behavior p(0) is between 0.375 and 0.625, then the final collective behavior p* becomes an intermediate value according to the trials and the initial collective behavior. If the initial collective behavior p(0) is 0.375, then the average final collective behavior p* becomes 0.05. If the initial

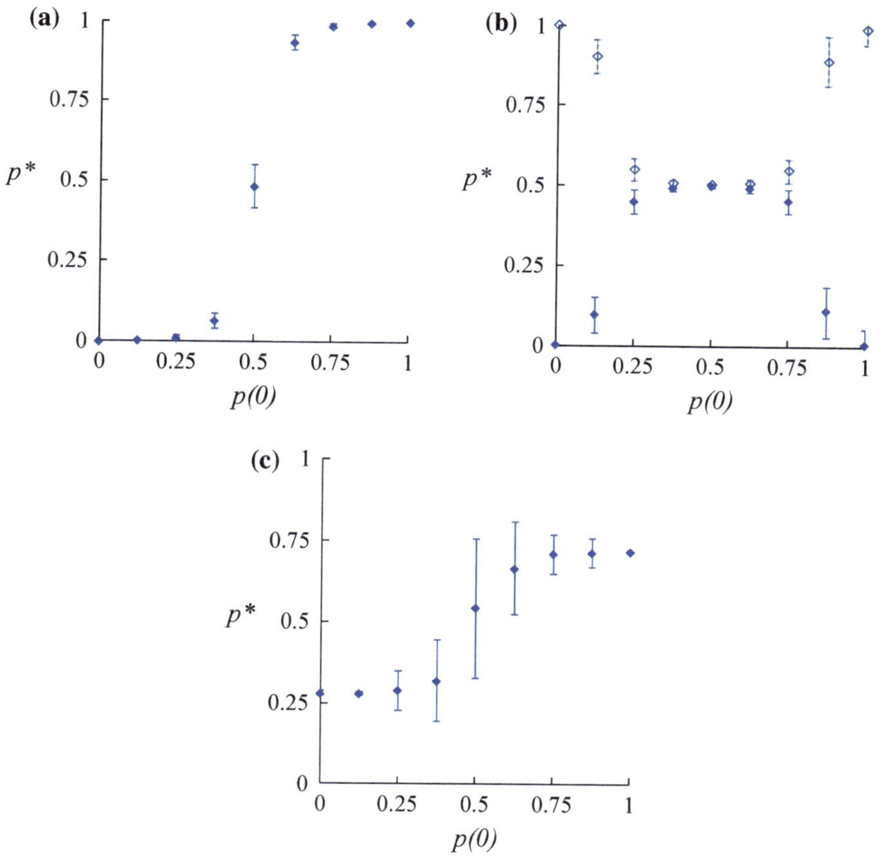

Fig. 5 The average final collective behavior p*, **a** Cooperative agents, **b** Perverse agents, **c** Cooperative and perverse agents

collective behavior p(0) is 0.5, then the average final collective behavior p* becomes 0.5. If the initial collective behavior p(0) is 0.675, then the average final collective behavior p* becomes 0.95. Behaviors S1 and S2 survive and coexist.

Figure 5b shows the results of all the perverse agents. If the initial collective behavior p(0) is between 0.375 and 0.625, then the final collective behavior p* becomes 0.5. Half of the agents choose S1 and the others choose S2. **Behaviors** S1 and S2 survive and coexist. We plot two sets of points—filled and outlined—as the final collective behavior p* oscillates. A filled point signifies a lower value, whereas an outlined point signifies a higher value. For example, the collective behavior is 0.0 (filled point) at one time and 1.0 (outlined point) at the next time, indicating that no agents choose S1 at one particular time, but all agents choose S1 at the next time. If the initial collective behavior p(0) is 0.25 or 0.75, then the final collective behavior p* oscillates between low and high values. A low value is between 0.35 and 0.5 while the high value is between 0.5 and 0.65. If the initial collective behavior p(0) is 0.125 or 0.875, then the final collective behavior p* oscillates between the low and high values. The low value is between 0.05 and 0.5 while the high value is between 0.5 and 0.95. If the initial collective behavior p(0) is 0.0 or 1.0, then the final collective behavior p* oscillates between 0.0 and 1.0.

For a mixed population of 50% cooperative and 50% perverse agents, the final collective behavior p* does not oscillate and becomes 0.28 or 0.72 according to the trials. If the initial collective behavior p(0) is less than 0.5, then the final collective behavior p* tends to be 0.28. If the initial collective behavior p(0) is more than 0.5, then the final collective behavior p* tends to be 0.72. The average final collective behavior p* is shown in Fig. 5c.

5.2 Decision Errors

Figure 6 shows a simulation result with a decision error of 0.001 for the population of cooperative agents. When the initial collective behavior p(0) is 0.0, then θi of all the agents becomes 1.0, as shown in Fig. 6a. When the initial collective behavior p(0) is 1.0, then θi of all the agents becomes 0.0, as shown in Fig. 6c. However, when the initial collective behavior p(0) is 0.5, then θi of half the agents becomes 0.0 and that of the other half becomes 1.0, as shown in Fig. 6b. Figure 7 shows a simulation result with a decision error of 0.001 for the population of perverse agents. When the initial collective behavior p(0) is 0.0, 0.5, or 1.0, then θi of half of the agents becomes 0.0 and that of the other half becomes 1.0.

Figure 8 shows the simulation results of the mixed population of cooperative and perverse agents. When the initial collective behavior p(0) is 0.0, then θi of almost all cooperative agents becomes 1.0, that of half the perverse agents becomes 0.0, that of the other half becomes 1.0, as shown in Fig. 8a. When the initial collective behavior p(0) is 1.0, then θi of almost all cooperative agents become 0.0 s, that of half the perverse agents becomes 0.0, and that of the other half becomes 1.0, as shown in Fig. 8c. When the initial collective behavior p(0) is 0.5, θi of almost

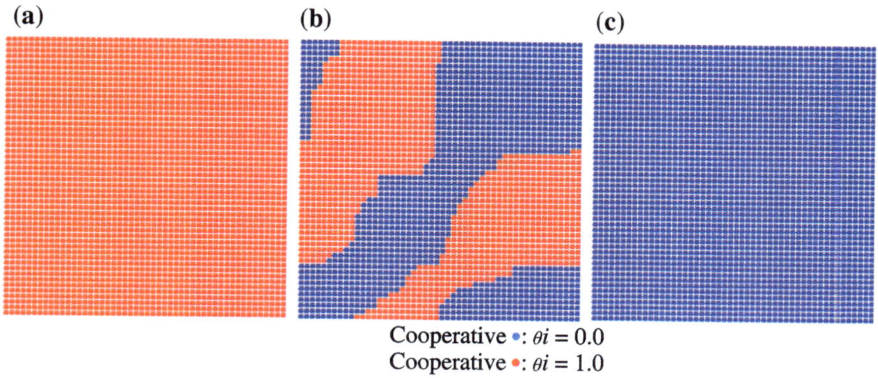

Fig. 6 Preferences of cooperative agents with error 0.001, **a** When p(0) = 0.0, **b** When p(0) = 0.5, **c** When p(0) = 1.0

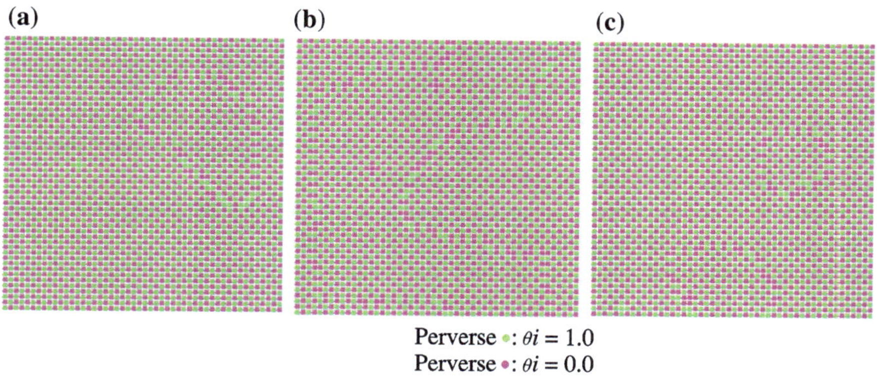

Fig. 7 Preferences of perverse agents with error 0.001, **a** When p(0) = 0.0, **b** When p(0) = 0.5, **c** When p(0) = 1.0

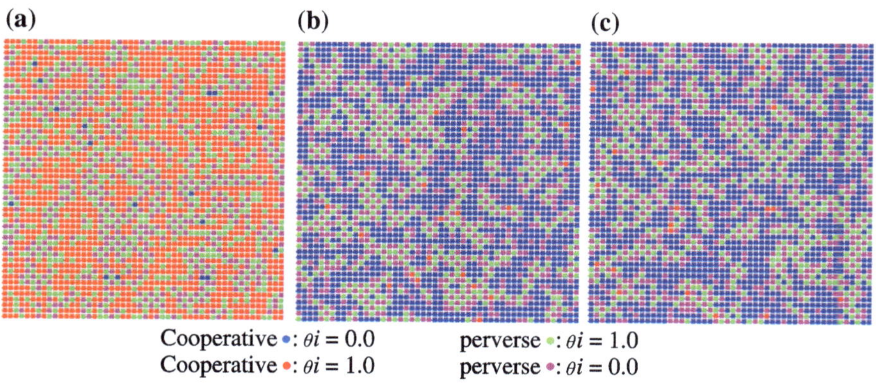

Fig. 8 Preferences of cooperative and perverse agents with error 0.001, **a** When p(0) = 0.0, **b** When p(0) = 0.5, **c** When p(0) = 1.0

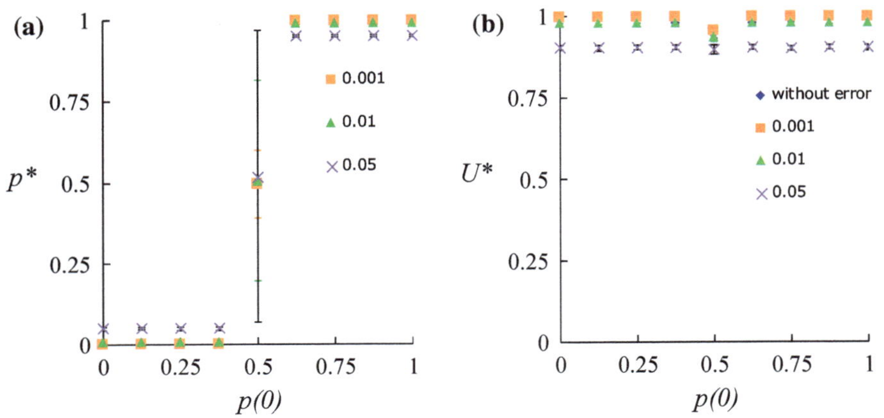

Fig. 9 Result in population of all cooperative agents, **a** The average final collective behavior p*, **b** The average utility

all cooperative agents becomes 0.0 according to the trials, since the final collective behavior depends on the trial. θi of half the perverse agents becomes 0.0 and that of the other half becomes 1.0, as shown in Fig. 8b. We found that preference diversity emerges from the existence of perverse agents.

Figure 9 shows the simulation results for the pure population of cooperative agents. The average final collective behavior p* is shown in Fig. 9a. The x-axis represents the initial collective behavior p(0) and the y-axis represents the average final collective behavior p*. Each plot shows the average of 100 trials and the error bar shows the standard deviation of the final collective behavior p*.

A decision error of 0.001 is equivalent to 0.1%. If the initial collective behavior p(0) is more than 0.5, then the collective behavior becomes 0.999 and almost all agents select S1 in any trial. S2 is chosen by 0.1% of the agents because of decision error. If the initial collective behavior p(0) is less than 0.5, then the collective behavior becomes 0.001 and almost all agents select S2 in any trial. S1 is chosen by 0.1% of the agents because of decision error. However, if the initial collective behavior p(0) is 0.5, then the final collective behavior p* becomes an intermediate value between 0.25 and 0.75 according to the trials. The final collective behavior is more radical than the behavior without error.

A decision error of 0.01 is equivalent to 1%. If the initial collective behavior p(0) is more than 0.5, then the final collective behavior p* becomes 0.99. Otherwise, if the initial collective behavior p(0) is less than 0.5, then the collective behavior becomes 0.01. However, if the initial collective behavior p(0) is 0.5, then the final collective behavior p* becomes an intermediate value between 0.01 and 0.99 according to the trials.

A decision error of 0.05 is equivalent to 5%. If the initial collective behavior p(0) is more than 0.5, then the collective behavior becomes 0.95. If the initial collective behavior p(0) is less than 0.5, then the collective behavior becomes 0.05. However, if the initial collective behavior p(0) is 0.5, then the final collective behavior p* becomes 0.05 or 0.95 according to the trials. With a decision error of 0.05, when the initial collective behavior p(0) is 0.5, then the final collective behavior p* is the most radical of the three cases when the decision error was 0.001, 0.01 and 0.05. It was difficult to anticipate the collective behavior when the initial collective behavior p(0) was 0.5.

The average utilities at the last timestep are shown in Fig. 9b. We found that the average utility with a decision error of 0.001 was the highest of the four cases. For any case, it was the lowest when an initial collective behavior p(0) = 0.5. Without error, the average utility was 0.93. With the decision errors of 0.001, 0.01, and 0.05, the average utility was 0.95, 0.93, and 0.88, respectively.

Figure 10 shows the simulation results for the pure population of all perverse agents. The average final collective behavior p* is shown in Fig. 10a. With a decision error of 0.001, if the initial collective behavior p(0) is between 0.125 and 0.875, then the final collective behavior p* becomes 0.5. If the initial collective behavior p(0) is 0.0 or 1.0, then the final collective behavior p* oscillates between low and high values. The low value is between 0.0 and 0.5 and the high value is between 0.5 and 1.0. The oscillation is suppressed by the decision error. With a decision error of 0.01, the collective behavior becomes 0.5 in any initial collective behavior. With a decision error of 0.05, the collective behavior also becomes 0.5. The oscillation is suppressed again by the decision error.

The average utilities at the last timestep are shown in Fig. 10b. We found that in almost all initial collective behaviors, the average utility with a decision error of 0.001 was the highest of the four cases. The average utilities are highest when initial collective behavior p(0) = 0.5 in any initial collective behavior. Without error, the average utility is 0.93. With decision errors of 0.001, 0.01 and 0.05, the average utility is 0.95, 0.93, and 0.68.

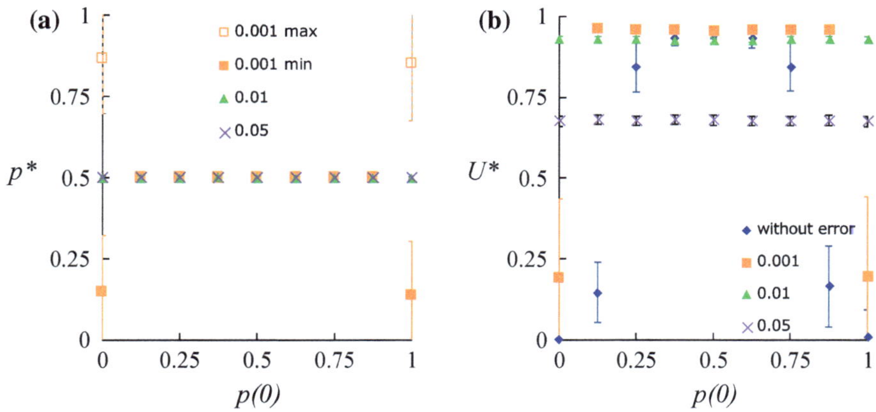

Fig. 10 Result in population of all perverse agents, **a** The average final collective behavior p*, **b** The average utility

Figure 11 shows the simulation results of a mixed population with 50% cooperative agents. The average final collective behavior p* is shown in Fig. 11a. With a decision error of 0.01, the average final collective behavior p* is also similar to that without error. If the initial collective behavior p(0) is less than 0.5, then the final collective behavior p* tends to be 0.32. If the initial collective behavior p(0) is more than 0.5, the final collective behavior p* tends to be 0.68. However, with a decision error of 0.05, the final collective behavior p* is different from the other results and takes on a value between 0.36 and 0.64 according to the trial in any initial collective behavior p(0). Therefore, the average final collective behavior p* is about 0.5. The average utilities at the last timestep are shown in Fig. 11b. They are almost the same in any initial collective behavior p(0). Without error, the average utility is 0.89. With decision errors of 0.001, 0.01 and 0.05, the average utility is 0.75, 0.71 and 0.32, respectively. We found that the average utility without error is the highest in the four cases.

Figure 12 shows the simulation results for populations consisting of 25 and 75% of cooperative agents without error. The average final collective behavior p* is shown in Fig. 12a. In the population with 25% cooperative agents, in any initial collective behavior, the final collective behavior p* becomes 0.4 or 0.6 according to the trials. Therefore, the average final collective behavior p* is 0.5. In the population of 75% cooperative agents, if the initial collective behavior p(0) is lower than 0.5, then the final collective behavior p* becomes 0.15. If the initial collective behavior p(0) is higher than 0.5, then the final collective behavior p* becomes 0.85. If the initial collective behavior p(0) is 0.5, then the final collective behavior p* is a random value between 0.15 and 0.85 according to the trials.

The average utilities at the last timestep are shown in Fig. 12b. In a population of 25 or 75% cooperative agents, the average utility is 0.92. In the population of 50% cooperative agents, the average utility is 0.88, which is the lowest of the three

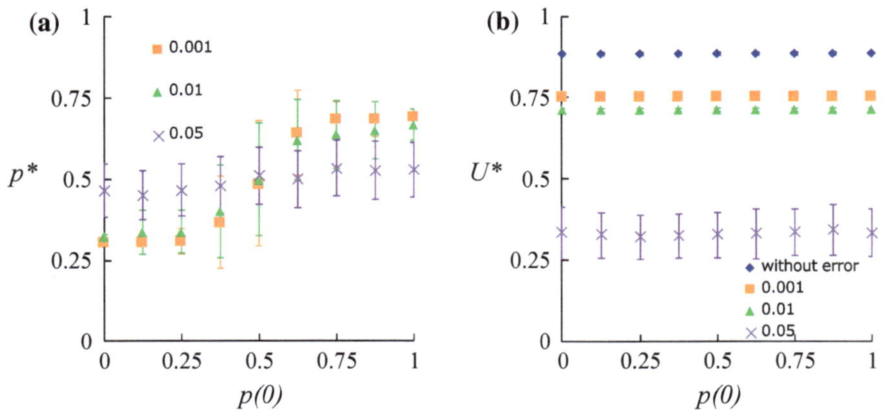

Fig. 11 Result in mixed (half and half) population of agents, **a** The average final collective behavior p*, **b** The average utility

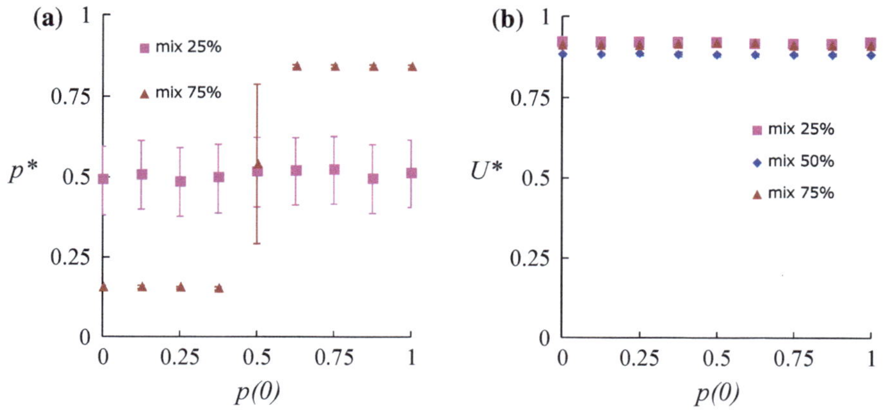

Fig. 12 Result in the mixed population, **a** The average final collective behavior p*, **b** The average utility

populations. As for preference diversity, in any case, agents appear with 0.0 and 1.0 of θi. We found that preference diversity emerged in the mixed population.

5.3 *Changes of Interaction Types*

Figure 13 shows a simulation result for a population in which all agents first adopted the cooperative game. When the initial collective behavior p(0) is 0.25, then θi of

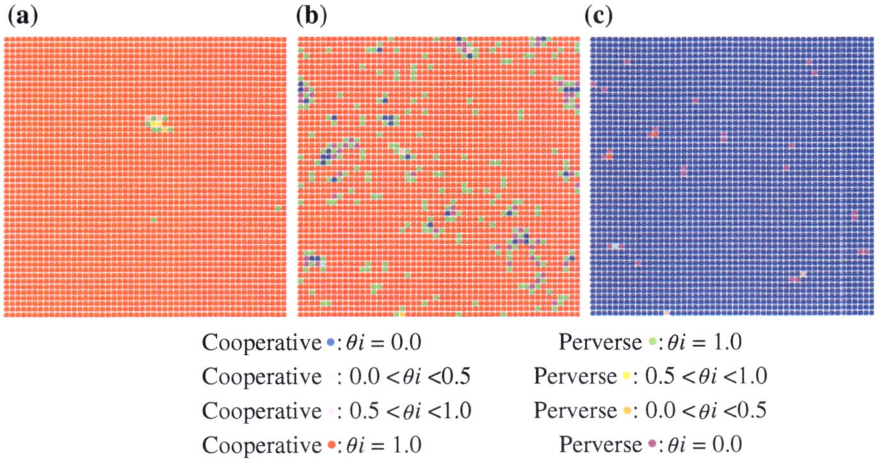

Fig. 13 A result: Preferences of cooperative agents at first, **a** When p(0) = 0.25, **b** When p(0) = 0.5, **c** When p(0) = 0.75

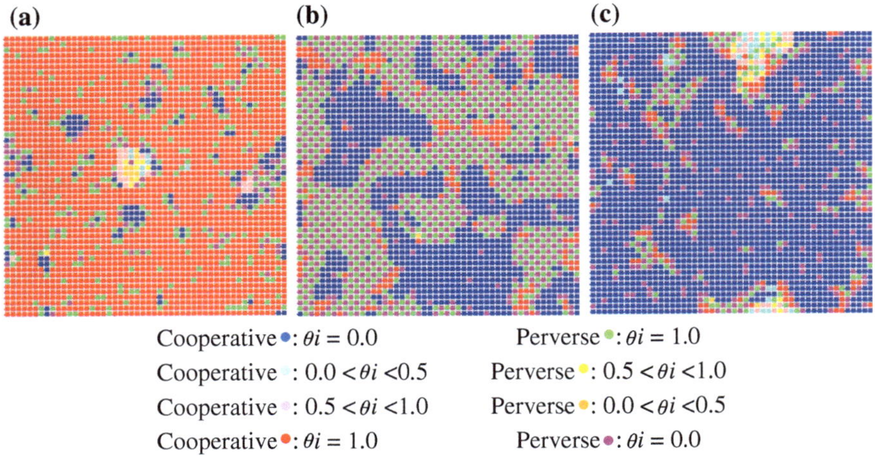

Fig. 14 A result: preferences of perverse agents at first, **a** When p(0) = 0.25, **b** When p(0) = 0.5, **c** When p(0) = 0.75

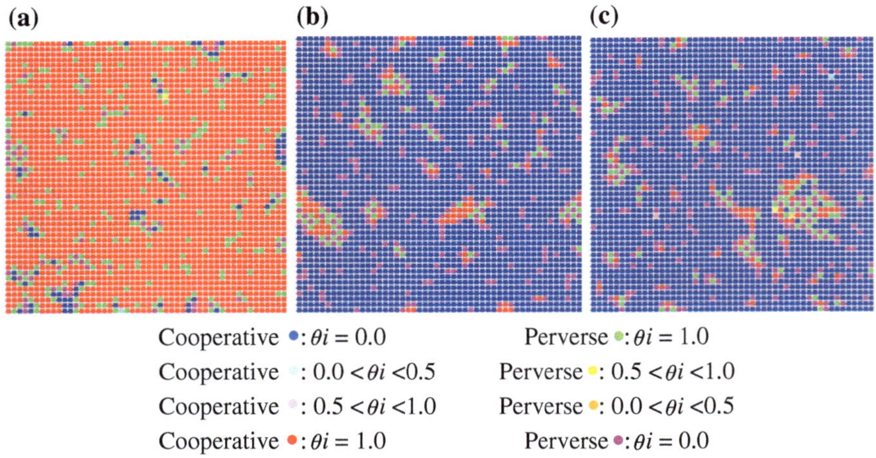

Fig. 15 A result: Preferences of cooperative and perverse agents at first, **a** When p(0) = 0.25, **b** When p(0) = 0.5, **c** When p(0) = 0.75

almost all cooperative agents becomes 1.0, as shown in Fig. 13a. Here some perverse agents appear. When the initial collective behavior p(0) is 0.75, then θi of almost all cooperative agents becomes 0.0, as shown in Fig. 13c. Here, some perverse agents also appear. However, when the initial collective behavior p(0) is 0.5, then θi of almost all cooperative agents becomes 1.0, as shown in Fig. 13b, according to the trials, since the final collective behavior depends on the trials.

Figure 14 shows a simulation result for a population in which all agents first adopted the cooperative game. When the initial collective behavior p(0) is 0.25, then many cooperative agents appear and θi of the majority becomes 1.0, as shown in Fig. 14a. When the initial collective behavior p(0) is 0.75, then many cooperative agents also appear and θi of the majority becomes 0.0, as shown in Fig. 14c. However, when the initial collective behavior p(0) is 0.5, then half of the agents become cooperative agents, whereas the other half become perverse agents, as shown in Fig. 14b. θi of the majority of the cooperative agents becomes 0.0 according to the trials, since the final collective behavior depends on the trials. Here, θi of half the perverse agents become 0.0, whereas that of the other half becomes 1.0.

Figure 15 shows a simulation result of a population in which 50% of the agents first adopted the cooperative game. When the initial collective behavior p(0) is 0.25, then many cooperative agents appear and θi of the majority becomes 1.0, as shown in Fig. 15a. When the initial collective behavior p(0) is 0.75, then many cooperative agents also appear and θi of the majority becomes 0.0, as shown in Fig. 15c. When the initial collective behavior p(0) is 0.5, many cooperative agents appear again, as shown in Fig. 15b. However, θi of the majority becomes 0.0 according to the trials, since the final collective behavior depends on the trials.

Figure 16 shows the simulation results for a population in which all agents first adopted the cooperative game. The average final collective behavior p* is shown in Fig. 16a. The x-axis represents the initial collective behavior p(0) and the y-axis represents the average final collective behavior p*. Each plot shows the average of 100 trials and the error bar shows the standard deviation of the final collective behavior p*.

Without error, the final collective behavior p* depends on the initial collective behavior p(0). When the initial collective behavior p(0) is low, then the final collective behavior p* becomes 0.0, but when the initial collective behavior p(0) is high, then the final collective behavior p* becomes 1.0. The final collective behavior p* is radical in all the results of this study. A decision error of 0.001 is equivalent to 0.1%. The final collective behavior p* depends on the initial collective behavior p(0). When the initial collective behavior p(0) is low, the final collective behavior p* becomes about 0.4, but when the initial collective behavior p(0) is high, then the final collective behavior p* becomes about 0.6 and radical collective behavior is suppressed. A decision error of 0.01 is equivalent to 1%. The final collective behavior p* does not depend on the initial collective behavior p(0) and becomes about 0.5.

The average utility U is shown in Fig. 16b. The x-axis represents the initial collective behavior p(0) and the y-axis represents the average utility U. Without error, when the initial collective behavior p(0) is 0.5, then the average utility U is about 0.8. The average utility U is about 0.6 with a decision error.

The diversity of the types of games is shown in Fig. 16c. The x-axis represents the initial collective behavior p(0) and the y-axis represents the average final ratio of agents adopting the cooperative game q*. Without error, when the initial collective behavior p(0) is low or high, the average ratio q* is about 1.0, but when the initial collective behavior p(0) is 0.5, then the average ratio q* is about 0.8. With a decision

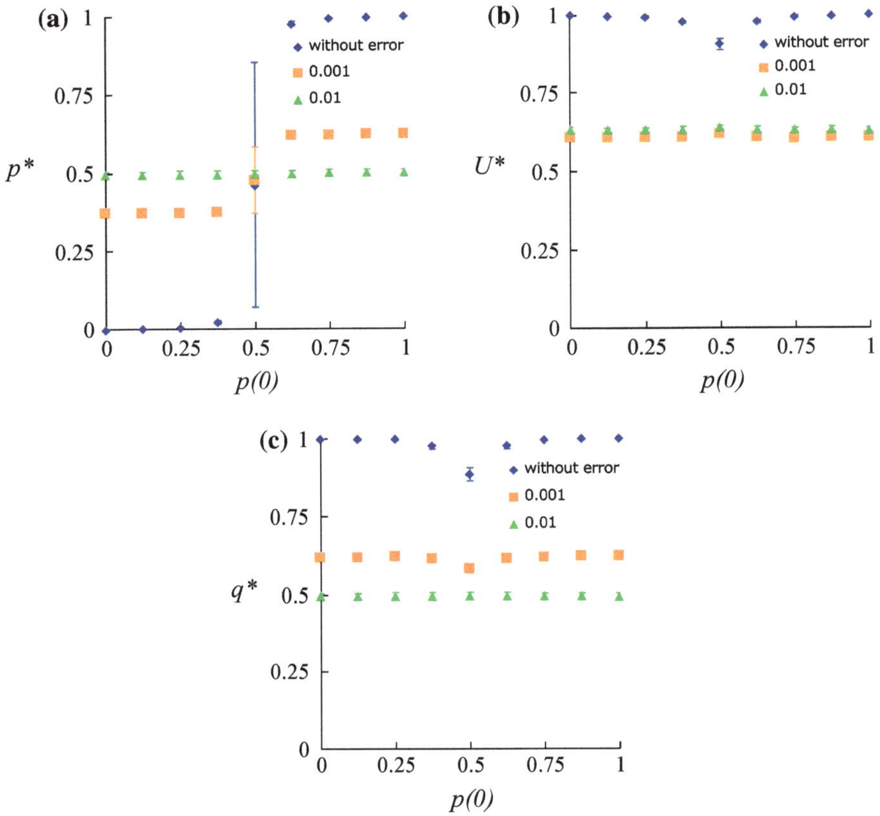

Fig. 16 100% of agents were cooperative agents at first, **a** The average final collective behavior p*, **b** The average utility, **c** The average final ratio of cooperative agents

error of 0.001, the average ratio q* is about 0.6. With a decision error of 0.01, the average ratio q* is about 0.5.

Figure 17 shows the simulation results for a population in which none of the agents first adopted the cooperative game. The average final collective behavior p* is shown in Fig. 17a. Without error, the final collective behavior p* takes on various values according to the trials. With a decision error of 0.001, a variety of final collective behavior p* is suppressed. With a decision error of 0.01, the final collective behavior p* is about 0.5. The average utility is shown in Fig. 17b. Without error, the average utility U is 0.8 in almost all initial collective behaviors p(0), but the average utility U is 0.2 when the initial collective behavior p(0) is 0.0 or 1.0. The average utility U is about 0.6 in any initial collective behavior p(0) with a decision error. The diversity of the types of interactions is shown in Fig. 17c. Without error, the final ratio of cooperative agents is about 0.8 when the initial collective behavior p(0) is between 0.25 and 0.375 or between 0.75 and 0.875. With a decision error, the final ratio of the cooperative agents is about 0.5.

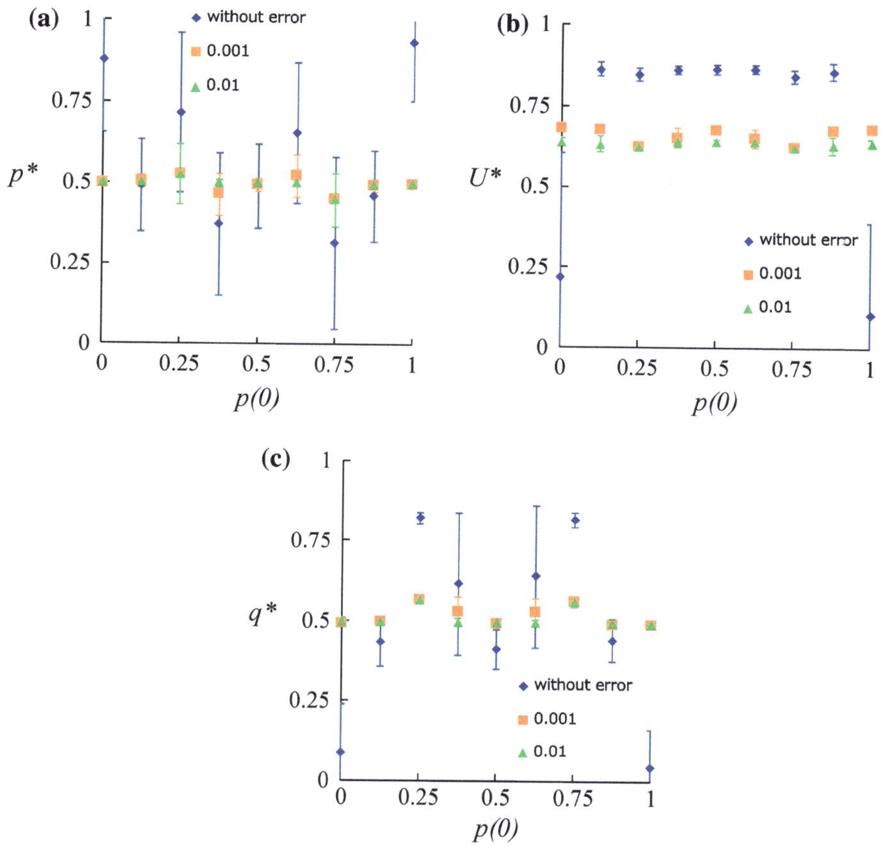

Fig. 17 0% of agents were cooperative agents at first, **a** The average final collective behavior p*, **b** The average utility, **c** The average final ratio of cooperative agents

Figure 18 shows the simulation results for a population in which 50% of the agents first adopted the cooperative game. The average final collective behavior p* is shown in Fig. 18a. Without error, the final collective behavior p* is about 0.2 or 0.8 according to the trials. With a decision error of 0.001, the final collective behavior p* is about 0.4 or 0.6 according to the trials. With a decision error of 0.01, the collective behavior is about 0.5. The diversity of the types of interactions is shown in Fig. 18c. As shown in Fig. 18b, the average utility with a decision error of 0.001 is about 0.6. The average utility without error is 0.8. Without error, the final ratio of the cooperative agents is about 0.8, as shown in Fig. 18c.

We found that, by the heterogeneity of interaction type, preference diversity had emerged as in the last simulation and the ratio of the cooperative agents tended to be 80%.

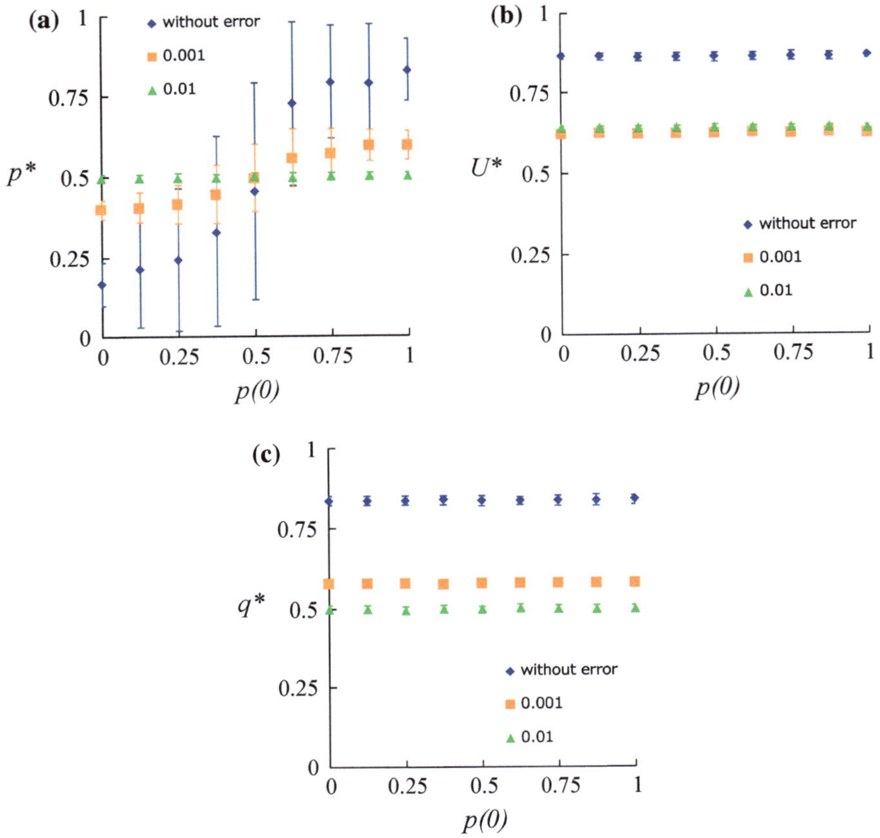

Fig. 18 50% of agents were cooperative agents at first, **a** The average final collective behavior p*, **b** The average utility, **c** The average final ratio of cooperative agents

6 Discussion

In this study, we conducted three types of simulations. At first, we examined self-reinforcement and found that the final collective behavior depended on the initial collective behavior and was heavily unstable against the initial collective behavior in the population in which all the cooperative agents played coordination games. In the population in which all the perverse agents played complementary games, we found that the final collective behavior depended on the initial collective behavior and oscillated heavily. In both populations, which we term pure populations of interaction type, behaviors S1 and S2 do not coexist, i.e., it is difficult to make the behaviors heterogeneous. On the other hand, in the mixed population of cooperative agents and perverse agents, the final collective behavior p* does not oscillate and becomes 0.28 or 0.72 according to the trials. Behaviors S1 and S2 coexist and become heterogeneous autonomously.

Second, we examined the effects of decision errors on collective behavior. In the coordination game or complementary game, we show that only a small decision error of 0.001 diversifies collective behavior and preferences while increasing the average utility. On the other hand, in the mixed population of cooperative agents and perverse agents, the cooperative agent decides differently from the perverse agent. The opposite type of agents is considered to play a role in decision error of behavior, which is too much for a mixed population. We found that the absence of errors is good for the mixed population. As for the mixed rates, we showed that a ratio of 25 or 75% cooperative agents is good for high utility.

Finally, we examined a population in which each agent can change its interaction type. We found that small decision errors make the final collective behavior p* and the final ratio of agents adopt the cooperative game q*, which is stable against the initial collective behavior p(0) but reduces the average utility in the population of all cooperative agents at first. This is true in the population of all perverse agents at first, i.e., the absence of errors is good for both populations. On the other hand, in the mixed population of cooperative agents and perverse agents, we found that the average utility is high at first. Behaviors S1 and S2, as well as cooperative agents and perverse agents, coexist. Behaviors and games become heterogeneous autonomously, the ratio of cooperative agents becomes 80%, and the average utility is stable high.

7 Conclusion

We deal with these coordination games and asymmetric coordination games as a type of interaction and each agent can change the interaction type. At first, we examined three populations: a pure population in which all agents play the coordination game, a pure population in which all agents play the complementary game, and a mixed population in which one half play the coordination game, whereas the other half play the complementary game. We found that in the mixed population, about 28 or 72% of the agents choose S1 at last according to the trials. Behaviors become heterogeneous autonomously. Next, we focused on the effects of decision errors on collective behavior. We showed that small decisions error make collective behavior stable in pure populations. Finally, we examined populations in which each agent can change the interaction type and showed that in the mixed population, behaviors and games become heterogeneous autonomously. The ratio of cooperative agents becomes 80% and the average utility is stable high.

References

Axelrod R (1985) The evolution of cooperation. Basic Books, New York
Axelrod R, Cohen MD (2001) Harnessing complexity. The free press, London
Huberman B, Glance N (1993) Diversity and collective action. In: Haken H, Mikhailov A (eds) Interdisciplinary approaches to nonlinear complex systems (Springer Series in Synergetics). Springer, Berlin, pp 44–64
Iwanaga S, Namatame A (2002) The complexity of collective decision. Nonlinear Dyn Psychol Life Sci 6(2):137–158
Iwanaga S, Namatame A (2003) Emergent division of labor and evaluation of its efficiency and equity. Int J Knowl-Based Intell Eng Syst 7(3):115–123
Iwanaga S, Namatame A (2017) Influence of partner selection on functional differentiation: emergence of diversity by isolated interaction and preference change. J Robot Netw Artif Life 3(4):257–264
Kirman AP (1997) The economy as an interactive system. In: Arthur WB, Durlauf SN, Lane DA (eds) The economy as an evolving complex system II. Addison Wesley, Reading, MA, pp 491–533
Kubo M, Kakazu Y (1994) Evaluation of coordinated motions of multi-agent systems on competition for food between ant colonies [in Japanese]. IPSJ J 35(8):1555–1566
Nowak MA (2006) Evolutionary dynamics: exploring the equations of life. Belknap Press, Cambridge
Reynolds C (1987) Flocks, herds, and schools: a distributed behavioral model, computer graphics. In: Proceedings of the 14th annual conference on computer graphics and interactive techniques, SIGGRAPH'87, pp 25–34
Rubenstein M, Cornejo A, Nagpal R (2014) Programmable self-assembly in a thousand-robot swarm. Science 345(6198):795–799
Sandholm WH (2001) Preference evolution two-speed dynamics and rapid social change. Rev Econ Dyn 4(3):637–679
Schelling T (1978) Micromotives and macrobehavior. W W Norton & Company, New York
Selten R (1975) Reexamination of the perfectness concept for equilibrium points in extensive games. Int J Game Theory 4(1):25–55
Sims K (1994) Evolved virtual creatures. In: Proceedings of the 21st annual conference on computer graphics and interactive, SIGGRAPH'94, pp 15–22
Terzopoulos D, Tu X, Grzeszczuk R (1996) Artificial fishes: autonomous locomotion, perception, behavior, and learning in a simulated physical world. Artif Life 1(4):327–351

Price Distortions and Public Information: Theory, Experiments, and Simulations

Alba Ruiz-Buforn, Simone Alfarano and Eva Camacho-Cuena

Abstract This paper studies the effects on the asset price of the introduction of a public signal in the presence of asymmetric private information in a decentralized market. We introduce an artificial market model populated by bounded rational agents with heterogeneous levels of reasoning: sophisticated and naive traders. The model captures the main impacts of public information analyzed in the laboratory experiments reported by Ruiz-Buforn et al. (Higher-order beliefs and overweighting of public information in a laboratory financial market, 2019). Public information, when correct, coordinates market activity, improving price convergence to the fundamentals. By contrast, unwarranted public information pushes prices away from fundamentals. This strong influence of public information on prices is primarily driven by its common knowledge property.

1 Introduction

The idea that a price system based on competitive markets is able to aggregate dispersed information in the economy dates back at least to Hayek (1945). A detailed description of the ability of markets in efficiently aggregate information, and the conditions under which this might take place, is found in the theoretical literature of rational expectations and market structure. Grossman and Stiglitz (1980) demonstrate that a paradox exists when in a competitive and efficient market the production of information is costly. Informed traders have no incentive to reveal their private information into the market if not properly compensated by the costs of producing

A. Ruiz-Buforn · S. Alfarano (✉) · E. Camacho-Cuena
Universitat Jaume I, Av. Vicent Sos Baynat, s/n, 12071 Castellón de la Plana, Spain
e-mail: alfarano@uji.es

A. Ruiz-Buforn
e-mail: buforn@uji.es

E. Camacho-Cuena
e-mail: camacho@uji.es

A. S. Chakrabarti et al. (eds.), *Network Theory and Agent-Based Modeling in Economics and Finance*, https://doi.org/10.1007/978-981-13-8319-9_4

information, and therefore it does not exists an equilibrium price. This problem is solved when prices only partially reveal the information.

There is a large experimental literature dealing with the aggregation and dissemination of information in laboratory financial markets to test the theoretical predictions of rational expectation equilibrium models. Many experimental contributions have shown that centralized asset markets can disseminate private information held by agents (Plott and Sunder 1982). However, the ability of markets to disseminate (free allocated) information is limited (Plott and Sunder 1988; Corgnet et al. 2015; Camerer and Weigelt 1991).

There is another set of models that study the aggregation of information in decentralized markets. These markets are characterized by their opaqueness, where the details of the contracts are only known by the two parties (Duffie 2012). Several theoretical studies suggest that decentralized markets are able to aggregate dispersed information in the market, although the aggregation process is slowed down with respect to centralized markets (Duffie et al. 2005, 2015; Duffie and Manso 2007).

Despite the extensive literature on aggregation of information, not much attention has been paid to the interplay between private and public information and its potential adverse effects on market performance. In addition to the information held privately by traders, one might assume the existence of a disciplining institution that releases public information to improve market efficiency. For instance, the European Central Bank employs the *forward guidance* to manage the expectations of investors and consumers, providing information about future monetary policy targets. Thus, the forward guidance can influence current financial and economic conditions. However, the central bank announcements might influence too much the informativeness of prices and create an overweighting phenomenon, enhancing the volatility of markets. Public information, in fact, provides common priors for the market and "significant market events generally occur only if there is similar thinking among large groups of people..." (Shiller 2002).

Taking stock of that, it is not trivial to predic19999t the effect of public announcements on market performance. Beyond the information on fundamentals, public announcements provide information about the beliefs of the other market participants. Morris and Shin (2005) state that "The central bank cannot manipulate prices and, at the same time, hope that prices yield informative signals." Another example is the sovereign bonds market where prices are closely tracked to assess the probability of debt default of a country. However, prices may become uninformative when some unwarranted information is publicly announced. This public information may allow self-fulfilling beliefs. Allen et al. (2006) prove that there is an excessive reliance on a public signal even when there is not an explicit coordination motive. Agents use that information to better forecast the aggregate demand and, therefore, they overweight its value above and beyond its role in predicting the liquidation value of the asset.

We address the overweighting of public information phenomenon within a simple trading model. We formalize a decentralized asset market with heterogeneous agents who differ in their level of reasoning and information.[1] Using Monte Carlo simulations and comparing them with the observed experimental data in Ruiz-Buforn et al. (2019), we establish two conjectures. Our first conjecture suggests that the presence of more traders with higher levels of reasoning increases the impact of public information in the aggregate transaction prices. Second, the common knowledge of the public signal is the main responsible of the distortive effect of a misleading public signal. Ruiz-Buforn et al. (2019) study the effects of releasing public information in a laboratory financial market where traders have access to private information.[2] Our model reproduces qualitatively the main patterns observed in the laboratory experiment. Prices are strongly biased toward the public signal independently of its realization, i.e., correct or incorrect prediction on fundamentals. However, the impact of mistaken information lessens when the released signal, even if it is observed by all traders, is not common knowledge.

The rest of the chapter is organized as follows. Section 2 introduces the behavioral trading model and its results. Section 3 describes the laboratory experiment. Section 4 illustrates the model calibration and the finite sample properties of the model, implementing Monte Carlo simulations and comparing them to the observed data. Finally, Sect. 6 concludes.

2 The Model

2.1 *Information Set*

The market is populated by N agents who are endowed with risky assets and cash.[3] The asset is essentially an Arrow–Debreu security, which can take two possible values $D \in \{0, 1\}$ with equal probability. At the beginning of the market, all agents observe a binary public signal $y \in \{-1, 1\}$ that predicts the value of the asset with probability $q \in [\frac{1}{2}, 1]$. A public signal $y = -1$ indicates that $D = 0$, whereas a signal $y = 1$ indicates that $D = 1$. Moreover, each agent receives two binary private signals that predict the value D, each one with probability $p \in [\frac{1}{2}, 1]$. Agent i's private information can take three values: (i) $x_i = 2$ if the agent receives two private signals predicting $D = 1$; (ii) $x_i = 0$ if they receive two opposite signals; and (iii) $x_i = -2$ if they receive two private signals predicting $D = 0$. Thus, there are three possible information levels depending on the realization of the private signals. Each level of information is denoted by "i", which indicates *high*, *medium*, and *low* $i \in \{H, M, L\}$.

[1]Cognitive hierarchical models represent stock markets where some traders believe, incorrectly and over-confidently, that their strategy is the most sophisticated. In such situations, "the players are not in equilibrium because some players' beliefs are mistaken" (Camerer et al. 2004).

[2]Ruiz-Buforn et al. (2018) test the overweighting phenomenon when traders can acquire costly private information.

[3]The amount of cash is a loan that they must give back at the closing of the market.

Note that y is common knowledge to all agents whereas x_i is private information for each agent and, therefore, not observable by the other agents.

According to the Bayesian inference, agent i's expected dividend is

$$E[D=1|x_i, y] = \frac{1}{1+\left(\frac{1-p}{p}\right)^{x_i}\left(\frac{1-q}{q}\right)^{y}}. \tag{1}$$

Given the informational levels, there are three possible expected dividend values in the market $D_i \in \{D_H, D_M, D_L\}$.[4] Agent i's expected dividend is *high* (D_H) when his private information is $x_i = 2$. If agent i observes $x_i = 0$, he is privately uninformed and his expected dividend is *medium* (D_M). Finally, if he observes $x_i = -2$, he has a *low* expected dividend D_L.

2.2 Agents' Decisions

Once private and public information is revealed, agents decide whether to be sellers or buyers and the price of their offer. Agents have one chance to decide their offer and bargaining is not allowed. Each agent's offer involves one randomly chosen agent as a counterpart. Thus, an agent who observes the offer of another agent in the market decides whether to accept or reject it.

We assume that all agents are risk-neutral and bounded rational since they are not fully aware of the strategic implications of their actions in an asymmetric information environment. Using the concepts of cognitive hierarchy theory, there are two types of agents $\tau \in \{N, S\}$ according to their level of reasoning. A fraction $\theta \in [0, 1]$ of the agents' population is sophisticated (S) while a fraction $1-\theta$ is constituted by naive traders (N). Agents desire to maximize expected payoffs,[5] using their information. Naive traders only consider the information they have about the fundamentals. Sophisticated traders, on the other hand, make use of the public information in order to forecast other agents' beliefs, considering that it also carries information on the asset liquidation value. Unlike naive traders, sophisticated traders compute the probability of acceptance for each offer. Essentially, our market population is characterized by agents trading based on their first-order beliefs (naive) and agents trading based on their second-order beliefs (sophisticated).

An important point should be clarified here. Our market is populated by heterogeneous agents with different time-invariant trading strategies. This means that the agents do not learn from their trading activity, but they follow the same strategy. The market is not centralized since we implement a bilateral trading mechanism between two agents. We use the average price as a measure of central tendency of the whole transactions distribution.

[4]Hereafter, we will denote the expected dividend as $D_i \equiv E[D=1|x_i, y]$.

[5]The expected payoff denotes the income of the trader after dividend payment.

Naive Traders

A naive trader acts as prior information trader considering only his own information without taking into account the zero-sum nature of the game and the strategic implications of his actions.

Naive Proposer: First, we define the features of naive traders' bidding behavior. If he submits a buy offer at price b and it is accepted, he gets an additional unit of the asset and his expected payoff is $\pi^N(b|D_i) = 2D_i - b$. If his sell offer at price a is accepted, he trades his unit and gets a payoff[6] $\pi^N(a|D_i) = a$. Finally, if he does nothing, i.e., there is no trade (nt), his expected payoff is $\pi^N(nt|D_i) = D_i$.

A naive trader takes the action that provides him with the highest expected payoff:

$$s_i = \underset{s \in \{a,b,nt\}}{\text{argsup}}\ \pi^N(s|D_i)\ , \tag{2}$$

where a, b, and nt refers to every possible action s of a naive trader: submitting a sell offer at price a, submitting a buy offer at price b, and doing nothing, respectively.

Comparing the three possible strategies—submitting a bid, an ask, or doing nothing—he prefers submitting bids below his expected dividend and asks above it ($b < D_i < a$). Since he only considers the information on fundamentals, we assume he estimates that the probability of an offer being accepted is exponentially decreasing with the gains from trading. So, he submits bids and asks close to his expected dividend D_i. The naive proposer i, therefore, submits bids $b_i = D_i - \varepsilon$ and asks $a_i = D_i + \varepsilon$ with the same probability, since both actions provide him with the same expected payoff, which is strictly higher than doing nothing. Note that he earns the extra profit ε with respect to doing nothing, which is independent of his type i. The parameter ε is exogenously fixed. We assume that $0 < \varepsilon < min\{D_i\}$. So that all bids and asks are within the range [0, 1] independent of i.

The expected payoff of a naive proposer when submits his optimal offer according to Eq. (2) is

$$\pi^N(a|D_i) = D_i + \varepsilon$$

and

$$\pi^N(b|D_i) = D_i + \varepsilon\ .$$

Since it is the same, he randomizes between the two strategies.

Naive Receiver: Similarly, we assume that a naive trader accepts offers that provide him with a higher expected payoff than no accepting them. If a naive trader receives a bid, the expected payoff of acceptance is b. If he receives an ask and accepts it,

---

[6]The proposer knows with certainty the gains of his action since they do not depend on the liquidation value of the asset.

he gets an additional asset and his expected payoff is $2D_i - a$. Thus, a naive trader accepts buy offers below and sells offers above his expected dividend:

$$\pi^N(b, D_i) = \begin{cases} b & \text{if he accepts} \\ D_i & \text{if he rejects} \end{cases}$$

and

$$\pi^N(a, D_i) = \begin{cases} 2D_i - a & \text{if he accepts} \\ D_i & \text{if he rejects} \end{cases}$$

In conclusion, he accepts a bid if $b > D_i$ and an ask if $a < D_i$.

Sophisticated Traders

We assume that a sophisticated trader acts with certain level of strategic reasoning. When deciding her strategy, a sophisticated trader uses her information set (x_i, y) and considers the trading motives of the counterpart to decide her optimal action. We assume that sophisticated traders consider their second-order beliefs based on the assumption that all other traders in the market are naive. As a consequence, sophisticated traders take into account how information (private and public) is distributed across traders in the market. The bounded rationality of this kind of traders stems from the fact that they do not contemplate higher order beliefs, i.e., they believe that all other traders are naive, without further iteration levels.

In this framework, the public signal enters in the information set of all traders in the market. The public nature of this signal allows sophisticated traders to better characterized other traders' expectations.

Sophisticated Proposer: When a sophisticated trader submits an offer, her expected payoff depends on the selling price a or the buying price b, her information D_i and the probability that her offer is accepted. She faces a trade-off between the transaction payoff and the probability of closing such transaction. If she submits an ask a, her expected payoff is[7]

$$\pi^S(a|D_i) = D_i + (a - D_i) \sum_j Pr[a < D_j|D_i] \ ,$$

where $Pr[a < D_j|D_i]$ denotes the probability that a sophisticated trader with expected dividend D_i sells her asset at price a. In other words, $Pr[a < D_j|D_i]$ represents the probability to be matched with a trader with an expected dividend $D_j > a$ given her information. Similarly, when submitting a bid b her expected payoff is

[7] See Appendix 7.1 for the extended functions.

$$\pi^S(b|D_i) = D_i + (D_i - b) \sum_j Pr[b > D_j|D_i] ,$$

where $Pr[b > D_j|D_i]$ denotes the probability that a sophisticated trader with expected dividend D_i buys her asset at price b. In case that the sophisticated trader decides to do nothing, her expected payoff is D_i.

A sophisticated trader takes the action that provides her with the highest expected payoff:

$$s_i^* = \underset{s\in\{a,b,nt\}}{\text{argsup}} \ \pi^S(s|D_i) , \tag{3}$$

where a, b, nt denotes every possible trader's action: selling at price a, buying at price b, and doing nothing, respectively. She faces a trade-off between maximizing her payoff and maximizing the potential market demand.

Solving Eq. (3), the optimal action for a sophisticated trader with an expected dividend D_H is submitting buy offers at price $b_H^* = D_M + \varepsilon$. If her expected dividend is D_L, she will submit sell offers at price $a_L^* = D_M - \varepsilon$. The trade-off between her transaction payoff and the potential market demand is optimized at the medium-price level. To give some intuition, at that price, she satisfies the demand of two out of three types of naive traders. Finally, if she is privately uninformed (D_M), submitting a bid $b_M^* = D_L + \varepsilon$ or an ask $a_M^* = D_H - \varepsilon$ provides her with the highest expected payoff. Note that an uninformed sophisticated trader is not able to exploit the difference between her private information and the public signal. Given that her information is the public signal, it does not help her to characterized other traders' expectations. So, her optimal choice is selling at the highest possible price or buying at the lowest possible price that a trader is willing to accept.

Sophisticated Receiver: In case a sophisticated trader receives an offer, it provides her with new information to be updated. Indeed, the received offer carries information about the proposer's private information. A sophisticated trader knows that traders submit offers that provide them with positive expected payoffs. This means that no trader will submit sell offers below his expected dividend nor buy offers above his expected dividend.[8] The expected payoff of a sophisticated trader when she receives a bid b or an ask a is

$$\pi^S(b, D_i) = \begin{cases} b & \text{if she accepts} \\ \sum_j D_{ij} \ \ Pr[D_j|D_j > b] & \text{if she rejects} \end{cases}$$

and

[8] For example, she identifies the proposer as type H when she observes a bid $b > D_M$. In case she observes a buy offer at $b > D_L$, she infers the probability that the expected dividend of the proposer is D_H or D_M. A bid $b < D_L$ does not carry additional information since any trader makes positive expected payoffs buying at a very low price.

$$\pi^S(a, D_i) = \begin{cases} -a + 2\sum_j D_{ij} \;\; Pr[D_j|D_j < a] & \text{if she accepts} \\ \sum_j D_{ij} \;\; Pr[D_j|D_j < a] & \text{if she rejects} \end{cases}$$

where D_{ij} denotes the updated expected dividend of the trader i when she infers from the offer that the proposer's expected dividend is D_j.[9] D_{ij} is computed by adding the proposer's private signals $x_j \in \{-2, 0, 2\}$ to her own information set $D_{ij} \equiv E[D = 1|x_i, x_j, y]$. Finally, $Pr[D_j|D_j > b]$ and $Pr[D_j|D_j < a]$ define the probability of the proposer's level of information given the observed bid and ask, respectively.[10] The optimal action depends on which of the two expected payoffs is higher.

2.3 Endogenous Order Flow

Having defined the optimal strategy for each type of trader, we now compute the optimal number of offers submitted by them. In order to do so, we introduce a cost function in the number of offers. This cost function is composed by a quadratic term[11] $c \cdot (z_i^\tau)^2$ and an opportunity cost term D_i. The parameter c is a constant of proportionality. The variable z_i^τ denotes the number of offers submitted by a trader of type τ and with expected dividend D_i. The quadratic term can be understood as some costs to manage all information related to the offers. The opportunity cost denotes trader's expected payoffs if he would not submit any offer. Proposer's cumulative payoff function is

$$\Pi^\tau(s|D_i) = \left[\pi^\tau(s|D_i) - c\, z_i^\tau - D_i\right] z_i^\tau \,, \tag{4}$$

where $\pi^\tau(s|D_i)$ denotes the proposer's expected payoffs when playing his or her optimal strategy s given her information. The order flow is the optimal number of offers per unit of time and it is computed maximizing Eq. (4) with respect to z_i^τ:

$$z_i^\tau = \frac{\pi^\tau(s|D_i) - D_i}{2c} \,.$$

For naive traders, we denote the order flow with $z_i^N = \nu = \frac{\varepsilon}{2c}$, which is independent of their information level. The order flow of sophisticated traders, on the other hand, changes based on their information level. The relative order flow of sophisticated traders with respect to the naive traders is

---

[9]Table 3 in Appendix 7.2 describes the information revealed in every offer, together with some illustrative examples to explain the computing process of the expected payoffs.

[10]Recall that sophisticated traders believe all other traders are naive.

[11]The quadratic nature of the costs is necessary for having an optimal value for the number of offers.

Table 1 Transactions

Proposer (τ_i)	Offer	Price	Receiver (τ'_j)
S_H	b^*_H	$D_M + \varepsilon$	N_L, N_M, S_L
S_M	a^*_M	$D_H - \varepsilon$	N_H
	b^*_M	$D_L + \varepsilon$	N_L
S_L	a^*_L	$D_M - \varepsilon$	N_H, N_M, S_H
N_H	a_H	$D_H + \varepsilon$	*No trade*
	b_H	$D_H - \varepsilon$	N_M, N_L, S_L
N_M	a_M	$D_M + \varepsilon$	N_H, S_H
	b_M	$D_M - \varepsilon$	N_L, S_L
N_L	a_L	$D_L + \varepsilon$	N_H, N_M, S_H
	b_L	$D_L - \varepsilon$	*No trade*

$$\mu_i = \frac{z_i^S}{z_i^N} = \frac{\pi^S(s|D_i) - D_i}{\varepsilon} .$$

Without loss of generality, we can set the value of c in such a way that the order flow of the naive traders is $\nu = 1$, so that $c = \frac{\varepsilon}{2}$.

2.4 Transactions

What we have characterized so far the traders' behavior given their level of reasoning and information. In general, a trade occurs because traders differ in endowments, preferences or beliefs. The latter element takes place in our framework. In this section, we study the tendency of transaction prices as a function of the fraction of sophisticated traders in the market $\theta \in [0, 1]$.

Without loss of generality, Table 1 describes the market transactions assuming $D = 1$. The first column lists the proposer's type according to their level of reasoning and information, the second and the third columns show the optimal strategy of every type of trader. Finally, the last column lists what types of traders accept the offer.

Let us define with $f(\tau_i, \tau'_j)$ the probability per unit of time that a given match between two traders, a proposer τ_i and a receiver τ'_j, turns out in a transaction.[12] If we sum over the index τ' and j, $\sum_{\tau' j} f(\tau_i, \tau'_j)$, we obtain the probability per unit of time that a trader τ_i closes a transaction according to her or his optimal strategy. Introducing the order flow $\omega_i^\tau = \begin{cases} \mu_i \text{ if } \tau = S \\ 1 \text{ if } \tau = N \end{cases}$ we have

$$t(\tau_i) = \omega_i^\tau \sum_{\tau' j} f(\tau_i, \tau'_j) , \tag{5}$$

[12]In order to compute it, we refer to Table 1. Note that $f(\tau_i, \tau'_i) = 0 \;\; \forall i$.

Table 2 Probability of transaction per unit of time

Proposer (τ_i)	T
S_H	$\mu_H\theta[(1-\theta)2p^3(1-p)+p^2(1-p)^2]$
S_M	$\mu_M\theta(1-\theta)p^3(1-p)$
	$\mu_M\theta(1-\theta)p(1-p)^3$
S_L	$\mu_L\theta[(1-\theta)2p(1-p)^3+p^2(1-p)^2]$
N_H	$\nu(1-\theta)[(1-\theta)p^3(1-p)+0.5p^2(1-p)^2]$
N_M	$\nu(1-\theta)[p(1-p)^3+p^3(1-p)]$
N_L	$\nu(1-\theta)[0.5p^2(1-p)^2+(1-\theta)p(1-p)^3]$

where $t(\tau_i)$ is the expected number of transactions of trader τ_i per unit of time.

Table 2 shows the expected number of transactions per unit of time for each trader of type τ_i. Note that it is already included the corresponding order flow. To give an example, the probability of observing a transaction of a N_H is computed as the probability that a naive with high expected dividend is matched with any trader who is willing to accept his offer $b_H = D_H - \varepsilon$: a sophisticated trader S_L with low expected dividend, an uninformed naive trader N_M or a naive trader N_L with low expected dividend. All the sum is multiplied by the order flow $\nu = 1$. The explicit calculation is the following:

$$t(N_H) = \nu\sum_{\tau' j} f(N_H, \tau'_j) = \nu[\frac{1}{2}(1-\theta)\theta p^2(1-p)^2$$
$$+(1-\theta)^2 p^3(1-p)] + \frac{1}{2}(1-\theta)^2 p^2(1-p)^2] \,.$$

Let us define a vector $\mathbf{T}$ whose components are $t(\tau_i)$. As a proxy for the tendency of transaction prices, we compute the average price $\overline{P}$ in the market. The vector of prices $\mathbf{P} = (D_M, D_H, D_L, D_M, D_H, D_M, D_L)$ is defined according to the proposer's offers in Table 1.[13] Finally, the average price $\overline{P}$ is computed as the weighted sum of each possible value of the price according to the probability of observing such price over the total feasible transactions, $F = \sum_{\tau_i} T(\tau_i)$. Defining $\mathbf{T}'$ as the transpose of $\mathbf{T}$, the average price is

$$\overline{P} = \frac{\mathbf{T}'}{F}\mathbf{P} \,. \tag{6}$$

[13] We omit the ε parameter for notational convenience.

3 Laboratory Experiment

In this section, we sketch briefly the experimental design of Ruiz-Buforn et al. (2019). The experiment took place in the LEE (*Laboratori d'Economia Experimental*) at University Jaume I in Castellón. Each session consists of ten independent markets lasting 3 min each. The asset market is implemented as a double auction where subjects are free to introduce their bids and asks or directly accept other trader's outstanding bid or/and ask. Every bid and ask concerns only one unit of the asset, but subjects can handle as many as desired as long as they have enough cash or assets (no short sale is allowed).

Each market is populated by 15 subjects who are endowed with 1000 units of experimental currency (ECU)[14] and 10 one-period life units of a risky asset. The dividend takes the value 0 or 10 with a 50% probability, which is common knowledge to all subjects. At the beginning of each market, the dividend is randomly determined by the experimenter, but not revealed to the subjects until the end of the same, when the dividend is paid. Additionally, subjects receive noisy signals on the dividend value. Signals are partially but not totally informative and they are presented to the subjects taking the value 10 or 0. If a subject observes a signal that results to be 10 (0), he can infer that the dividend is expected to be 10 (0) with probability 0.8 and 0 (10) with probability 0.2.

The experiment consists of three treatments depending on the source of information in the market. In the baseline treatment (B), subjects receive two noisy private signals. In the public information treatment (PS), all subjects observe an identical noisy public signal besides the two private signals. In the common information treatment (CS), subjects observe three signals and one of them is identical to all subjects in the market. However, unlike in the PS treatment, this signal is not common knowledge.[15,16]

At the end of each market, dividends are paid out and the subjects' profits are computed as the difference between their initial endowment and the money held at the end of the market. Essentially, profits consist of the gains or losses generated by the trading activity and the dividend. Each subject's final payoff is computed as the accumulated profit in all markets.

---

[14]During the experiment, earnings and dividends are designated in experimental currency units (ECU) and converted into Euro at the end of the session.

[15]Within treatments, we differentiate between two types of markets. Markets with a correct public or common signal are labeled "Correct PS" and "Correct CS", respectively. Markets with an incorrect released signal are labeled as "Incorrect PS" and "Incorrect CS".

[16]In treatment B, one group of subjects participate and, therefore, there are 10 markets. In treatments PS and CS, two groups of subjects participate; we have therefore 20 markets for each treatment.

4 Model Calibration

In this section, we calibrate our model to compare the theoretical mean price $\overline{P}$ and the experimental data (Ruiz-Buforn et al. 2019). It is important to emphasize that this model provides a *post hoc* interpretation of the impact of public information on traders' behavior. Indeed, we did not design the experiment to test this model.

Morris and Shin (2002) claims that public information is a double-edged instrument that simultaneously provides information about the fundamentals and information about other traders' beliefs. The second *edge* is due to the common knowledge of that signal, and it is the reason for the emergence of overreliance on public information above and beyond its information on fundamentals. Our study aims at providing some theoretical insights into the overweighting mechanism. We disentangle the dual role of public information by comparing the mean price when traders observe a public signal and the mean price when they observe an identical signal, which is not common knowledge among traders. In this case, the identical signal carries information on the fundamentals but does not reveal information about the other traders. We refer to that signal as *common signal*. If an incorrect public signal distorts prices more than an incorrect common signal, we can state that the public nature of information is the main determinant of traders' overreaction. Stated differently, we find evidence on the overweighting phenomenon if the mean price is biased toward the public signal regardless of its realization, namely whether it is correct or incorrect. Conversely, the mean price should never be biased toward the common signal, independently of its realization. Onward, we will refer to the following scenarios: markets with a public signal are labeled as PS scenario, and markets with an identical signal that is not common knowledge (common signal) as CS scenario.[17] Additionally, we introduce a baseline scenario (B) where there is no identical signal released to the market. Thus, each trader only observes two private signals.

The public information benchmark (PB) represents the theoretical expected dividend considering only the public information and is computed by the following formula:

$$E[D = 1|\hat{y}] = \frac{1}{1 + \left(\frac{1-q}{q}\right)^{\hat{y}}}, \tag{7}$$

where $\hat{y}$ takes values 1 or -1 if the signal is public and $\hat{y} = 0$ if the signal is common.[18] $PB = 0.8$ when the public signal predicts $D = 1$ and $PB = 0.2$ when that signal predicts $D = 0$. Finally, $PB = 0.5$ in the B and CS scenarios.

Recall that, for simplicity, we focus our attention on the case $D = 1$ since the model is symmetric in the two states of the world. Thus, the fraction of traders that

---

[17] In the CS scenario, the procedure for the resolution of the theoretical model is explained in Appendix 8.

[18] We denote now the released signal by $\hat{y}$ instead of y like in Eq. (1) to unify the three scenarios (benchmark, public signal, and common signal) into a single equation.

receive two signals pointing to the dividend and, then, have a high expected dividend D_H is p^2. A fraction of $2p(1-p)$ are uninformed traders whose expected dividend is D_M and a fraction of $(1-p)^2$ are misinformed traders whose expected dividend is D_L. Considering those probabilities, instead of the corresponding frequencies in the population, implies that we neglect the fluctuations in the configuration of the population due to the finite number of traders.[19] Just like the experimental design, we fix the quality of every private, public and common signal at $p = q = 0.8$.

4.1 Results

Without loss of generality, Fig. 1 shows the mean price in the three scenarios (B, PS, and CS) when $D = 1$ as a function of the proportion of sophisticated traders in the market population. The mean price is computed separately according to the correctness of the released signal. We use the B scenario as a benchmark for evaluating the impact of releasing a public signal. One can see that the mean price in the B scenario (dashed-dotted line) is biased toward the dividend, although without converging to it. The presence of sophisticated traders drives prices away from the dividend.

Looking at the bottom lines of the figure, it is evident that an incorrect released signal pushes prices away from the dividend $D = 1$. In particular, an incorrect public signal (thick-solid line) has a stronger distorting impact on price performance than an incorrect common signal. An incorrect public signal (thick-solid line) has a stronger distorting impact on price performance. The mean price quickly drops when there are sophisticated traders in the market. In fact, a small fraction of sophisticated traders ($\theta = 0.2$) is sufficient to observe that the mean price clearly tends to the incorrect public signal ($PB = 0.2$), getting closer to the PB as θ increases. The maximum level of overweighting is eventually reached when $\theta = 1$. It is worth noting that this phenomenon is quite a robust outcome. The distance between mean prices and the public signal is, in fact, almost unchanged in the interval $\theta \in [0.2, 0.7]$. This means that, in order to observe the price biased toward the incorrect public signal, it is not necessary a process of fine-tuning the value of θ.

On the other hand, the price behavior in the CS scenario is markedly different. Even though an incorrect common signal distorts the mean price (thin-solid line), this negative effect is less harmful than the negative impact of an incorrect public signal. Interestingly, the presence of sophisticated traders has no impact on the mean price until they make up the majority of the market population. The mean price starts from the middle of range values and remains constant until sophisticated traders reach a percentage close to 80%, which is the large majority of the population.

The top lines of the Fig. 1 show the mean price when the released signal points toward the dividend. Mean prices show a lower sensitivity to the presence of sophisticated traders in both scenarios with respect to the case with an incorrect signal.

[19] We assume that the number of traders is sufficiently large that the fluctuations around the mean can be neglected.

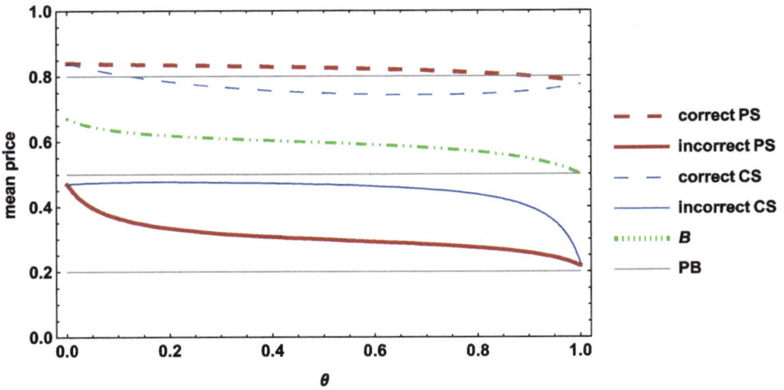

Fig. 1 Mean price over the proportion of sophisticated traders, assuming $\varepsilon = 0.025$

The convergence of the mean price to the $PB = 0.8$ is largely independent of the fraction of sophisticated traders. Surprisingly, there is almost no difference between markets with a public and a common signal. The mean price takes similar values in both scenarios. The mean price gets closer to the dividend with respect to the B scenario. We can claim that releasing an identical signal improves market performance, moving traders' activity at levels closer to fundamentals.

We can deduce from our model that public information is beneficial *per se* when it is correct. Both information on fundamentals and on other traders' expectations help market price to converge to the dividend value. The mean price is almost indistinguishable in the public or common scenario. However, we observe a different impact when there is an identical misleading signal in the market. The fact that traders are aware that they observe an identical signal reinforces the distorting effects. The public signal is overweighted in the aggregation of information into prices due to the overreliance of the traders on public information. Differences between mean prices in scenarios PS and CS, when the released signal is incorrect, indicate the importance of the common knowledge of the public announcements.

We evaluate now whether the released signal has different impacts on the mean price in aggregate terms. Figure 2 plots the mean price for every scenario computed as the weighted probability of occurrence of the signal. Stated differently, the aggregate mean price is computed by the sum of two terms: (i) the mean price when the released signal predicts dividend 1 weighted by the probability of being correct, p and (ii) the mean price when the released signal predicts dividend 0 weighted by the probability of being incorrect $(1 - p)$. On aggregate, releasing a signal into the market improves mean price performance. The mean prices in both PS and CS scenarios are closer to the dividend than the mean price in the baseline scenario. Figure 2 shows that the effect of the common knowledge is almost indistinguishable in aggregate terms.

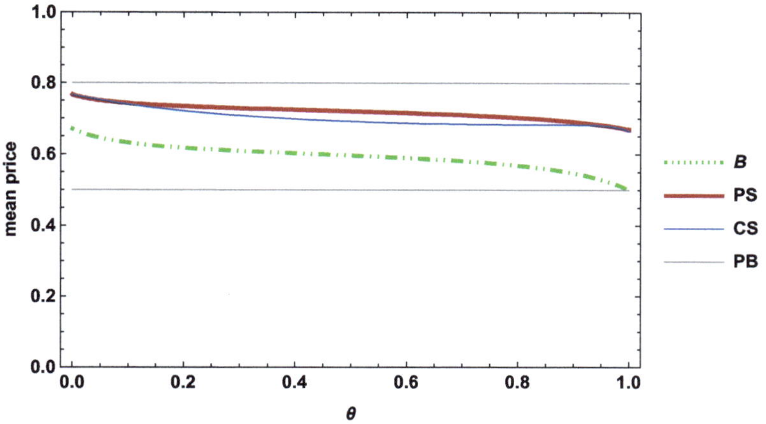

Fig. 2 Mean price over the proportion of sophisticated traders, assuming $\varepsilon = 0.025$ in aggregate terms

4.2 *Monte Carlo Simulations*

Equation (6) assumes a very large number of traders and encounters, since we replace the frequencies with probabilities. We study now the finite sample properties of our model. We run Monte Carlo simulations based on our theoretical model, assuming 15 heterogeneous traders who have different levels of reasoning. We run 100 market configurations for every realization of the public or common signal given $D = 1$. In each market configuration, 30 private signals are drawn using a binomial distribution and allocated to the traders. Once we fix the distribution of signals among traders, the simulations are initialized with $\theta = 0$, progressively increasing the value of θ in steps of 0.1 until $\theta = 1$. One trader of the whole pool is randomly chosen with equal probability. The probability of being sophisticated or naive depends on the value of θ. Moreover, the number of submitted offers changes depending on the type of the proposer (S,N) and his or her expected dividend. Every offer of a given trader is associated with a counterpart, which is randomly chosen among the rest of traders. The receiver may accept or reject the offer depending on his or her level of reasoning and information. For each value of θ, this operation is repeated 100 times. Finally, the average price of transactions is computed in each case.

Figure 3 shows the mean price obtained in Monte Carlo simulations for PS cases on the panels (a, c) and CS cases on the panels (b, d). We also differentiate between markets where the released signal is correct ($y = 1$) on the panels (a, b) and those where it is incorrect ($y = -1$) on the panels (c, d). One can see that mean price of simulations closely follow the theoretical predictions in all cases. We note further that the mean price dispersion is larger in CS scenario than in PS scenario. This finding indicates that the price is more sensitive to the distribution of signals in CS scenario rather than in the PS scenario.

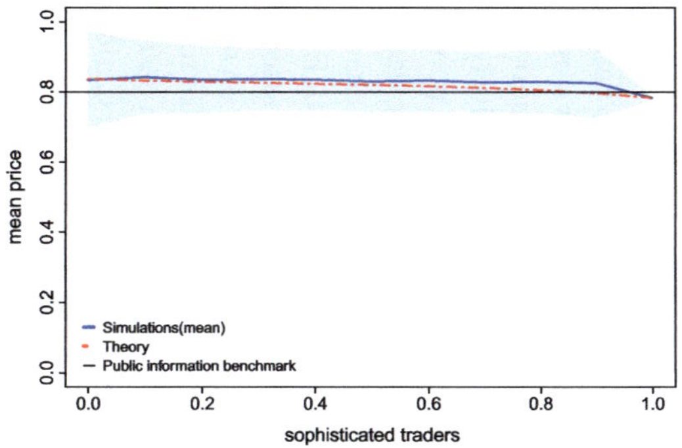

(a) PS scenario with $y = 1$

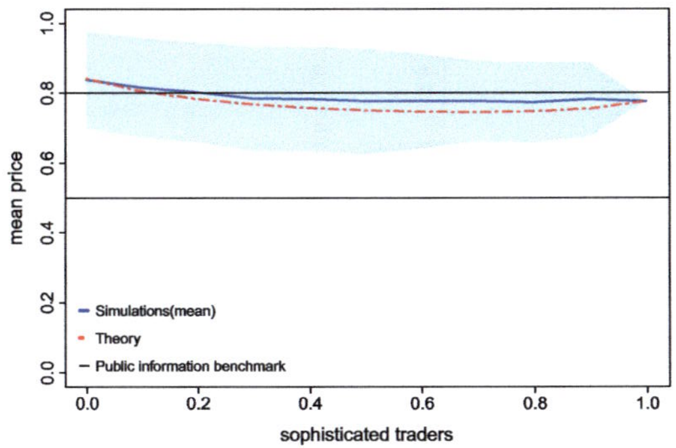

(b) CS scenario with $y = 1$

Fig. 3 The results of 100 Monte Carlo simulations with a public signal on the left panels (**a**, **c**), and a common signal on the right panels (**b**, **d**). The X-axis denotes the proportion of sophisticated traders in the market and Y-axis denotes mean prices. Dashed-blue line describes the theoretical mean price, the solid line represents the average of simulated mean market prices and shaded area shows two standard deviations. Horizontal lines represent the $PB \in \{0.2, 0.5, 0.8\}$ depending on the value of the released signal

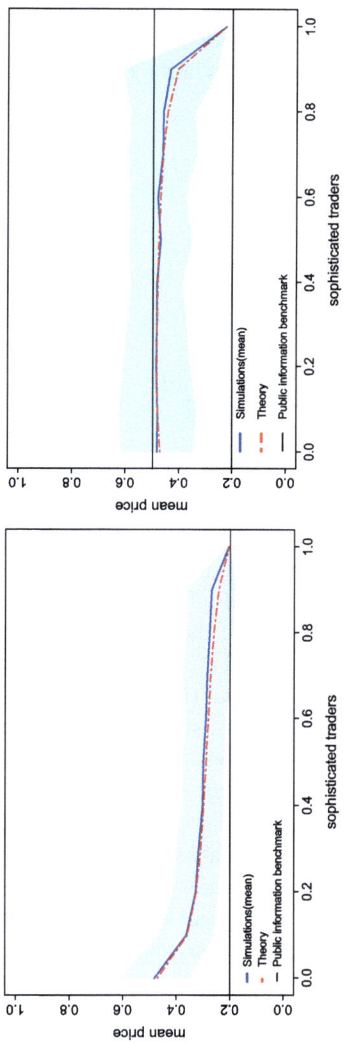

(c) PS scenario with $y = -1$

(d) CS scenario with $y = -1$

Fig. 3 (continued)

5 Corroborating Evidence: Observed Versus Simulated Data

This section compares the computational with the experimental data. The computational data are generated following the process explained in Sect. 4.2 with a fixed value of $\theta = 0.2$. In order to compare the impact of public information on market prices, we evaluate how public information pushes prices away or toward the dividend. We compute the mean absolute deviation of transaction prices PR_{tr} from the dividend value in the laboratory markets:

$$DP_e = \frac{1}{T_r} \sum_{tr=1}^{T_r} \frac{|D - PR_{tr}|}{10}, \tag{8}$$

where T_r is the total number of transactions. For the computational data, the formula is

$$DP_s = \frac{1}{M} \sum_{m=1}^{M} |D - \overline{P}_m|, \tag{9}$$

where $\overline{P}_m$ refers to the mean price of every simulated market (m), and M denotes the number of Monte Carlo runs. When $DP_e = 0$ or $DP_s = 0$, prices or mean prices converge to the dividend value.

Figure 4a plots the DP_s indicator and Fig. 4b plots the DP_e indicator. At a first glance, similarities are clear although a higher dispersion is present in the experimental data. Comparing to the B treatment, the release of a correct public signal helps prices to converge toward the dividend (Correct PS). However, an incorrect public signal drives prices far from the dividend (Incorrect PS). The impact of an incorrect common signal is strongly attenuated in some laboratory markets. This result suggests that subjects are able to learn from prices in the laboratory, even when they receive incorrect signals. Note, however, that an incorrect public signal seems to drag this learning process out. We have tested the effect of misleading information assuming that simulated traders are not able to learn. Although this is a weakness of our model, the main results of the laboratory experiment are reproduced.

After analyzing the impact of a correct and incorrect public signal, it remains to answer to the question: What is the aggregate impact of public information? Fig. 5 plots the data averaging over the different realizations of the signal. One can see that the release of information, public or common, improves price convergence. Therefore, we can conclude that public information is beneficial for market dynamics in the experimental as well as simulated markets.[20]

In conclusion, our model is able to reproduce qualitatively the patterns observed in the experiment, which are (i) prices are biased toward the fundamentals when an

[20]From Fig. 3 one can infer that we would obtain similar results if the fixed proportion of sophisticated traders lies in the interval $\theta \in [0.2, 0.7]$.

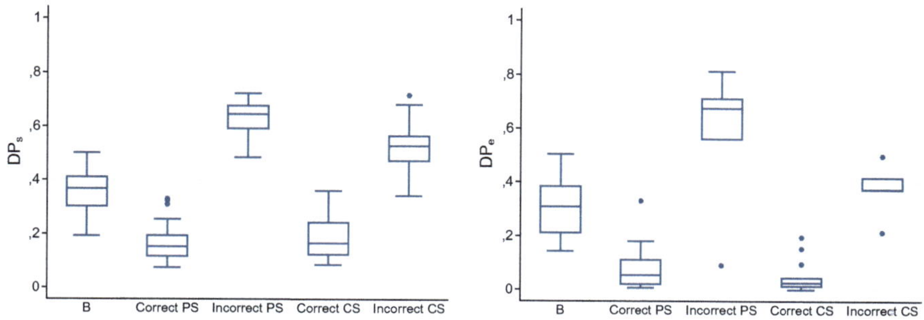

(a) Simulations ($\theta = 0.2$)　　(b) Experimental results

Fig. 4 Distribution of DP across markets, considering whether the released signal is correct or incorrect

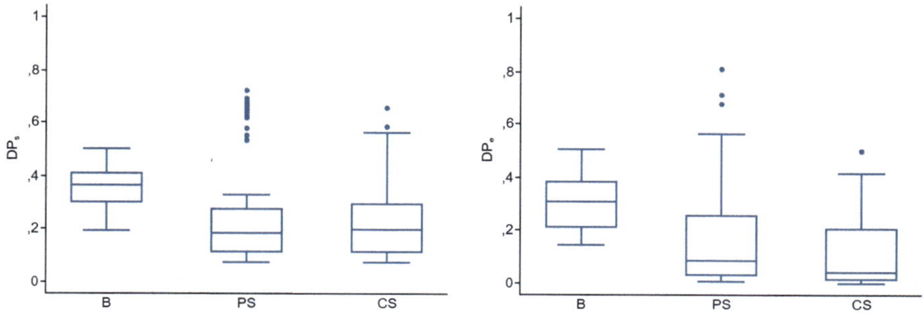

(a) Simulations ($\theta = 0.2$)　　(b) Experimental results

Fig. 5 Distribution of DP across markets

additional signal is released, (ii) the presence of price distortion when the released signal is incorrect, and (iii) a limited distortion if traders observe the same signal without being common knowledge.

6 Conclusions

We propose a simple decentralized asset market with asymmetric information populated by naive and sophisticated traders. The model aims at identifying the principal effects of unwarranted or mistaken public information on prices when it interplays with noisy private information. Under bounded rationality, public information differently affects traders' behavior. Whereas, naive traders only consider their own information, sophisticated traders make use of public information to infer the distribution of aggregate demand. We find that a noisy public signal pushes prices away from fundamentals when it predicts the wrong state of the world. A low proportion of sophisticated traders is sufficient to observe that the mean transaction price follows a mistaken public signal.

We also perform Monte Carlo simulations with a finite sample of traders and calibrate the key parameters to match the ones in the laboratory experiment. We compute three scenarios: markets where there is no public information, markets with public information, and markets where one of the signals is observed by all traders but they are not aware of it. An interesting result emerges: the price is biased toward the incorrect public signal rather than the dividend value. Whereas the distorting impact of unwarranted public information emerges, this effect is much lower under the assumption of non-common knowledge about the released signal. In our behavioral model, the common knowledge nature of public information makes traders overrely on public information.

Our simple model qualitatively reproduces the aggregate behavior observed in the laboratory asset markets of Ruiz-Buforn et al. (2019). Heterogeneity combined with bounded rationality generates similar findings to those of the experimental study. Finally, future work should relax some strong assumptions as learning capacity of traders and implement different market architectures.

Acknowledgements The authors are grateful to the Universitat Jaume I under the project UJI-B2018-77 and the Generalitat Valenciana for the financial support under the project AICO/2018/036. Alba Ruiz Buforn acknowledges the Spanish Ministry of Science and Technology under an Formación de Profesorado Universitario (FPU14/01104) grant.

7 Public Signal Scenario

7.1 Sophisticated Proposers

The expected payoff function of a sophisticated proposer has two components: (i) the expected payoff if the offer is accepted multiplied by the probability of acceptance and (ii) the expected payoff if the offer is rejected, i.e., D_i. Submitting sell offers:

$$\pi^S(a|D_i) = \sum_j \left(a\ Pr[a < D_j|D_i] + D_i\ Pr[a \geq D_j|D_i]\right) .$$

Submitting buy offers:

$$\pi^S(b|D_i) = \sum_j \left((2D_i - b)\ Pr[b > D_j|D_i] + D_i\ Pr[b \leq D_j|D_i]\right) .$$

Table 3 Sophisticated receivers' inference about the expected dividend of the proposer

Observed offer		Type of the proposer
bid	$b \geq D_M$	N_H
	$D_L \leq b < D_M$	N_H, N_M
	$b < D_L$	N_H, N_M, N_L
ask	$a > D_H$	N_H, N_M, N_L
	$D_M < a \leq D_H$	N_M, N_L
	$a \leq D_M$	N_L

7.2 *Sophisticated Receivers*

This section provides some illustrative examples to clarify the computation of expected payoffs when a sophisticated trader receives an offer. Table 3 lists all inferences that a sophisticated trader can make observing a particular offer, assuming all offers are submitted by naive traders.

Receiving Buy Offers: An Example[21]

Let us suppose that a sophisticated trader S_L, whose expected dividend is D_L observes a bid. She updates her beliefs and decides whether accepting or rejecting the offer. For instance, in case she observes a bid $b_{H^-} = D_H - \varepsilon$, she infers the type of proposer is a naive whose expected dividend is D_H.[22]

$$\pi^S(b_{H^-}, D_L) = \begin{cases} b_{H^-} & \text{accepting the bid} \\ \sum_j D_{Lj}\ Pr[D_j | D_j > b_{H^-}] & \text{rejecting the bid} \end{cases}$$

where $D_j = D_H$ since a naive trader with a high expected dividend is the only trader submitting this offer without incurring in losses. Thus $D_{Lj} = D_{LH}$ refers to the updated expected dividend, where subindex L means her prior expected dividend and H is the guessed proposer's expected dividend. Her updated expected dividend is

$$D_{LH} \equiv E[D = 1 | x_L, x_H, y] = \frac{1}{1 + \left(\frac{1-p}{p}\right)^{-2+2} \left(\frac{1-q}{q}\right)^{y}}.$$

In case she observes a bid $b_{M^-} = D_M - \varepsilon$, the type of proposer might be M or H.

[21] The intuition to follow when a sophisticated trader receives a sell offer is similar to a buy offer.

[22] We adopt the following notation throughout the examples of received offers. b and a indicate whether the received offer is a buy or a sell offer, respectively; subindex $\{H, M, L\}$ stands for the level of the price, which is equivalent to the expected dividend level; H^- and H^+ are used to denote that the price is slightly below or above the level D_H, namely $D_H - \varepsilon$ and $D_H + \varepsilon$, respectively.

$$\pi^S(b_{M^-}, D_L) = \begin{cases} b_{M^-} & \text{accepting the bid} \\ \quad D_{LM}\ Pr[D_M|D_M > b_{M^-}] & \text{rejecting the bid} \\ +\ D_{LH}\ Pr[D_H|D_H > b_{M^-}] & \end{cases}$$

where the updated expected dividend is given by

$$D_{LM} \equiv E[D = 1|x_L, x_M, y] = \frac{1}{1 + \left(\frac{1-p}{p}\right)^{-2+0} \left(\frac{1-q}{q}\right)^{y}}$$

and

$$D_{LH} \equiv E[D = 1|x_L, x_H, y] = \frac{1}{1 + \left(\frac{1-p}{p}\right)^{-2+2} \left(\frac{1-q}{q}\right)^{y}}\ .$$

The probability assigned to a proposer of type M given that the receiver has an expected dividend D_L is computed by

$$\begin{aligned} Pr[D_M|D_M > b_{M^-}] &= \frac{Pr[b_{M^-}|D_M]\ \ Pr[D_M|D_L]}{Pr(b_{M^-}|D_L)} \\ &= \frac{\frac{1}{4}2pq[(1 - D_L) + D_L]}{\frac{1}{4}[D_L(p^2 + 2pq) + (1 - D_L)(q^2 + 2pq)]} \end{aligned}$$

Conversely, she cannot update her beliefs when she observes a bid $b_{L^-} = D_L - \varepsilon$ because any type of trader could submit that offer.

$$\pi^S(b_{L^-}, D_L) = \begin{cases} b_{L^-} & \text{accepting the bid} \\ D_L & \text{rejecting the bid} \end{cases}$$

8 Common Signal Scenario

The analysis of the common signal scenario follows the same structure as the case of PS scenario. The main difference with the PS scenario lies in the sophisticated traders' strategies. Nonetheless, the lack of common knowledge does not change naive traders' behavior since they evaluate signals according to their precision about fundamentals. This Appendix explains the main differences in the CS scenario and the results of the model.

8.1 Sophisticated Traders

Sophisticated traders consider the distribution of information in order to assess market demand. However, contrary to public signal, the common signal does not allow them to better characterized the potential market demand. They estimate the potential demand assuming each trader possesses three independent private signals $\{x_i, y_i\}$ because they are not aware that y_i is identical to all traders. We must redefine, therefore, the expected dividend for a trader of type i as:

$$E[D = 1|x_i, y_i] = \frac{1}{1 + \left(\frac{1-p}{p}\right)^{x_i} \left(\frac{1-q}{q}\right)^{y_i}} \tag{10}$$

where $x_i = \{-2, 0, 2\}$ refers to private signals and $y_i = \{-1, 1\}$ refers to the common signal. Notwithstanding the common signal is unique for all traders in the market, the sophisticated traders classify traders in four groups according to the four possible expected dividends $\{D_H, D_{\overline{M}}, D_{\underline{M}}, D_L\}$, corresponding to all the possible combinations of $\chi_i = (x_i + y_i)$.[23] We introduce the notation $\overline{M}$ and $\underline{M}$ to denote the low and high intermediate levels. The variable i takes the values $\{H, \overline{M}, \underline{M}, L\}$. It is important to stress, however that only three are the levels effectively present in the market. For instance, if the common signal is $y_i = 1$, existing types of traders are $\{H, \overline{M}, \underline{M}\}$ and the types of traders are $\{\overline{M}, \underline{M}, L\}$ when common signal is $y_i = -1$. The optimal offer is computed by following the process explained in Sect. 2.2.

In case a sophisticated trader receives an offer, it provides her with new information to be updated. Unlike markets in the PS scenario, she identifies four possible type of proposers $j \in \{H, \overline{M}, \underline{M}, L\}$, although one of them does not actually exist.

8.2 Transactions

Tables 4 and 5 list the market transactions when the dividend is $D = 1$ and the common signal is correct or incorrect, respectively. The first column denotes the proposer's type according to his level of reasoning and expected dividend. The second and the third columns show the optimal offer of each trader while the last column shows the counterpart of every transaction.

In order to compare the results between common and public signal, one should consider that when the common signal indicates dividend 1, $j = \overline{M}$ corresponds to the M and $j = \underline{M}$ corresponds to L. If the common signal indicates dividend 0, $j = \overline{M}$ corresponds to the H and $j = \underline{M}$ corresponds to M. We rename the type of traders and offers for each prediction of the common signal $y_i = \{-1, 1\}$ for an easier comparison between markets with common signal and markets where

---

[23] In CS, privately uninformed traders are absent, therefore $\chi_i \in \{-3, -1, 1, 3\}$. Remember that in PS scenario, traders might be informed $x_i \in \{-2, 2\}$ or uninformed $x_i = 0$.

Table 4 Transactions when the common signal is 1

Proposer (τ_i)	Order	Price	Receiver (τ_j)	
S_H	b^*_H	$D_{\overline{M}} + \varepsilon$	$N_{\underline{M}}, N_{\overline{M}}$	
		$D_{\underline{M}} + \varepsilon$	$N_{\underline{M}}$	
$S_{\overline{M}}$	$b^*_{\overline{M}}$	$D_{\underline{M}} + \varepsilon$	$N_{\underline{M}}$	
$S_{\underline{M}}$	$a^*_{\underline{M}}$	$D_{\overline{M}} - \varepsilon$	$N_H, N_{\overline{M}}$	S_H
	a_H	$D_H + \varepsilon$	*No trade*	
N_H	b_H	$D_H - \varepsilon$	$N_{\overline{M}}, N_{\underline{M}},$	$S_{\underline{M}}$
$N_{\overline{M}}$	$a_{\overline{M}}$	$D_{\overline{M}} + \varepsilon$	N_H	S_H
	$b_{\overline{M}}$	$D_{\overline{M}} - \varepsilon$	$N_{\underline{M}},$	$S_{\underline{M}}$
$N_{\underline{M}}$	$a_{\underline{M}}$	$D_{\underline{M}} + \varepsilon$	$N_H, N_{\overline{M}},$	$S_H, S_{\overline{M}}$
	$b_{\underline{M}}$	$D_{\underline{M}} - \varepsilon$	*No trade*	

Table 5 Transactions when the common signal is -1

Proposer (τ_i)	Order	Price	Receiver (τ_j)	
$S_{\overline{M}}$	$b^*_{\overline{M}}$	$D_{\underline{M}} + \varepsilon$	$N_L, N_{\underline{M}},$	S_L
$S_{\underline{M}}$	$a^*_{\underline{M}}$	$D_{\overline{M}} - \varepsilon$	$N_{\overline{M}}$	
S_L	a^*_L	$D_{\overline{M}} - \varepsilon$	$N_{\overline{M}}$	
		$D_{\underline{M}} - \varepsilon$	$N_{\overline{M}}, N_{\underline{M}},$	
$N_{\overline{M}}$	$a_{\overline{M}}$	$D_{\overline{M}} + \varepsilon$	*No trade*	
	$b_{\overline{M}}$	$D_{\overline{M}} - \varepsilon$	$N_{\underline{M}}, N_L$	$S_{\underline{M}}, S_L$
$N_{\underline{M}}$	$a_{\underline{M}}$	$D_{\underline{M}} + \varepsilon$	$N_{\overline{M}}$	$S_{\overline{M}}$
	$b_{\underline{M}}$	$D_{\underline{M}} - \varepsilon$	N_L	S_L
N_L	a_L	$D_L + \varepsilon$	$N_{\overline{M}}, N_{\underline{M}},$	$S_{\overline{M}}$
	b_L	$D_L - \varepsilon$	*No trade*	

the released signal is public. Considering only private signals, the possible types of traders are $\{H, \overline{M}, \underline{M}\}$ if the common signal predicts dividend 1 (Table 4); otherwise $j \in \{\overline{M}, \underline{M}, L\}$ (Table 5). Considering the previous changes, we define a vector of market prices following the proposer's type offer in Table 4, $\mathbf{P} = (D_M, D_L, D_L, D_M, D_H, D_M, D_L)$. The vector of transaction prices when the common signal predicts dividend 0 is $\mathbf{P} = (D_M, D_H, D_H, D_M, D_H, D_M, D_L)$, which is listed in Table 5.

Finally, the expected number of transactions per unit of time is listed in Table 6. The mean price is computed by Eq. (6).

Table 6 Expected number of transactions per unit of time for every type of trader, given $D = 1$ in CS scenario

Proposer (τ_i)	T (if $y_i = 1$)	T (if $y_i = -1$)
S_H	$\mu_H\theta(1-\theta)p^2(1-p)(p+0.5(1-p))$	$\mu_H\theta p^2 q[(1-\theta)2p+(1-p)]$
	$\mu_H\theta(1-\theta)0.5p^2(1-p)^2$	
S_M	$\mu_M\theta(1-\theta)2p(1-p)^3$	$\mu_M 2\theta(1-\theta)p^3(1-p)$
S_L	$\mu_L\theta(1-p)^2 p[p+(1-\theta)2(1-p)]$	$\mu_L 0.5\theta(1-\theta)(1-p)^2p^2$
		$\mu_L\theta(1-\theta)(1-p)^2p[0.5p+(1-p)]$
N_H	$\nu(1-\theta)[(1-\theta)p^3(1-p)+0.5p^2(1-p)^2]$	$\nu(1-\theta)[p^3(1-p)+0.5p^2(1-p)^2]$
N_M	$\nu(1-\theta)[p(1-p)^3+p^3(1-p)]$	$\nu(1-\theta)[p(1-p)^3+p^3(1-p)]$
N_L	$\nu(1-\theta)[0.5p^2(1-p)^2+p(1-p)^3]$	$\nu(1-\theta)p(1-p)^2[0.5p+(1-\theta)(1-p)]$

9 Robustness: Does Market Configuration Matter?

This subsection aims at testing the relevance of the distribution of signals in markets with public information. Intuitively, the proportion of informed traders in the aggregation and dissemination of information matters. For example, an incorrect public signal might largely distort prices when the proportion of informed traders is small. However, an incorrect public signal should be harmless when most of the traders are informed. Since the most concerning case is the impact of an incorrect public signal, we restrict our attention to the PS scenario to assess the importance of market configuration. We define three market configurations based on observed distributions of information across traders in the laboratory experiment. (i) Config. 1, markets are populated by 5 uninformed and 10 informed traders. (ii) Config. 2, markets are populated by 1 misinformed trader, 7 uninformed traders and 7 informed traders. (iii) Config. 3 where markets are populated by 2 misinformed, 5 uninformed and 8 informed traders.

Figure 6 shows that mean prices change depending on the distribution of private information. When the public signal is correct, one can see that the computational mean takes similar values to the theoretical prediction in markets where uninformed and misinformed traders have a large presence (Config. 2 and Config. 3). For the markets with an incorrect public signal, the public signal always dominates the mean price. The impact is larger when the proportion of informed traders is small (Config. 2 and Config. 3).

Altogether, we can claim that the market configuration can generate systematic deviations from the theoretical prediction, however "not too large", i.e., the general conclusions still hold. A special case seems to be the configuration where there is absence of misinformed traders. The mean price is noticeably higher than the other market configurations, independently of the prediction of the released signal $y =$

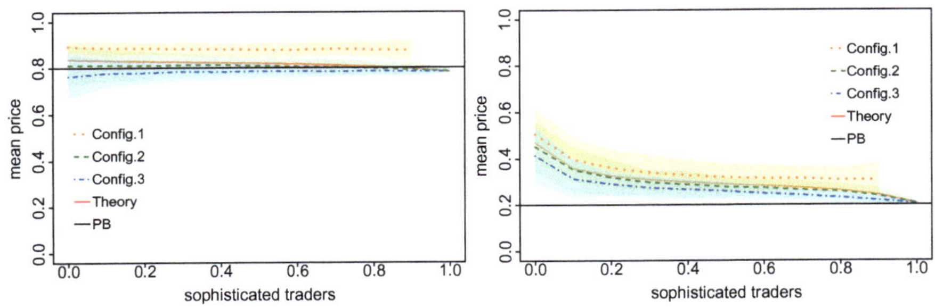

(a) Correct public signal $y = 1$ (b) Incorrect public signal $y = -1$

Fig. 6 Mean price of the market configurations assuming dividend $D = 1$. Shaded area shows 1 standard deviation of the Monte Carlo simulations

$\{1, -1\}$. Besides, it is interesting to note that there are no transactions when $\theta = 1$. Therefore, if a market where all traders are sophisticated and none is misinformed, we have no transactions.

References

Allen F, Morris S, Shin HS (2006) Beauty contests and iterated expectations in asset markets. Rev Finan Stud 19(3):719–752

Camerer C, Weigelt K (1991) Information mirages in experimental asset markets. J Bus 64:463–493

Camerer CF, Ho T-H, Chong J-K (2004) A cognitive hierarchy model of games. Q J Econ 119(3):861–898

Corgnet B, DeSantis M, Porter D (2015) Revisiting information aggregation in asset markets: reflective learning & market efficiency. ESI Working Papers

Duffie D (2012) Dark markets: asset pricing and information transmission in over-the-counter markets. Princeton University Press

Duffie D, Gârleanu N, Pedersen LH (2005) Over-the-counter markets. Econometrica 73(6):1815–1847

Duffie D, Malamud S, Manso G (2015) Reprint of: information percolation in segmented markets. J Econ Theory 158:838–869

Duffie D, Manso G (2007) Information percolation in large markets. Am Econ Rev 97(2):203–209

Grossman SJ, Stiglitz JE (1980) On the impossibility of informationally efficient markets. Am Econ Rev 70:393–408

Hayek FA (1945) The use of knowledge in society. Am Econ Rev 35(4):519–530

Morris S, Shin HS (2002) Social value of public information. Am Econ Rev 92(5):1521–1534

Morris S, Shin HS (2005) Central bank transparency and the signal value of prices. Brook Papers Econ Act 2005(2):1–66

Plott CR, Sunder S (1982) Efficiency of experimental security markets with insider information: an application of rational-expectations models. J Polit Econ 90(4):663–698

Plott CR, Sunder S (1988) Rational expectations and the aggregation of diverse information in laboratory security markets. Econometrica 56:1085–1118

Ruiz-Buforn A (2019) On the impact of public information in financial markets: an experimental approach (Doctoral dissertation, Universitat Jaume I, Castellón de la Plana, España). Retrieved from http://hdl.handle.net/10803/667025

Ruiz-Buforn A, Alfarano S, Camacho-Cuena E, Morone A (2018) Crowding out effect and traders' overreliance on public information in financial markets: a lesson from the lab. MPRA Paper 88866, University Library of Munich, Germany
Shiller R J (2002) Irrational exuberance in the media. The Right to Tell, p 83

Order Book on Financial Networks

Alessio Emanuele Biondo

Abstract This chapter presents results from a class of agent-based models describing a realistic financial order book, aiming to discuss some aspects related to market stability. Major empirical regularities of data are correctly replicated and simulations are used to infer possible policy implications. After a detailed explanation of model features, an application to networks is presented to advance intuitions about the role of social interaction on induced imitation and herding phenomena. Thus, the initial market structure is augmented by a dynamic multiplex with two layers devoted to information and trading. The first one, representing social interactions, is designed according to different topologies in order to show how investors decide their behavior by following perceived informative flows. The second one, where the central hub is the market maker (i.e., the owner of the venue holding the order book) is devoted to the execution of all transactions. Some policy implications oriented to foster market stability are finally provided.

Keywords Order book · Imitation · Agent based models · Time series · Financial Markets

JEL classification: C63 · C15 · G41 · E71

1 Introduction

Financial markets represent one of the most intriguing example of economic interaction among independent individuals with heterogeneous attributes and non-necessarily cooperative goals. The cooperation among them would somehow imply an explicit coordination of choices, all following a given criterion. In the absence of such a superior direction, each market participant feels *peripheral* with respect to the core dynamics of the whole market and assumes the strategy that max-

A. E. Biondo (✉)
Department of Economics and Business, University of Catania, Corso Italia 55, 95129 Catania, Italy
e-mail: ae.biondo@unict.it

A. S. Chakrabarti et al. (eds.), *Network Theory and Agent-Based Modeling in Economics and Finance*, https://doi.org/10.1007/978-981-13-8319-9_5

imizes her individual target. In other words, by means of her own transactions, each trader participates to a global process that involves many other actors, each focused on his own convenience, without a complete awareness of causes and targets and, specially, without a measure of the impact of individual choices on the aggregate market equilibrium. Normally, an investor does not even know who is her counterpart when is negotiating a transaction.

In such a context, informative differentials play a very relevant role, both in terms of single contracts and of global market behavior. The order book governing a financial market is the backbone of its structure, for it registers all terms of contracting agreements (such as price, quantity, type of order, counterparts, conditions, timing, etc.) for each negotiation occurred in that market. Thus, the order book dynamics is, as Mitchell (2009) has underlined, a valid example of a complex system, because it emerges as a global result of local individual interactions among traders. It is not straightforward to define complexity, and many different approaches to this concept can be pursued. It is useful to focus on some of its most relevant ingredients and say, following Gallegati and Richiardi (2009), that it emerges when the interaction among elements of a system generate spontaneous self-organized structures at different layers of a hierarchical configuration, so that the aggregate outcome results to be qualitatively different from the features of its constituents. In other words, the forces of mutual interaction among individual components of the system count more than the set of attributes of each element in determining the overall aggregate dynamics. Such a rationale explains fairly well the reason why it is rather strange to challenge the sense of human logic by looking for forecasts in complex systems, where deterministic predictions, have not reasonable foundations (Prigogine 1997).

Macroeconomics is a domain of complexity, since it deals with aggregate dynamics generated by means of the interactions among individual entities, consumers and firms, who match locally within a networked series of relations mapped beyond the awareness of single individuals. The microeconomic approach cannot therefore be extended to describe and study such non-individual effects, despite the fact that this is the mainstream of the economic profession. Economic systems with emergent properties should be studied by means of specific statistical tools, as explained in Mantegna and Stanley (2000) with reference to financial markets, and by new models designed in such a way that they can replicate empirical data and true experience. Such a new paradigm, explained in Delli Gatti et al. (2008, 2011), is rooted on agent-based models. Once the lens of complexity is used in economic modelling, the understanding of macroeconomics improves, as discussed in Sornette (2009), Kirman (2010) and Ladyman et al. (2013), among many others. Such a consciousness is not, indeed, a novelty and dates back to pioneering contributions by Keynes (1936) and Majorana (1942), after which many other relevant economists focused on implications of complexity as, among others, Schumpeter (2003), von Hayek (2015), Leijonhufvud (1993), Simon (1957), Kahneman and Tversky (1974, 1979), Camerer (2003), Barberis and Thaler (2003), Colander (2009).

A very wide literature on financial order books exists, scrupulously surveyed in Parlour and Seppi (2008) and Chakraborti et al. (2011). Different approaches can be recognized. One can be defined "trader-centric", because in some papers authors aim to derive fully-rational trading strategies, more or less directly referred to agents' utility maximization (e.g.: the optimal choice between limit and market orders), as in Copeland and Galai (1983), Glosten and Milgrom (1985), Kyle (1985), Glosten (1994), Chakravarty and Holden (1995), Bertsimas and Lo (1998), Parlour (1998), Foucault (1999), Almgren and Chris (2001), Rosu (2009, 2016). Another can be defined "facts-centric" because some contributions are addressed to study the statistical features of the market as a dynamic process, regardless of a realistic design of market participants, as in Bak et al. (1997), Maslov (2000), Daniels et al. (2003), Farmer et al. (2005), Mike and Farmer (2008), Bouchaud et al. (2009), Farmer and Foley (2009), Cont et al. (2010), Cont and de Larrard (2013). Finally, a third stream of literature, which in some cases overlaps with previous ones, proposes the computational approach of agent-based models (ABMs) in economics. Such models, developed since the Nineties, have shown to be able to describe many aspects of true markets by simulating a global environment where interactions among heterogeneous traders give rise to emergent phenomena at the aggregate level, as explained in Tesfatsion (2006). Examples are, among others, Brock and Hommes (1997, 1998), Chiarella (1992), Chiarella and He (2001), Day and Huang (1990), Franke and Sethi (1998), Hommes (2001), Lux (1995, 1998), Lux and Marchesi (1999). The heterogeneity of individuals and the global properties emerging from their interaction can be analyzed by means of agent-based models, thus assuming a determinant descriptive role in models of financial markets, as in Hommes (2006), LeBaron (2006), and in models of order books, as in Raberto et al. (2001), Chiarella and Iori (2002), Consiglio et al. (2005), Gil-Bazo et al. (2007), Chiarella et al. (2009), Anufriev and Panchenko (2009), Tedeschi et al. (2009, 2012).

Focusing on financial markets, the main element of the analysis is the interdependence of individual opinions, since the process of decision-making of investors strictly depends on strong informative connections among them. Many papers deal specifically with such aspects, among which, Orléan (1995), Stauffer and Sornette (1999), Cont and Bouchaud (2000), Cooper et al. (2001), Iori (2002), Markose et al. (2004), Tedeschi et al. (2009, 2012), and Zhao et al. (2014). In general, opinion dynamics models are based on network theory, given that the topological displacement (connection properties) of people (nodes) affects both the speed and the magnitude of the diffusion (contagion) of the signal/phenomenon being studied. Networks literature has proliferated so much that the number of related contributions has become almost uncountable in few years. Very comprehensive surveys can be found in Boccaletti et al. (2006, 2014).

In economics, the terms contagion and imitation are not necessarily synonyms. The first is often referred to circumstances in which the instability arises from the interaction among banks and institutions, as for example in Allen and Gale (2000), Kiyotaki et al. (1997), Lagunoff et al. (2001), Rochet et al. (1996), Leitner (2005), Elliot et al. (2014) and Chong and Küppelberg (2018). The second is, instead, frequently used to describe the situation in which a person copies the behavior assumed by someone else. Causes of imitation, which may also reflect an individual predisposition, are mainly related to social influence, in terms of both diversity—as discussed, for example by Lorenz et al. (2011)—, and dimension of groups—as explained by

Kao and Couzin (2014). The emergence of behavioral avalanches is the visible effect of such a complex interaction, for people think it is convenient to follow the crowd, Bikhchandani et al. (1992), revealing the existence a common root in very different contexts, Moussaid et al. (2009). The first and very fundamental step of any herding process is the choice of a leader to follow, specially in financial markets, where the reputation of market participants are inferred by looking at their performances (Cooper et al. 2001; Hirshleifer and Hong Teoh (2003); Clement and Tse (2005); Booth et al. (2014), among many others).

This chapter presents results from a class of agent-based models describing a realistic financial order book, aiming to discuss some aspects related to market stability. In particular, traders are displaced over networks, in order to provide some specific intuitions relating informative flows—spreading by means of social interaction—with market instability. While preserving the ability to replicate stylized facts of empirical data and to confirm results of existing related literature on the topic, the model shows the correlation between imitative processes (triggered by three different criteria) and volatility of the market and presents some policy implications. In particular, the proposed approach is related to Tedeschi et al. (2009, 2012) but, since imitation is a social concept, based on the acknowledgement of success of others, it has been designed to operate by means of information networks. Successful traders to be imitated are selected on the basis of different criteria, namely, wealth, performance, and reputation. The microstructural part of the model draw substantially on Biondo (2018a, b, c), and Biondo (2019), and the network, describing the consequences of imitation on market stability, is designed as a two-layer multiplex. More precisely, the market structure is represented by a dynamic multiplex with two layers devoted, respectively, to information and trading. The first one, representing social interactions, is designed according to different topologies in order to show how investors decide their behavior by following perceived informative flows, thus evolving because of the social interaction and the imitation. The second one could be represented as a simple star network, where the central hub is the market maker (i.e., the owner of the venue holding the order book) and all transactions are executed, thus dynamically changing according to negotiations.

The chapter is organized as follows: Sect. 2 contains the model description and the results of simulations used to test some stylized facts of empirical data; Sect. 3 reports the effect of relevant parameters of the model on market stability; Sect. 4 presents the effects of the imitation-driven dynamics, on different network topologies; Sect. 5 contains conclusions and some policy suggestions.

2 The Model

Consider a population of heterogeneous traders, each endowed with a portfolio, $W_{i,t} = m_{i,t} + p_t a_{i,t}$, composed by an amount of money $m_{i,t}$ and a quantity of shares $a_{i,t}$, which evolves in time according to negotiated transactions and to the market price p_t. At $t = 0$, all traders receive equal initial amounts of money $m_{i,0}$ and of shares $a_{i,0}$. Each of them participates to a single-asset artificial stock market and presents one of two possible behavioral root, namely fundamentalists and chartists, as in many other models of related literature. Such traders submit their orders to a market maker, who owns the venue, set the rules and updates the order book, which eventually regulates the possible negotiations.

2.1 *Simulative Steps of Microstructure*

Each simulative cycle (1 complete run) consists of five steps, which are continuously repeated: (1) expectations setting; (2) strategy setting; (3) orders setting (price and quantity) and, in case of imitative behavior, imitation; (4) order book setting; (5) transactions. Since at the end of the last step contracts are possibly executed and a new price is thus established, the length of a simulation is determined by setting the desired length of the simulated price series.

2.1.1 Expectations

Fundamentalists are usually described as competent traders, who form their opinions by looking at informations regarding the economic activity of the issuer. Because of their knowledge-oriented approach to the market, fundamentalists presume to know the "correct" value for the asset, i.e., its so-called fundamental value. Thus, they form their expectations on the market price by referring to variables that are exogenous with respect to the past financial price dynamics, such as the quality of the production, the competition and market power played by the issuer in the sector etc. In simulative terms, this can be replicated by letting fundamentalists refer to a stochastic process (totally disconnected to past prices). Thus, the Fundamental Value, which is assumed to vary in time as in Grinblatt and Han (2005), is computed as:

$$FV_t = FV_{t-1} \pm \Theta_t \tag{1}$$

where Θ_t is a bounded random variable, drawn with uniform distribution from $[-\sigma_\Theta, \sigma_\Theta]$. Each fundamentalist i's expectation is built on the idea that information is imperfectly distributed and that each trader may have her own perception of variables. Then, it is computed as:

$$_F p_{i,t}^{\exp} = FV_t \pm \Phi_{i,t} \tag{2}$$

where Φ_t is a bounded random variable, drawn with uniform distribution from $[-\sigma_F, \sigma_F]$, representing the mistake in perception done by i.

Chartists are, instead, defined as trend followers who refer to the (more or less) recent past values assumed by the asset price. They are depicted as market analysts, who try to infer the next price by comparing a reference value computed by using past prices with the last observed price in order to infer the future trend. Thus, first of all, each chartist i computes a Reference Value as an average of past prices included in a time window of heterogeneous length τ_i:

$$RV_{i,t} = \sum_{j=(t-\tau)}^{t} p_j/\tau_i \tag{3}$$

Then, their individual expectations are given by:

$$_C p_{i,t}^{\exp} = p_t + \frac{p_t - RV_{i,t}}{\tau_i - 1} \pm \Lambda_{i,t} \tag{4}$$

where Λ_t is a random variable with uniform distribution drawn from $[-\sigma_C, \sigma_C]$, representing an individual factor of differentiation (i.e., it ensures that two chartists choosing an equally long time window may still have different expectations). Expectations of agents, above defined in Eqs. (2) and (4), are well known in related literature. Similar approaches can be found also in Lux and Marchesi (1999), Takayasu et al. (2006), Alfi et al. (2006, 2007).

2.1.2 Strategy

The strategy setting of each trader i is devoted to maximize the value of her portfolio, according to the available information on the variables relevant for the individual behavioral root. Thus, the strategy adopted by each investor is based on the comparison between her expectations and the current market price, as reported in Table 1. If the expected price is greater (smaller) than the current one, the investor will have the convenience to buy (sell) because, in case the expectation is correct, the value of the portfolio will increase (decrease). The parameter κ, which can be used to model transaction costs and other rigidities, will be shown to affect the market volatility.

Table 1 Expectations-based individual strategy setting, where $j = F, C$ is trader's type and κ is the transaction cost (possibly) levied by the market maker, which will be discussed later

Expectations	Strategy	Order
$^i_j p_t^{\exp} > p_t + \kappa$	Buy	Bid
$p_t - \kappa \leq {}^i_j p_t^{\exp} \leq p_t + \kappa$	Hold	N/A
$^i_j p_t^{\exp} < p_t - \kappa$	Sell	Ask

2.1.3 Orders Setting: Price and Quantity

The definition of the elements of orders (price and quantity) follows two phases. The first one defines the price. It the trader i is a bidder, she decides the price for her order (${}_Bp_{i,t}$) according to her willingness to pay, $w_{i,t}$, randomly drawn with uniform distribution from (0, 1) and to her bargaining ability, $p_{i,t}^{\text{rev}}$, which consists in her personal prediction of asker's willingness to accept, computed as a random variable, drawn with uniform distribution, from the interval $({}_Ap_{t-1}^{\text{best}}, p_t)$, where ${}_Ap_{t-1}^{\text{best}}$ is the lowest ask price (see below). Thus, the price of a bid order is given by:

$$ {}_Bp_{i,t} = p_{i,t}^{\text{rev}} + w_{i,t}({}_jp_{i,t}^{\text{exp}} - p_t) \pm z_{i,t}\Delta\pi_{t-1} \tag{5} $$

where ${}_jp_{i,t}^{\text{exp}}$ is i's expected price determined as shown in either Eq. (2) or Eq. (4), according to the trader's type, indicated by $j = F, C$; the term $\Delta\pi_{t-1}$ is the market imbalance, resulting from the order book (shown later) at the previous instant, weighted by the individual perception threshold $z_{i,t} \in [0, 1]$.

If, instead, the trader i is an asker, the price setting rule is correspondingly symmetric: the price depends on her willingness to accept, $w_{i,t} \in (0, 1)$, and on her prediction of bidders' willingness to pay, $p_{i,t}^{\text{rev}} \in (p_t, {}_Bp_{t-1}^{\text{best}})$, with ${}_Bp_{t-1}^{\text{best}}$ being the highest bid price (see below). An ask order is priced as:

$$ {}_Ap_{i,t} = p_{i,t}^{\text{rev}} - w_{i,t}(p_t - {}_jp_{i,t}^{\text{exp}}) \pm z_{i,t}\Delta\pi_{t-1} \tag{6} $$

where other variables and parameters follow the same definitions given above.

The second phase determines the quantity for bid ($q_{i,t}^B$) and ask ($q_{i,t}^A$) orders. Since traders can be imagined as utility/profit maximizers, each of them will bear in mind their own target function to optimized under their budget constraints, represented by their endowments, $W_{i,t}$. In simulative terms, such a process has been depicted without imposing any specific utility function but, instead, modelling the choice of each trader as a random extraction from the set of her possible decisions. Thus, defined the feasible interval, in such a way that neither short selling, nor loans are allowed, the quantity of each order is defined as:

$$ q_{i,t}^B = \zeta \quad and \quad q_{i,t}^A = \xi \tag{7} $$

where $\zeta \in [1, m_{i,t}/p_t]$ and $\xi \in [1, a_{i,t}]$ are uniformly distributed random variables. Thus, a bidder can decide to buy, at most, the highest number of shares she can pay, and an asker can decide to sell, at most, all shares she has.

2.1.4 Order Book Setting

All active traders continuously submit their orders to the market maker who registers them into the two sections of the order book, one for bid orders (decreasingly ranked

Table 2 Limit and market orders

Orders	Bid	Ask
Limit	${}_{B}^{i}p_t < {}_{A}p_t^{\text{best}}$	${}_{A}^{i}p_t > {}_{B}p_t^{\text{best}}$
Market	${}_{B}^{i}p_t \geq {}_{A}p_t^{\text{best}}$	${}_{A}^{i}p_t \leq {}_{B}p_t^{\text{best}}$

by price) and the other for ask orders (increasingly ranked by prices). Thus, the bid order with the highest price, i.e., the best bid (${}_{B}p_t^{\text{best}}$), is the first in the bid side and the trader who posted it, i.e. the best bidder, will have the priority. Symmetrically, the ask order with the lowest price, i.e., the best ask (${}_{A}p_t^{\text{best}}$), is the first in the ask side and the trader who posted it, i.e. the best asker, will have the priority. Following Chiarella et al. (2009), only limit and market orders will be considered, defined as shown in Table 2.

Each limit order has a finite time validity, chosen by the trader at submission and then it is automatically cancelled. While pending to be executed, limit orders are assumed, as in the approach of Lux and Marchesi (1999, 2000), to express a signal about the "market sentiment". In other words, traders are designed to perceive the excess demand or excess supply and adaptively adjust their price settings rules, by means of the market pressure term $\Delta\pi_{t-1}$, as in Eqs. (5) and (6). It is computed as the product of two factors: the first one is the absolute value of the bid-ask spread, $\Delta S_{t-1} = |\, {}_{A}p_{t-1}^{\text{best}} - {}_{B}p_{t-1}^{\text{best}}|$; the second one is the ratio between the greatest amount of orders on one side of the order book over the amount of orders of the other, $\Delta n_{t-1} = ({}_{A}n_{t-1}/{}_{B}n_{t-1})$ if ${}_{A}n_{t-1} > {}_{B}n_{t-1}$, and $\Delta n_{t-1} = ({}_{B}n_{t-1}/{}_{A}n_{t-1})$ vice versa. Thus, finally, the market pressure term is computed as:

$$\Delta\pi_{t-1} = \Delta S_{t-1}\Delta n_{t-1} \tag{8}$$

when a market order is posted, the transaction is immediately executed at the best price of the counter side of the book. The following quantity-matching mechanism operates:

(a) if $q_t^B = q_t^A$, the negotiation is executed by assigning to the bidder (asker) the corresponding increase (decrease) in the owned asset quantity, and the related decrease (increase) in the quantity of money;
(b) if $q_t^B \neq q_t^A$, the traded quantity will be the "shortest side of the market" (i.e., the smallest quantity). The partly unsatisfied trader remains queueing in the order book, waiting for a new counterpart. The trader who has entirely negotiated her order is erased and the successive one becomes the new best trader to continue the transaction, at the new best price. This process is repeated either until the unmatched quantity has been entirely negotiated or if the allowed order book length is reached. The order book length is defined as the maximum number of new counterparts that a trader is allowed to match in a single turn of transactions.

2.1.5 Transactions

The execution of an order completes the negotiation and establishes the new market price (which will have a different value of the previous one if the minimum tick size has been surpassed). The model here presented is using a concept of time that is referred to ticks, as explained in Chakraborti et al. (2011). A real time version of the model, with a comparative presentation of some stylized facts of data at four different time resolutions (ticks, 1 s, 1 min, 5 min) replicating the time schedule of Borsa Italiana S.p.A, is contained in Biondo (2019). After an order (either limit or market) is executed, portfolios of involved traders are updated. If no transactions happen, then all traders update their expectations and the whole set of steps (above described) starts again. In such a way, the market is truly driven by actual contracts dealt among traders.

2.2 *Stylized Facts of Financial Markets*

Empirical data of financial returns exhibit some statistical regularities, which are considered *stylized facts*, as shown in Cont (2001) and Chakraborti et al. (2011). The present model is able to generate simulative returns that replicate some of such facts. In particular, referring to: (1) the existence of fat tails in returns distribution; (2) the lack of autocorrelation of returns; (3) the presence of volatility clustering. All simulated series have effective length of 15,000 entries, net of a transient of 2000 entries, which has always been drop out. Returns have been obtained from the price series generated by the model, according to the canonical definition: $r_t = (p_t - p_{t-1})/p_{t-1}$.

The first stylized fact of financial data, described by Mandelbrot (1963), Gopikrishnan et al. (1999), among others, is the existence of fat tails in the returns probability density functions. It shows that the probability to find values which are distant from the average is greater than the Gaussian case. This is a strong regularity in financial series, whose theoretical stochastic generator process would manifest infinite variance (Mantegna and Stanley 2000). A major implication of this is that no predictions can be effectively done on future values: errors in forecasts are not bounded in any standard deviation known a priori. In the left panel of Fig. 1, true and simulated returns distributions are shown in log-linear plots. Simulated data exhibit a fat tailed leptokurtic PDF of returns, which mimics true data very closely. In order to compare different assets among them and true data with simulated one, normalized returns have been used. They have been computed, for all series, as $r_t^{\text{norm}} = (r_t - r_{av})/\sigma_r$, where r_{av} and σ_r are the respective mean and standard deviation.

The second stylized fact is the lack of autocorrelation in the returns time series, as reported by Chakraborti et al. (2011), Pagan (1996) and Cont et al. (1997), among

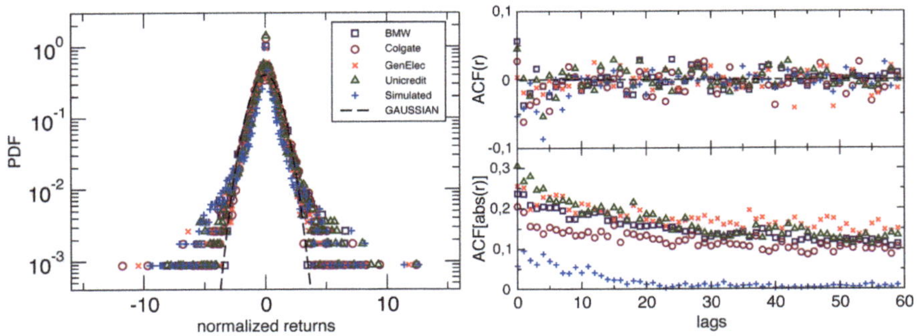

Fig. 1 Simulated data and main statistical regularities of empirical time series. Left panel: density functions of empirical returns of four financial time series (daily closings ranging from 2 January 73 to 30 June 16) are compared with the one of simulated returns. Data exhibit a fat-tailed leptokurtic distribution almost completely overlapping to those of true time series. Right panel: (top) the ACF of returns series reveals the absence of autocorrelation for all plotted series; (bottom) the ACF of absolute returns shows a positive and decaying autocorrelation. Despite simulated data show a faster decay than empirical data, it is still appreciable for more than 20 lags

others. The autocorrelation function has been computed, by reporting lag-time variation of the theoretical autocorrelation index $\rho(k)$ for up to 60 lags, as:

$$\rho(k) = \text{corr}(r_t, r_{t-k}) = \frac{\sigma_{r_t r_{t-k}}}{\sigma_{r_t}\sigma_{r_{t-k}}} \tag{9}$$

The ACF functions of true series are similar to the one of simulated returns (top-right panel of Fig. 1).

The third stylized fact is frequently referred to as the volatility clustering property (Mandelbrot 1963). It means that periods with high volatility are followed by periods with high volatility, whereas periods with low volatility are followed by periods with low volatility. In other words, the existence of autocorrelation in absolute values of returns shows that a correlation is present, but that it is not linear. This can be shown by computing the autocorrelation function, defined as in Eq. (9), of absolute values of returns: all series exhibit a positive and decreasing ACF (bottom-right panel of Fig. 1). Absolute values of simulated returns are lower than true data, but the ACF of absolute values of simulated returns is still positive and decreasing for a significant number of lags.

3 Model Parameters and Market Stability

The model presented in this chapter has been tested in previous studies (Biondo 2018a, b, c) to discuss the possibility of stabilization policies. In the present section, such a possibility will be discussed referring to the possibility influence some behavioral attributes of traders (corresponding to some parameters of the model) in such a way that the volatility of the market could be dampened. In the next section, it will be referred to the responsibility of imitation, instead. Results will be shown by reporting decumulative distribution functions of simulated returns, as defined by Yaari (1987)—i.e., the probability to have a return, in absolute value, greater than the plotted values.

3.1 *Subjective Factors*

The behavioral heterogeneity of traders, within each group, is a very relevant ingredient for market stability. The differentiation in expectations of traders can, instead, effectively reduce fat tails of returns distributions. It can derive from the personal perception of the relevant variable—either the fundamental value or the past reference value—considered by traders. In simulative terms, it can be tested by modifying the two specific parameters of the model regulating the differentiation of traders expectations, i.e., σ_F for fundamentalists and σ_C for chartists, respectively. Three tests have been run in order to compare results to the standard null case (in which both parameters have been set to 1): firstly, the intervals from which values are extracted have been reduced for both groups ($\sigma_F = \sigma_C = 0.1$); then the mixed cases have been tried ($\sigma_F = 1$, $\sigma_C = 0.1$ and $\sigma_F = 0.1$, $\sigma_C = 1$). Results shown in the left panel of Fig. 2, confirm that the behavioral synchronization generates fatter tails, because of a longer queue of orders, all convergent to similar prices, which push (up or down) prices more effectively than dispersed ones can do. Such a result is introductory of next discussion on imitation. One of the most evident features of imitation is that every involved trader behaves the same (that, in other words is equivalent to the hypothetical case in which everybody has the very same expectations).

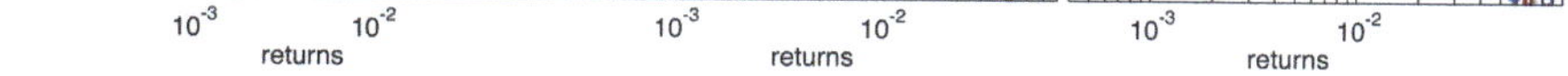

Fig. 2 Left panel: effect of behavioral heterogeneity. The probability of high volatility is higher in case of strong behavioral homogeneity (F 0.1–C 0.1 case). Mixed cases actually overlaps (F 0.1–C 1 and F 1–C 0.1 cases) and stay above of the heterogeneity case F 1–C 1. Middle panel: effect of reinforcing behaviors: traders acting in the same direction expressed by the market (i.e., bidders willing to pay more and askers willing to accept less) increase fat tails. Right Panel: the effect of market pressure on volatility can be very dangerous; as long as the strength of market signals perceived by traders increases, fat tails become fatter

Equations (5) and (6) describe pricing of orders in a quite standard fashion: each bidder (asker) will set a price positively (negatively) proportional to her willingness to pay (accept). Then, parameter $w_{i,t}$ regulates the weight of the difference between the market price and the expected price on the price of orders. The individual value of $w_{i,t}$ for each trader is drawn with uniform distribution from the interval $[0, \epsilon_1]$. In the null setting, $\epsilon_1 = 0.75$. Different values of ϵ_1 have been tested and results of performed simulations are reported in the middle panel of Fig. 2: higher willingness to pay/accept, ($w_{i,t} \in [0, 1.25]$ case) induces fatter tails. Such a result may seem paradoxical, but in behavioral terms it can be interpreted, once again, as the result of feeling peripheral with respect to the market dynamics: it would suggest to "obey" to market signal and reinforce its dynamics, i.e., buy (sell) as much as you can when the market is buying (selling). In general, however, the fact that traders suffer consistently from the influence of market signals has very bad effect on market stability. This can be shown even before analyzing the perverse effect of mutual contagion passing through social networks. Indeed, the impact of information on pricing functions is weighted by the parameter $z_{i,t}$, whose value is drawn, with uniform distribution, from the interval $[0, \epsilon_2]$ (and in the null model, $\epsilon_2 = 0.75$). Results reported in the right panel of Fig. 2 dramatically show what happens when, starting from $\epsilon_2 = 0.25$ and increasing it by steps of 0.25 points, the parameter approaches the value $\epsilon_2 = 1.5$: when traders give greater importance to information, i.e., their individual propensity to follow market signals grows, fat tails increase. In other words, the individual reaction to information is a source of aggregate volatility: a possible rationale is that induced adjustments lead to aggregate oscillations that are greater than needed ones.

3.2 Objective Factors

Other aspects playing a role in dampening market volatility are related to orders negotiations. In particular, for limit orders, the time they are allowed to remain pending in the book if not executed is directly correlated to fat tails, as shown in the left panel of Fig. 3. In simulative terms, the parameter ϑ, is a random variable drawn, with uniform distribution, from the interval $[0, \epsilon_3]$ (and in the null model it was set to 5). Results derive from different values of ϵ_3.

The rationale derives from the fact that a greater amount of orders standing in the book have a signalling role that affects pricing strategies of traders, and thus, the more they last, the more they contribute to the market pressure signal. Also the order book length can influence the volatility of the market. Being the maximum number of counterparts that a market order can match in order to satisfy its quantity requirements, such parameter regulates the penetrative ability of such orders. In the null model, this parameter has been fixed to the value of 3. As shown by the middle panel of Fig. 3, the greater the parameter governing the length is, the fatter the tails.

The null setting of the model consider the hypothetical case in which the market maker does not apply transaction costs to financial contracts negotiated in his trading venue. In true markets, instead, the existence of transaction costs distorts observed

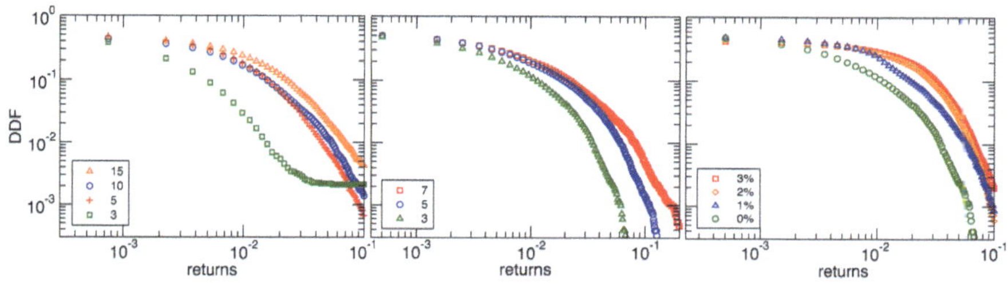

Fig. 3 Left panel: the validity of orders constitute a proxy of excess demand/supply and influence traders' pricing strategies. Middle panel: the order book length parameter reduces the penetrative potential of market orders because the number of successive matching counterparts is reduced. Right panel: transaction costs and, more broadly, market frictions in general, exacerbate fat tails

and used prices, inducing traders towards adjusted decisions, according to expectations requirements. In the right panel of Fig. 3, transaction costs are shown to have a dangerous effect in exacerbating the fat tails of returns. Three possible configurations (1, 2, and 3%) have been compared with the commissions-free case (0%) of the null model. Commissions/transaction costs act as a wedge that induces traders to wait for a ticker expected price variation (because, otherwise, any strategy would not be profitable enough and they would "hold"). It is worth noticing that the explained result is generally applicable to several other cases of market frictions/threshold levels in prices-expectations comparisons. This supports the empirical evidence of possible perverse effects caused by Tobin taxes, as shown in CME (2010).

4 Market Signals, Pressure and Financial Networks

In the previous section, the effect of market influence has been showed to be dangerous for the market instability even without a specific topological structure regulating communication among investors. This section is devoted to provide support to such a finding, by further investigations related to a more detailed specification of social interactions.

4.1 *Information, Imitation and Social Topology*

Let us imagine to fit traders of the model within a realistic community. This would imply that they could "know each other", at least when they are "close enough". Such concepts of knowledge and proximity have been used for they have a very natural meaning in network theory. From this point of view, traders would have now more than a unique reason to get in contact among them. In fact, they continue to have relations because of their negotiations (whether and when they trade), but they would have also a relation based on social interaction: each one could observe the others, learn what happens to their investments, to their portfolios, whether they have a good success and raise a lot of money or if they lay down because of misfortune. Then, the enrichment gained by the model from a networked framework is that traders have now two reasons to be "in relation with" someone else: the first is, as before, the execution of transactions (matched, as before, by the order book mechanism); the second is the desire to imitate successful behaviors of other investors (as usual with the aim to maximize the value of the own portfolio). In both cases, this dynamically adaptive set of relations can be described by means of a dynamic network configuration. In particular, with regards to the first reason, the network will represent the venue where the negotiations occur, where the market maker is central and allows each trader to exchange with anybody else. The social interaction, instead, will be modelled with different topologies, in such a way that traders will have personal contacts with their neighbours, thus obtaining information about both the local community close to them and the global market. As a whole, the model is a very simplified dynamic multiplex: a 2-layer network $(\mathcal{M})_t = [(\mathcal{G})_t, (\mathcal{H})_t]$, where $(\mathcal{G})_t$ is a family of two undirected graphs $G_\alpha = (N_\alpha, [L_\alpha]_t)$ and $G_\beta = (N_\beta, [L_\beta]_t)$, with static nodes sets $N_\alpha = N_\beta = N$ and no interlayer connections (except that each trader is linked with herself on both layers), i.e., $(\mathcal{H})_t = \mathcal{H} = \{L_{\alpha\beta} \subseteq N \times N\}$ with $L_{\alpha\beta} = \{(n, n); n \in N\}$, and dynamic links sets $[L_\alpha]_t$ and $[L_\beta]_t$, representing interactions as explained below. The layer where transactions will be regulated will be called the "market layer", while the layer where social contacts are developed will be called the "social layer". In both layers links change in time.

In the market layer, the market maker is a sort of super-node representing the unique path through which two traders can be linked together. When, a trader submits her order, a link between her and the market maker is created. When two orders are matched by means of the order book mechanism, each of the two counterparts involved in its execution are disconnected from the market maker and temporarily connected to each other until the negotiation has been completed. When the regulation has occurred, they are finally disconnected. Thus, in the market layer the link represent the order submission: an activity that all traders do with the market maker who, in turn, creates the contact between traders whose compatible orders can be submitted to each other. In case there exists a commission on transactions (as a transaction cost), the market maker obtains a profit from this "coordination" activity, as a function of the number of links created between couples of traders.

The social layer represents, instead, the informative exchange among agents. This layer can be designed according to different topologies, following the way in which local communities are modelled. Here, three very general topological configurations have been chosen, as it will be described below, in order to investigate different significant settings. Each agent i engages a sort of game of persuasion with herself, in which imitation is modelled in probabilistic terms. Being $\rho_i \in (0, 1)$ the individual propensity to imitate of trader i, every time she has to decide her trading strategy a number $\zeta \in (0, 1)$ is extracted randomly (with uniform distribution): if $\zeta < \rho_i$, i will copy the behavior of someone else and the only choice is to select *whom* to imitate, the so-called "master". Such a selection requires two ingredients: the first is

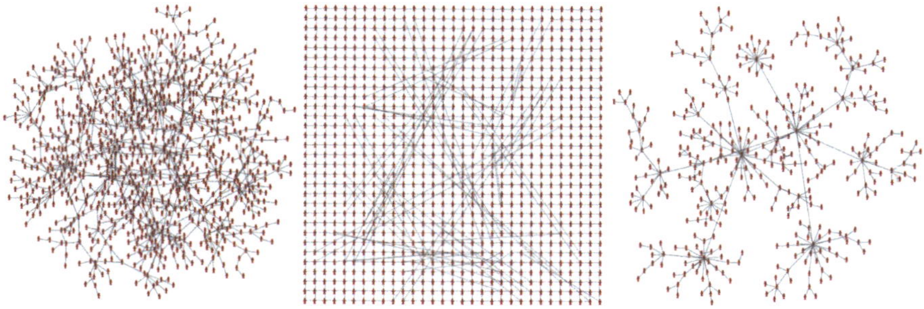

Fig. 4 The community of traders organized as a random graph (left panel), as a small-world (middle panel), and as a scale-free tree (right panel)

a criterion of preference; the second is a definition for the set of other traders to apply the chosen criterion to. With regards to the first element, the imitation is modelled to follow the perceived success of other market participants, defined in three alternative ways: (1) performance, measured as the gain yielded in the last transaction; (2) wealth, measured as the current value of the portfolio; (3) reputation, measured as the number of other traders already following him. With regards to the second element, the information disclosure has been designed in two possible configurations: global and local. In the global case, agents are assumed to have access to needed information about the entire community; in the local case agents have a clusterized knowledge for they can look up to their second degree of separation (i.e., know details of their nearest neighbors and of the nearest neighbors of them). It is finally worth noting that, differently from other existing contributions—as, e.g. Tedeschi et al. (2009, 2012), among others, where the imitation is referred to expectations—here, the imitation refers to the trading choice instead (whether to buy, hold, or sell), without modifying the group (fundamentalist of chartist) which the trader belongs to.

The community of investors has been designed on three alternative topologies, as shown in the three panels of Fig. 4: the first one (left), is a random graph, as introduced by Erdős-Rényi (1959); the second one (middle), is a $2D$ lattice version of the small-world presented in Watts-Strogatz (1998); and, finally, the third one (right) is the scale-free tree, as presented by Barabási-Albert (1999).

4.1.1 Dimension of Herding and Fat Tails

Simulations have been devoted to obtain two different results: first of all, to show the weight of the existing linkage between the amplitude of the imitative process and the market instability; secondly, to compare the effects of different criteria in order to gain an intuition on which one appears more prone to reinforce negative effects of imitation.

The first set of simulations has been performed to compare, for all social configurations, the effect of global and local amplitude of informative sets of traders. Results are shown in northern panels of Fig. 5. It is evident that global imitation (i.e., the imitation of the behavior of a master chosen from the whole community) is much more dangerous of the local one (i.e., the imitation of the behavior of a master chosen from the local community of the first two degrees of separation of the decision-maker). Further, quite unsurprisingly, the higher the individual imitation probability of investors (tested values have been, as indicated in legends 0.3, 0.5, and 0.7,), the stronger the effect on market stability is (tails become fatter). Reported results have obtained by using a unique selection criterion for the master, based on his reputation (and other two ones—performance and wealth—provide same results, with global imitation always worse than local one). Such an evidence is also consistent with another aspect of financial networks: the most severe crises are those involving the highest number of traders. Biondo et al. (2015, 2016) have successfully compared financial crises to earthquakes, proposing a model explaining self-organized criticality (SOC) in financial markets, by means of an adaptation of the OFC model (Olami et al. 1992) to a simple agent-based model without a true order book. The juxtaposition of such contributes is evident, once one focuses on the fact that the amplitude of the herding here induced by the deliberated synchronization (i.e., the imitative choice) is *corresponding* to the spontaneous involvement of agents in the financial-quakes deriving from the SOC in action. Forthcoming research, based on recent studies (Biondo 2018d), provides a model of the financial market, based on the order-book model here presented, in which the imitation is acknowledged as the *inner* ingredient of SOC. One of the main results advanced here and in previous studies (Biondo 2018a, b, c) is that heterogeneity of behaviors is the root of market stability: if a unique investor (i.e., a unique strategy) is considered as "the master" by the whole community, then many traders will behave as him, at the same time. Consequently, a bigger synchronization occurs and this unavoidably widens fat tails, increasing ultimately the market volatility.

The second set of simulations, whose results are reported in the southern panels of Fig. 5, investigated the effect of global and local imitation (based on all three criteria) in different topologies, when all agents have the same probability to imitate someone else. Regardless of the network configuration, the final result is fully consistent with the previous finding that imitation and any form of behavioral synchronization leads to fatter tails, i.e., a more unstable market.

4.1.2 Market Liquidity and Prices of Orders

The financial activity on true markets is done by a number of traders interacting on common frameworks where exchange are administrated and prices are shown. Sometimes, the liquidity condition of a market is intended as a measure of how easy is to trade an asset exchanged in it. Thus, if in a market purchases and sales can be completed without difficulties, at any time, and with a negligible impact on the equilibrium price, then that market is considered as liquid, (Foucault et al. 2013). Motivated by the main finding that the behavioral heterogeneity is the key factor in ensuring a reduction of market instability, a final set of simulations has been

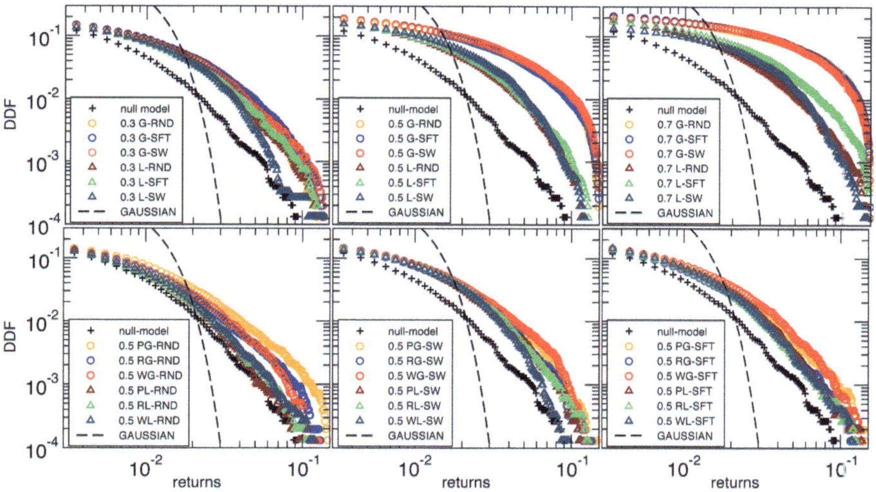

Fig. 5 Northern panels: global imitation reveals its dangerous potential in comparison with local one, independently of topological configurations, although both the scale-free and the small-world topologies exhibit slightly fatter tails. Southern panels: simulations show that, unsurprisingly, a higher individual propensity to imitate will increase market instability. Legends report the value of the individual imitation propensity and the criterion being tested, namely P (performance), R (reputation), and W (wealth). The tested scenario for the social interaction is labelled by the letter G, which stands for global, or L for local. Acronyms RND, SFT, and SW stand for, respectively, the random graph, the scale-free tree, and the 2D small-world network

performed in order to investigate how strongly the market dynamics is influenced by a reduction in the number of traders participating to the market activity, N^{tot}. The reason why an effect was expected is that the number of independent traders (thus neglecting imitative behaviors) is per se an indicator of how differentiated trading behaviors are. Indeed, in terms of market stability and fat tails, since a reduction in the number of active traders reduces the number of different orders (each heterogeneously priced), fat tails increase. The reason is that an involuntary synchronization operates, due to the increase in the probability to find groups of orders, barely differentiated within in terms of price, and distant instead among themselves. Simulations for the null model (where $N_t^{\text{tot}} = N$, i.e., all agents can become active and trade) have been compared with alternative ones, run for the same community, when part of traders, $N_{\iota,t}$, do not trade and remain inactive (thus $N_t^{\text{tot}} = N - N_{\iota,t} < N$, with $\iota \in (0, 1)$). Such a reduction has been obtained by randomly selecting traders, regardless of their behavioral root (i.e., whether fundamentalists or chartists), in two ways: a first hypothesis is that, chosen a given proportion of traders involved in market activity, it remains always the same, i.e., $N_{\iota,t} = \iota N \; \forall t$; a second hypothesis is that the proportion of active traders may vary in time without surpassing a chosen proportion, i.e., $N_{\iota,t} \leq \iota N$. Further, the $N_{\iota,t}$ traders are not always the same individuals. Results reported in Fig. 6 show that the reduction in the number of active traders effectively

Fig. 6 Liquidity conditions affect market volatility according to the number of active traders

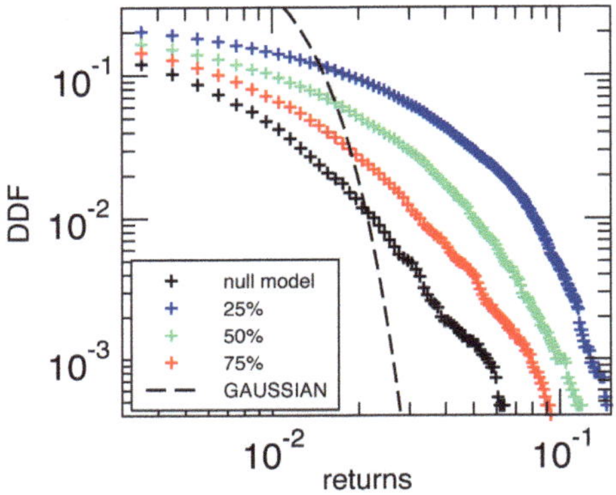

worsens fat tails. Such finding is counter-intuitive with respect to the idea that a less populated market is necessarily more stable.

5 Conclusive Remarks

This chapter presented a collection of results obtained by simulations performed with an agent-based model of financial order book. Compared to other existing models, it relies on some novelties, oriented to realism. A peculiar advantage of the presented framework is that simulated data is exclusively generated by transactions among simulated traders. The reliability of the model has been commented by showing its compliance with some of the most relevant stylized facts of true financial time series. In order to discuss the relevance of network theory in studying financial stability, the model has been build over different topology settings in order to show the effect of social interaction on market volatility.

A dynamic multiplex configuration, consisting of two layers, has been presented, where agents have been displaced in order to negotiate transactions within a venue held by a market maker. Particular emphasis has been placed on the so-called "social layer" of the network, where links transmit information about evaluation criteria used to determine which trader can be selected to be imitated. The effects of imitation have been investigated, by focusing on different topologies and comparing the tails of the returns density functions from different cases.

First of all, some subjective factors have been discussed to advance the idea that market instability can be seen as a consequence of individual conducts. The main relevant result provided by the model, which remains the major finding even when the model is operated without implementing the use of social networks, is that the behavioral heterogeneity of traders is the determinant key factor that matters in order to reduce market instability. Consistently with previous results in existing literature, homogeneity and synchronization are shown to be the very core problem of financial markets, whose dynamics appear dominated by self organized criticality. In turn, this poses the question about the credibility that should be attributed to models built on the assumption that market participants have perfectly complete informative sets and, thus, rational expectations. It has been shown that a very useful reduction of fat tails can be obtained by relying on the restriction of the imitation potential to local communities. Such a result could be implemented by following policy suggestions advanced in previous contributions (Biondo 2018b, c) where a reduction of information is suggested in order to reduce market instability. The rationale is that market instability emerges as a result of independent decisions assumed under the storming effect of a multitude of informative signals: each trader attempts to tune her individual conduct according to the most credible news, but informative sources provide more than the truly needed amount of information and this leads the market, at aggregate level, to dangerous fluctuations. A paradoxical reduction of information can reveal useful in order to reduce fat tails of returns distributions. From the most microeconomic point of view, the determinants of each transaction is the willingness to pay and to accept of the buyer and the seller. Such components are measured by demand and supply curves respectively and, as such, are linked to their elasticities, i.e., the sensitivity of variation in demanded/supplied quantities caused by price variations. The personal willingness to pay/accept of investors is directly related to market volatility. A possible interpretation of such a finding is that they act as an amplifier of the global dynamics. Fat tails are instead reduced if traders behave aggressively and price their orders lower. Further, the sensitivity to the market environment, as modelled by means of the parameter regulating the weight of external sources of information on the individual pricing of orders, is also a symptom of scarce ability and becomes common if traders have not competence in trading. Thus, the financial education of market participants is essential and can contribute to reduce market fluctuations.

Secondly, the model confirms that market volatility is linked with objective factors regarding the microstructure of the order book. A reduction of the time validity of limit orders has proven effective in improving market stability. The number of pending limit orders has been shown to have a role in inducing the adaptive attitude of investors to the context from which their decisions depend. This suggests that the microstructure, can be tuned in sight of a reduction of market volatility. The same applies to the limitation of the number of allowed counterparts that a market order can progressively match to trade the desired quantity. The price impact of orders could be, in principle, limited by imposing temporal and quantitative limits to orders. Finally, any transaction costs, as commissions levied by the market maker or any kind of wedge affecting prices—as Tobin taxes—are dangerous in terms of volatility and can destabilize the market.

A third finding is that imitation causes fatter tails. Three criteria have been used to choose the trader to imitate, the so-called "master", and they basically manifest same results. Synchronization of behaviors is not surprisingly correlated to fat tails and higher volatility. Simulations have been presented to argue to what extent effects of imitation can depend on the informative width of traders. Reported results show evidence that, also in this case, a perverse use of information globally available can worsen fat tails because the whole community select the same master and such a coordination generates much more volatility. The number of existing different order prices is a value driver to reduce instability. Such a result comes from simulations in which the liquidity of the market, defined as the circumstance that purchases and sales of an asset are easy and do not imply strong impact on the price, has been shown to be a relevant factor in reducing fat tails.

Previous studies, Biondo et al. (2015, 2016), have challenged a comparison between financial crises and earthquakes. The analysis of conditions for the emergence of self-organized criticality (SOC) in simulated dynamics goes beyond the scope of the present chapter. However, the amplitude of the herding here induced by the imitative choice can be correctly interpreted as *corresponding* to the spontaneous involvement of agents in avalanches generated by self-organized criticality. Forthcoming research, based on recent studies (Biondo 2018d), provides a model of the financial market, based on the order-book model here presented, in which the existence of imitation is a condition inducing criticality in the dynamics of the system.

References

Alfi V, Coccetti F, Marotta M, Pietronero L, Takayasu M (2006) Hidden forces and fluctuations from moving averages: a test study. Physica A 370:30–37

Alfi V, DeMartino A, Tedeschi A, Pietronero L (2007) Detecting the traders'strategies in minority-majority games and real stock-prices. Physica A 382:1–8

Allen F, Gale D (2000) Financial contagion. J Polit Econ 108:1–33

Almgren R, Chriss N (2001) Optimal execution of portfolio transactions. J Risk 3:5–40

Anufriev M, Panchenko V (2009) Asset prices, traders' behavior and market design. J Econ Dyn Control 33(5):1073–1090

Bak P, Paczuski M, Shubik M (1997) Price variations in a stock market with many agents. Physica A 246:430–453

Banerjee AV (1992) A simple model of herd behavior. Q J Econ 107(3):797–817

Banerjee AV (1993) The economics of rumours. Rev Econ Stud 60(2):309–327

Barabási AL, Albert R (1999) Emergence of scaling in random networks. Science 286(5439):509–512

Barberis N, Thaler R (2003) A survey of behavioural finance. In: Constantinides GM, Harris M, Stulz R (eds) Handbook of the Economics of Finance. Elsevier Science B.V

Ben-Elia E, Shiftan Y (2010) Which road do I take? A learning-based model of route-choice behavior with real-time information. Transp Res Part A 44:249–264

Bertsimas D, Lo AW (1998) Optimal control of execution costs. J Financ Mark 1(1):1–50

Bikhchandani S, Hirshleifer D, Welch I (1992) A theory of fads, fashion, custom, and cultural change as informational cascades. J Polit Econ 100(5):992–1026. https://doi.org/10.1086/261849

Biondo AE (2018b) Learning to forecast, risk aversion, and microstructural aspects of financial stability. Economics 12(2018–20):

Biondo AE (2018d) Self-organized criticality in financial markets and order book modeling. Working Paper, *mimeo*

Biondo AE (2019) Information versus imitation in a real-time agent-based model of financial markets. J Econ Interacti Coord. https://doi.org/10.1007/s11403-019-00249-2

Biondo AE (2018a) Order book microstructure and policies for financial stability. Stud Econ Financ 35(1):196–218. https://doi.org/10.1108/SEF-04-2017-0087

Biondo AE (2018c) Order book modeling and financial stability. J Econ Interact Coord. https://doi.org/10.1007/s11403-018-0227-6
Biondo AE, Pluchino A, Rapisarda A (2015) Modeling financial markets by self-organized criticality. Phys Rev E 92(4):042814
Biondo AE, Pluchino A, Rapisarda A (2016) Order book, financial markets, and self-organized criticality. Chaos Solitons Fractals 88:196–208. https://doi.org/10.1016/j.chaos.2016.03.001
Boccaletti S, Latora V, Moreno Y, Chavez M, Hwang DU (2006) Complex networks: structure and dynamics. Phys Rep 424(4):175–308
Boccaletti S, Bianconi G, Criado R, del Genio CI, Gómez-Gardeñes J, Romance M, Sendiña-Nadal I, Wang Z, Zanin M (2014) The structure and dynamics of multilayer networks. Phys Rep 544(1):1–122
Booth L, Chang B, Zhou J (2014) Which analysts lead the herd in stock recommendations? J Account Audit Financ 29(4):464–491. https://doi.org/10.1177/0148558X14537825
Bouchaud JP, Farmer JD, Lillo F (2009) How markets slowly digest changes in supply and demand. In: Hens T, Schenk-Hoppe KR (eds) Handbook of financial markets: dynamics and evolution. Handbooks in Finance, North-Holland, San Diego, pp 57–160
Brock WA, Hommes CH (1997) A rational route to randomness. Econometrica 65:1059–1095
Brock WA, Hommes CH (1998) Heterogeneous beliefs and routes to chaos in a simple asset pricing model. J Econ Dyn Control 22(8–9):1235–1274
Camerer C (2003) Behavioral game theory: Experiments in strategic interaction. Princeton University Press, Princeton
Chakraborti A, Toke IM, Patriarca M, Abergel F (2011) Econophysics review: I. Empirical facts. Quant Financ 11(7):991–1012
Chakravarty S, Holden CW (1995) An integrated model of market and limit orders. J Financ Intermed 4:213–241
Chiarella C (1992) The dynamics of speculative behavior. Ann Oper Res 37(1):101–123
Chiarella C, He XZ (2001) Asset price and wealth dynamics under heterogeneous expectations. Quant Finance 1(5):509–526
Chiarella C, Iori G (2002) A simulation analysis of the microstructure of double auction markets. Quant Finance 2:346–353
Chiarella C, Iori G, Perello J (2009) The impact of heterogeneous trading rules on the limit order book and order flows. J Econ Dyn Control 33(3):525–537
Chong C, Küpelberg, (2018) Contagion in financial systems: a bayesian network approach. SIAM J Financ Math 9(1):28–53
Clement MB, Tse SY (2005) Financial analyst characteristics and herding behavior in forecasting. J Financ 60(1):307–341. https://doi.org/10.1111/j.1540-6261.2005.00731.x
CME Group (2010), Impact of Tobin Taxes, Executive Summary
Colander D, Goldberg M, Haas A, Juselius K, Kirman A, Lux T, Sloth B (2009) The financial crisis and the systemic failure of the economics profession. Critical Review 21(2/3):249–267
Consiglio A, Lacagnina V, Russino A (2005) A simulation analysis of the microstructure of an order driven financial market with multiple securities and portfolio choices. Quant Finance 5(1):71–87
Cont R, Bouchaud JP (2000) Herd behavior and aggregate fluctuations in financial markets. Macroecon Dyn 4(2):170–196
Cont R, De Larrard A (2013) Price dynamics in a Markovian limit order market. SIAM J Financ Math 4(1):1–25
Cont R, Potters M, Bouchaud JP (1997) Scaling in stock market data: stable laws and beyond. In: Dubrulle B, Graner F, Sornette D (eds) Scale invariance and beyond. Springer, Berlin, Heidelberg
Cont R, Stoikov S, Talreja R (2010) A stochastic model for order book dynamics. Oper Res 58(3):549–563
Cooper RA, Day TE, Lewis CM (2001) Following the leader: a study of individual analysts' earnings forecasts. J Financ Econ 61(3):383–416. https://doi.org/10.1016/S0304-405X(01)00067-8
Copeland TE, Galai D (1983) Information effects on the bid-ask spread. J Finance 38(5):1457–1469

Daniels M, Farmer JD, Gillemot L, Iori G, Smith E (2003) Quantitative model of price diffusion and market friction based on trading as a mechanistic random process. Phys Rev Lett 90:108102
Day RH, Huang W (1990) Bulls, bears and market sheep. J Econ Behav Organ 14(3):299–329
Delli Gatti D, Gaffeo E, Gallegati M, Giulioni G, Palestrini A (2008) Emergent macroeconomics an agent-based approach to business fluctuations. Springer, Milan
Delli Gatti D, Desiderio S, Gaffeo E, Cirillo P, Gallegati M (2011) Macroeconomics from the Bottom-up. Springer Science+Business Media
Elliot M, Golub B, Jackson MO (2014) Financial networks and contagion. Am Econ Rev 104(10):3115–3153
Erdős P, Rényi A (1959) On random graphs. I. Publicationes Mathematicae, Debrecen, pp 290–297
Farmer JD, Patelli P, Zovko II, (2005) The predictive power of zero intelligence in financial markets. Proc Natl Acad Sci USA 102:2254–2259
Farmer JD, Foley D (2009) The economy needs agent-based modelling. Nature 460:685–686
Foucault T (1999) Order flow composition and trading costs in a dynamic limit order market. J Financial Mark 2:99–134
Foucault T, Pagano M, Roell A (2013) Market liquidity: theory, evidence, and policy. Oxford Scholarship Online: September 2013. https://doi.org/10.1093/acprof:oso/9780199936243.001.0001, Print ISBN-13: 9780199936243
Franke R, Sethi R (1998) Cautious trend-seeking and complex asset price dynamics. Res Econ 52(1):61–79
Gallegati M, Richiardi M (2009) Agent-based modelling in economics and complexity. In: Meyer RA (ed) Encyclopedia of complexity and system science. Springer, New York, pp 200–224
Gil-Bazo J, Moreno D, Tapia M (2007) Price dynamics, informational efficiency, and wealth distribution in continuous double-auction markets. Comput Intell 23(2):176–196
Glosten LR (1994) Is the electronic open limit order book inevitable? J Finance 49:1127–1161
Glosten LR, Milgrom PR (1985) Bid, ask and transaction prices in a specialist market with heterogeneously informed traders. J Financ Econ 14:71–100
Gopikrishnan P, Plerou V, Amaral LA, Meyer M, Stanley HE (1999) Scaling of the distribution of fluctuations of financial market indices. Phys Rev E 60:5305–5316
Grinblatt M, Han B (2005) Prospect theory, mental accounting, and momentum. J Financ Econ 78:311–339
Hirshleifer D, Hong TS (2003) Herd behaviour and cascading in capital markets: a review and synthesis. Eur Financ Manag 9(1):25–66. https://doi.org/10.1111/1468-036X.00207
Hommes CH (2001) Financial markets as nonlinear adaptive evolutionary systems. Quant Finance 1(1):149–467
Hommes CH (2006) Heterogeneous agent models in economics and finance. In: Tesfatsion L, Judd KL (eds) Handbook of computational economics, vol 2. North-Holland, Amsterdam
Iori G (2002) A microsimulation of traders activity in the stock market: the role of heterogeneity, agents' interactions and trade frictions. J Econ Behav Org 49(2):269–285
Kahneman D, Tversky A (1974) Judgment under uncertainty: heuristics and biases, Science (New York, N.Y.), Vol. 185 No. 4157, pp. 1124–1131
Kahneman D, Tversky A (1979) Prospect theory: an analysis of decision under risk. Econometrica 47(2):263–291
Kao AB, Couzin ID (2014) Decision accuracy in complex environments is often maximized by small group sizes. Proc Royal Soc B 281(1784). https://doi.org/10.1098/rspb.2013.3305
Keynes JM (1936) The general theory of unemployment, interest and money. MacMillan, London
Kiyotaki N, Moore J (1997) Credit chains. Working Paper, University of Minnesota and London School of Economics
Kyle AS (1985) Continuous auctions and insider trading. Econometrica 53:1315–1335
Lagunoff R, Schreft SL (2001) A model of financial fragility. J Econ Theory 99:220–264
LeBaron B (2006) Agent-based computational finance. In: Tesfatsion L, Judd KL (eds) Handbook of computational economics, vol 2. North-Holland, Amsterdam

Leijonhufvud A (1993) Towards a not-too-rational macroeconomics. Southern Economic Journal 60(1):1–13
Leitner Y (2005) Financial networks: contagion, commitment, and private sector bailouts. J Financ IX(6):2925–2953
Lorenz J, Rauhut H, Schweitzer F, Helbing D (2011) How social influence can undermine the wisdom of crowd effect. PNAS 108(22):9020–9025. https://doi.org/10.1073/pnas.1008636108
Lux T (1995) Herd behavior, bubbles and crashes. Econ J 105:881–896
Lux T (1998) The socio-economic dynamics of speculative markets: interacting agents, chaos, and the fat tails of return distributions. J Econ Behav Organ 33(2):143–165
Lux T, Marchesi M (1999) Scaling and criticality in a stochastic multi-agent model of a financial market. Nature 397(6719):498–500
Lux T, Marchesi M (2000) Volatility clustering in financial markets: a microsimulation of interacting agents. Int J Theor Appl Financ 3(4):675–702
Majorana E (1942) Il valore delle leggi statistiche nella fisica e nelle scienze sociali, Scientia, Quarta serie, Febbraio-Marzo 1942, pp. 58-66. English translation in Majorana E (2005) The value of statistical laws in physics and social sciences. Quant Finance 5:133–140
Mandelbrot B (1963) The variation of certain speculative prices. J Bus 36(4):394–419
Mantegna RN, Stanley HE (2000) Introduction to econophysics: correlations and complexity in finance. Cambridge University Press, Cambridge
Markose SM, Alentorn A, Krause A (2004) Dynamic learning, herding and guru effects in networks. University of Essex Department of Economics Discussion Papers. http://repository.essex.ac.uk/id/eprint/3732
Maslov S (2000) Simple model of a limit order-driven market. Physica A 278:571–578
Mike S, Farmer JD (2008) An empirical behavioral model of liquidity and volatility. J Econ Dyn Control 32(1):200–234
Mitchell M (2009) Complexity: a guided tour. Oxford University Press, New York
Moussaid M, Garnier S, Theraulaz G, Helbing D (2009) Collective information processing and pattern formation in swarms, flocks, and crowds. Top Cogn Sci 1(3):469–497. https://doi.org/10.1111/j.1756-8765.2009.01028.x
Olami Z, Feder HJS, Christensen K (1992) Self-organized criticality in a continuous, nonconservative cellular automaton modeling earthquakes. Phys Rev Lett 68(8):1244
Orlean A (1995) Bayesian interactions and collective dynamics of opinion: herd behavior and mimetic contagion. J Econ Behav Org 28(2):257–274
Pagan A (1996) The econometrics of financial markets. J Empir Financ 3:15–102
Parlour CA (1998) Price dynamics in limit order markets. Rev Financ Stud 11:789–816
Parlour CA, Seppi DJ (2008) Limit order markets: a survey. In: Thakor A, Boot A (eds) Handbook of financial intermediation and banking. Elsevier, Amsterdam
Raberto M, Cincotti S, Focardi SM, Marchesi M (2001) Agent-based simulation of a financial market. Phys A Stat Mech Appl 299(1):319–327
Rochet J-C, Tirole J (1996) Interbank lending and systemic risk. J Money Credit Bank 28:733–762
Rosu I (2009) A dynamic model of the limit order book. Rev Financ Stud 22:4601–4641
Rosu I (2016) Liquidity and information in orden driven markets. http://webhost.hec.fr/rosu/research/info_lob.pdf. Accessed on 22 Aug 2018
Schumpeter JA (2003) How does one study social science? Society 40(3):57–63
Simon HA (1957) Models of man: social and rational. John Wiley and Sons, New York
Sornette D (2009) Why stock markets crash: critical events in complex financial systems. Princeton University Press, Princeton
Stauffer D, Sornette D (1999) Self-organized percolation model for stock market fluctuations. Phys A Stat Mech Its Appl 271(3–4):496–506
Takayasu M, Mizuno T, Takayasu H (2006) Potential force observed in market dynamics. Physica A 370:91
Tedeschi G, Iori G, Gallegati M (2009) The role of communication and imitation in limit order markets. Eur Phys J B 71(4):489

Tedeschi G, Iori G, Gallegati M (2012) Herding effects in order driven markets: the rise and fall of gurus. J Econ Behav Organ 81(1):82–96
Tesfatsion L (2006) Agent-based computational economics: a constructive approach to economic theory. In: Tesfatsion L, Judd KL (eds) Handbook of computational economics, vol II. North-Holland, Amsterdam
Von Hayek FA (2015) The pretence of knowledge, Nobelprize.org. Nobel Media AB 2014, Web. Accessed on 1 Jul 2015
Watts DJ, Strogatz SH (1998) Collective dynamics of small-world networks. Nature 393(6684):440
Yaari M (1987) The dual theory of choice under risk. Econometrica 55(1):95–115
Zhao Z, Zhang Y, Feng X, Zhang W (2014) An analysis of herding behavior in security analysts' networks. Physica A 413:116–124. https://doi.org/10.1016/j.physa.2014.06.082

Detection of Factors Influencing Market Liquidity Using an Agent-Based Simulation

Isao Yagi, Yuji Masuda and Takanobu Mizuta

Abstract Recently, investors have become more interested in market liquidity, which is regarded as a measure of a booming financial market. When market liquidity is high, market participants are able to smoothly buy and sell their intended amount at a price close to the market mid-price. When discussing market liquidity in empirical studies, researchers have defined liquidity indicators that are consistent with their research objectives. However, it has not been clarified which market factors affect these indicators. In the present paper, we investigated which market factors affect major liquidity indicators, including Volume, Tightness, Resiliency, and Depth, using an artificial market, which is a type of agent-based simulation system. As a result, market liquidity based on Volume is completely opposite to market liquidity based on Tightness, Resiliency, or Depth. Moreover, we confirmed the price decline rate from the fundamental price and the price convergence periods to the fundamental price as a measure of the convergence speed, which is the original meaning of Resiliency, from the price level, which has been brought about by random price changes. Therefore, the trades of fundamentalists have the effect of shortening the convergence period, i.e., causing market liquidity to increase.

1 Introduction

These days, many investors attach importance to market liquidity. In general, a highly liquid market is a market in which market participants can quickly sell (or buy) in the quantities that they want to sell (or buy) and at prices close to the market price observed at that time (Kurosaki et al. 2015).

I. Yagi (✉) · Y. Masuda
Kanagawa Institute of Technology, Atsugi, Kanagawa 243-0292, Japan
e-mail: iyagi2005@gmail.com

Y. Masuda
e-mail: s1885003@cce.kanagawa-it.ac.jp

T. Mizuta
SPARX Asset Management Co., Ltd., Tokyo 108-0075, Japan
e-mail: mizutata@gmail.com

A. S. Chakrabarti et al. (eds.), *Network Theory and Agent-Based Modeling in Economics and Finance*, https://doi.org/10.1007/978-981-13-8319-9_6

Many empirical studies on market liquidity have produced various beneficial findings. For example, Kyle (1985) proposed the evaluation of market liquidity using a number of concepts, including (1) tightness, (2) depth, and (3) resiliency. Amihud (2002) proposed the measurement of illiquidity, which is the ratio of a stock absolute daily return to its daily dollar volume, averaged over some period (ILLIQ) and showed the proposition that asset expected returns were increasing in illiquidity using ILLIQ. However, the definition of liquidity often depends on the purpose of the research. For example, a market in which the purchases and sales of each market participant have little impact on the market price and a market having low volatility are both often regarded as high-liquidity markets. As a result, the indicators of liquidity often differ between studies, so it is difficult to analyze whether the findings on liquidity obtained in one study are consistent with the findings obtained in other studies.

One method for dealing with situations that defy analysis by previous empirical research methods is an artificial market. An artificial market, which is a type of agent-based simulation system, is an agent-based model of financial markets (Chiarella et al. 2009; Chen et al. 2012; Mizuta et al. 2014). Each agent is assigned a specific trading (i.e., buying or selling) rule and is then set to trade financial assets as an investor. The market can then be observed to determine how the agents behave. At the same time, by modeling and incorporating certain restrictions on the market side (e.g., limitations to ensure market stability and efficiency such as a short selling regulation), it is possible to examine how investors behave, as well as what kinds of effects their behaviors induce in the market.

Some artificial market simulations have been fruitful in terms of producing useful findings (Yamamoto et al. 2013; Yagi et al. 2017). However, it seems that no one has been discussing in detail the effects on market liquidity of different definitions by using an artificial market.

Thus, in the present study, we set out to investigate which market factors affect four typical liquidity indicators—Volume, Tightness, Resiliency, and Depth—by changing the artificial market factors. We also confirmed the convergence speed, which is the original meaning of Resiliency but cannot be measured in the real world, from the price level, which has been brought about by random price changes. First, in Sect. 2, we explain the definition of market liquidity and review some empirical studies on market liquidity. Next, we explain the proposed artificial market used in the present study in Sect. 3. Then, in Sect. 4, we describe the results of simulations performed using our artificial market and the mechanism thereof and discuss the difference between the relationship among liquidity indicators in the present paper and the relationship in empirical studies. In Sect. 4.5, we investigate the price decline rate from the fundamental price and the price convergence periods to the fundamental price as a measure of the convergence speed and discuss the relationship between market liquidity and investors. Finally, we discuss our conclusions regarding the present study and our future work in Sect. 5.

2 Market Liquidity

2.1 *Market Liquidity and the Four Liquidity Indicators*

Market liquidity has not been defined clearly. However, generally it is said that a liquid market is a market in which participants can rapidly execute large-volume transactions with a small impact on price (Bank for International Settlements 1999). In a related empirical study (Nishizaki et al. 2013), four indicators—Volume, Tightness, Resiliency, and Depth—were used to examine and evaluate market liquidity (Fig. 1).

2.1.1 Volume

The liquidity indicator Volume is the amount of trading in the market. The turnover, which is the number of shares traded in an entire market during a given period of time, can be used to measure Volume. The reason for this is as follows. We can see that frequent orders and a large amount of orders in a short period are easier to match when the turnover is large. Therefore, it can be said that the higher the turnover (Volume), the higher the liquidity. In the present paper, we observe trading volume per simulation run as the turnover.

2.1.2 Tightness

Tightness refers to the lowness of transaction costs, such as the difference between buy and sell order prices. It is said that the bid-ask spread, which is the amount by

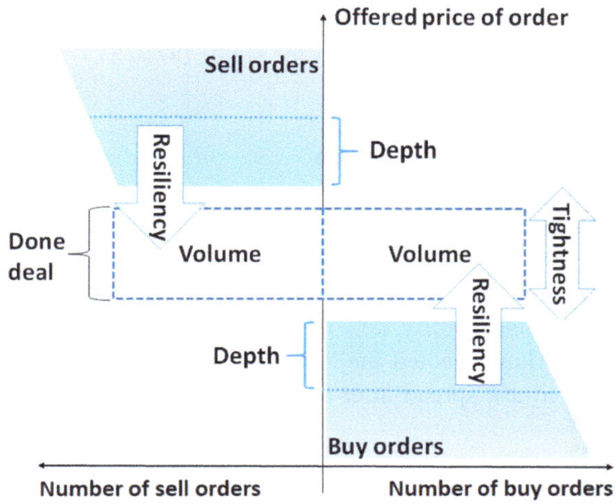

Fig. 1 Indicators of market liquidity in view of buy and sell orders

which the ask (sell) price exceeds the bid (buy) price for an asset in the market, can be used to measure Tightness. Therefore, it can be said that the smaller the bid-ask spread, the higher the liquidity. Thus, we observe the average of the bid-ask spreads in simulation runs as Tightness.

2.1.3 Resiliency

Resiliency is a characteristic of markets in which new orders flow quickly to correct order imbalances, which tend to move prices away from what is warranted by the fundamentals.

The daily price range to turnover ratio (Nishizaki et al. 2013), which is the difference between the highest and lowest transaction prices of the day divided by turnover in a trading day, can be used to measure Resiliency. This is because a smaller ratio can be interpreted to mean that market conditions are more resilient, and traders might be able to execute transactions smoothly and rapidly. We observe the average of the daily price range to turnover ratio (Nishizaki et al. 2013) in simulation runs as Resiliency, and hereinafter refer to this as the P/T ratio for simplicity. A lower P/T ratio corresponds to a higher liquidity.

2.1.4 Depth

Depth is the amount of orders around the highest range of buy orders and the lowest range of sell orders. Depth refers to the abundance of orders, either actual or easily uncovered, of potential buyers and sellers, both above and below the price at which a security currently trades. Therefore, it can be said that the larger the Depth, the higher the liquidity. Since the information on the trading board indicates market Depth, the sum of the amount of sell orders from the lowest price to 50 ticks higher than the lowest price and the amount of buy orders from the highest price to 50 ticks lower than the highest price is regarded as the indicator of Depth in the present paper, where a tick is the minimum unit for prices (Sect. 3).

2.1.5 Related Research

Muranaga (1999) studied the dynamic aspects of market liquidity of Japanese stocks, namely, the first section of the Tokyo Stock Exchange (TSE). Muranaga and his colleagues checked the relationship between trade frequency and the three indicators of market liquidity, i.e., Tightness, Depth, and Resiliency, by a cross-sectional analysis. They selected the bid-ask spread, market impact, and convergence speed of the bid-ask spread after trades corresponding to Tightness, Depth, and Resiliency, respectively. Note that market impact refers to quote changes triggered by trade execution divided by the corresponding trade volume. As a result, they found that all three indicators correlated with trade frequency. They also analyzed whether the reduction in tick size of the TSE affected the market liquidity of the TSE. They found that the trade frequency increased and the bid-ask spread and volatility decreased with the reduction in tick size, which indicates clearly that a reduction in tick size affects market liquidity.

Chung et al. (2005) confirmed the effect of tick size on market quality by using unique data from a market with multiple tick sizes. They revealed how the minimum price variation affects market liquidity by analyzing both the cross-sectional and inter-temporal relation between tick size and market liquidity. They demonstrated that Depth has a positive correlation with volume and turnover rate and a negative correlation with market price level and volatility.

3 Artificial Market Model

In the present study, a new artificial market model based on the artificial market model of Mizuta et al. (2015) is proposed.

In the proposed model, only one risk asset is available for trading. Hereinafter, we refer to risk assets as simply assets, whereas we refer to non-risk assets as cash, because non-risk assets are equivalent to cash in the proposed artificial market.

There are n agents, agents $j = 1, \ldots, n$, each of which places an order in sequence, i.e., from agent 1 to agent n. After the final agent, agent n, has placed an order, the first agent, agent 1, places the next order. The time t is incremented by 1 each time an agent places an order. Thus, the process moves forward one step even when a trade does not occur and this new order is placed on the order book.

The mechanism for determining the price in this model being a continuous double auction (continuous trading session) means that if there are sell (buy) order prices in the order book that are lower (higher) than the buy (sell) order price of the agent, then the agent's order is immediately matched to the lowest sell order (highest buy order) in the order book. We refer to this as a market order. If there are no such orders in the order book, then the order does not match any other order and remains in the order book. We refer to this order as a limit order. The remaining orders in the order book are canceled at time t_c (order effective period) after the order was placed. The tick size, which is the minimum unit for the price, is ΔP, and when orders are sell orders, fractional values smaller than ΔP are rounded up. On the other hand, when orders are buy orders, they are rounded down. Each agent sends one order each time. An agent can possess assets indefinitely because the quantity of cash of the agent is set indefinitely. Agents can also short sell.

3.1 Order Process

The order prices of agent j by transaction are determined as shown below. The rate of change of the price expected by agent j at time t (the expected return) $r_{e_j}^t$ is given by

$$r_{e_j}^t = \frac{1}{w_{1,j}^t + w_{2,j}^t + u_j}(w_{1,j}^t r_{e1,j}^t + w_{2,j}^t r_{e2,j}^t + u_j \epsilon_j^t) \tag{1}$$

where $w_{i,j}^t$ is the weight of the ith term for agent j at time t and is set according to the uniform distribution between 0 and $w_{i,\max}$ at the start of the simulation and then varied using the learning process described later herein. Furthermore, u_j is the weight of the third term and is set according to the uniform distribution between 0 and $u_{\max}$ at the start of the simulation and is kept constant thereafter.

The first term on the right-hand side of Eq. (1) represents the fundamental strategy, which indicates that an agent expects a positive (negative) return when the market price is lower (higher) than the fundamental price. The term $r_{e1,j}^t$ is the expected return of the fundamental strategy for agent j at time t, given by $r_{e1,j}^t = \log(P_f / P^{t-n})$, where P_f is the fundamental price, which is constant over time, and P^t is the market price at time t. The market price is set to the most recent price at the time if no trading is occurring. The initial market price is set to the fundamental price, i.e., $P^0 = P_f$.

The second term represents the technical strategy, which indicates that an agent expects a positive (negative) return when the historical return is positive (negative). Here, $r_{e2,j}^t$ is the expected return of the technical strategy for agent j at time t, given by $r_{e2,j}^t = \log(P^{t-n} / P^{t-n-\tau_j})$, where τ_j is set according to the uniform distribution between 1 and $\tau_{\max}$ at the start of the simulation.

The third term represents the noise strategy. Here, ϵ_j^t is a normally distributed random error with mean zero and standard deviation σ_ϵ.

Based on the expected return $r_{e_j}^t$, the expected price $P_{e_j}^t$ is found using the following equation:

$$P_{e_j}^t = P^{t-1} \exp(r_{e_j}^t) \tag{2}$$

The order price $P_{o_j}^t$ is a normally distributed random number with mean $P_{e_j}^t$ and standard deviation P_σ^t given by

$$P_\sigma^t = P_{e_j}^t \cdot Est \tag{3}$$

where $Est (0 < Est \leq 1)$ is referred to as the variation coefficient of the order price for expedience. The choice between buying and selling is determined by the relative sizes of the expected price $P_{e_j}^t$ and the order price $P_{o_j}^t$.

- An agent places a buy order for one share if $P_{e_j}^t > P_{o_j}^t$.
- An agent places a sell order for one share if $P_{e_j}^t < P_{o_j}^t$.

3.2 Learning Process

Previous studies using an artificial market implemented various kinds of learning processes. For example, agents switched strategies and/or tuned their strategy parameters based on their performance, market price, etc. (Arthur et al. 1997; Lux and Marchesi 1999; Nakada and Takadama 2013). The learning process in the present study is implemented to switch the strategy between fundamental and technical strategies.

We modeled the learning process as follows based on Yagi et al. (2017). For $r_{e_{i,j}^t}$, learning is performed by each agent immediately before the agent places an order. That is, when $r_{e_{i,j}^t}$ and $r_l{}^t = \log(P^t/P^{t-t_l})$ are of the same sign, $w_{i,j}$ is updated as follows:

$$w_{i,j}^t \leftarrow w_{i,j}^t + k_l |r_l{}^t| q_j^t (w_{i,\max} - w_{i,j}^t) \tag{4}$$

where k_l is a constant, and q_j^t is set according to the uniform distribution between 0 and 1. When $r_{e_{i,j}^t}$ and $r_l{}^t$ have opposite signs, $w_{i,j}$ is updated as follows:

$$w_{i,j}^t \leftarrow w_{i,j}^t - k_l |r_l{}^t| q_j^t w_{i,j}^t \tag{5}$$

Separately from the process for learning based on past performance, $w_{i,j}^t$ is reset with a small probability m, according to the uniform distribution between 0 and $w_{i,\max}$.

4 Simulation Results and Discussion

4.1 Overview

In Sect. 4.3, we show how the four liquidity indicators change when one parameter is changed. Moreover, we check the difference between Resiliency based on the P/T ratio and Resiliency based on the convergence speed in Sect. 4.5. We set the initial values of the model parameters as listed in Table 1. In the simulations, various parameters are then changed as listed in Table 2. For each simulation, only one parameter is changed, whereas the others are maintained at their initial values. We ran simulations from $t = 0$ to $t = 1{,}000{,}000$, and the four liquidity indicator values were calculated as averages of 20 simulation runs. We define 20,000 time steps as 1 day because the number of trades for 20,000 time steps is approximately the same as that in actual markets in 1 day.

Table 1 Parameters

Parameters	Initial values
n	1,000
$w_{1,\max}$	1
$w_{2,\max}$	10
$u_{\max}$	1
$\tau_{\max}$	10,000
σ_ϵ	0.06
Est	0.003
t_c	20,000
ΔP	0.1
P_f	10,000
k_l	4
m	0.01

Table 2 Parameters and settings

Parameters	Values				
ΔP	0.01	0.1	1.0	10	100
$w_{1,\max}$	1.0	3.0	5.0	8.0	10.0
$w_{2,\max}$	1.0	3.0	5.0	8.0	10.0
σ_ϵ	0.02	0.04	0.06	0.08	0.1
Est	0.003	0.005	0.01	0.02	0.03
t_c	10,000	15,000	20,000	25,000	30,000

4.2 Validation of Proposed Artificial Market

As many empirical studies have mentioned (Sewell 2006; Cont 2001), a fat tail and volatility clustering appear in actual markets. Therefore, we set the artificial market parameters so as to replicate these features.

Table 3 shows the statistics for stylized facts, which are averages of 20 simulation runs, for which we calculated the price returns at intervals of 100 time units. Table 3 shows that both kurtosis and autocorrelation coefficients for squared returns with several lags are positive, which means that all runs replicate a fat tail and volatility clustering. This indicates that the model replicates long-term statistical characteristics observed in real financial markets.

Table 3 Stylized facts

Kurtosis		0.802
	Lag	
Autocorrelation coefficients for squared returns	1	0.206
	2	0.132
	3	0.103
	4	0.088
	5	0.078

4.3 Results

4.3.1 Case that Tick Size is Changed

Table 4 shows how the four liquidity indicators change with the tick size ΔP. When the tick size becomes large, Volume becomes small, but Tightness, Resiliency, and Depth become large.

The reasons for these results are as follows.

We discuss the reason why Volume has become smaller, taking the tick size of $\Delta P = 1$ as an example. Assume that the expected price is 10,000 and that the order price is P. At this time, if P becomes a sell order at a price greater than 10,000 and less than 10,001, then the fraction of P smaller than ΔP is rounded up and the sell order price is 10,001. Otherwise, if P becomes a buy order at a price greater than 9,999 and less than 10,000, then the fraction of P less than ΔP is truncated and the buy order price is 9,999. As a result, there are no orders for which the price is 10,000, but there are orders for which the price is 9,999 or 10,001. This means that there is a range in which orders are entered into the market. This range expands as the tick size increases. Therefore, it becomes difficult to trade shares, and Volume becomes smaller.

The reason why Tightness becomes larger is that, as the tick size becomes larger, the bid-ask spread becomes larger due to the minimum bid-ask spread being equal to the tick size.

Table 4 Four liquidity indicators when ΔP is changed

ΔP	Volume	Tightness	Resiliency	Depth
0.01	265,884	12.015	0.0104	2,362
0.1	267,309	12.174	0.0106	2,326
1.0	265,894	12.557	0.0106	2,369
10	266,892	17.579	0.0118	2,431
100	253,092	98.393	0.0190	3,114

The reason why Resiliency becomes larger is as follows. The minimum of the numerator of the P/T ratio, which is the difference between the highest and lowest transaction prices of the day, is equal to the tick size. As the tick size becomes larger, the minimum of the numerator of the P/T ratio also becomes larger. Note that the denominator of the P/T ratio, which is the turnover, becomes smaller (Table 4). Therefore, Resiliency becomes larger.

When the turnover decreases, the amount of limit orders on the trading board increases. Therefore, Depth also increases.

4.3.2 Case that Maximum Fundamental Strategy Weight is Changed

Table 5 shows how the four liquidity indicators change when the maximum fundamental strategy weight of the expected return $w_{1,\max}$ changes. As shown in the table, as it becomes larger, Volume, Tightness, and Resiliency become smaller, but Depth becomes larger. Figure 2 shows the mechanism when the fundamental strategy weight of the expected return becomes larger.

We explain the reasons for these results as follows.

The reason why Volume becomes smaller is as follows. When the maximum fundamental strategy weight of the expected return becomes larger, the expected prices tend to converge to the fundamental price (Fig. 2). Therefore, the order prices

Table 5 Four liquidity indicators when $w_{1,\max}$ is changed

$w_{1,\max}$	Volume	Tightness	Resiliency	Depth
1.0	265,894	12.557	0.0106	2,369
3.0	247,682	12.014	0.0094	2,807
5.0	230,026	11.772	0.0084	3,283
8.0	207,453	11.171	0.0074	3,896
10.0	19,6318	10.675	0.0070	4,201

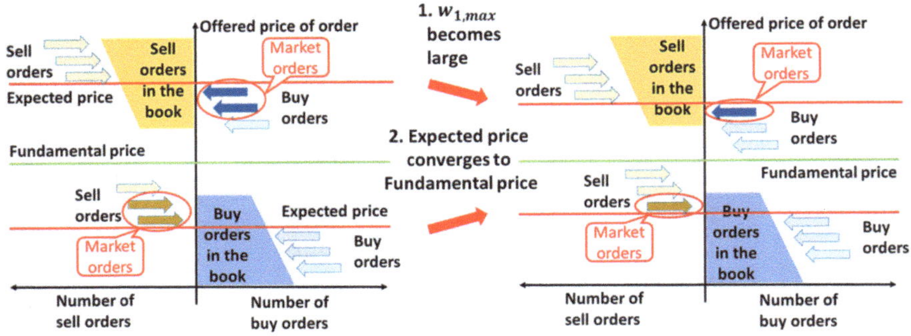

Fig. 2 Mechanism when the fundamental strategy weight of the expected return becomes larger

and the market price also tend to converge to the fundamental price. The amount of sell (buy) orders in the order book that are lower (higher) than the expected price decreases, and the amount of the market buy (sell) orders also decreases. Thus, Volume becomes smaller.

The reason why Tightness becomes smaller is as follows. As described above, when the market price converges to the fundamental price, the limit orders, the prices of which become close to the market price, increase. Then, the best-bid (best-ask) price tends to decrease (increase) and the difference between the best-bid price and the best-ask price decreases.

The reason why Resiliency becomes smaller is as follows. When the maximum fundamental strategy weight of the expected return becomes larger and the order prices are close to the fundamental price, the difference between the highest and lowest market prices of the day is smaller. Thus, the numerator of the P/T ratio decreases. Note that the denominator of the P/T ratio becomes smaller. Therefore, Resiliency becomes larger.

When the turnover, i.e., Volume, decreases, the limit orders in the order book increase. Therefore, Depth also increases.

4.3.3 Case that Maximum Technical Strategy Weight is Changed

Table 6 shows how the four liquidity indicators change when the maximum technical strategy weight of the expected return $w_{2,\max}$ changes. As $w_{2,\max}$ becomes larger, Volume, Tightness, and Resiliency become smaller, but Depth becomes larger. Figure 3 shows the mechanism when the technical strategy weight of the expected return becomes larger.

We explain the reasons for these results as follows.

The reason why Volume becomes smaller is as follows. When the maximum technical strategy weight of the expected return becomes larger, the expected prices tend to converge to the market price (Fig. 3). Therefore, the amount of sell (buy) orders in the order book that are lower (higher) than the expected price decreases, and the amount of market buy (sell) orders also decreases. Thus, Volume becomes smaller.

The reason why Tightness becomes smaller is as follows. As described above, when the order prices converge to the market price, the limit orders, the prices of which become close to the market price, increase. Then, the best-bid (best-ask) price tends to decrease (increase), and the difference between the best-bid price and the best-ask price decreases.

The reason why Resiliency becomes smaller is as follows. When the maximum technical strategy weight of the expected return becomes larger and the order prices are close to the market price, the difference between the highest and lowest market prices of the day is smaller. Thus, the numerator of the P/T ratio decreases. Note that the denominator of the P/T ratio becomes smaller. Therefore, Resiliency becomes larger.

Table 6 Four liquidity indicators when $w_{2,max}$ is changed

$w_{2,max}$	Volume	Tightness	Resiliency	Depth
1.0	372,930	78.297	0.0518	223
3.0	348,262	31.672	0.0267	645
5.0	320,592	21.473	0.0180	1,080
8.0	285,224	14.869	0.0123	1,840
10.0	265,894	12.557	0.0106	2,369

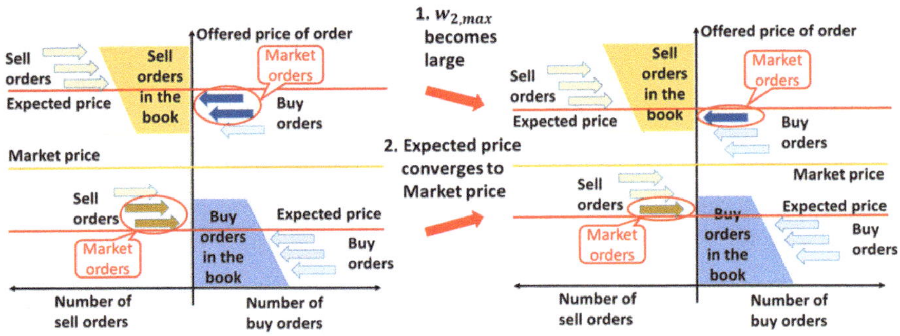

Fig. 3 Mechanism when the technical strategy weight of the expected return becomes larger

When the turnover, i.e., Volume, decreases, the limit orders in the order book increase. Therefore, Depth also increases.

4.3.4 Case that SD of Noise Factor of Expected Return is Changed

Table 7 shows how the four liquidity indicators change when the standard deviation of the noise factor of the expected return σ_ϵ changes. When the standard deviation becomes large, Volume, Tightness, and Resiliency also become large, but Depth becomes small. Figure 4 shows the mechanism when the standard deviation of the noise factor of the expected return becomes larger.

The reasons for these results are as follows.

We first explain the reason why Volume becomes larger. When the standard deviation of the noise factor of the expected return becomes large, the expected prices of each agent begin to diffuse. As the expected prices diversify from the market price, the amount of limit orders in the order book between the market price and the expected price increases (Fig. 4). Therefore, any new order is more likely to be a market order, and the possibility of trade matching to an order in the order book increases. Thus, Volume becomes larger.

Table 7 Four liquidity indicators when σ_ϵ is changed

σ_ϵ	Volume	Tightness	Resiliency	Depth
0.02	151,109	5.308	0.0061	5,655
0.04	222,479	8.975	0.0081	3,644
0.06	265,894	12.557	0.0106	2,369
0.08	296,523	15.928	0.0137	1,648
0.1	318,035	18.821	0.0169	1,268

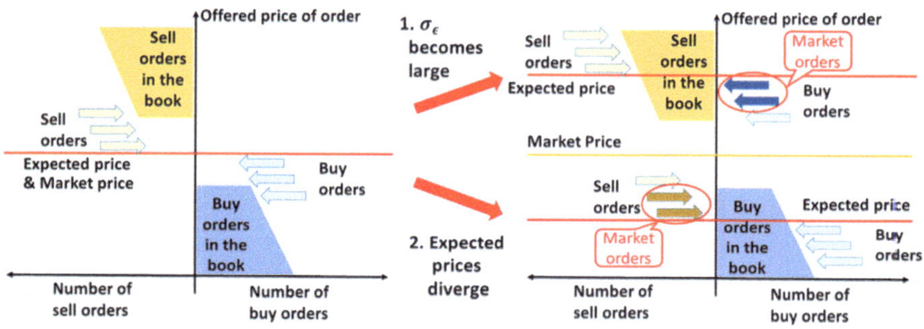

Fig. 4 Mechanism when the standard deviation of the noise factor of the expected return becomes larger

The reason why Tightness becomes larger is that, in addition to the above mechanism, the bid-ask spread becomes large as the possibility of trade matching to an order in the order book increases.

The reason why Resiliency becomes larger is as follows. The numerator of the P/T ratio increases when the order prices that each agent calculates based on the expected price diffuse. Note that the denominator of the P/T ratio also becomes larger. However, the growth rate of the turnover, i.e., Volume, is no greater than that of the numerator of the P/T ratio (Table 7).

When the turnover, i.e., Volume, increases, the limit orders on the trading board decrease because current orders being sent tend to become market orders. Therefore, Depth also decreases.

4.3.5 Case that Variation Coefficient of Order Price is Changed

Table 8 shows how the four liquidity indicators change when the variation coefficient of the order price Est changes. As Est becomes larger, Volume and Depth become smaller, and Tightness and Resiliency become larger. Figure 5 shows the mechanism when the variation coefficient of the order price becomes larger.

The reasons for these results are as follows.

Table 8 Four liquidity indicators when Est is changed

Est	Volume	Tightness	Resiliency	Depth
0.003	265,894	12.557	0.0106	2,369
0.005	206,866	12.937	0.0125	2,405
0.01	131,180	13.605	0.0191	1,821
0.02	73,462	14.208	0.0338	1,088
0.03	51,052	14.598	0.0491	754

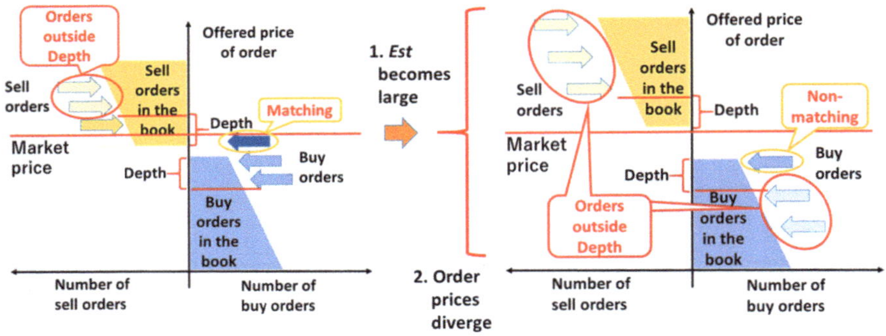

Fig. 5 Mechanism when the variation coefficient of the order price becomes larger

The reason why Volume becomes smaller is as follows. As the variation coefficient of the order price becomes larger, the order prices also diverge (Fig. 5). Therefore, the amount of new buy (sell) orders that are lower (higher) than the market price increases and the turnover decreases.

When the variation coefficient of the order price becomes larger, the order prices also diverge, and the amount of limit buy (sell) orders that are lower (higher) than the best-bid (best-ask) price increases. Thus, the difference between the highest buy price and the lowest sell price, i.e., Tightness, also increases.

Volume rapidly decreases as the variation coefficient of the order price increases, as described above. Therefore, the P/T ratio increases, and Resiliency becomes larger.

The reason why Depth becomes smaller is as follows. As the variation coefficient of the order price become larger, the order prices diverge and the amount of limit orders that are placed outside the measurement range of Depth increases. On the other hand, the limit orders that have been placed in the order book inside the measurement range of Depth decrease because these orders match the orders that are placed around the best quotations. Therefore, Depth becomes smaller.

Table 9 Four liquidity indicators when t_c is changed

t_c	Volume	Tightness	Resiliency	Depth
10,000	319,649	8.461	0.0229	1,166
15,000	268,288	11.719	0.0124	1,937
20,000	265,894	12.557	0.0106	2,369
25,000	265,865	12.480	0.0102	2,806
30,000	265,303	12.417	0.0099	3,308

Table 10 Changes of four liquidity indicators and liquidities (in parentheses) when each parameter increases

Parameter	Volume	Tightness	Resiliency	Depth
ΔP	Smaller (↓)	Larger (↓)	Larger (↓)	Larger (↑)
$w_{1,\max}$	Smaller (↓)	Smaller (↑)	Smaller (↑)	Larger (↑)
$w_{2,\max}$	Smaller (↓)	Smaller (↑)	Smaller (↑)	Larger (↑)
σ_ϵ	Larger (↑)	Larger (↓)	Larger (↓)	Smaller (↓)
Est	Smaller (↓)	Larger (↓)	Larger (↓)	Smaller (↓)
t_c	Smaller (↓)	Larger (↓)	Smaller (↑)	Larger (↑)

4.3.6 Case that Order Effective Period in Order Book is Changed

Table 9 shows how the four liquidity indicators change when the order effective period in the order book t_c changes. When t_c becomes larger, Volume and Resiliency become smaller, but Tightness and Depth become larger.

4.4 Relationship Among Four Liquidity Indicators

Table 10 shows the changes of the liquidity indicators when the value of a parameter increases. The arrows in parentheses indicate the directions of change in liquidity.

Table 10 indicates that market liquidity based on Volume is completely opposite to liquidity based on one of the other indicators in some situations (Sects. 4.3.2 through 4.3.4).

We discuss the results in view of order types again. First, when the parameters are changed, the order price variation is assumed to become large. Then, since the amount of market orders that match the orders in the order book becomes large, Volume also increases. On the other hand, as the amount of orders in the order book decreases, the other three indicators decrease. Next, suppose that the order price variation becomes small. Then, since the amount of market orders that match the orders in the order book becomes small, Volume also decreases. However, as the amount of orders in the order book increases because many orders do not match and remain in the order book, the other three indicators increase.

From the above, in principle, it seems unlikely that both Volume and Depth rise (or fall), although some empirical studies (Muranaga 1999; Chung et al. 2005) showed that Depth also increased when Volume increased. This difference in results may indicate that we need to implement additional mechanisms in the proposed model. For example, we could implement a mechanism whereby, when Volume increases, some investors who place many limit orders participate in the proposed market.

Based on the results in Sect. 4.3.5, it might seem that Est could help to implement this mechanism. However, this is not correct, because Est is not a parameter that decides the amount of the order according to the amount of the turnover, but rather Est expresses how confident investors are in their own expected prices.

4.5 Market Liquidity Based on Original Meaning of Resiliency

As mentioned previously, although the P/T ratio traditionally corresponds to Resiliency, it was pointed out that the P/T ratio cannot measure market resiliency exactly[1] The original meaning of Resiliency is the convergence speed from the price level brought about by random price changes (Muranaga 1999). However, in the empirical studies, the convergence speed cannot be measured. Therefore, the existing Resiliency, such as the P/T ratio, has been used involuntarily.[2]

In this section, we investigate the convergence period when the price that has diverged from the fundamental price due to market impacts converges to the fundamental price by changing the parameters as shown in Table 2. Concretely, we measure the price decline rate from the fundamental price and the convergence period to the neighborhood of the fundamental price when large turbulence caused by large erroneous orders occurs. We set the neighborhood of the fundamental price to five percent less than the fundamental price. We modeled large erroneous orders as follows. The erroneous orders existence time is set to the period between 300,001 and 330,000, i.e., 30,000. Among the erroneous order existence time, each agent places a sell order at the best bid price with a probability 20%, rather than the expected price that was calculated. Figure 6 shows the type of price transition for 20 simulation runs with large erroneous orders.

Note that the order price P_{oj}^{t} is determined randomly according to a uniform distribution between $P_{oj}^{t} - P_d$ and $P_{oj}^{t} + P_d$, where $P_d = 1{,}000$.

[1]For example, if the difference between the highest and lowest prices of the trading day is smaller, then the P/T ratio tends to become lower (Kurosaki et al. 2015).

[2]For example, the P/T ratio may be small, although the price does not return from the price level that has been brought about by random price changes if the price fluctuates randomly far from the original price level.

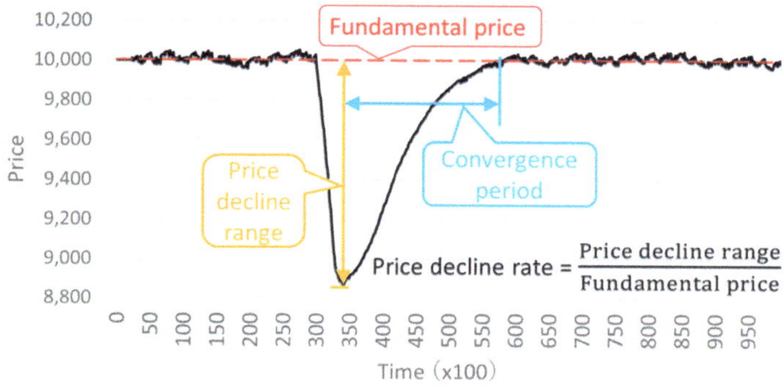

Fig. 6 Price transition with erroneous orders

4.5.1 Simulation Results and Discussion of Resiliency

Table 11 shows how the price decline rate and the price convergence period change when the tick size ΔP changes. When the tick size increases, the price decline rate becomes large and the price convergence period becomes small. Note that the price does not converge to the neighborhood of the fundamental price for the case of $\Delta P = 100$. This is because the price converges to 9,900 or 9,800 when the tick size is large. When the tick size becomes large, the range of the price decline per transaction also becomes large and the price decline rate increases. On the other hand, when the price rebounds after the decline, the range of an increase in price is large. Therefore, the convergence period tends to become small.

Table 12 shows how the price decline rate and the price convergence period change when the maximum fundamental strategy weight $w_{1,\max}$ changes. When the maximum fundamental strategy weight increases, the price decline rate and the price convergence period become small. When the maximum fundamental strategy weight becomes large, the force that causes the price to converge to the fundamental price becomes large, because the influence of the expected return of the fundamental strategy term on the expected return increases (Eq. 1). Therefore, the range of the price decline becomes smaller, and the price decline rate also becomes small. On the

Table 11 Price decline rate and convergence period when ΔP is changed

ΔP	Price decline rate (%)	Convergence period
0.01	11.461	197,206
0.1	11.502	195,811
1	11.364	193,921
10	11.602	187,271
100	11.759	N/A

Table 12 Price decline rate and convergence period when $w_{1,\max}$ is changed

$w_{1,\max}$	Price decline rate (%)	Convergence period
1.0	11.364	193,921
3.0	10.072	93,843
5.0	9.229	65,763
8.0	8.216	48,494
10.0	7.757	41,512

Table 13 Price decline rate and convergence period when $w_{2,\max}$ is changed

$w_{2,}$	Price decline rate (%)	Convergence period
1.0	10.784	56,108
3.0	10.834	90,916
5.0	11.149	125,155
8.0	11.361	167,895
10.0	11.364	193,921

other hand, once the price rebounds after the decline, the price rises rapidly, and the convergence period becomes small.

Table 13 shows how the price decline rate and the price convergence period change when the maximum technical strategy weight $w_{2,\max}$ changes. When the maximum technical strategy weight increases, the price decline rate and the price convergence period become large. When the maximum technical strategy weight becomes large, the force that causes the expected price to converge to the neighborhood of the market price increases. Once the market price starts to fall, the force causing the price fall increases, and the price decline continues, so the price decline rate increases. Moreover, the price trend does not reverse easily due to the force causing the price decrease Thus, the convergence period becomes large.

Table 14 shows how the price decline rate and the price convergence period change when the standard deviation of the noise factor of the expected return σ_ϵ changes. When the standard deviation of the noise factor of the expected return increases, the price decline rate increases, but the price convergence period decreases. When the standard deviation of the noise factor of the expected return becomes large, the expected price of agents diffuses, and lower sell orders tend to increase (Fig. 4). As the lower sell orders tend to be market orders, the price decline rate increases when the price decreases. On the other hand, when the market price reverses, higher buy orders tend to increase, and the buy orders easily become market orders. Therefore, the price rises rapidly, and the convergence period decreases.

Table 15 shows how the price decline rate and the price convergence period change when the order effective period in the order book t_c changes. When the order effective period in the order book increases, the price decline rate decreases, but the price convergence period increases. When the order effective period in the order book becomes large, the amount of the orders in the order book tends to increase. Then,

Table 14 Price decline rate and convergence period when σ_ϵ is changed

σ_ϵ	Price decline rate (%)	Convergence period
0.02	10.870	213,717
0.04	11.183	204,126
0.06	11.364	193,921
0.08	11.747	187,771
0.1	12.060	182,204

Table 15 Price decline rate and convergence period when t_c is changed

t_c	Price decline rate (%)	Convergence period
10,000	96.910	40,298
15,000	21.365	118,922
20,000	11.364	193,921
25,000	8.096	235,371
30,000	6.151	254,249

Table 16 Changes of price decline rate and convergence period and liquidities (in parentheses) when each parameter increases

Parameter	Price decline rate	Convergence period
ΔP	Larger (↓)	Smaller (↑)
σ_ϵ	Larger (↓)	Smaller (↑)
$w_{1,}$	Smaller (↑)	Smaller (↑)
$w_{2,}$	Larger (↓)	Larger (↓)
t_c	Smaller (↑)	Larger (↓)

there remain a number of orders that are around the best quotation in the order book, and the market price does not decline very much even if the market sell orders match the buy orders in the order book. Therefore, the price decline rate decreases. On the other hand, when the market price reverses, the market buy orders have to match the sell orders in the order book that were placed during the previous price decline. Therefore, the price rises slowly, and the price convergence period becomes large.

Table 16 shows the changes of the price decline rate and the price convergence period when the value of a parameter increases. The arrows in parentheses indicate the directions of changes in liquidity.

As described in Sect. 4.3, when the market volatility is large, market liquidity based on Resiliency in the empirical studies is small. However, when the maximum fundamental strategy weight $w_{1,\max}$ increases, the liquidity based on the original meaning of Resiliency also increases. The result may indicate that the liquidity is large, because the convergence period is small when the expected return when the price approaches the fundamental price is large. This means that the existence of fundamentalists in the market affects strengthening of Resiliency; that is, an increase in trading by fundamentalists may lead to greater liquidity. On the other hand, when the maximum technical strategy weight $w_{2,\max}$ increases, the liquidity based on the original meaning of Resiliency decreases. This indicates that trades by chartists may diverge the market price from the fundamental price and weaken Resiliency, i.e., that an increase in trading by chartists may lead to smaller liquidity.

5 Conclusion

In the present paper, we set out to investigate the relationship among four liquidity indicators—Volume, Tightness, Resiliency, and Depth—by changing the parameters of an artificial market, such as the tick size of the market price. As a result, we found that when the maximum fundamental strategy weight and maximum technical strategy weight of the expected return and the standard deviation of the noise factor of the expected return change, the tendency of market liquidity based on Volume is completely opposite that based on Tightness, Resiliency, or Depth. However, some empirical studies show that Volume correlates with Depth. This may indicate that we need to add mechanisms that are not currently implemented in the proposed model. For example, we could add a mechanism whereby, when Volume increases, some investors who place many limit orders participate in the proposed market. On the other hand, market liquidity based on Volume correlates with liquidity based on Tightness, Resiliency, or Depth when the variation coefficient of the order price changes. Moreover, when the tick size and the order effective period in the order book change, we confirmed different results when the other parameters are changed. Clarifying the mechanism of the order effective period in the order book change is a subject for future study.

Moreover, we confirmed that the price decline rate and the price convergence period can be used to measure the convergence speed, which could not be measured in empirical studies, based on the price level brought about by random price changes, by changing the artificial market parameters. Thus, when the maximum fundamental strategy weight increases, the price decline rate decreases but the price convergence period increases. Otherwise, when the maximum technical strategy weight increases, the price decline rate increases but the price convergence period decreases.

Generally, market liquidity based on Resiliency used in the empirical studies is small when the market volatility is large. However, when we focus on the convergence speed as Resiliency, as shown by the above results, the existence of fundamentalists in the market appears to affect strengthening of Resiliency, i.e., an increase in trading by fundamentalists leads to greater liquidity. On the other hand, trades by chartists appear to diverge the market price from the fundamental price and to weaken Resiliency; that is, an increase in trading by chartists leads to smaller liquidity.

Disclaimer Note that the opinions expressed herein are solely those of the authors and do not necessarily reflect those of SPARX Asset Management Co., Ltd.

References

Amihud Y (2002) Illiquidity and stock returns: cross-section and time-series effects. J Financ Mark 31–56

Arthur W, Holland J, Lebaron B, Palmer R, Tayler P (1997) Asset pricing under endogenous expectations in an artificial stock market. The economy as an evolving complex system II. Addison-Wesley, pp 15–44

Chen SH, Chang CL, Du YR (2012) Agent-based economic models and econometrics. Knowl Eng Review 27(2):187–219

Chiarella C, Iori G, Perellò J (2009) The impact of heterogeneous trading rules on the limit order book and order flows. J Econ Dyn Control 33(3):525–537

Chung KH, Kim KA, Kitsabunnarat P (2005) Liquidity and quote clustering in a market with multiple tick sizes. J Financ Res 28(2):177–195. https://doi.org/10.1111/j.1475-6803.2005.00120.x, https://onlinelibrary.wiley.com/doi/abs/10.1111/j.1475-6803.2005.00120.x

Cont R (2001) Empirical properties of asset returns: stylized facts and statistical issues. Quant Financ 1(2):223–236

Kurosaki T, Kumano Y, Okabe K, Nagano T (2015) Liquidity in JGB markets: an evaluation from transaction data. Bank of Japan Working Paper Series 15-E-2, Bank of Japan. https://ideas.repec.org/p/boj/bojwps/wp15e02.html

Kyle AS (1985) Continuous auctions and insider trading. Econometrica 53:1315–1336

Lux T, Marchesi M (1999) Scaling and criticality in a stochastic multiagent model of a financial market. Nature 397:498–500

Mizuta T, Izumi K, Yagi I, Yoshimura S (2014) Regulations' effectiveness for market turbulence by large erroneous orders using multi agent simulation. In: 2014 IEEE conference on computational intelligence on for financial engineering and economics (CIFEr), pp 138–143

Mizuta T, Izumi K, Yagi I, Yoshimura S (2014) Investigation of price variation limits, short selling regulation, and uptick rules and their optimal design by artificial market simulations. Electron Commun Jpn 98(7):13–21

Muranaga J (1999) Dynamics of market liquidity of Japanese stocks: an analysis of tick-by-tick data of the Tokyo Stock Exchange. In: Bank for International Settlements (ed) Market liquidity: research findings and selected policy implications, vol 11. Bank for International Settlements, pp 1–25. https://EconPapers.repec.org/RePEc:bis:biscgc:11-13

Nakada T, Takadama K (2013) Analysis on the number of XCS agents in agent-based computational finance. In: IEEE conference on computational intelligence on for financial engineering and economics (CIFEr), pp 8–13

Nishizaki K, Tsuchikawa A, Yagi T (2013) Indicators related to liquidity in JGB markets. Bank of Japan Review Series 13-E-3, Bank of Japan. https://EconPapers.repec.org/RePEc:boj:bojrev:13-e-3

Bank for International Settlements (1999) Recommendations for the design of liquid markets. Bank for International Settlements. https://www.bis.org/publ/cgfs13.htm

Sewell M (2006) Characterization of financial time series. http://finance.martinsewell.com/stylized-facts/

Yagi I, Nozaki A, Mizuta T (2017) Investigation of the rule for investment diversification at the time of a market crash using an artificial market simulation. Evol Inst Econ Rev 14(2):451–565. https://link.springer.com/article/10.1007%2Fs40844-017-0070-9

Yamamoto R, Hirata H (2013) Strategy switching in the Japanese stock market. J Econ Dyn Control 37(19):2010–2022. http://EconPapers.repec.org/RePEc:eee:dyncon:v:37:y:2013:i:10:p:2010-2022

Macroscopic Properties in Economic System and Their Relations

Atushi Ishikawa, Shouji Fujimoto and Takayuki Mizuno

Abstract Econophysics, which reconsiders economics from the perspective and approach of physics, has many stylized facts about firms and clarifies the relationships among them. To build a microscopic behavioral model of firms as economic agents in an economic system, such stylized facts and their relationships must be organized to determine their minimum required properties. For this reason, in this chapter, we consider the properties observed in the aggregate of a large number of firm variables that represent the size of firms, including total sales, total revenue, net income, number of employees, tangible fixed assets, total assets, net assets. We describe the properties observed at a fixed time, in a short-term period of two consecutive years, and in a long-term period over several decades and examine how they are related. These properties and relationships are universal macroscopic structures that are not influenced by the microscopic details of individual firms and the microscopic interactions among firms.

1 Introduction

A firm is defined as an economic entity or an agent in an economic system consisting of firms, households, and government. In a system composed of many firms, various statistical properties are observed in terms of quantities that represent the size of firms, such as sales, profits, and number of employees (herein referred to as firm-size variables). The universal structure observed in natural science also exists in economic

A. Ishikawa (✉) · S. Fujimoto
Kanazawa Gakuin University, 10 Sue, Kanazawa, Ishikawa 920-1154, Japan
e-mail: at.ishikawa@gmail.com

S. Fujimoto
e-mail: sjfjmt@gmail.com

T. Mizuno
National Institute of Informatics, 2-1-2 Hitotsubashi, Chiyoda-ku, Tokyo 101-8430, Japan
e-mail: mizuno@nii.ac.jp

Department of Informatics, The Graduate University for Advanced Studies, 2-1-2 Hitotsubashi, Chiyoda-ku, Tokyo 101-8430, Japan

A. S. Chakrabarti et al. (eds.), *Network Theory and Agent-Based Modeling in Economics and Finance*, https://doi.org/10.1007/978-981-13-8319-9_7

science. In recent years, econophysics has recaptured economic phenomena from the viewpoint and method of physics. Many stylized facts about firms and their relationships are also being clarified.

In the history of physics, we extracted various universal structures from nature and described them logically and mathematically, including Newtonian mechanics in the 17th century, electromagnetism, relativity and thermodynamics in the 19th century, and quantum mechanics in the 20th century.

Here we focus on thermodynamics, a system that describes the relationship between macroscopic matter and heat based on various observations and experiments codified in the mid-19th century. Next came statistical mechanics and molecular theory, and it became clear that matter's micro structure is dominated by quantum mechanics. Surprisingly, despite the development of statistical and quantum mechanics, thermodynamics didn't require any corrections, suggesting that its universal structure is not influenced by the unfocused details of microscopic structure and unknown elements. In other words, a universal structure exists that describes the macroscopic world independently of microscopic theory (Tazaki 2000). In addition to thermodynamics, the motion of rigid bodies, critical phenomena in magnetic materials and fluids, etc. are universal structures of the macroscopic world that are also independent of system details. Furthermore, statistical mechanics that discusses microscopic multi-body systems is a theoretical system that is designed to be consistent with thermodynamics.

The above physical viewpoint should be a guide to clarify the structure of economic science. Using an analogy to physics, the macroscopic statistical properties of socio-economic systems, obtained as a result of the actions of individual firms, provide critical guidance for creating an agent-based model. Especially if we construct an agent-based model from the minimum essential properties, various properties, which are derived from the basic properties, are automatically incorporated. In this chapter from this point of view, we discuss various statistical properties and their relationships that we identified in a number of firm-size variables.

It is organized as follows. In Sect. 2, we refer to data that can confirm various properties discussed in the following sections. Section 3 describes the power-law distribution in the large-scale range of firm-size variables at fixed times and log-normal distribution in the mid-scale range.

Section 4 describes three types of short-term properties observed in the changes of firm-size variables in two consecutive years. The first type is time-reversal symmetry that describes a system's equilibrium state in the short term, and quasi-time-reversal symmetry that describes changes in its quasi-static equilibrium state. The second type is a property in short-term growth, Gibrat's law in large-scale firm-size variables, and non-Gibrat's property in the mid-scale. The third type is the regular dependence of the inactive rate of firms, which is the proportion of firms that stop their operations, on the firm's age or its size.

Section 5 describes the long-term properties of firms observed in a long-term period comprised of several decades. As a long-term growth property, the geometric mean of firm-size variables with less than ten years of existence obeys rapid power-law growth first and follows a moderate exponential growth. It also describes the properties of firm-age distribution.

After describing the statistical properties of a large number of firm-size variables at fixed times and in short- and long-term periods, we discuss the relations among them in Sect. 6. First, we derive power-law and log-normal distributions at fixed times from the short-term properties of firms that continue to operate. Second, we derive the long-term firm growth properties and firm-age distribution from the short-term properties of the firms that continue to operate and those that discontinue their activities. Finally, we show that in the short-term period, the Gibrat's law and non-Gibrat's properties of the firms that continue their activities are related to the properties of the inactive rate of firms.

Section 7 summarizes this chapter and discusses future issues.

2 Database Statistical Properties

Although this chapter does not analyze actual data, the various properties covered in it have been confirmed by previous researches.

We obtained the total sales, net income, and number of employees of Japanese firms from Tokyo Shoko Research's database, CD-Eyes 50 (Tokyo Shoko Research, Ltd. 2019), which gathers information of the top 500,000 firms with the largest total sales. From it, we can obtain information on firms with large-scale sales.

In addition, using a database called Orbis compiled by Bureau van Dijk Electronic Publishing KK (2019), we can access the following information of approximately 300 million worldwide firms: total sales, gross income, net income, number of employees, tangible fixed assets, total assets, net assets, establishment year of firms, corporate activity status, etc. This incredibly comprehensive database lists data from all of the world's leading countries. International comparisons are also easy because the data are organized in the same format, and the various statistical properties described in the following sections are confirmed for many years and in various countries including a sufficient amount of data.

Although this article does not describe such data analysis, it can be confirmed by the past studies referenced in each section.

3 Properties at Fixed Time

As mentioned in the introduction, in the large amount of sales, profits, assets, and number of employees that represent the size of firms (firm-size variables), properties are universally observed, which are unaffected by individual microscopic structures and firm behaviors. In this section, we describe the power-law and log-normal distributions because they are the oldest known properties of firms. Both distributions are observed in the large- and mid-scale ranges of firm-size variables at fixed times. The properties of the small-scale range have not been established because the amount of data is too scarce.

3.1 Power-Law Distribution

The power-law distribution has been observed in the large-scale range of various firm-size variables (Pareto 1897; Newman 2005; Clauset et al. 2009):

$$P(x) \propto x^{-\mu-1} \quad \text{for} \quad x > x_0. \tag{1}$$

Here x is such firm-size variable as sales, $P(x)$ is the probability density function (PDF), μ is a parameter that represents the spread of the power-law distribution, and x_0 is the lower limit of the large-scale range. This is also called Pareto's law, which was discovered in UK's personal income distribution by Italian economist Pareto. Power-law index μ is also called the Pareto index. Although individual firm-size variable changes annually, this property is universally observed in the large-scale range regardless of the type of firm-size variables, the country to which the firms belong, and the year that was measured. The Pareto index is generally around 1, although it varies depending on the observation variables.

3.2 Log-Normal Distribution

Similar to the large-scale power law in the large-scale range of firm-size variables, the log-normal distribution is also found in the mid-scale range (Gibra 1932; Badger 1980; Montroll and Shlesinger 1983):

$$P(x) \propto x^{-\mu-1} \exp\left[-\alpha \ln^2 \frac{x}{x_0}\right] \quad \text{for} \quad x < x_0. \tag{2}$$

Here α is a parameter that represents the spread of the log-normal distribution, and x_0 is the upper limit of the mid-scale range, which is a boundary with a large-scale range. Equation (2) can be connected smoothly with Eq. (1) in the large-scale range by setting $\alpha = 0$. Therefore, we rewrite the notation of the log-normal distribution from the standard one and express it as Eq. (2). Similar to the power law in the large-scale range, although the firm-size variables change from year to year, the log-normal distribution is observed universally in the mid-scale range regardless of the type of firm-size variables, the country to which the firms belong, and the year that was measured.

4 Properties in Short-Term Period

Next we introduce three types of properties are identified in the amount of firm-size variables in a short-term period for two consecutive years. The first is time-reversal symmetry, which indicates that a firm-size variable system is in equilibrium for two consecutive years, and quasi-time-reversal symmetry, which describes how a system changes quasi-statically. The second is a property that is found in the growth of firm-size variables from one year to the next, Gibrat's law in the large-scale range, and non-Gibrat's property in the mid-scale range. The third is a property in the dependence of the rate at which firms cease to operate (referred to as the inactive rate of firms) on the firm-age and firm-size variables for two consecutive years.

4.1 Equilibrium or Quasi-equilibrium State of System

In this chapter, we use T as a symbol for the calendar year. Denote (x_T, x_{T+1}) as firm-size variables for two consecutive years and focus on the joint PDF $P(x_T, x_{T+1})$. This is observed as a scatter plot of (x_T, x_{T+1}) in data analysis. Two types of symmetries are observed in the joint PDF when the firm-size variables are in an equilibrium state and change quasi-statically.

4.1.1 Time-Reversal Symmetry

Even though individual firm-size variables vary, when their aggregates are in equilibrium, the system's symmetry with respect to a time reversal: $x_T \leftrightarrow x_{T+1}$ is studied. This is called time-reversal symmetry, which is expressed using the joint PDF (Fujiwara et al. 2003, 2004):

$$P_J(x_T, x_{T+1}) = P_J(x_{T+1}, x_T). \tag{3}$$

Note that its function form is identical on both sides of Eq. (3). The system is symmetrical with respect to a straight line: $\log_{10} x_{T+1} = \log_{10} x_T$ in both logarithmic axes. In Eq. (3), measure $dx_T dx_{T+1}$ is omitted because it is shared by both sides. Time-reversal symmetry, which is a concept that describes the equilibrium state of a system in thermodynamics, is also called the detailed balance.

4.1.2 Quasi-time-reversal Symmetry

When a system composed of a large number of firm-size variables changes quasi-statically, it is symmetric with respect to time inversion: $a{x_T}^{\theta} \leftrightarrow x_{T+1}$, which is a quasi-static extension of time-reversal symmetry. We call this quasi-time-reversal symmetry, which is expressed using the following joint PDF (Ishikawa 2006a, 2009a):

$$P_J(x_T, x_{T+1})\, dx_T\, dx_{T+1}$$
$$= P_J\left(\left(\frac{x_{T+1}}{a}\right)^{1/\theta},\ a\, {x_T}^{\theta}\right) d\left(\left(\frac{x_{T+1}}{a}\right)^{1/\theta}\right) d\left(a\, {x_T}^{\theta}\right). \tag{4}$$

Here θ and a are parameters of this symmetry, and the system is symmetrical with respect to a straight line: $\log_{10} x_{T+1} = \theta\ \log_{10} x_T + \log_{10} a$ on a logarithmic axis. That is, parameter θ is the slope of the symmetry axis, and $\log_{10} a$ is its intercept. An interesting nature of θ will be revealed in Sects. 6.1.2 and 6.1.4 in which the short-term properties and those at fixed times are related.

4.2 Short-Term Growth Properties

In Sect. 4.1, we considered the system's equilibrium or quasi-equilibrium state, and here we address the short-term growth rate of firm-size variables. We define the growth rate of the firm-size variables for two consecutive years as $R = x_T / x_{T+1}$. At this time, conditional PDF $Q(R|x_T)$ of growth rate R, which is conditioned by the firm-size variable of first year x_T, has different characteristics in the large- and mid-scale ranges (Hart and Oulton 1996; Lotti et al. 2009).

4.2.1 Gibrat's Law

Interestingly, $Q(R|X_T)$ does not depend on the firm-size variable in first year x_T in large-scale range $x_T > x_0$ (Gibra 1932; Sutton 1997):

$$Q(R|x_T) = Q(R) \quad \text{for} \quad x_T > x_0. \tag{5}$$

This is called Gibrat's law, and lower bound x_0 corresponds to the lower limit of the large-scale range in which the power law (1) holds. As will be described in Sect. 4.2.2, there are two types of growth rate distributions depending on the type of firm-size variables. One is linear in both the logarithmic axes, and the other has a downward convex curvature. In both cases, $Q(R|x_T)$ does not depend on x_T in the large-scale range. This is a quite interesting property where the growth of the large-scale firm-size variables beyond threshold x_0 is not dependent on its initial value. This universal structure has been identified in different firm-size variables, different countries, and different years.

4.2.2 Non-Gibrat's Properties

On the other hand, in the mid-scale range, the PDF $Q(R|x_T)$ of growth rate R conditioned on initial value x_T is regularly dependent on x_T. The dependency is different between the cases where the shape of $Q(R|x_T)$ is linear on the bilogarithmic axis and has a downward convex curvature. From previous studies, the growth rate distribution has a downward convex curvature for firm-size variables without negative values, such as sales and the number of employees, and the growth rate distribution is linear for firm-size variables that can take a negative value, such as profit (Ishikawa 2009b; Ishikawa et al. 2011a, b).

To simplify the discussion, we treat only firms with positive x_T and x_{T+1} and use logarithmic growth rate $r = \log_{10} R$. When the logarithm of the conditional PDF $q(r|x_T)$ is linear with respect to r, the x_T dependency is expressed (Ishikawa 2006b, 2007):

$$\log_{10} q(r|x_T) = c - \left(C_1 + \alpha \ln \frac{x_T}{x_0} \right) r \quad \text{for} \quad r > 0, \tag{6}$$

$$\log_{10} q(r|x_T) = c + \left(C_2 - \alpha \ln \frac{x_T}{x_0} \right) r \quad \text{for} \quad r < 0. \tag{7}$$

Here c, C_1, C_2, α, and x_0 are the parameters. We refer to this dependency as the first non-Gibrat's property. Interestingly, it can be analytically shown that, when the logarithm of $q(r|x_T)$ is linear with respect to r, the only first non-Gibrat's property is permitted without contradiction under time-reversal symmetry (3) or quasi-time-reversal symmetry (4). The proof is performed in Sects. 6.1.3 and 6.1.4.

On the other hand, when the logarithm of $q(r|x_T)$ has a convex curvature downward with respect to r, approximating the curvature by a quadratic function of r, the x_T dependency is expressed as follows (Ishikawa 2006b; Ishikawa et al. 2011a):

$$\log_{10} q(r|x_T) = c - \left(D_1 + 2\alpha \ln \frac{x_T}{x_0} \right) r + \ln 10\, D_3\, r^2 \quad \text{for} \quad r_c > r > 0, \tag{8}$$

$$\log_{10} q(r|x_T) = c + D_2\, r + \ln 10\, (D_3 + \alpha)\, r^2 \quad \text{for} \quad - r_c < r < 0. \tag{9}$$

As above, c, D_1, D_2, D_3, α, and x_0 are parameters here as well. r_c is also a cut-off for not diverging the probability. We call this the second non-Gibrat's property. We can also analytically show that, when the logarithm of $q(r|x_T)$ is expressed by a quadratic function of r and does not depend on x_T in $r < 0$, which is confirmed with real data, only the second non-Gibrat's property is consistent with the time-reversal symmetry (3) or the quasi-time-reversal symmetry (4). This proof will also be performed in Sect. 6.1.3.

In the non-Gibrat's property, mid-sized firm-scale variables below threshold x_0 are more likely to grow, the smaller the initial scale is. Also, the smaller the initial scale is, the smaller the negative growth (the first non-Gibrat's property) is, or it does not depend on the initial scale (the second non-Gibrat's property). This property is also a universal structure that has been identified in different types of firm-size variables, in different countries, and in different years, as in the case of Gibrat's law. However, the property is divided into first and second types, depending on the type of the firm-size variables. As shown in Sect. 6.1.3, the origins of the two types of non-Gibrat's properties are shown in the figure of the growth rate distribution. The mechanism of the difference in the shape of the growth rate distribution between the cases where the firm-size variables take only positive values and when they take positive and negative values is shown using a simulation (Ishikawa 2009b).

4.3 *Inactive Rate of Firms*

In Sects. 4.1 and 4.2, we considered the properties observed in firms with firm-size variables for two consecutive years, that is, firms that continue to operate. In contrast, we deal with the properties of firms that stop their activities and found that the inactive rate of firms depends on their age (Jovanovic 1982; Audretsch et al. 2000; Thompson 2005) and size (Audretsch and Mahmood 1995).

4.3.1 Firm-Age Dependence of Inactive Rate

First, we consider the firm-age dependency of the inactive rate of firms. We used T to denote the calendar year and here after use t to denote the number of years the firm has existed. Here we measured firm-age t from the year of its establishment (the firm's first year of establishment is represented as $t = 1$). In many countries, young firms tend to be more prone to be inactive as their age t decreases. On the other hand, for firms that have grown up, a stable period can be identified where the inactive rate becomes constant independently of the firm's age. Expressing these properties in a formula, firm-age t dependence of inactive rate D can be expressed (Ishikawa et al. 2015a, b, 2017a):

$$D(t) = \lambda_1 + \lambda_2 e^{-\kappa t}. \tag{10}$$

Here λ_1, λ_2, and κ are positive parameters. Although this property is universally observed in countries, each parameter differs from country to country. For example, in Japan and Spain, $\lambda_2 \sim 0$ is found. In case $\lambda_2 \neq 0$, the value of κ varies depending on the country. In many countries $\lambda_2 \neq 0$ is found, indicating that the lower the firm's age, the smaller its size and the higher the inactive rate of firms. After they age and reach a certain size, the inactive rate does not depend on the firm's age. As mentioned above, in Japan and Spain, we found no effect of the second term of Eq. (10). The reason may be that government support measures have been implemented for young firms in those countries, or perhaps a culture or mechanism exists in which firms with a high possibility of ceasing their operations never start (Klapper et al. 2006, Autio 2011; Lee et al. 2011; Honjo 2015).

4.3.2 Firm-Size Dependence of Inactive Rate

Next we consider the dependence of the inactive rate of firms on the scale of firms. In many countries, mid-sized firms tend to be more easily suspended, as the smaller their size are (Audretsch and Mahmood 1995). On the other hand, in large-sized firms, the inactive rate is generally constant regardless of the firm size. Expressing these in formula, the firm-size variable x dependence of inactive rate D is expressed as follows (Ishikawa et al. 2017b, 2018):

$$D(x) \sim b \quad \text{for} \quad x > x_0, \tag{11}$$

$$D(x) \propto x^{-D_2} \quad \text{for} \quad x < x_0. \tag{12}$$

Here b and D_2 are positive parameters.

The inactive rate of firms, whose firm size exceeds threshold x_0, is constant, and below it the inactive rate decreases as the firm size increases. Firms must have a certain size to stably continue their activities. In addition, Sect. 4.3.1 suggests that the larger a firm's age, the larger its size. This idea will be discussed in the long-term properties of the next section. Equation (10) in Sect. 4.3.1 is an exponential function, and Eq. (12) is a power function based on observation. In Sect. 6.3, a mechanism will be shown through which Eq. (12) takes an exponential function.

5 Long-Term Properties

With regard to firm-size variables, we described the properties that can be observed at fixed times in Sect. 3 and the properties in two consecutive years in Sect. 4. Long-term properties were also found in a long-term period of over several decades. In this section, we discuss the properties of the firm-age distribution and the long-term growth of firm-size variables.

5.1 Firm-Age Distribution

In some countries firm-age distribution follows an exponential distribution (Coad 2010a, b; Miura et al. 2012). Our recent research concluded that the age distribution of young firms has a downward convex curvature in some countries (Ishikawa et al. 2015b, 2017a). These properties can be described uniformly:

$$P(t) \propto \exp\left[-\Lambda_1\, t - \Lambda_2\, t^2\right]. \tag{13}$$

Here $P(t)$ is a PDF of age t, and Λ_1 and Λ_2 are positive parameters. This property is universally seen in many countries, but its parameters vary from country to country. For example, in Japan $\Lambda_2 \sim 0$ is observed.

5.2 Long-Term Growth Properties

Within an economic system, although each firm changes its size in various ways, there are also properties of firm-size variables in the long-term period. In the correlation between firm-size variables and firm age in a given year, the geometric mean of a large number of firm-size variables follows two universal growth properties with respect to firm age. First, from their establishment until around ten years, young firms grow rapidly under power-law functions. Then their growth slows under an exponential function with small exponents.

In this observation, although we do not directly follow the growth of individual firms, we must capture the statistical growth property by investigating millions of pieces data from one country. Because of this observation, no sampling bias only deals with firms that have become large. This property is found in firms that continue their activities, as opposed to the age distribution formed by firms that have become inactive in Sect. 5.1.

5.2.1 Power-Law Growth

The age dependence of the firm-size variables in the long-term period shows that geometric mean $\langle x_t \rangle$ of young firms less than around ten years ($t < 10$) grows rapidly under a power law of t (Ishikawa et al. 2016a, b):

$$\langle x_t \rangle \propto t^{\gamma} \quad \text{for} \quad t < 10. \tag{14}$$

Here γ is a positive parameter. This phenomenon is known as a rocket-start for start-up firms. This power-law growth is a universal property found in many countries.

5.2.2 Exponential Growth

The geometric mean of the size variables of firms, which have been established for at least ten years ($t \geq 10$), enters a period of gradual stable growth following an exponential function (Ishikawa et al. 2016a, b):

$$\langle x_t \rangle \propto e^{\beta\, t} \quad \text{for} \quad t \geq 10. \tag{15}$$

Here β is a positive parameter. Since index β takes a very small value, its growth is slower than the power-law growth (14). This exponential growth is also a universal property observed in many countries.

6 Relationships Among Fixed-Time, Short-Term, and Long-Term Properties

We described the fixed-time properties observed in a number of firm-size variables in Sect. 3, the short-term properties in two consecutive years in Sect. 4, and the long-term properties observed over several decades in Sect. 5. This section explains how these properties are related.

First, we explain that the fixed-time properties described in Sect. 3 are derived from the short-term properties in Sect. 4. Next, the long-term properties in Sect. 5 are derived from the short-term properties in Sect. 4. Finally, in the short-term properties in Sect. 4, we connect the properties of firms that have been operating continuously for two consecutive years to the properties of firms that have become inactive.

6.1 Derivation of Fixed-Time Properties from Short-Term Properties

Here we show that the PDF (power-law and log-normal distributions) of the firm-size variables in a fixed time is derived from the short-term growth property (Gibrat's law and non-Gibrat's properties) in two consecutive years under the system's equilibrium state (time-reversal symmetry) or its quasi-equilibrium state (quasi-time-reversal symmetry).

First, the power law, which is in the equilibrium state and whose index does not change, is derived by combining Gibrat's law and the time-reversal symmetry. Next the power law where the exponent changes annually is derived by combining Gibrat's law and the quasi-time-reversal symmetry. By combining the non-Gibrat's property and the time-reversal symmetry, we derive a log-normal distribution that is in equilibrium and whose parameters do not change. Finally, by combining the non-Gibrat's property and the quasi-time-reversal symmetry, we derive a log-normal distribution with varying parameters.

6.1.1 Static Power Law

First, as the simplest case, we derive static power law (1) at a fixed time from Gibrat's law (5) where the conditional growth rate distribution does not depend on the value of the first year in the large-scale range under the equilibrium state of firm-size variables (Fujiwara et al. 2003, 2004). When the system is in equilibrium, time-reversal symmetry (3) holds. Using growth rate $R = x_{T+1}/x_T$, time-reversal symmetry is rewritten by variables x_T, R as follows:

$$P_J(x_T, R) = R^{-1}\, P_J(R x_T, R^{-1}). \tag{16}$$

Using conditional PDF $Q(R|x_T) = P_J(x_T, R)/P(x_T)$ and Gibrat's law (5), this can be reduced:

$$\frac{P(x_T)}{P(R x_T)} = \frac{1}{R}\frac{Q(R^{-1}|R x_T)}{Q(R|x_T)} = \frac{1}{R}\frac{Q(R^{-1})}{Q(R)}. \tag{17}$$

Since the system has time-reversal symmetry, Gibrat's law (5) is also established under the transformation of $x_T \leftrightarrow R x_T\ (= x_{T+1})$. The right side of Eq. (17) is only a function of R, and so we signify it by $G(R)$ and expand Eq. (17) by $R = 1 + \epsilon$ ($\epsilon \ll 1$) in ϵ. The 0-th order term of ϵ is a trivial expression, and the 1-st order term yields the following differential equation:

$$G'(1) P(x_T) + x_T \frac{d}{dx_T} P(x_T) = 0. \tag{18}$$

Here $G'(\cdot)$ means the R differentiation of $G(\cdot)$. No more useful information can be obtained from the second and higher order terms of ϵ. The solution to this differential equation is uniquely given:

$$P(x_T) \propto {x_T}^{-G'(1)}. \tag{19}$$

This solution satisfies Eq. (17) even if R is not near $R = 1$, when $Q(R) = R^{-G'(1)-1}$ $Q(R)$ holds (this is called a reflection law). Reflection law has been confirmed with various actual data (Fujiwara et al. 2004). Finally, if we set $G'(1) = \mu + 1$, Eq. (19) is derived from Gibrat's law under time-reversal symmetry and matches the power law (1) observed at time T. Since this system has symmetry $x_T \leftrightarrow x_{T+1}$, the power law holds for the same Pareto index μ even at time $T + 1$.

6.1.2 Quasi-static Power Law

Next we show that, in a large-scale range where Gibrat's law (5) holds, a power index change can be derived from the quasi-time-reversal symmetry where firm-size variables change quasi-statically in two consecutive years (Ishikawa 2006a, 2009a). When the system changes quasi-statically, quasi-time-reversal symmetry (4) is found. Using extended growth rate $R = x_{T+1}/a{x_T}^{\theta}$ in the quasi-time-reversal system, this is rewritten by variables x_T, R:

$$P_J(x_T, R) = R^{1/\theta - 2}\, P_J(R^{1/\theta} x_T, R^{-1}). \tag{20}$$

Equation (20) is reduced to Eq. (16) at $\theta = 1$. Using the conditional PDF $Q(R|x_T)$ and Gibrat's law (5), this is reduced to

$$\frac{P(x_T)}{P(R^{1/\theta}x_T)} = R^{1/\theta-2}\frac{Q(R^{-1}|R^{1/\theta}x_T)}{Q(R|x_T)} = R^{1/\theta-2}\frac{Q(R^{-1})}{Q(R)}. \tag{21}$$

Here we assume that Gibrat's law (5) holds under a transformation: $x_T \leftrightarrow R^{1/\theta}x_T$ $(= (x_{T+1}/a)^{1/\theta})$. This is valid in a system that has quasi-time-reversal symmetry. Since the last term in Eq. (21) is only a function of R, we signify it by $G_\theta(R)$ and expand Eq. (21) to R near 1 as $R = 1+\epsilon$ $(\epsilon \ll 1)$. The 0-th order of ϵ is a trivial expression, and the 1-st order term yields the following differential equation:

$$G_\theta{}'(1)P(x_T) + \frac{x_T}{\theta}\frac{d}{dx_T}P(x_T) = 0. \tag{22}$$

Here $G_\theta{}'(\cdot)$ denotes the R differentiation of $G_\theta(\cdot)$. No more useful information can be obtained from the second and higher order terms of ϵ. The solution to this differential equation is uniquely given:

$$P(x_T) \propto x_T{}^{-\theta G_\theta{}'(1)}. \tag{23}$$

Similar to Sect. 6.1.1, this solution satisfies Eq. (21) even if R is not near $R = 1$, when $Q(R) = R^{-G_\theta{}'(1)-1}\ Q(R)$ holds.

Next, in quasi-static system (x_T, x_{T+1}), we identify distribution $P(x_{T+1})$. Actually, we should write $P_{x_T}(x_T)$, $P_{x_{T+1}}(x_{T+1})$; however, because function forms are complicated, they are collectively written as P. From Eq. (23) and $P(x_T)dx_T = P(x_{T+1})dx_{T+1}$, $P(x_{T+1})$ can be expressed:

$$P(x_{T+1}) = P(x_T)\frac{dx_T}{dx_{T+1}} \propto x_{T+1}{}^{-G_\theta{}'(1)+1/\theta-1}. \tag{24}$$

Here we signify Pareto indices at T, $T+1$ by μ_T, μ_{T+1} and represent $P(x_T)$, $P(x_{T+1})$ as follows:

$$P(x_T) \propto x_T{}^{-\mu_T-1}, \quad P(x_{T+1}) \propto x_{T+1}{}^{-\mu_{T+1}-1}. \tag{25}$$

Comparing Eqs. (23) and (24) to Eq. (25), we obtain $\theta G_\theta{}'(1) = \mu_T + 1$, $G_\theta{}'(1) - 1/\theta + 1 = \mu_{T+1} + 1$ and conclude the relation among μ_T, μ_{T+1}, and θ as follows:

$$\theta = \frac{\mu_T}{\mu_{T+1}}. \tag{26}$$

From this, we understand that parameter θ in the quasi-time-reversal symmetry represents the rate of the change of Pareto indices μ_T, μ_{T+1} at time T, $T+1$. This is geometrically consistent with that the width of the power law at time T can be expressed as $1/\mu_T$ and the width at time $T+1$ as $1/\mu_{T+1}$ on the logarithmic axis.

6.1.3 Static Log-Normal Distribution

In Sects. 6.1.1 and 6.1.2, a power law is derived from Gibrat's law and time-reversal or quasi-time-reversal symmetries that hold in the large-scale range of firm-size variables. Here we show that a static log-normal distribution is derived from a non-Gibrat's property that holds in the mid-scale range under time-reversal symmetry. In the process, the expression of the non-Gibrat's property is only permitted in the form of a first non-Gibrat's property or a second non-Gibrat's property. First, we discuss when the logarithm of the conditional growth rate distribution is linear with respect to the logarithmic growth rate and consider a case where it is represented by a quadratic function.

When the logarithm of the PDF $q(r|x_T)$ of logarithmic growth rate r (which is conditioned on initial firm size x_T) is linear with respect to r, it is described as follows (Ishikawa 2006b, 2007):

$$\log_{10} q(r|x_T) = c - t_+(x_T)\, r \quad \text{for} \quad r > 0, \tag{27}$$

$$\log_{10} q(r|x_T) = c + t_-(x_T)\, r \quad \text{for} \quad r < 0. \tag{28}$$

Using the relation between $q(r|x_T)$ and $Q(R|x_T)$,

$$\log_{10} q(r|x_T) = \log_{10} Q(R|x_T) + r + \log_{10}(\ln 10), \tag{29}$$

$Q(R|x_T)$ is described as follows:

$$Q(R|x_T) = d\ R^{-1-t_+(x_T)} \quad \text{for} \quad r > 0, \tag{30}$$

$$Q(R|x_T) = d\ R^{-1+t_-(x_T)} \quad \text{for} \quad r < 0. \tag{31}$$

Although c and d are the functions of x_T in general, here we concentrate on the case where the x_T dependency can be ignored. In this case, the expression up to the second equation of Eq. (21) becomes the following for $R > 1$:

$$\frac{P(x_T)}{P(Rx_T)} = R^{1+t_+(x_T)-t_-(Rx_T)}. \tag{32}$$

Expanding this equation around $R = 1$, we obtain the following differential equations:

$$\left\{1 + t_+(x) - t_-(x)\right\} P(x) + x \frac{dP(x)}{dx} = 0, \tag{33}$$

$$\frac{dt_+(x)}{dx} + \frac{dt_-(x)}{dx} = 0, \tag{34}$$

$$\frac{dt_+(x)}{dx} + x \frac{d^2 t_+(x)}{d^2 x} = 0. \tag{35}$$

Here x_T is expressed as x for simplicity. The same differential equation is obtained for $R < 1$. The solutions to these differential equations are generally given:

$$t_+(x) = C_1 + \alpha \ln \frac{x}{x_0}, \tag{36}$$

$$t_-(x) = C_2 - \alpha \ln \frac{x}{x_0}, \tag{37}$$

$$P(x) \propto x^{-\mu-1} \exp\left[-\alpha \ln^2 \frac{x}{x_0}\right]. \tag{38}$$

Here C_1, C_2, and α are integration constants, $\mu = C_1 - C_2$, and x_0 is a parameter that is introduced to connect with the power law in the large-scale range. These solutions satisfy Eq. (32) except in the neighborhood of $R = 1$. Therefore, they are both necessary and sufficient conditions.

Combining Eqs. (36) and (37), and Eqs. (27) and (28), when the logarithm of conditional growth rate distribution $q(r|x_T)$ is linear with respect to r, only the first non-Gibrat's property Eqs. (6) and (7) is permitted to be consistent with time-reversal symmetry. Equation (38) is also the best log-normal distribution (2). Since this system has $x_T \leftrightarrow x_{T+1}$ symmetry, log-normal distribution has identical parameters α and μ at time x_{T+1}.

Next when the logarithm of conditional growth rate distribution $q(r|x_T)$ is a quadratic function of r and not dependent on x_T in $r < 0$, we show that only the second non-Gibrat's property is permitted to be consistent with the time-reversal symmetry. When the logarithm of the PDF $q(r|x_T)$ of logarithmic growth rate r (which is conditioned by initial firm size x_T) is a quadratic equation with respect to r, it is represented as Tomoyose et al. (2009), Ishikawa et al. (2011a):

$$\log_{10} q(r|x_T) = c - t_+(x_T)\, r + \ln 10\; u_+(x_T) r^2 \quad \text{for} \quad r > 0, \tag{39}$$

$$\log_{10} q(r|x_T) = c + t_-(x_T)\, r + \ln 10\; u_-(x_T) r^2 \quad \text{for} \quad r < 0. \tag{40}$$

Here ln 10 is a coefficient for simplifying the notation later. $Q(R|x_T)$ is given by:

$$Q(R|x_T) = d\; R^{-1-t_+(x_T)+u_+(x_T)\ln R} \quad \text{for} \quad r > 0, \tag{41}$$

$$Q(R|x_T) = d\; R^{-1+t_-(x_T)+u_-(x_T)\ln R} \quad \text{for} \quad r < 0. \tag{42}$$

The expression up to the second equation of Eq. (21) is rewritten as follows for $R > 1$:

$$\frac{P(x_T)}{P(Rx_T)} = R^{1+t_+(x_T)-t_-(Rx_T)-\left\{u_+(x_T)-u_-(Rx_T)\right\}\ln R}. \tag{43}$$

As before, expanding this around $R = 1$ gives the following differential equations:

$$\left\{1 + t_+(x) - t_-(x)\right\} P(x) + x\frac{dP(x)}{dx} = 0, \tag{44}$$

$$x\left\{\frac{dt_+(x)}{dx} + \frac{dt_-(x)}{dx}\right\} + 2\left\{u_+(x) - u_-(x)\right\} = 0, \tag{45}$$

$$2\frac{dt_+(x)}{dx} + \frac{dt_-(x)}{dx} + 6\frac{du_+(x)}{dx} + x\left\{2\frac{d^2t_+(x)}{d^2x} + \frac{d^2t_-(x)}{d^2x}\right\} = 0, \tag{46}$$

$$\frac{dt_+(x)}{dx} + \frac{dt_-(x)}{dx} + 3x\left\{\frac{d^2t_+(x)}{d^2x} + \frac{d^2t_-(x)}{d^2x}\right\} + x^2\left\{\frac{d^3t_+(x)}{d^3x} + \frac{d^3t_-(x)}{d^3x}\right\} = 0, \tag{47}$$

$$\frac{dt_+(x)}{dx} + 7x\frac{d^2t_+(x)}{d^2x} + 6x^2\frac{d^3t_+(x)}{d^3x} + x^3\frac{d^4u_+(x)}{d^4x} = 0. \tag{48}$$

The same differential equations are obtained for $R < 1$. As solutions of these differential equations, $t_\pm(x)$ and $u_\pm(x)$ can be expressed as follows:

$$t_+(x) = \frac{D_{-3}}{3}\ln^3\frac{x}{x_0} + \frac{D_{-2}}{2}\ln^2\frac{x}{x_0} + 2\alpha\ln\frac{x}{x_0} + D_1, \tag{49}$$

$$t_-(x) = -\frac{D_{-3}}{3}\ln^3\frac{x}{x_0} + \frac{D_{+2} - D_{-2}}{2}\ln^2\frac{x}{x_0} + (D_{+1} - 2\alpha)\ln\frac{x}{x_0} + D_2, \tag{50}$$

$$u_+(x) = -\frac{D_{-3}}{6}\ln^2\frac{x}{x_0} - \frac{D_{+2} + D_{-2}}{6}\ln\frac{x}{x_0} + D_3, \tag{51}$$

$$u_-(x) = -\frac{D_{-3}}{6}\ln^2\frac{x}{x_0} + \frac{2D_{+2} - D_{-2}}{6}\ln\frac{x}{x_0} + \alpha + D_3, \tag{52}$$

$$P(x) \propto x^{-\mu-1}\exp\left[-\frac{D_{-3}}{6}\ln^4\frac{x}{x_0} + \frac{D_{+2} - 2D_{-2}}{6}\ln^3\frac{x}{x_0} - \frac{4\alpha - D_{+1}}{2}\ln^2\frac{x}{x_0}\right]. \tag{53}$$

Here D_{-3}, $D_{\pm2}$, D_{+1}, α, D_1, D_2, and D_3 are integration constants, $\mu = D_1 - D_2$, and x_0 is a parameter introduced to connect with the power law in the large-scale range. Since these solutions satisfy Eq. (43) except in the neighborhood of $R = 1$, they are necessary and sufficient conditions. Furthermore, when $q(r|x)$ does not depend on x for $r < 0$, $t_-(x)$, $u_-(x)$ does not depend on x_T, and therefore, we obtain $D_{-3} = D_{-2} = D_{+2} = 0$ and $D_{+1} = 2\alpha$.

Combining Eqs. (49)–(52) and Eqs. (39) and (40), when the logarithm of the conditional growth rate distribution $q(r|x_T)$ is quadratic function of r and is not dependent on x_T in $r < 0$, only the second non-Gibrat's property (8) and (9) is permitted to be consistent with time-reversal symmetry. At this time, from Eq. (53) $P(x)$ follows log-normal distribution (2). Since this system has $x_T \leftrightarrow x_{T+1}$ symmetry, the log-normal distribution also remains for identical parameters μ and α at time x_{T+1}.

From the above, in both cases where the growth rate distribution of the firm-size variables is linear or convex downward on both logarithmic axes in two consecutive years, the distribution function in the mid-scale range at a fixed time is a static log-normal distribution.

6.1.4 Quasi-static Log-Normal Distribution

At the end of this section, we show that in the mid-scale range where non-Gibrat's property is observed, the change in the parameters of the log-normal distribution is derived from the quasi-time-reversal symmetry where firm-size variables change quasi-statically in two consecutive years (Ishikawa 2009a).

When the growth rate distribution can be expressed as Eqs. (27) and (28), the expression up to the second equation of the quasi-static time-reversal symmetry (21) can be transformed as follows for $R > 1$:

$$\frac{P(x_T)}{P(R^\theta x_T)} = R^{1+t_+(x_T)-t_-(R^\theta x_T)}. \tag{54}$$

Expanding this equation around $R = 1$, the following differential equations are obtained:

$$\theta\Big\{1 + t_+(x) - t_-(x)\Big\}P(x) + x\frac{dP(x)}{dx} = 0, \tag{55}$$

$$\frac{dt_+(x)}{dx} + \frac{dt_-(x)}{dx} = 0, \tag{56}$$

$$\frac{dt_+(x)}{dx} + x\frac{d^2t_+(x)}{d^2x} = 0. \tag{57}$$

Here x_T is expressed as x for simplicity. The same differential equations are obtained for $R < 1$. Equations (56) and (57) from which $t_\pm(x)$ is obtained are the same as the case when there is time-reversal symmetry. The solutions of these differential equations are given by Eqs. (36) and (37) and the following:

$$P(x) \propto x^{-\mu-1} \exp\Big[-\theta\alpha \ln^2 \frac{x}{x_0}\Big]. \tag{58}$$

These solutions satisfy Eq. (54) except in the neighborhood of $R = 1$. Therefore, these solutions are not only necessary but also sufficient conditions. From the above, when the logarithm of the conditional growth rate distribution $q(r|x_T)$ is linear with respect to r, only the first non-Gibrat's property is permitted to be consistent with the quasi-time-reversal symmetry. At the same time, $P(x_T)$ also follows the log-normal distribution.

Next in quasi-static system (x_T, x_{T+1}), we identify distribution $P(x_{T+1})$ as in the case of the large-scale range. From Eq. (58) and $P(x_T)dx_T = P(x_{T+1})dx_{T+1}$, $P(x_{T+1})$ can be expressed:

$$P(x_{T+1}) = P(x_T)\frac{dx_T}{dx_{T+1}} \propto {x_{T+1}}^{-\frac{\mu_1}{\theta}-1} \exp\left[-\frac{\alpha}{\theta}\ln^2\frac{x_{T+1}}{a x_0}\right]. \tag{59}$$

Using Eq. (26) and signifying the parameters that change quasi-statically at time T and $T+1$ by α_T and α_{T+1}, $P(x_T)$ and $P(x_{T+1})$ can be written:

$$P(x_T) \propto {x_T}^{-\mu_T-1} \exp\left[-\alpha_T \ln^2\frac{x}{x_0}\right], \tag{60}$$

$$P(x_{T+1}) \propto {x_{T+1}}^{-\mu_{T+1}-1} \exp\left[-\alpha_{T+1} \ln^2\frac{x}{a x_0}\right]. \tag{61}$$

In this case, $\alpha_T = \theta\alpha$, $\alpha_{T+1} = \alpha/\theta$, and therefore the relation among α_T, α_{T+1}, and θ is concluded:

$$\theta^2 = \frac{\alpha_T}{\alpha_{T+1}}. \tag{62}$$

In the same manner, a similar result is obtained for the second non-Gibrat's property. Since α is proportional to the square of the reciprocal of logarithmic standard deviation σ, Eq. (62) can be rewritten:

$$\theta = \frac{\sigma_{T+1}}{\sigma_T}. \tag{63}$$

This is geometrically consistent because the width of the log-normal distribution at time T can be expressed as σ_T and the width of the log-normal distribution at time $T+1$ can be expressed as σ_{T+1} on the logarithmic axis.

6.2 Derivation of Long-Term Properties from Short-Term Properties

In Sect. 6.1, we showed that the properties of firm-size variables at fixed times are derived from short-term properties. Here we show that the properties observed in the long-term period are also derived from short-term properties. First, we show that the short-term growth properties in Sect. 4.2 lead to the long-term growth properties in Sect. 5.2. Next, we show that the short-term inactive rate of firms in Sect. 4.3 leads to long-term firm-age distribution in Sect. 5.1.

6.2.1 Derivation of Long-Term Growth from Short-Term Growth

As described in Sect. 4.2, we observed two properties in the growth of firm-size variables in two consecutive years. One is Gibrat's law (5) where the growth rate distribution does not depend on the initial value, which is observed in the large-scale range. The other is non-Gibrat's property where the growth rate distribution changes regularly depending on the initial value, which is observed in the mid-scale range. One characteristic of a non-Gibrat's property is that the smaller the growth rate of mid-sized firms, the greater the possibility of growth in a positive direction and the smaller the possibility of growth in a negative direction (the first non-Gibrat's property) or the possibility of growth in a negative direction is constant (the second non-Gibrat's property). Such short-term growth properties lead to the rapid power growth of young firms and the gradual exponential growth of firms that entered a stable period. This can be shown analytically using rough approximations (Ishikawa et al. 2016a). On the other hand, considering the following numerical simulation, long-term growth property can be explicitly derived from short-term growth law (Ishikawa et al. 2016b).

Here we signify the firm size of age t as x_t. First, we define the growth rate R_t of firm size variables x_t as

$$R_t = \frac{x_{t+1}}{x_t}. \tag{64}$$

We assume that short-term properties, which are confirmed for two successive years $(T,\ T+1)$, also hold in two successive ages $(t,\ t+1)$. In other words, we postulate that calendar year T in Sect. 4.2 can be replaced by firm age t.

Under this assumption, we interpret Eq. (64) as the following stochastic process:

$$x_{t+1} = R_t\, x_t. \tag{65}$$

Here R_t is a random variable that depends on initial value x_t, and we assume that R_t follows the non-Gibrat's property and Gibrat's law for $x_t < x_0$ and $x_t > x_0$. This model has no additive noise because no system stability is required (Takayasu et al. 1997). Using this model, we develop a system stochastically and investigate the growth of firm-size variables x_t. The aim of this simulation is the confirmation of the initial power-law growth and the subsequent exponential growth of geometric mean values $\langle x_t \rangle$ that were described in Sect. 5.2.

In the literature Ishikawa et al. (2016b), we first prepared 1000 initial values x_1, which were randomly sampled from the sales data of Japanese firms in 2011. In addition, multiplicative noise R_t was also randomly sampled from the growth rate distribution from 2011 to 2012. Then we iteratively developed the system 49 times by Eq. (65) and obtained x_t $(t = 1, 2, \cdots, 50)$. In this process, x_t evolves in time based on the second non-Gibrat's property when the scale is small and according to Gibrat's law when the size is large. We observed geometric mean $\langle x_t \rangle$ and confirmed that it grew under a power law for the first ten years and under mild exponential growth for the next forty years. Note that when multiplicative noise R_t was set to be independent of the initial value, no initial power growth was observed; only exponential growth was observed.

From the above with respect to firm-size geometric average $\langle x_t \rangle$, we numerically confirmed that young, small firms achieved rapid growth in their first ten years and then entered a stable period of moderate exponential growth.

6.2.2 Derivation of Long-Term Age Distribution from Short-Term Inactive Rate

From the firm age dependency of the short-term inactive rate of firms (10) described in Sect. 4.3.1, the long-term firm-age distribution (13) can be derived (Ishikawa et al. 2015a, b, 2017a). Denoting the number of firms of age t as $N(t)$, the following equation is established from Eq. (10):

$$\frac{N(t+1)}{N(t)} = 1 - \lambda_1 - \lambda_2 e^{-\kappa t}. \tag{66}$$

This can be transformed into the following differential equation:

$$\frac{\frac{dN(t)}{dt}}{N(t)} = -\lambda_1 - \lambda_2 e^{-\kappa t}. \tag{67}$$

The solution can be uniquely expressed:

$$N(t) = N_0 \exp\left[-\lambda_1 t + \frac{\lambda_2}{\kappa}\left(e^{-\kappa t} - 1\right)\right]. \tag{68}$$

In the case of $\kappa t \ll 1, \kappa t \gg 1$, age distribution $P(t)$ is given by the following equations:

$$P(t) \propto \exp\left[-(\lambda_1 + \lambda_2)\, t - \frac{\lambda_2 \kappa}{2} t^2\right] \quad \text{for} \quad \kappa t \ll 1, \tag{69}$$

$$P(t) \propto \exp\left[-\lambda_1 t\right] \quad \text{for} \quad \kappa t \gg 1. \tag{70}$$

This results in Eq. (13) by the definition of parameters Λ_1, Λ_2.

As described above, assuming that the inactive rate of firms (10) observed in two consecutive years does not change for a long-term period over several decades, the firm-age distribution is derived in long-term periods. In fact, in Japan, a discontinuity in firm-age distribution was observed before and after World War II, and the value of λ_1 also changed before and after World War II (Ishikawa et al. 2015a). As mentioned in Sect. 4.3.1, in Japan since there is no parameter λ_2 of Eq. (10), perhaps the value of parameter λ_1 of the inactive rate of firms changed before and after World War

II. This approach might be able to estimate the past inactive rate of firms, which is difficult to confirm using data, by linking the short-term inactive rate to the long-term firm-age distribution.

6.3 Relationship Between Short-Term Growth and Inactive Rate of Firms

In Sect. 6.1, we derived a property observed at fixed times from the short-term property. In that process, we showed that, under time-reversal symmetry that represents the system's equilibrium state in the short term or under quasi-time-reversal symmetry that represents a quasi-static system, the short-term growth of the non-Gibrat's property is uniquely determined in the mid-scale range of firm-size variables. This is the relationship among the short-term properties of firms that continue their activities. Here we discuss the relationship between the short-term property observed in firms that continue their activities and in those that stop their activities (Ishikawa et al. 2017b, 2018).

First, in the large-scale range, consider the relation between Gibrat's law (5) observed in the growth rate of firms continuing their activities and the property (11) where the inactive rate of firms does not depend on their firm size. In the large-scale range, firm-size variables x_T at time T follow the power law (1). Interestingly, among these firms, the distribution of firms that stop operating at $T+1$ also follows the power law, and power-law index μ equals the original one. Therefore, inactive rate of firms D, given by the logarithmic difference between these two power-law distributions, becomes constant regardless of the firm size, as in Eq. (11). As described in Sect. 6.1.1, the power-law distribution is derived from Gibrat's law. In connection with these discussions, the relationship between Gibrat's law and a property where the inactive rate does not depend on the firm size is a front-back relationship between firms that continue to operate and those that stop. In fact, we confirmed from the data that lower limit x_0 of the power law agrees with lower limit x_0 at which the inactive rate is constant regardless of the firm size.

Next, in the mid-scale range, consider the relation between the non-Gibrat's property observed in the growth rate of firms continuing their activities and a property (12) where the inactive rate of firms decreases under the power law as firm size increases. For analyzing the data and identifying the non-Gibrat's property, we investigated the conditional PDF $q(r|x_T)$ of logarithmic growth rate r for each initial firm-size variable x_T. By scrutinizing them, we found that in the mid-scale range, $-\infty < r < -\log_{10} x_T$ data systematically lacks $r < 0$, probably reflecting a lower limit on the firm size to maintain their activities and that negative growth below that lower limit is not permitted. As a result, firms that have ceased operating are excluded from the data of firms that maintained their activities in two consecutive years. These firms are observed as the data loss of range $-\infty < r < -\log_{10} x_T$ of $q(r|x_T)$. Instead, they are denoted as firms that suspended their activities.

From the above data loss range, $-\infty < r < -\log_{10} x_T$, ignoring the r^2 term in the case for the second non-Gibrat's property, the inactive rate of such firms is approximated:

$$D(x_T) \propto \int_{-\infty}^{-\log_{10} x_T} q(r|x_T)\, dr \sim \int_{-\infty}^{-\log_{10} x_T} 10^{c+D_2 r}\, dr \propto {x_T}^{-D_2}. \tag{71}$$

This is Eq. (12) in Sect. 4.3.2 itself. Thus, there is a close relationship between the data loss in the non-Gibrat's property in the mid-scale range and the dependence of the inactive rate of firms on the firm size. In the mid-scale range as well, the property observed in firms that continue to operate for two consecutive years and the property of firms that stop their activities in the second year share a front-back relationship. In fact, as in the case of the large-scale range, we confirmed from the data that upper limit x_0 of non-Gibrat's property agrees with upper limit x_0 of the range in which the inactive rate of firms depends on the firm size.

7 Conclusion and Discussion

In this chapter, to provide a guide to build a micro behavior model of a firm as an economic agent in an economic system, we discussed various macroscopic properties observed in a system composed of multiple firms and their relationships. We analyzed the total sales, the gross and net income, the number of employees, the tangible fixed assets, and the total and net assets, etc. that represent the size of firms (firm-size variables) and their activity.

First, power-law distribution in the large-scale range and log-normal distribution in the mid-scale range are properties observed at a fixed time. As properties observed in a short-term period of two consecutive years, we described the property of the system's equilibrium state, the property of the growth rate of the firm-size variables, and the property of the inactive rate of firms. As properties observed in a long-term period over several decades, we also mentioned the firm-age distribution and the growth of firm-size variables.

Then we discussed how these properties, which we observed in fixed times, in a short-term period, and a long-term period, are related. These properties and their relationships are macroscopic properties that are not influenced by the microscopic detailed structure of individual firms and their shared micro interactions. In the history of the development of physics, a theoretical system dealing with macroscopic properties of matter called thermodynamics was constructed without identifying the microscopic structure of matter. After that, micro-statistical and quantum mechanics were constructed with a guiding principle that was consistent with macro thermodynamics. Since this guideline is effective even in understanding economic systems, the statistical properties observed in many firms discussed in this chapter can act as a guide to construct a micro model with firms as agents.

The time-reversal symmetry discussed in this chapter can simultaneously be extended to the inversion symmetry between different firm-size variables (Mizuno et al. 2012; Ishikawa et al. 2013, 2014). This corresponds to an extension in the spatial direction, which is orthogonal to the time-oriented properties of essentially one type of firm-size variable discussed in this chapter. In economics, production, assets, and labor are deemed typical firm-size variables that characterize firm activities, and discussions have been developed based on the Cobb-Douglas production function observed among them (Cobb and Douglass 1928). Even though the Cobb-Douglas production function is one of the foundations of economics, few studies in economics led to its analytical derivation (Houthakker 1955). We are examining whether we can extend the methodology of this chapter to derive it and will report our results in the near future. The result is crucial for identifying an important property to be incorporated into a micro model in which firms are agents.

Acknowledgements The authors thank Dr. Tsutomu Watanabe. Much of this work is based on discussions with him. The authors also thank the 23rd Annual Workshop on Economic Science with Heterogeneous Interacting Agents. Discussions during the workshop were useful to complete this work. This study was supported by JSPS KAKENHI Grant Numbers 17K01277, 16H05904, and 18H05217, and the Obayashi Foundation.

References

Audretsch DB, Mahmood T (1995) New firm survival: new results using a hazard function. Rev Econ Stat 77:97–103

Audretsch DB, Houweling P, Thurik AR (2000) Firm survival in the Netherlands. Rev Ind Organ 16:1–11

Autio E (2011) High-aspiration entrepreneurship. In: Minniti M (ed) The dynamics of entrepreneurship: evidence from global entrepreneurship monitor data. Oxford University Press, Oxford, pp 251–275

Badger WW (1980) An entropy-utility model for the size distribution of income. In: West BJ (ed) Mathematical models as a tool for the social science. Gordon and Breach, New York, pp 87–120

Bureau van Dijk Electronic Publishing KK. https://www.bvdinfo.com/en-gb

Clauset A, Shalizi CR, Newman MEJ (2009) Power-law distributions in empirical data. SIAM Rev 51:661–703

Coad A (2010a) Investigating the exponential age distribution of firms. Economics 4(2010–17):

Coad A (2010b) The exponential age distribution and the Pareto firm size distribution. J Ind Compet Trade 389–395

Cobb CW, Douglass PH (1928) A theory of production. Am Econ Rev 18:139–165

Fujiwara Y, Souma W, Aoyama H, Kaizoji T, Aoki M (2003) Growth and fluctuations of personal income. Physica A 321:598–604

Fujiwara Y, Guilmi CD, Aoyama H, Gallegati M, Souma W (2004) Do Pareto-Zipf and Gibrat laws hold true? An analysis with European firms. Physica A 335:197–216

Gibra R (1932) Les Inégalités Économique. Sirey, Paris

Hart PE, Oulton N (1996) Growth and size of firms. Econ J 106:1242–1252

Honjo Y (2015) Why are entrepreneurship levels so low in Japan? Jpn World Econ 36:88–101

Houthakker HS (1955) The Pareto distribution and the Cobb-Douglas production function in activity analysis. Rev Econ Stud 23:27–31
Ishikawa A (2006a) Annual change of Pareto index dynamically deduced from the law of detailed quasi-balance. Physica A 371:525–535
Ishikawa A (2006b) Derivation of the distribution from extended Gibrat's law. Physica A 367:425–434
Ishikawa A (2007) The uniqueness of firm size distribution function from tent-shaped growth rate distribution. Physica A 383:79–84
Ishikawa A (2009a) Quasistatically varying log-normal distribution in the middle scale region of Japanese land prices. Prog Theor Phys Supple No 179(2009):103–113
Ishikawa A (2009b) Power-law and log-normal distributions in temporal changes of firm-size variables. Econ E-J Spec Issues Reconstr Macroecon 3(2009–11):
Ishikawa A, Fujimoto S, Mizuno T (2011a) Shape of growth rate distribution determines the type of non-Gibrat's property. Physica A 390:4273–4285
Ishikawa A, Fujimoto S, Tomoyose M (2011b) The difference of growth rate distributions between sales and profits. In: Proceedings of the APFA7: applications of physics in financial analysis & Tokyo Tech—Hitotsubashi interdisciplinary conference. Journal of physics: conference series, vol 221, p 012007
Ishikawa A, Fujimoto S, Watanabe T, Mizuno T (2013) The emergence of different tail exponents in the distributions of firm size variables. Physica A 392:2104–2113
Ishikawa A, Fujimoto S, Mizuno T, Watanabe T (2014) Analytical derivation of power laws in firm size variables from Gibrat's law and quasi-inversion symmetry: a geomorphological approach. J Phys Soc Jpn 83:034802
Ishikawa A, Fujimoto S, Mizuno T, Watanabe T (2015a) Firm age distributions and the decay rate of firm activities. In: Takayasu H, Ito N, Noda I, Takayasu M (eds) Proceedings of the international conference on social modeling and simulation, plus econophysics colloquium 2014, pp 187-194
Ishikawa A, Fujimoto S, Mizuno T, Watanabe T (2015b) The relation between firm age distributions and the decay rate of firm activities in the United States and Japan. IEEE Int Conf Big Data 2015:2726–2731. https://doi.org/10.1109/BigData.2015.7364073
Ishikawa A, Fujimoto S, Mizuno T, Watanabe T (2016a) Firm growth function and extended-Gibrat's property. Adv Math Phys 2016, Article ID 9303480. https://doi.org/10.1155/2016/9303480
Ishikawa A, Fujimoto S, Mizuno T, Watanabe T (2016b) Long-term firm growth properties derived from short-term laws of sales and number of employees in Japan and France. Evol Inst Econ Rev 409–422. https://doi.org/10.1007/s40844-016-0055-0
Ishikawa A, Fujimoto S, Mizuno T, Watanabe T (2017a) Dependence of the decay rate of firm activities on firm age. Evol Inst Econ Rev 351–362. https://doi.org/10.1007/s40844-017-0084-3
Ishikawa A, Fujimoto S, Mizuno T, Watanabe T (2017b) Transition law of firms' activity and the deficit aspect of non-Gibrat's law. JPS Conf Proc 16:011005
Ishikawa A, Fujimotoo S, Mizuno T (2018) Statistical law observed in inactive rate of firms. Evol Inst Econ Rev 1–12. https://doi.org/10.1007/s40844-018-0119-4
Jovanovic B (1982) Selection and the evolution of industry. Econometrica 50:649–670
Klapper L, Laeven L, Rajan R (2006) Entry regulation as a barrier to entrepreneurship. J Financ Econ 82:591–629
Lee SH, Yamakawa Y, Peng MW, Barney JB (2011) How do bankruptcy laws affect entrepreneurship development around the world? J Bus Ventur 26:505–520
Lotti F, Santarelli E, Vivarelli M (2009) Defending Gibrat's Law as a long-run regularity. Small Bus Econ 32:31–44
Miura W, Takayasu H, Takayasu M (2012) Effect of coagulation of nodes in an evolving complex network. Phys Rev Lett 108:168701
Mizuno T, Ishikawa A, Fujimoto S, Watanabe T (2012) Power laws in firm productivity. Prog Theor Phys Supple No 194:122–134
Montroll EW, Shlesinger MF (1983) Maximum entropy formalism, fractals, scaling phenomena, and 1/f noise: a tale of tails. J Stat Phys 32:209–230

Newman MEJ (2005) Power laws, Pareto distributions and Zipf's law. Contemp Phys 46:323–351
Pareto V (1897) Cours d'Économie Politique. Macmillan, London
Sutton J (1997) Gibrat's legacy. J Econ Lit 35(1997):40–59
Takayasu H, Sato A, Takayasu M (1997) Stable infinite variance fluctuations in randomly amplified Langevin system. Phys Rev Lett 79:966–969
Tazaki H (2000) Thomodynamics. Baifukan, Tokyo (in Japanese)
Thompson P (2005) Selection and firm survival: evidence from the shipbuilding industry, 1825–1914. Rev Econ Stat 87:26–36
Tokyo Shoko Research, Ltd. http://www.tsr-net.co.jp/en/outline.html
Tomoyose M, Fujimono S, Ishikawa A (2009) Non-Gibrat's law in the middle scale region. Prog Theor Phys Supple No 179:114–122

How Much Income Inequality is Fair? Nash Bargaining Solution and Its Connection to Entropy

Venkat Venkatasubramanian and Yu Luo

Abstract The question about fair income inequality has been an important open question in economics and in political philosophy for over two centuries with only qualitative answers such as the ones suggested by Rawls, Nozick, and Dworkin. We provided a quantitative answer recently, for an ideal free market society, by developing a game-theoretic framework that proved that the ideal inequality is a lognormal distribution of income at equilibrium. In this paper, we develop another approach, using the Nash bargaining solution (NBS) framework, which also leads to the same conclusion. Even though the conclusion is the same, the new approach, however, reveals the true nature of NBS, which has been of considerable interest for several decades. Economists have wondered about the economic meaning or purpose of the NBS. While some have alluded to its fairness property, we show more conclusively that it is all about fairness. Since the essence of entropy is also fairness, we see an interesting connection between the Nash product and entropy for a large population of rational economic agents.

1 Introduction

Extreme economic inequality is widely seen as a serious concern to the future of stable and vibrant capitalist democracies. In 2015, the World Economic Forum in Davos identified deepening income inequality as the number one challenge of our time. As many political observers remarked, the social and political consequences of extreme economic inequality and the uneven sharing of prosperity seem to have played a role in the outcome of the U.S. presidential election in 2016. Many in the U.S. feel that the nation's current level of economic inequality is unfair and that

V. Venkatasubramanian (✉)
Department of Chemical Engineering, Columbia University, New York City, NY 10027, USA
e-mail: venkat@columbia.edu

Y. Luo
Department of Chemical and Biomolecular Engineering, University of Delaware, Newark, DE 19716, USA
e-mail: yuluo@udel.edu

A. S. Chakrabarti et al. (eds.), *Network Theory and Agent-Based Modeling in Economics and Finance*, https://doi.org/10.1007/978-981-13-8319-9_8

capitalism is not working for 90% of the population (Piketty and Goldhammer 2014; Stiglitz 2015; Reich 2015).

Yet some income inequality is inevitable, even desirable and necessary for a successful capitalist society. As different people have different talents and skills, and different capacities for work, they make different contributions in a society, some more, others less. Therefore, it is only fair that those who contribute more earn more. *But how much more?* In other words, what is the fairest inequality of income? This critical question is at the heart of the inequality debate. The debate is not so much about inequality *per se* as it is about fairness.

Consider a simple example to illustrate this point. John is hired as a temporary worker to perform a job for one hour and makes $100. Lilly also performs the same job, at the same level of quality, but works for two hours and makes $200. Is there inequality in their incomes? Of course, there is. But is the inequality *fair*? Of course, it is. Lilly earned more because she contributed more. Their incomes are *not equal*, but *equitable*. They are both paid at the same rate per hour of work, which is the basis for fairness here.

In this simple case, it was easy to ensure equity, but how do we accomplish this, in general, in a free market society consisting of millions of workers of varying degrees of talent, skill, and capacity for work? Is there a measure of fairness that can guide us to accomplish this? Is there an income distribution that ensures equity? Given the complexity of the problem, one might anticipate the answer to be no for both questions. But, surprisingly, the answer is yes, under certain conditions.

The first author has shown, in a series of papers (Venkatasubramanian 2009, 2010; Venkatasubramanian et al. 2015) and a book (Venkatasubramanian 2017), that the measure of fairness we are seeking is entropy and the equitable income distribution is lognormal at equilibrium in an ideal free market. These results were arrived at by analyzing the problem in two different, but related, perspectives: statistical mechanics and potential game theory.

In this paper, we demonstrate yet another approach, namely, the Nash bargaining solution (NBS) formalism (Nash 1950; Muthoo 1999). We consider this paper to be valuable in two respects. One, this is the first time one has proposed the NBS formalism for the income distribution problem. Even though the NBS formalism has been well-known for nearly 70 years, and has been used extensively in many fair allocation problems (see, for example, Muthoo 1999), we find it quite surprising that it has not been used to address the central question of fair distribution of income in a free market economy. Thus, we consider our NBS formulation to this problem as an important contribution to economic literature. Second, while the final result of our analysis in itself is not new, this new approach, however, reveals something unexpected, namely, the true meaning of the Nash product and its connection with entropy.

In the NBS formalism, one arrives at the solution by maximizing the *product* of utilities, known as the *Nash product* (more on this in Sect. 4). Over the years, economists have wondered about the true meaning, i.e., about the economic content, of the Nash product. The *sum* of utilities of different agents makes economic sense as it gives us the total utility of the system, but why a *product* of utilities? What does it stand for?

As Trockel (2003) observed:

> While all other characterizations, via axioms or via support by equilibria of noncooperative games appear to reflect different aspects of this solution and to open new ways of interpretation, its most simple description via the Nash product seems to have escaped up to now a meaningful interpretation …Yet, concerning its direct interpretation the situation is best described by the quotation of Osborne and Rubinstein (1994, p. 303): "Although the maximization of a product of utilities is a simple mathematical operation it lacks a straightforward interpretation; we view it simply as a technical device."

Most people use it as a convenient mathematical device, but what is it really?

Interestingly, the meaning of potential (P^*) in game theory also had posed a similar puzzle as Monderer and Shapley (1996) had pointed out:

> This raises the natural question about the economic content (or interpretation) of P^*: What do the firms try to jointly maximize? We do not have an answer to this question.

This is again related to a similar fundamental question raised by Samuelson (1972) decades ago in his Nobel lecture:

> What it is that Adam Smith's "invisible hand" is supposed to be maximizing?

We showed in our earlier work (Venkatasubramanian et al. 2015; Venkatasubramanian 2017) that Samuelson, and Monderer and Shapley, were right in suspecting that something quite interesting and deep was missing in our understanding of these economic theories—what was missing was the understanding of how the concept of fairness was intimately connected with all this. We showed that both potential and entropy stand for the concept of fairness in a distribution and that this is what the "invisible hand" is maximizing.

Building on this insight, we offer, in this paper, a novel interpretation of the Nash product and its connection with entropy and fairness. We show how both employ the same mathematical device to accomplish the same objective. Our earlier work showed the deep connection between potential game and statistical mechanics via the concept of entropy. In this paper, we show how these two frameworks are deeply connected with the NBS framework via the connection between entropy and the Nash product. Thus, all three puzzles—the "invisible hand," the potential, and the Nash product—are related in a deep and interesting manner to each other, and to entropy, and all these are related to the same critical economic concept, namely, fairness.

2 Potential Game-Theoretic Framework: Summary of Past Work

Before we can proceed, we need to recall the central ideas and results from our earlier work (Venkatasubramanian et al. 2015; Venkatasubramanian 2017), and so we summarize them here for the benefit of the reader.

Let us first recall the expression we derived for the utility of a rational agent employed in a company. We arrived at this by seeking to answer the basic question of why people seek employment. At the most fundamental, *survival*, level, it is to be able to pay bills now so that they can make a living, with the hope that the current job will lead to a better future. One hopes that the present job will lead to a better one next, acquired based on the experience from the current job, and to a series of better jobs in the future, and hence to a better life. Thus, the utility derived from a job is made up of two components: the *immediate* benefits of making a living (i.e., "present" utility) and the prospects of a better *future* life (i.e., "future" utility). There is, of course, the cost or *disutility* of effort or contribution to be accounted for as well.

Hence, we proposed that the *effective* utility from a job is determined by these three dominant elements: (i) utility from salary, (ii) disutility of effort or contribution, and (iii) utility from a fair opportunity for future prospects. By effort, we do not mean just the physical effort alone, even though it is a part of it.

Thus, the effective utility for an agent is given by

$$h_i(S_i, N_i) = u_i - v_i + w_i \tag{1}$$

where h_i is the effective utility of an employee earning a salary S_i by expending an appropriate effort, while competing with $(N_i - 1)$ other agents in the same job category i for a fair shot at a better future. $u(\cdot)$ is the utility derived from salary, $v(\cdot)$ the disutility from effort, and $w(\cdot)$ is the utility from a fair opportunity for a better future. Every agent tries to maximize its effective utility by picking an appropriate job category i.

2.1 Utility of a Fair Opportunity for a Better Future

The first two elements are rather straightforward to appreciate, but the third requires some discussion. Consider the following scenario. A group of freshly minted law school graduates (totaling N_i) have just been hired by a prestigious law firm as associates. They have been told that one of them will be promoted as partner in 8 years depending on her or his performance. Let us say that the partnership is worth $\$Q$. So any associate's chance of winning the coveted partnership goes as $1/N_i$, where N_i is the number of associates in her peer group i, her *local competition*. Therefore, her expected value for the award is Q/N_i, and the utility derived from it goes as $\ln(Q/N_i)$ because of diminishing marginal utility. Diminishing marginal utility is a well-known economics concept that states that the incremental benefit derived by consuming an additional unit of a resource decreases as one consumes more and more of it. This property is usually modeled by a logarithmic function. It is important to recognize here that the benefit ($\ln Q$) lies in the future, but its cost or disutility ($\ln(1/N_i)$) is paid in the present, in the daily competition with one's peers toward the partnership. This is akin to buying a lottery ticket. The cost (say, \$1) of the ticket is incurred right away, right at the purchase, but the benefit of potentially winning a large sum lies in the future. For the time being, one is out \$1—this disutility is to be accounted for right away.

Therefore, the disutility incurred toward a fair opportunity for career advancement in a competitive environment is:

$$w_i(N_i) = -\gamma \ln N_i \tag{2}$$

where γ is a constant parameter. This equation models the effect of *competitive interaction* between agents. Considering the society at large, this equation captures the notion that in a fair society, an appropriately qualified agent with the necessary education, experience, and skills should have a fair shot at growth opportunities irrespective of her race, color, gender, and other such factors—i.e., it is a *fair competitive* environment. This is the cost or the disutility incurred for *equality of access* or *equality of opportunity* for a better life, for *upward mobility*. The category i would correspond to her qualification category in the society. The other agents in that category are the ones she will be competing with for these growth opportunities.

2.2 Modeling the Disutility of a Job

For the utility derived from salary, we again employ the commonly used logarithmic utility function:

$$u_i(S_i) = \alpha \ln S_i \tag{3}$$

where α is another constant parameter. As for the effort component, every job has certain disutility associated with it. This disutility depends on a host of factors such as the investment in education needed to qualify oneself for the job, the experience to be acquired, working hours and schedule, quality of work, work environment, company culture, relocation anxieties, etc. To model this, one can combine u and v to compute

$$u_{\text{net}} = au - bv \quad (a \text{ and } b \text{ are positive constant parameters}) \tag{4}$$

which is the net utility (i.e., net benefit or gain) derived from a job after accounting for its cost. Typically, net utility will increase as u increases (because of salary increase, for example). However, generally, after a point, the cost has increased so much, due to personal sacrifices such as working overtime, missing quality time with family, giving up on hobbies, job stress resulting in poor mental and physical health, etc., u_{net} begins to decrease after reaching a maximum. The simplest model of this commonly occurring inverted-U profile is a quadratic function, as in

$$u_{\text{net}} = au - bu^2. \tag{5}$$

Since, $u \sim \ln$ (Salary), we get Eq. (6):

$$v_i(S_i) = \beta(\ln S_i)^2. \tag{6}$$

2.3 Effective Utility from a Job

Combining all three, we have

$$h_i(S_i, N_i) = \alpha \ln S_i - \beta(\ln S_i)^2 - \gamma \ln N_i \tag{7}$$

where $\alpha, \beta, \gamma > 0$.

In general, α, β, and γ, which model the relative importance an agent assigns to these three elements, can vary from agent to agent. However, we examined the ideal situation where all agents have the same preferences and hence treat these as constant parameters.

2.4 Equilibrium Income Distribution

We then used the potential game-theoretic framework to prove (Venkatasubramanian et al. 2015; Venkatasubramanian 2017) that a large population of agents with this utility function will reach Nash equilibrium. In potential games (Rosenthal 1973; Sandholm 2010; Easley and Kleinberg 2010), there exists a single scalar-valued global function, called a *potential*, that captures the necessary information about utilities. The *gradient* of the potential is the *utility*. For such games, Nash equilibrium is reached when the potential is maximized.

So, using the potential game formalism, we have an employee's utility as the gradient of potential $\phi(\boldsymbol{x})$, i.e.,

$$h_i(\boldsymbol{x}) \equiv \frac{\partial \phi(\boldsymbol{x})}{\partial x_i} \tag{8}$$

where $x_i = N_i/N$ denotes the fraction of population at category i and $\boldsymbol{x}$ is the population vector. Therefore, by integration (we replace partial derivative with total derivative because $h_i(\boldsymbol{x})$ can be reduced to $h_i(x_i)$ expressed in (1)–(6)),

$$\phi(\boldsymbol{x}) = \sum_{i=1}^{n} \int h_i(\boldsymbol{x}) \mathrm{d}x_i. \tag{9}$$

We observe, using (7), that our game is a potential game with the potential function

$$\phi(\boldsymbol{x}) = \phi_u + \phi_v + \phi_w + \text{constant} \tag{10}$$

where

$$\phi_u = \alpha \sum_{i=1}^{n} x_i \ln S_i \tag{11}$$

$$\phi_v = -\beta \sum_{i=1}^{n} x_i (\ln S_i)^2 \tag{12}$$

$$\phi_w = \frac{\gamma}{N} \ln \frac{N!}{\prod_{i=1}^{n} (N x_i)!} \tag{13}$$

where we have used Stirling's approximation in Eq. (13).

One can see that $\phi(\boldsymbol{x})$ is strictly concave:

$$\frac{\partial^2 \phi(\boldsymbol{x})}{\partial x_i^2} = -\frac{\gamma}{x_i} < 0. \tag{14}$$

Therefore, *a unique Nash equilibrium for this game exists*, where $\phi(\boldsymbol{x})$ is maximized, as per the well-known theorem in potential games (Sandholm 2010, p. 60).

Thus, the self-organizing free market dynamics, where employees switch jobs, and companies switch employees, in search of better utilities or profits, ultimately reaches an equilibrium state, with an *equilibrium income distribution*. This happens when the potential $\phi(\boldsymbol{x})$ is maximized. The equilibrium income distribution is the following *lognormal distribution*:

$$x_i = \frac{1}{S_i D} \exp\left[- \frac{\left(\ln S_i - \frac{\alpha+\gamma}{2\beta} \right)^2}{\gamma/\beta} \right] \tag{15}$$

where $D = N \exp\left[\lambda/\gamma - (\alpha + \gamma)^2/4\beta\gamma \right]$ and λ is the Lagrange multiplier used in maximizing $\phi(\boldsymbol{x})$ with the $\sum_{i=1}^{n} x_i = 1$ constraint.

We also proved that the effective utility, h^*, at equilibrium is given by

$$h^* = \gamma \ln Z - \gamma \ln N \tag{16}$$

where $Z = \sum_{j=1}^{n} \exp\left([\alpha \ln S_j - \beta (\ln S_j)^2]/\gamma \right)$ resembles the partition function seen in statistical mechanics (it is easy to show that $\lambda = h^*$). *At equilibrium, all agents enjoy the same effective utility or effective welfare, h^**. Everyone is *not* making the same salary, of course, but they all enjoy the same effective utility. This is an important result, for it proves that all agents are treated *equally* with respect to the economic outcome, namely, effective utility. This proves that the ideal free market exhibits *outcome fairness*.

We also proved that this distribution is socially optimal. A socially optimal distribution is one where the effective utility of the entire population (i.e., the *total effective utility* of society, H) is maximized. This is the outcome desired by utilitarians such as Jeremy Bentham and John Stuart Mill. This *maximum* total effective utility (H^*) is given by

$$H^* = \sum_{i=1}^{n} N_i h_i^* = \sum_{j=1}^{N} h_j^* = N h^* \tag{17}$$

(since $h_i^* = h_j^* = h^*$ at equilibrium), subject to the constraints

$$\sum_{i=1}^{n} N_i S_i = M \tag{18}$$

$$\sum_{i=1}^{n} N_i = N \tag{19}$$

Note that the index i covers the n different salary levels S_i, whereas j covers the N employees. M is the total salary budget and N is the total number of employees in a company.

Now that we have covered the necessary background work, we are ready to forge ahead to develop the Nash bargaining solution framework and its connection to maximum entropy. Toward that, let us first provide an intuitive explanation of the maximum entropy principle.

3 Entropy as a Measure of Fairness: An Intuitive Explanation of $S = -\sum_{i=1}^{n} p_i \ln p_i$

As we know, entropy is maximized at statistical equilibrium. We also know that the equivalent defining criterion for statistical equilibrium is the *equality* of all accessible phase space cells, which, in fact, is the fundamental postulate of statistical mechanics. In other words, a given molecule is *equally likely* to be found *in any one* of the accessible phase space cells, at any given time.

For example, consider a very large number of identical gas molecules enclosed in a chamber. To say that a gas molecule, in the long run, is more likely to be found in the left half of the chamber than the right, and assign it a higher probability (say, $p(\text{left}) = 0.6$ and $p(\text{right}) = 0.4$), would be a *biased* and an *unfair* assignment of probabilities. This assignment *presumes* some information that has not been given. What does one know about the molecule, or its surrounding space, that would make one prefer the left chamber over the right? There is no basis for this preference. The unequal assignment of probabilities is thus arbitrary and unwarranted. Therefore, the *fairest* assignment of probabilities is one of *equality* of outcomes, i.e., $p(\text{left}) = 0.5$ and $p(\text{right}) = 0.5$.

Let us examine this example further. Let us say that the chamber is divided into n imaginary partitions of equal volume such that the molecules are free to move about from partition to partition. As noted, the essence of statistical equilibrium is the *equality* of all accessible phase space cells. But how is this connected to the maximization of entropy as defined by $S = -\sum_{i=1}^{n} p_i \ln p_i$ (not to be confused with salary S)? The connection is not obvious from this equation. However, there is a fascinating connection with an important implication on the true nature of entropy, as well as for the Nash product.

Now, the criterion for equilibrium in our chamber example is

$$p_1 = p_2 = \ldots = p_n = p \tag{20}$$

where p_i is the probability of finding a given molecule in the partition i.

But how do we recognize and enforce this criterion in practice?

This is easy to do, for example, if there are only a few partitions in phase space, e.g., two partitions, left and right, as we did above. We would immediately recognize a nonequilibrium situation if one had $p(\text{left}) = 0.6$ and $p(\text{right}) = 0.4$. And we would also recognize with equal ease the equilibrium situation, $p(\text{left}) = 0.5$ and $p(\text{right}) = 0.5$.

But this gets trickier when there are a large number of partitions or cells in phase space. For example, what if there are 1000 partitions, as opposed to just two, and we are given the following two situations to evaluate?

$$p_1 = 0.0015,\ p_2 = 0.001,\ p_3 = 0.0009, \ldots,\ p_{1000} = 0.0011$$

$$p_1 = 0.0011,\ p_2 = 0.0008,\ p_3 = 0.0012, \ldots,\ p_{1000} = 0.0009$$

Since all the partitions are of equal volumes, at equilibrium,

$$p_1 = p_2 = \ldots = p_{1000} = 0.001$$

How do we compare these two configurations and determine which one is *closer* to equilibrium? In general, how do we make this comparison among billions of such alternative configurations? How do we enforce this equality criterion?

One could compute the sum of squared residuals for comparison, or other such methods, but this gets to be really tedious and messy, when the number of molecules and the number of configurations run into billions as is typical in statistical mechanics. Is there a simpler, more elegant way of accomplishing this? It turns out there is, one that exploits a wonderful result regarding the product of a set of positive numbers that sum to a fixed amount. This is what is at the heart of the entropy equation, and the Nash product as we show later.

While we recognize the equilibrium state by the maximum entropy criterion, hidden in its mathematical form is the *equality* of all accessible cells criterion given by

Eq. (20). To see this, we first observe that maximizing $\{-\sum_{i=1}^{n} p_i \ln p_i\}$ is the same as maximizing $\{-\ln \prod_{i=1}^{n} p_i^{p_i}\}$. And this is the same as minimizing $\{\ln \prod_{i=1}^{n} p_i^{p_i}\}$, which is the same as minimizing $\{\prod_{i=1}^{n} p_i^{p_i}\}$. Now, according to two well-known results (which are related), the arithmetic mean-geometric mean inequality theorem (AM–GM theorem) Lohwater (1982), and the Jensen's inequality, this product is minimized if and only if

$$p_1 = p_2 = ... = p_n$$

where $0 < p_i < 1$ and $\sum_{i=1}^{n} p_i = 1$ (one can also demonstrate this using the Lagrangian multiplier method, but, we believe, the geometric intuition is more transparent using the AM-GM theorem or the Jensen's inequality). This equality is, of course, the fundamental criterion for statistical equilibrium. Therefore, when we maximize entropy, what we are really doing is *enforcing this equality constraint* buried in it. This is a clever mathematical trick for enforcing equality, by exploiting this important relationship between the product and the sum of a set of positive numbers. Therefore, enforcing equality and hence fairness, is the objective buried in the mathematical trick of maximizing the product, $\{-\prod_{i=1}^{n} p_i^{p_i}\}$, which is the same as minimizing $\{\prod_{i=1}^{n} p_i^{p_i}\}$. We will now see in the next section how the same clever mathematical device of maximizing a product to test and enforce equality is employed for the Nash product.

We can see clearly now how entropy maximization is related to enforcing equality, which is the same as enforcing *fairness*—treating equal things equally. This insight that entropy really is a measure of fairness in a distribution has never been clearly recognized in statistical mechanics, in information theory, or in economics literature until the first author discussed its importance in his 2009–2010 papers (Venkatasubramanian 2009, 2010). It is a historical accident that the concept of entropy was first discovered in the context of thermodynamics and, therefore, it has generally been identified as a measure of randomness or disorder. However, the true essence of entropy is fairness, which appears with different masks in different contexts (for a detailed discussion of this, see Ray 2017).

4 Nash Bargaining Solution

We now derive the Nash bargaining solution to our problem. Following Nash's original formulation (Nash 1950) for the two-player bargaining problem there has been extensive literature on this topic, which has been generalized to the n-player case using both cooperative and noncooperative approaches with many applications (we cite a few select papers here Harsanyi (1963), Rubinstein (1982), Binmore and Wolinsky (1986), Mazumdar and Douligeris (1991), Chatterjee et al. (1993), Krishna and Serrano (1996), Muthoo (1999), Yaïche and Rosenberg (2000), Ray (2007), Okada (2010), Compte and Jehiel (2010)).

We consider N players labeled $j = 1, \ldots, N$ (use j instead of i for the index of individual players), and we consider only the grand coalition involving all N players because all N players are required for the company to succeed. The surplus that this coalition generates is nonnegative. Once this team forms, the game stops.

Suppose that players $1, \ldots, N$, have complementary skills which can be combined to produce a divisible pie of some total utility. The pie is produced and divided among the N players *only* when all players reach an agreement on their shares $(h_1, h_2, \ldots, h_N)$, where h_j is the share or the utility of the jth player.

The following description of the NBS is adapted from Mazumdar and Douligeris (1991):

> Consider a cooperative game of N players (e.g., employees in a company). Let each individual player j have an utility function $h_j(\boldsymbol{x}) : \mathbb{X} \to \mathbb{R}$ where $\mathbb{X}$ is a convex, closed, and bounded set of $\mathbb{R}^N$. For example, in communication networks $\mathbb{X}$ denotes the space of throughputs. Let $\boldsymbol{h}_\mathrm{d} = [h_{\mathrm{d},1}, h_{\mathrm{d},2}, \ldots, h_{\mathrm{d},N}]$ where $h_{\mathrm{d},j} = h_j(\boldsymbol{x}_\mathrm{d})$ for some $\boldsymbol{x}_\mathrm{d} \in \mathbb{X}$ denote a disagreement point which all the players agree to as a starting point for the game. In general, $\boldsymbol{h}_\mathrm{d}$ can be thought of as the vector of individual default utilities which the users would like to achieve, at least, if they enter the game. It is also referred to as the threat point. Let $[\mathbb{H}, \boldsymbol{h}_\mathrm{d}]$ denote the game defined on $\mathbb{X}$ with initial disagreement point $\boldsymbol{h}_\mathrm{d}$ where $\mathbb{H}$ denotes the image of the set $\mathbb{X}$ under $h(\cdot)$, i.e., $h(\mathbb{X}) = \mathbb{H}$. Let $F[\cdot, \boldsymbol{h}_\mathrm{d}] : \mathbb{H} \to \mathbb{H}$ be an arbitration strategy. Then F is said to be an NBS if it satisfies the four axioms below.
>
> 1. Let $\psi(\boldsymbol{h}) = \boldsymbol{h}'$ where $h'_j = a_j h_j + b_j$ for $j = 1, 2, \ldots, N$ and $a_j > 0, b_j$ are arbitrary constants. Then
> $$F[\psi(\mathbb{H}), \psi(\boldsymbol{h}_\mathrm{d})] = \psi(F[\mathbb{H}, \boldsymbol{h}_\mathrm{d}]). \tag{21}$$
> This states that the operating point in the space of strategies is *invariant* with respect to *linear* utility transformations.
> 2. The solution must satisfy
> $$(F[\mathbb{H}, \boldsymbol{h}_\mathrm{d}])_j \geq h_{\mathrm{d},j} \quad (j = 1, 2, \ldots, N) \tag{22}$$
> and furthermore there exists no $\boldsymbol{h} \in \mathbb{H}$ such that $h_j \geq (F[\mathbb{H}, \boldsymbol{h}_\mathrm{d}])_j$ for all $j = 1, 2, \ldots, N$. This is the notion of *Pareto optimality* of the solution.
> 3. Let $[\mathbb{H}_1, \boldsymbol{h}_\mathrm{d}]$ and $[\mathbb{H}_2, \boldsymbol{h}_\mathrm{d}]$ be two games with the same initial agreement point such that:
> $$\mathbb{H}_1 \subset \mathbb{H}_2 \tag{23}$$
> $$F[\mathbb{H}_2, \boldsymbol{h}_\mathrm{d}] \in \mathbb{H}_1. \tag{24}$$
> Then $F[\mathbb{H}_1, \boldsymbol{h}_\mathrm{d}] = F[\mathbb{H}_2, \boldsymbol{h}_\mathrm{d}]$. This is called the *independence of irrelevant alternatives* axiom. This states that the NBS of a game with a larger set of strategies is the same as that of the smaller game if the arbitration point is a valid point for the smaller game. The additional strategies are superfluous.
> 4. Let $\mathbb{H}$ be symmetrical with respect to a subset $\mathbb{J} \subseteq \{1, 2, \ldots, N\}$ of indices (i.e., let $j, k \in \mathbb{J}$ and $j < k$), then
> $$\{h_1, h_2, \ldots, h_{j-1}, h_k, h_{j+1}, \ldots, h_{k-1}, h_j, h_{k+1}, \ldots, h_N\} \in \mathbb{H}. \tag{25}$$
> If $h_{\mathrm{d},j} = h_{\mathrm{d},k}$, then $(F[\mathbb{H}, \boldsymbol{h}_\mathrm{d}])_j = (F[\mathbb{H}, \boldsymbol{h}_\mathrm{d}])_k$ for $j, k \in \mathbb{J}$. This is the axiom of *symmetry* which states that if the set of utilities is symmetric then, for any two players, if the initial agreement point corresponds to equal performance, then their arbitrated values are equal.

Nash proved that the unique solution, i.e., the NBS, of the game that satisfies the four axioms is given by the point which maximizes the expression $\prod_{j=1}^{N}(h_j(\boldsymbol{x}) - h_{\mathrm{d},j})$, known as the *Nash product*. This can be written as

$$\max \prod_{j=1}^{N}(h_j(\boldsymbol{x}) - h_{\mathrm{d},j}) = \max \prod_{j=1}^{N}(h_j - h_{\mathrm{d},j}) \tag{26}$$

where h_j denotes both the jth player's utility function and its utility.

5 NBS of the Income Distribution Game

In this section, we define the set of feasible utilities of the income game and show that it is a convex set. By maximizing the product of utilities over such set, we obtain the NBS of the income game and it is identical to the Nash equilibrium solution we had reported earlier in Venkatasubramanian et al. (2015).

Recall the utility function $h_i(S_i, N_i)$ defined in (7). Suppose the n salary levels are predetermined. Then the utility for agents at each salary level solely depends on its occupancy, that is

$$h_i(N_i) = h_{0,i} - \gamma \ln N_i \tag{27}$$

where $h_{0,i} \equiv \alpha \ln S_i - \beta(\ln S_i)^2$ is a constant unique to the ith salary level.

We now define the set of population $\mathbb{N}$

$$\mathbb{N} \equiv \{N_i \in \mathbb{Z}_+ : \sum_{i=1}^{n} N_i \leq N\} \tag{28}$$

where the total number of agents does not exceed N. It is easy to verify that $\mathbb{N}$ is convex. Next, we construct the set of utilities $\mathbb{H}$ through the following mapping:

$$\mathbb{H} \equiv \{h_i = h_{0,i} - \gamma \ln N_i : \sum_{i=1}^{n} N_i \leq N\}. \tag{29}$$

To show that $\mathbb{H}$ is also convex, let $\boldsymbol{h}_1 = \boldsymbol{h}_0 - \gamma \ln \boldsymbol{N}_1$ and $\boldsymbol{h}_2 = \boldsymbol{h}_0 - \gamma \ln \boldsymbol{N}_2$ be any two vectors of utilities in $\mathbb{H}$. They correspond to two vectors of occupancies $\boldsymbol{N}_1 \in \mathbb{N}$ and $\boldsymbol{N}_2 \in \mathbb{N}$. Define $\boldsymbol{h}_3$ as the convex combination of $\boldsymbol{h}_1$ and $\boldsymbol{h}_2$:

$$\begin{aligned} \boldsymbol{h}_3 &\equiv t\boldsymbol{h}_1 + (1-t)\boldsymbol{h}_2 \\ &= \boldsymbol{h}_0 - \gamma[t \ln \boldsymbol{N}_1 + (1-t)\ln \boldsymbol{N}_2] \\ &= \boldsymbol{h}_0 - \gamma \ln \boldsymbol{N}_3 \end{aligned} \tag{30}$$

where $0 \leq t \leq 1$ is the ratio of the convex combination and $\boldsymbol{N}_3 \equiv e^{t \ln \boldsymbol{N}_1 + (1-t) \ln \boldsymbol{N}_2}$. We have

$$\begin{aligned}
\sum_{i=1}^{n} N_{i,3} &= \sum_{i=1}^{n} e^{t \ln N_{i,1} + (1-t) \ln N_{i,2}} \\
&\leq \sum_{i=1}^{n} \left[t e^{\ln N_{i,1}} + (1-t) e^{\ln N_{i,2}} \right] \\
&= \sum_{i=1}^{n} \left[t N_{i,1} + (1-t) N_{i,2} \right] \\
&= t \sum_{i=1}^{n} N_{i,1} + (1-t) \sum_{i=1}^{n} N_{i,2} \\
&\leq t N + (1-t) N \\
&= N
\end{aligned} \tag{31}$$

where $N_{i,j}$ is the ith component of the vector $\boldsymbol{N}_j$. The first inequality

$$e^{t \ln N_{i,1} + (1-t) \ln N_{i,2}} \leq t e^{\ln N_{i,1}} + (1-t) e^{\ln N_{i,2}} \tag{32}$$

follows Jensen's inequality where

$$f\big(t x_1 + (1-t) x_2\big) \leq t f(x_1) + (1-t) f(x_2) \tag{33}$$

if the function $f(x)$ is convex; and the second follows the definition of $\mathbb{N}$. $\boldsymbol{h}_3$ is therefore also in $\mathbb{H}$ because N_3 is in $\mathbb{N}$.

Recall that the NBS is obtained by maximizing the Nash product in (26). By grouping players with the same amount of utility together (i.e., they are in the same income level), we convert the original NBS problem to the following problem:

$$\max_{\boldsymbol{x} \in \mathbb{X}} \prod_{j=1}^{N} (h_j(\boldsymbol{x}) - h_{\mathrm{d},j}) = \max_{\boldsymbol{h} \in \mathbb{H}} \prod_{i=1}^{n} h_i^{N_i} \tag{34}$$

where we have set the disagreement point or the threat point of the jth player $h_{\mathrm{d},j}$ to be zero (i.e., a player agrees to enter the game as long as her effective utility is greater than zero). Note that the first product is over all the players (N) whereas the second is over all the levels (n).

Therefore, (34) is equivalent to solving

$$\begin{aligned} \max \quad & g(\boldsymbol{N}) = \sum_{i=1}^{n} N_i \ln h_i \\ \text{s.t.} \quad & l(\boldsymbol{N}) = \sum_{i=1}^{n} N_i - N \leq 0 \end{aligned} \tag{35}$$

because logarithm is continuous and monotonic. The Karush–Kuhn–Tucker (KKT) necessary conditions of (35) are as follows:

$$\nabla g(\boldsymbol{N}^*) = \mu \nabla l(\boldsymbol{N}^*) \tag{36}$$

$$l(\boldsymbol{N}^*) \leq 0 \tag{37}$$

$$\mu \geq 0 \tag{38}$$

$$\mu l(\boldsymbol{N}^*) = 0 \tag{39}$$

where μ is a KKT multiplier. Expanding (36), we have

$$\ln h_i^* - \frac{\gamma}{h_i^*} = \mu \quad (i = 1, \ldots, n). \tag{40}$$

There exists a unique solution to (40) where $h_i^* = h^* = \gamma / W(\gamma e^{-\mu})$ and W denotes the Lambert-W (product log) function. Since $W(x)$ monotonically increases as $x \geq 0$ increases, h^* is maximized when $\mu > 0$. From the complementary slackness in (39), we have $l(\boldsymbol{N}^*) = 0$ or

$$\sum_{i=1}^{n} N_i = N. \tag{41}$$

From (27), we have

$$h^* = \alpha \ln S_i - \beta (\ln S_i)^2 - \gamma \ln N_i^*. \tag{42}$$

The NBS of the income game therefore requires

$$N_i^* = \frac{N}{S_i D} \exp \left[-\frac{\left(\ln S_i - \frac{\alpha+\gamma}{2\beta} \right)^2}{\gamma/\beta} \right] \tag{43}$$

where $D = \exp\left[h^*/\gamma - (\alpha + \gamma)^2/4\beta\gamma\right]$. Equation (43) is identical to the Nash equilibrium solution (15) since $\lambda = h^*$. This corresponds to all agents enjoying the same effective utility h^* at equilibrium given by Eq. (16).

6 Summary and Conclusions

We have presented a Nash bargaining solution to the problem of fair income distribution in an ideal free market economy. As noted, this is the first time the NBS formalism has been proposed for this problem even though the formalism itself has been well-known for about seven decades and the fair inequality question has been open for over two centuries.

In addition, since the NBS outcome results in a lognormal distribution, which we have proved in our earlier work as the fairest outcome, we see the connection between NBS and fairness. By maximizing the Nash product, we are indeed maximizing fairness. We also showed that by maximizing entropy one was essentially maximizing fairness in the probability distribution by enforcing the equality of all accessible cells through the mathematical device of maximizing the product $-\ln\prod_{i=1}^{n} p_i^{p_i}$, which is the same as minimizing $\prod_{i=1}^{n} p_i^{p_i}$, by exploiting the arithmetic mean–geometric mean inequality theorem or Jensen's inequality.

In a similar manner, in NBS one again invokes the mathematical device of maximizing a product, this time the Nash product. Both techniques achieve the same desired result—the enforcement of equality. In the case of maximum entropy, we achieve the equality of the probabilities of all accessible phase space cells—$p_1 = p_2 = ... = p_n$. That is, we achieve maximizing fairness at equilibrium. Similarly for NBS, we achieve the equality of effective utilities for all agents—$h_1 = h_2 = ... = h_N = h^*$. That is, we achieve maximizing fairness at equilibrium.

Thus, the true economic objective of maximizing the Nash product is to treat all agents fairly subject to the Pareto optimality constraint. Since the fairness objective is buried deeply in the mathematical device of maximizing the product, just as it is buried in the formulation of maximum entropy principle, the fairness property is not obvious even when you probe it. So, naturally, people are surprised when they find the fairness outcome in the final results. That is why we have economists somewhat mystified, making observations such as "Nash product seems to have escaped up to now a meaningful interpretation," "Although the maximization of a product of utilities is a simple mathematical operation it lacks a straightforward interpretation; we view it simply as a technical device," and "The Unreasonable Fairness of Maximum Nash Welfare" (Caragiannis et al. 2016), as quoted earlier. Thus, our work reveals the deep and surprising connection between the Nash product, entropy, and fairness. Achieving maximum fairness is the purpose of the maximum entropy principle as well as for the maximum Nash welfare function. Enforcing equality and hence fairness, under the given constraints, is the true objective buried in the mathematical device of maximizing a product—we see this in entropy and in Nash product.

Acknowledgements This work is supported in part by the Center for the Management of Systemic Risk (CMSR) at Columbia University. The authors thank Resmi Suresh for her helpful comments and assistance with the preparation of this manuscript.

References

Binmore ARKG, Wolinsky A (1986) The nash bargaining solution in economic modelling. RAND J Econ 17:176–188

Caragiannis I, Kurokawa D, Moulin H, Procaccia AD, Shah N, Wang J (2016) Ec'16, in: Proceedings of the 2016 ACM Conference on Economics and Computation, pp 305–322

Chatterjee DRK, Dutta B, Sengupta K (1993) A noncooperative theory of coalitional bargaining. Rev Econ Stud 60:463–477

Compte O, Jehiel P (2010) The coalitional nash bargaining solution. Econometrica 78(5):1593–1623

Easley D, Kleinberg J (2010) Networks, crowds, and markets: reasoning about a highly connected world. Cambridge University Press

Harsanyi J (1963) A simplified bargaining model for the n-person cooperative game. Int Econ Rev 4:194–220

Krishna V, Serrano R (1996) Multilateral bargaining. Rev Econ Stud 63:61–80

Lohwater A (1982) Introduction to inequalities, Tech. rep., e-book online

Mazumdar LMRR, Douligeris C (1991) Fairness in network optimal flow control: optimality of product forms. IEEE Trans Commun 39:775–782

Monderer D, Shapley LS (1996) Potential games. Games and economic behavior 14(1):124–143

Muthoo A (1999) Bargaining theory with applications. Cambridge University Press, UK

Nash JF (1950) The bargaining problem. Econometrica 18(2):155–162

Okada A (2010) The nash bargaining solution in general n-person cooperative games. J Econ Theory 6(145):2356–2379

Osborne MJ, Rubinstein A (1994) A course in game theory. MIT Press, Cambridge, MA

Piketty T, Goldhammer A (2014) Capital in the twenty-first century. Belknap Press

Ray D (2007) A game theoretic perspective on coalition formation. Oxford University Press, U.K

Reich RB (2015) Saving capitalism: for the many. Knopf, Not the Few

Rosenthal RW (1973) A class of games possessing pure-strategy nash equilibria. Int J Game Theory 2(1):65–67

Rubinstein A (1982) Perfect equilibrium in a bargaining model. Econometrica 50:97–109

Samuelson PA (1972) Maximum principles in analytical economics. Am Econ Rev, 249–262

Sandholm WH (2010) Population games and evolutionary dynamics. MIT press

Stiglitz JE (2015) The great divide: unequal societies and what we can do about them, W. W. Norton & Company

Trockel W (2003) On the meaning of the nash product, Tech. rep., Bielefed: IMW-working paper Nr. 354

Venkatasubramanian V (2017) How Much inequality is fair?. Optimal, and Stable Capitalist Society, Columbia University Press, Mathematical Principles of a Moral

Venkatasubramanian V (2009) What is fair pay for executives? an information theoretic analysis of wage distributions. Entropy 11(4):766–781

Venkatasubramanian V (2010) Fairness is an emergent self-organized property of the free market for labor. Entropy 12(6):1514–1531

Venkatasubramanian V, Luo Y, Sethuraman J (2015) How much inequality in income is fair? a microeconomic game theoretic perspective. Phys A: Stat Mech Appl 435:120–138

Yaïche RRMH, Rosenberg C (2000) Game theoretic framework for bandwidth allocation and pricing in broadband networks. IEEE/ACM Trans Netw 8(5):667–678

A Limit Cycle View of Inequality

Anindya S. Chakrabarti and Ratul Lahkar

Abstract We consider the relationship between inequality and the business cycle. Empirical evidence suggests mutual causality between GDP and inequality over the business cycle, and the countercyclical nature of inequality. We construct an evolutionary game-theoretic model with two types of workers to explain these observations. Cognitive workers produce capital of varying quality, which determines the productivity of physical workers who obtain the capital through an exogenous search process. We apply the logit dynamic to the model. The model satisfies positive definiteness due to which, both GDP and inequality converge to limit cycles and, therefore, fluctuate over time. These cyclical fluctuations have a common origin in the underlying evolutionary dynamic process, which explains the empirical observation of mutual causality between these variables. Further, simulations suggest an inverse relationship between GDP and inequality in our model, which accounts for the countercyclical nature of inequality.

Keywords Inequality · Business cycles · Logit dynamic

JEL Classification: C63 · C72 · C73 · E32

1 Introduction

How is inequality related to the business cycle? This question is of interest in economics because the impact of a business cycle on different economic groups depends upon the relationship between inequality and business cycle. Krueger et al. (2010) summarize the findings of the analysis of the *Review of Economic Dynamics* (RED)

A. S. Chakrabarti
Economics Area, Indian Institute of Management Ahmedabad, Ahmedabad 380015, Gujarat, India
e-mail: anindyac@iima.ac.in

R. Lahkar (✉)
Economics Department, Ashoka University, Sonipat 131029, Haryana, India
e-mail: ratul.lahkar@ashoka.edu.in

A. S. Chakrabarti et al. (eds.), *Network Theory and Agent-Based Modeling in Economics and Finance*, https://doi.org/10.1007/978-981-13-8319-9_9

database, which contains extensive data on macroeconomic variables of OECD countries. While there are differences across countries depending upon institutional and policy characteristics, these findings show a countercyclical relationship between GDP and inequality measured in terms of earnings, disposable income, and consumption in most of the OECD countries. In a subsequent analysis of the same database, Maestri and Roventini (2012) also find that "inequalities in hours of work, earnings, market and net income are generally countercyclical ...." Bonhomme and Hospido (2017) analyze Spanish social security data and show that earning inequality is countercyclical. Such empirical evidence, therefore, suggests that the relationship between the business cycle and inequality is broadly countercyclical, although the precise nature of the relationship depends upon the details of countries and variables for which inequality is measured.[1]

Maestri and Roventini (2012) also analyze the direction of (Granger) causation between inequality and GDP.[2] They find evidence of a mutual causal relationship between GDP and inequality in most of the countries in the RED database. The fact that Granger causality is mutual between GDP and inequality suggests both variables might emerge from a common underlying dynamical economic process, thus appearing to explain each other's dynamics over time.

This paper provides a theoretical account of these two empirical observations—the mutual causality between GDP and inequality and the countercyclical nature of inequality. To examine these issues, we apply the evolutionary game-theoretic model used by Chakrabarti and Lahkar (2017) to analyze the question of productivity dispersion. In the present paper, we make appropriate amendments to our earlier model to analyze the evolution and relationship between aggregate income and inequality. We argue that both aggregate income and inequality will fluctuate in our model. Since both aggregate income and inequality are defined in terms of the same underlying state variable in the model, these fluctuations have a common origin thereby suggesting a mutual causal relationship between the two variables. Simulations then reveal that there is an inverse relationship between aggregate income and inequality in our model, making inequality countercyclical in nature.

Fundamentally, this paper is based on the canonical price dispersion model of Burdett and Judd (1983). Lahkar (2011) and Chakrabarti and Lahkar (2017) provide evolutionary game theoretic extensions of Burdett and Judd (1983) for the analysis price dispersion and productivity dispersion, respectively. The use of evolutionary game theory in these papers was motivated by empirical observations of not just dispersion in the relevant phenomenon, but the cyclical nature of such dispersion. Limit cycles are a natural phenomenon in evolutionary game theory and, therefore, provide

---

[1] We should note that there are certain variations in the conclusions obtained by Krueger et al. (2010) and Maestri and Roventini (2012). For example, while Krueger et al. (2010) find that consumption inequality in countercyclical in most countries, Maestri and Roventini (2012) find that consumption inequality is pro-cyclical. Neither paper shows a uniform relationship between business cycles and wage inequality.

[2] They also consider Granger causality between inequality and various other macroeconomic variables like unemployment, inflation rate, consumption, etc. Our interest is in the causality between inequality and GDP.

a convenient tool to model such a cyclical phenomenon. These considerations also prompt us to apply evolutionary game theory for the modeling of inequality, which is a dispersion of income levels, and the cyclical relationship between inequality and GDP. The paper, therefore, also contributes to the literature on the application of evolutionary game theory to macroeconomic phenomena.

Our model has two groups (or populations in the terminology of evolutionary game theory) of workers whom we call cognitive workers and physical workers. A strategy for a cognitive worker is to produce capital of some quality, which is then sold to physical workers. An important assumption is that the price of capital obtained by cognitive workers does not vary with the quality but the cost of producing capital is strictly increasing in quality.[3] Physical workers use the highest quality capital they find through a search process and use the capital to produce output.[4] We may interpret the distinction between cognitive and physical workers as follows. Each physical worker supplies one unit of labor inelastically. They combine this labor with the capital obtained from cognitive workers. As in Chakrabarti and Lahkar (2017), the quality of capital that a physical worker obtains then represents productivity of that worker. Hence, cognitive workers do not produce final output in our model. Instead, they determine the productivity of physical workers who then create the final output.

We fix the search behavior of physical workers exogenously. We assume that there are two types of workers, those who search just once and those who search twice. Further, the proportion of physical workers of each type is also given exogenously. In contrast, in Chakrabarti and Lahkar (2017), the search behavior of the corresponding population (called firms in that paper) arises endogenously. The reason behind this simplification in this paper is that we wish to focus on the more important variable which causes the emergence of limit cycles, which is the behavior of cognitive workers. The definition of payoffs of cognitive workers, along with our particular assumptions of exogenous search behavior suffices for the model to satisfy a technical condition called positive definiteness, which renders all mixed equilibria evolutionarily unstable and causes the emergence of limit cycles.[5]

The state variable in our model is, therefore, the behavior of cognitive workers. Following our earlier paper, we apply the logit dynamic Fudenberg and Levine (1998), which is one of the most widely used dynamic in evolutionary game theory, to this state variable. Positive definiteness then causes the state variable to converge

[3]This is the reverse of the situation in the price dispersion model of Burdett and Judd (1983), where the quality and production cost of a product sold by sellers is the same but the price may vary.

[4]There is a minor difference in terminology used in this paper and in Chakrabarti and Lahkar (2017). Here, we call the two populations, cognitive workers and physical workers, while in the earlier paper they are called innovators and firms. We introduce this difference because we also want to allow the possibility that business cycles may have a different impact on different economic groups. For this purpose, it seems more natural to classify the two populations as two different groups of workers. This difference in nomenclature does not have any implication on the formal analysis of the model.

[5]We know from Chakrabarti and Lahkar (2017) that the same condition of positive definiteness becomes relevant even with endogenous search behavior because even in that case, physical workers will search just once or twice in equilibrium. It is, therefore, not a drastic oversimplification to exogenously limit physical workers to searching just once or twice and, instead, focus on the behavior of cognitive workers which is the more important determinant of positive definiteness.

to a limit cycle. We compute aggregate income in our model, which represents GDP, as well as Gini coefficients for both groups of workers to measure inequality. The cyclical behavior of the state variable also causes cyclical fluctuations in aggregate income and the Gini coefficients. We present simulations to show some examples of these cycles. Fluctuations in aggregate income represent a business cycle in our model while fluctuations in the Gini coefficients represent changes in inequality. Furthermore, both GDP and inequality fluctuate due to the same underlying factor, which is the behavior of cognitive workers. This explains one of the empirical observations we referred to, which is the mutual causality between GDP and inequality.

Our simulations lead to an interesting observation. The aggregate income of physical workers is procyclical in our model, whereas the aggregate income of cognitive workers is countercyclical. This difference arises due to the different ways in which the behavior of cognitive workers affect the aggregate income of the two groups of workers. When the proportion of cognitive workers producing higher qualities of capital is high, then the average quality of capital that physical workers get to use is itself high. This, in turn, increases both aggregate output, which is produced by physical workers, and the income of such workers. Thus, the aggregate income of physical workers is procyclical. On the other hand, when cognitive workers produce higher qualities of capital, their payoffs are low since the price of capital is fixed while the production cost of capital is increasing in quality. Therefore, the income of cognitive workers is countercyclical. These observations have the added implication that the share of cognitive workers in aggregate income is countercyclical while the share of physical workers is procyclical. Hence, our analysis suggests that different types of workers may be affected differently in a business cycle.[6]

Recall our interpretation that the quality of capital determines the productivity of physical workers in our model. We can use this interpretation to relate our observation that the income of such workers is procyclical to empirical findings that real wages in an economy are generally procyclical Solon et al. (1994).[7] Our analysis suggests a reason why real wages may be procyclical. Both GDP and the aggregate income of physical workers in our model is high when the level of productivity is high, which, in turn, happens when cognitive workers produce a higher quality of capital.

The other empirical observation we sought to explain is the relationship between inequality and the business cycle. Our simulations suggest that the Gini coefficients of both types of workers are inversely related to the aggregate income in the economy. Hence, inequality in both populations is countercyclical. This is in line with the general message of Krueger et al. (2010) and Maestri and Roventini (2012). Once again, we may relate this conclusion to the average quality of capital produced by cognitive workers. Our simulations suggest that when this is high, the dispersion (measured by the standard deviation) in the quality of capital produced by cognitive

[6]We note in this context that Hoynes et al. (2012) find evidence that recessions affect different social groups differently. If we interpret the two groups of workers in our model as two different social groups, then our conclusion matches the empirical findings of Hoynes et al. (2012).

[7]Of course, there are also empirical studies that find that real wages in certain sectors of the economy are countercyclical. See, for example, Swanson (2004).

workers and, hence, used by physical workers is also low. But this would then imply that the dispersion in income levels in both groups will also below. Therefore, high GDP is accompanied by low dispersion of income in both population groups leading to the countercyclical behavior of inequality. If we combine this observation with the countercyclical nature of cognitive workers' income and procyclical nature of physical workers' income, we obtain an additional implication. When inequality is low among cognitive workers, it means a greater prevalence of low income among such workers whereas low inequality among physical workers imply that high income is more common among such workers.[8]

To complement our analysis in the time domain, we also examine the dynamics of our model through spectral analysis in the frequency domain. The purpose of this exercise is to understand how cyclical behavior of aggregate income and inequality in our model may change if we change the exogenously fixed distribution of search behavior of physical workers. We observe that if this distribution changes, the corresponding limit cycles also change both qualitatively and quantitatively. The economic implication is that both the nature of the business cycles as well as the spread in income change substantially across the search distribution.

To summarize, this paper provides an explanation of two important aspects of the relationship between GDP and inequality; namely, the common origin of fluctuations in GDP and inequality and the countercyclical nature of inequality over a business cycle. As we have explicitly noted earlier, the basic model in this paper is a simplification of the model analyzed in Chakrabarti and Lahkar (2017). The present work makes three contributions. First, this paper presents a dynamic relationship between GDP and income spread which was not evident in our earlier paper. One reason why this is important is that the countercyclical nature of income risk has been incorporated in a variety of business cycle models.[9] Our model provides an understanding of how such a countercyclicality may arise. Second, we provide a methodological innovation by introducing the tools of spectral analysis to analyze limit cycles in evolutionary game theory and showing that such tools can credibly differentiate between limit cycles of different frequencies. Third, in combination with our earlier work, this paper provides a common evolutionary game theoretic framework to explain the cyclical nature of a variety of important macroeconomic variables like productivity dispersion, aggregate income, and inequality.

The rest of the paper is as follows. Section 2 presents the basic model, characterizes Nash equilibria, and defines aggregate income and inequality in the model. Section 3 explains the instability of equilibria in our model under the logit dynamic. In Sect. 4, we present our simulations of the relationship between the business cycle

---

[8]We note here that the scope of this paper is restricted to the short-run relationship between GDP and inequality over the business cycle. We do not address the long-term issue of the relationship between economic growth and inequality, as emphasized by, for example, Piketty and Saez (2014).

[9]See, for example, Krebs (2007), Edmond and Veldkamp (2009), Braun and Nakajima (2012), Chen et al. (2015).

and inequality and provide an explanation of the relationship. Section 5 presents certain additional simulations using spectral analysis, where we change the search behavior of physical workers. Section 6 concludes.

2 The Model: Cognitive and Physical Workers

We now propose the basic model of this paper. This model is based on Chakrabarti and Lahkar (2017), which itself is an extension and adaptation of the price dispersion model of Burdett and Judd (1983) to the analysis of productivity dispersion. The original model of Burdett and Judd (1983) is in the setting of continuous strategy sets. However, the evolutionary game- theoretic analysis of a game-theoretic model with continuous strategy sets is technically difficult. Hence, as in Chakrabarti and Lahkar (2017), we consider a finite strategy adaptation of the Burdett and Judd (1983) model to our question of interest.

We consider a population of mass 1 consisting of a continuum of cognitive workers. A cognitive worker chooses to create capital of quality $t_i = \frac{i}{n}$ from the set $S^n = \left\{0, \frac{1}{n}, \frac{2}{n}, \ldots, \frac{n-1}{n}, 1\right\}$, where n is a strictly positive integer. The cost of producing capital of quality t_i is t_i.[10] We denote the set of states in the population of cognitive workers as $\Delta^n = \{x \in \mathbf{R}_+^{n+1} : \sum_{i=0}^n x_i = 1\}$. Given the state x, the scalar x_i represents the proportion of cognitive workers producing capital of quality $t_i \in S^n$.

There also exists a continuum of physical workers of mass 1. Cognitive workers sell capital to physical workers who then use the capital to produce output. Physical workers in our model are not strategic agents. Instead, we fix the behavior of physical workers exogenously in the following way. Each physical worker observes either one or two cognitive workers and then buys one unit of capital at a price of 1 from that cognitive worker whose quality is highest among the ones observed. In case a physical worker observes two cognitive workers and both the cognitive workers produce the same quality of technology, then we assume that the physical worker buys capital from either of the two cognitive workers with equal probability. Capital of quality t_i generates an output of $q(t_i)$, where we assume that $q : [0, 1] \rightarrow \mathbf{R}_+$ is a strictly increasing function with $q(0) \geq 1$.[11] We denote by $y_1 \in [0, 1]$ the proportion of physical workers who search just once. The remaining proportion of physical workers, $y_2 = 1 - y_1$, search twice.[12] As noted in the Introduction, we can interpret

---

[10] We can generalize the cost of producing capital to a strictly increasing cost function $c(t_i)$ without introducing any significant change in our conclusions.

[11] The assumption $q(0) \geq 1$ will ensure that the income of a physical worker will always be non-negative, which will allow us to calculate the Gini coefficient for such workers.

[12] The key difference between this particular adaptation of the Burdett and Judd (1983) model and the adaptation in Chakrabarti and Lahkar (2017) is that here, the search distribution $\{y_1, y_2\}$ is fixed exogenously. In contrast, the search behavior arises endogenously in our earlier paper. As noted in footnote 5, we introduce this difference in order to focus on the more important variable, x, which describes the behavior of cognitive workers.

physical workers as providing one unit of labor. The quality of capital then determines their productivity through the function $q(t)$.

Given the exogenous distribution $y = \{y_1, y_2\}$ of the behavior of physical workers, the payoff of a cognitive worker who chooses to produce capital of quality t_i at the population state x is

$$\pi_i(x; y) = (1 - t_i)\left[y_1 + 2y_2\left(\frac{x_i}{2} + \sum_{j<i} x_j\right)\right]. \tag{1}$$

To understand this payoff function, consider a physical worker who observes two cognitive workers, including this particular cognitive worker who is producing capital of quality t_i. Then, $\sum_{j<i} x_j$ represents the probability that the other cognitive worker is producing capital of a lower quality. In that case, this cognitive worker is able to sell his capital to the physical worker with certainty. On the other hand, x_i is the probability that the other cognitive worker also produces capital of quality t_i, in which case this particular cognitive worker is able to sell his capital with probability $\frac{1}{2}$. Therefore, $\frac{x_i}{2} + \sum_{j<i} x_j$ is the probability with which this particular cognitive worker is able to sell his capital to a physical worker who observes two cognitive workers. The expression inside the square bracket then represents the expected number of sales for a cognitive worker choosing strategy t_i while $1 - t_i$ is the profit per unit of sales. The payoff function (1), therefore, implies that a lower quality capital improves profit per unit of sale for the cognitive worker, but reduces the expected number of sales.

We describe the population game defined by (1) as π. A Nash equilibrium of π is a population state $x^* \in \Delta^n$ such that given y, if $x_i^* > 0$, then $i \in \text{argmax}_{j \in S^n} \pi_j(x; y)$. Thus, at a Nash equilibrium any strategy i that is used by a positive mass of cognitive workers must be a payoff maximizer. A pure equilibrium is then a Nash equilibrium in which all cognitive workers use the same strategy. A mixed equilibrium is a Nash equilibrium in which at least two different strategies are in use in the population of cognitive workers.

The following proposition characterizes the set of Nash equilibria of π. The proof of the result is in Appendix 7.1.

Proposition 2.1 *Consider the population game π defined by (1).*

1. *Let $y_1 = 1$. Then, for all n, the unique Nash equilibrium is the pure equilibrium in which $x_0 = 1$.*
2. *Let $y_1 = 0$. For any n, there always exist two Nash equilibria of π; one in which $x_n = 1$ and the other in which $x_{n-1} = 1$. If $n \geq 2$, then the state in which $x_{n-2} = 1$ is also a Nash equilibrium. Moreover, there does not exist any other Nash equilibria for any n. Hence, in this case, all possible Nash equilibria are pure equilibria.*
3. *Let $0 < y_1 < 1$. Then, for n sufficiently large, all Nash equilibria of π are mixed equilibria.*

To prove this proposition, we rely on a change of variable which allows us to apply results from Lahkar (2011). Part 1 follows because if $y_1 = 1$, then $t_i = 0$ becomes the dominant strategy for all cognitive workers. If $y_1 = 0$, then all physical workers search twice. The game then becomes akin to a Bertrand competition model where each cognitive worker produces close to the highest possible quality of capital and, hence, earns close to zero profit. Part 3 follows because if n is large, then at any pure state, a cognitive worker can increase quality slightly and, thereby, increase market share significantly. This would enable that worker to increase payoff.

2.1 Aggregate Income and Inequality

We now define aggregate income levels of both type of workers at a given population state x and exogenous distribution y. For this purpose, it will be useful to define

$$h_i(x;y) = y_1 x_i + 2y_2 x_i \left(\frac{x_i}{2} + \sum_{j<i} x_j \right). \tag{2}$$

Note that $\sum_i h_i(x;y) = y_1 \sum_i x_i + y_2(\sum_i x_i)^2 = y_1 + y_2 = 1$. Hence, $h(x;y)$ is a probability distribution. In fact, $h_i(x;y)$ represents the proportion of physical workers who use capital of quality t_i in their production process. This follows because $y_1 x_i$ is the mass of physical workers who search once and use capital t_i while $2y_2 x_i \left(\frac{x_i}{2} + \sum_{j<i} x_j\right)$ is the mass of such workers who search twice and use capital t_i.

The aggregate payoff of cognitive workers is then

$$\begin{aligned}
CY(x;y) &= \sum_{i=0}^{n} x_i \pi_i(x;y) \\
&= \sum_{i=0}^{n} x_i \left(y_1 + 2y_2 \left(\frac{x_i}{2} + \sum_{j<i} x_j \right) \right) - \sum_{i=0}^{n} x_i t_i \left(y_1 + 2y_2 \left(\frac{x_i}{2} + \sum_{j<i} x_j \right) \right) \\
&= \sum_{i=0}^{n} h_i(x;y) - \sum_{i=0}^{n} h_i(x;y) t_i \\
&= 1 - C(x;y),
\end{aligned} \tag{3}$$

where $C(x;y)$ is the aggregate cost that cognitive workers incur in producing capital.

To obtain the aggregate income of physical workers, note that since capital of quality t_i generates output $q(t_i)$, $h_i(x;y)$ is also the share of physical workers who produce this output level $q(t_i)$. After paying 1 as the price of this capital, the income of such a physical worker is $q(t_i) - 1$. Hence, the aggregate income of physical workers is

$$\begin{aligned} PY(x;y) &= \sum_{i=0}^{n} h_i(x;y)(q(t_i)-1) \\ &= \sum_{i=0}^{n} h_i(x;y)q(t_i) - \sum_{i=0}^{n} h_i(x;y) \\ &= Q(x;y) - 1, \end{aligned} \tag{4}$$

where $Q(x; y)$ is the aggregate output produced by physical workers. Hence, aggregate income in the economy is

$$\begin{aligned} AY(x;y) &= CY(x;y) + PY(x;y) \\ &= Q(x;y) - C(x;y) \\ &= \sum_{i=0}^{n} h_i(x,y)(q(t_i) - t_i), \end{aligned} \tag{5}$$

where (5) arises from the definitions of $C(x; y)$ and $Q(x; y)$, respectively, in (3) and (4). We interpret $AY(x; y)$ as GDP in our model.

The definitions of the three aggregate income levels in (3), (4), and (5) allow us to measure the share of GDP accruing to cognitive workers and physical workers. We denote these shares as $SC(x; y)$ and $SH(x; y)$, respectively, and define them as

$$SC(x;y) = \frac{CY(x;y)}{AY(x;y)} \qquad SH(x;y) = \frac{PY(x;y)}{AY(x;y)}. \tag{6}$$

Clearly, $SH(x; y) = 1 - SC(x; y)$.

We now consider inequality within the two groups. For this purpose, we compute the Gini coefficient for each group separately. We compute the Gini coefficient for a particular group as follows. Let z denote the income level of a worker in a particular group and let $f(z)$ denote the proportion of workers in that group who have income level z. The standard definition of the Gini coefficient requires that all possible income levels be nonnegative. For cognitive workers, an income level z is the payoff $\pi_i(x; y) \geq 0$ defined in (1) and the proportion of such workers with that income level is x_i. For physical workers, a particular income level is $q(t_i) - 1$. Our assumption about $q(t)$ ensure any such income level is nonnegative (see footnote 11 The proportion of physical workers who earn $q(t_i) - 1$ is the share of workers who use capital t_i is their production process, which is given by $h_i(x; y)$ defined in (2). Note that for each population, we have $n + 1$ different income levels $\{z_0, \ldots, z_n\}$ corresponding to the $n + 1$ different qualities of capital $\{t_0, \ldots, t_n\}$.

We sort z in ascending order so that we have $z_i \leq z_{i+1}$ for $i \leq n$. Then we compute the vector $S = \{S_{-1}, S_0, \ldots, S_n\} \in \mathbf{R}_+^{n+2}$ in which $S_{-1} = 0$ and $S_i = \sum_{k=0}^{i} f(z_k)z_k$ for $i \in \{0, 1, \ldots, n\}$. The Gini coefficient for the income series represented by z is then given by

$$G = 1 - \frac{\sum_{i=0}^{n} f(z_i)(S_{i-1} + S_i)}{S_n}. \tag{7}$$

The interpretation of the Gini coefficient is well known. The Gini coefficient takes a value in [0, 1] with larger values of the coefficient signifying greater inequality. Applying (7), we obtain two Gini coefficients, one for cognitive workers which we denote as $Gini_C$ and the other for physical workers which we denote as $Gini_P$.[13]

We note that all three definitions of aggregate income given by (3), (4), and (5), the two income shares given by (6) as well as both the Gini coefficients are functions of the state variable x. The dependence of $Gini_C$ on the state variable arises through the definition of income level $\pi_i(x; y)$ and the proportion x_i of such workers who receive this income level. For $Gini_P$, this dependence arises from the definition of the proportion $h_i(x; y)$. These considerations will be of importance when we discuss the dynamics of these variables. The dynamics of x will also determine the dynamics of aggregate income and inequality.

3 Logit Dynamic

For the evolutionary analysis of our model, we apply the logit dynamic to the population game π defined by (1). The logit dynamic is one of the standard dynamics in evolutionary game theory.[14] To define this dynamic, we fix the the distribution $y = \{y_1, 1 - y_1\}$. We then postulate that the resulting population game π is played repeatedly over time, which we denote as τ, by the same population of cognitive workers.[15] Cognitive workers behave myopically and update their strategies at random strategy revision opportunities according to the *logit best response*. The logit best response depends upon a perturbation factor $\eta > 0$ and is defined as the function $L_\eta : \Delta^n \to \text{int}\Delta^n$ whose value for strategy $i \in S^n$ takes the form

$$L_{\eta,i}(x; y) = \frac{\exp(\eta^{-1}\pi_i(x; y))}{\sum_{j \in S^n} \exp(\eta^{-1}\pi_j(x; y))}. \tag{8}$$

The fraction $L_{\eta,i}(x; y) \in (0, 1)$ describes the probability with which a cognitive worker, when faced with the chance to change strategy at state x, chooses to play

[13] It is not sensible to compute a single Gini coefficient for both populations combined in our model. This is because the definition of income in the two populations are not comparable and depend upon exogenous factors like the production function $q(t)$. If we arbitrarily change $q(t)$, that will also affect the income of physical workers and, hence, any combined Gini coefficient. Nevertheless, as we discuss in Sect. 4, we can use the two separate Gini coefficients to consider the overall relationship between GDP and inequality in our model.

[14] We refer the reader to Sandholm (2010) for a more detailed analysis of the logit dynamic.

[15] The population of physical workers also remain the same over time. In the present case, physical workers are not strategic agents. Hence, their behavior is not important in describing the logit dynamic.

strategy $i \in S^n$. Intuitively, the logit best response is an approximation of the best response to the population state x as for η small, $L_\eta(x; y)$ puts almost the entire probability mass on the set of best responses to x. Unlike the best response, the logit best response is uniquely defined for all x and, hence, is technically more convenient to apply than the best response.

Under the logit best response with parameter η, the population state x in the population game π evolves according to the logit(η) dynamic, which is an ODE system that takes the form

$$\dot{x}_i = L_{\eta,i}(x; y) - x_i. \tag{9}$$

For a fixed y, the logit dynamic generates a unique solution trajectory $\{x(\tau)\}_{\tau\geq 0}$ in Δ^n from every initial state $x(0) \in \Delta^n$. A rest point of the logit dynamic is called a *logit equilibrium*. We denote a logit equilibrium as $\tilde{x}_\eta$. It is well known that for η small, every logit equilibrium $\tilde{x}_\eta$ is close to a Nash equilibrium x^* of the underlying population game (McKelvey and Palfrey 1995).

In any application of the logit dynamic, the main issue of interest is whether a logit equilibrium is stable under the dynamic, particularly for small values of η. If not, then it will imply that the population will diverge from that logit equilibrium. Intuitively, it means that for small η, the Nash equilibrium which is approximated by that logit equilibrium is not evolutionarily robust. If we can establish that all logit equilibria are unstable, then that will mean that the population, instead of converging to an equilibrium, will converge to more complex attractors like a limit cycle.

Results from Chakrabarti and Lahkar (2017) show that this will indeed be the case for the population game π defined by (1). The basis of this conclusion is that π satisfies a condition called positive definiteness.[16] To introduce this condition, define the tangent space of Δ^n as $T\Delta^n = \{z \in \mathbf{R}^{n+1} : \sum_{i\in S^n} z_i = 0\}$. Then, given y, the population game π defined by (1) satisfies positive definiteness if for all $z \in T\Delta^n \setminus \mathbf{0}$ and all $x \in \Delta^n$,

$$zD\pi(x; y)z > 0.$$

Positive definiteness will imply that any logit equilibrium that approximates a mixed Nash equilibrium of π for small values of η will be unstable under the logit dynamic (9). We state this result in the following proposition. We refer the reader to Chakrabarti and Lahkar (2017) for details of the proof, which follows from a local stability analysis of rest points of the logit dynamic.

Proposition 3.1 *Let the distribution $y = \{y_1, 1 - y_1\}$ be such that $0 < y_1 < 1$. Then, the population game π defined by (1) is positive definite. Hence, for η sufficiently small, any logit equilibrium $\tilde{x}_\eta$ of π that approximates a mixed equilibrium x^* of π is unstable under the logit(η) dynamic (9).*

---

[16]Positive definiteness is the reverse of the condition called negative definiteness, which usually leads to stability of Nash equilibrium under evolutionary dynamic. See, for example, Hofbauer and Sandholm (2009) for further details of these conditions.

Proof. Follows from the proofs of Lemmas A.6 and A.7 in Chakrabarti and Lahkar (2017). ■

The relevant proofs in our earlier paper make it clear that once we fix y such that $0 < y_1 < 1$, positive definiteness in our model arises entirely from the payoff of cognitive workers as defined in (1).[17] It is in this sense that we argue that the role of cognitive workers is more important in determining positive definiteness. Hence, fixing y allows us to focus on the more important determinant of positive definiteness. This is important because, as we elaborate below, it is positive definiteness that will generate cyclical behavior in our variables of interest.

Intuitively, positive definiteness implies that if a small group of agents switch strategies, then the strategy which the agents have adopted becomes relatively better off than the strategy that they have left. This happens at any population state. But at a mixed equilibrium in which, by definition, all strategies in the support of the equilibrium have equal payoff, relative improvement translates into absolute improvement. Hence, at a mixed equilibrium, positive definiteness will make "mutants" absolutely better off than "non-mutants", which renders such an equilibrium unstable.[18]

For η sufficiently small, any logit equilibrium will be close to a Nash equilibrium. Hence, positive definiteness implies that any such logit equilibrium that approximates a mixed Nash equilibrium will also be unstable under the logit dynamic. This leads to the conclusion in Proposition 3.1. Also recall from Proposition 2.1 that if $y_1 \in (0, 1)$, then for n sufficiently large, all Nash equilibria of π are mixed. Hence, for such n, every logit equilibrium of π will be close to a mixed Nash equilibrium and, therefore, will be unstable under the logit dynamic if η is sufficiently small. Since logit equilibria are the only possible rest points of the logit dynamic, we conclude that if $y_1 \in (0, 1)$, n is sufficiently large and η is sufficiently small, then any solution trajectory of the logit dynamic $x(t)$ in the population game π defined by (1) will converge to a complex attractor like a limit cycle in Δ^n.

4 Dynamics of Income and Inequality

We have defined the logit dynamic (9) for the population game π defined by (1). We now apply this dynamic to consider the evolution of income and inequality in our model. The purpose of this exercise is to explain the two empirical observations noted in the Introduction—the common origin of fluctuations in GDP and inequality and the countercyclical nature of inequality. As noted at the end of Sect. 2, both aggregate income and inequality in our model are functions of the state variable x. Hence, the evolution of these variables will depend upon the evolution of x. Since we are interested in the relationship between GDP and inequality over a business cycle,

[17] See footnote 5 in this context.

[18] We refer the reader to Sect. 5.1 in Chakrabarti and Lahkar (2017) for a broader discussion of this point. More formally, positive definiteness implies that a mixed equilibrium will violate the well known condition of being an evolutionary stable state (ESS) (Maynard Smith and Price GR 1973).

we need to consider a situation where x itself converges to a limit cycle under the logit dynamic. For a nonlinear dynamic like the logit dynamic, it is not feasible to establish the existence of any such limit cycle analytically. Only simulations provide a feasible way to obtain such a limit cycle.

It is evident from (8) and (9) that the logit dynamic in π depends upon the exogenous distribution $y = \{y_1, 1 - y_1\}$. The two extreme cases of $y_1 = 1$ and $y_1 = 0$ are not very interesting from the point of view of obtaining limit cycles. As we know from Proposition 2.1, the only Nash equilibria in these two cases are pure Nash equilibria. Logit equilibria corresponding to such pure equilibria are generally locally stable. Hence, in such a case, limit cycles may not exist. Some simulations we ran with appropriate parameter values failed to reveal any limit cycle.

The more interesting case is where $y_1 \in (0, 1)$. We have noted in the last paragraph of Sect. 3 for any such y_1, all possible logit equilibria of π will be unstable under the logit dynamic for η sufficiently small and n sufficiently large. This will give rise to complex dynamics of income and inequality levels, which we now explore through simulations.

For our first set of simulations, we fix $y = \{y_1, y_2\} = \{0.5, 0.5\}$ and $n = 5$.[19] The strategy set is, therefore, $S^5 = \{0, 0.2, 0.4.0.6, 0.8, 1\}$. We let the production function of physical workers be $q(t) = 1 + 5\sqrt{t}$ and the cost function of cognitive worker by $c(t) = t$. We then apply the logit (0.001) dynamic to π from the the initial state $x(0) = (\frac{1}{6}, \frac{1}{6}, \ldots, \frac{1}{6})$. For such small η, the population state x converges to a limit cycle, which results in a cyclical pattern of fluctuations in aggregate income levels and inequality. We first provide a description of these fluctuations followed by a discussion in Sect. 4.1.[20]

The top-left panel of Fig. 1 presents the trajectories of $CY(x; y)$, $PY(x; y)$, and $AY(x; y)$ defined in (3), (4), and (5), respectively, when x evolves as described in the previous paragraph. These are, respectively, the aggregate payoff of cognitive workers, aggregate income of physical workers, and the aggregate income in the economy. The absolute values of these income aggregates are not of much significance as they are determined by exogenous factors like the price of capital, the distribution y, and the production and cost functions. Instead, we are interested in the relationships between these trajectories establish among the aggregate income measures and with the inequality measures.

We notice that all three income levels display cyclical fluctuations. Recall that AY is GDP in our model. Hence, we interpret the fluctuations in AY as a business cycle in our model. We also note that PY is broadly procyclical in the sense that it exhibits a similar pattern of fluctuation as AY. On the other hand, CY is countercyclical in the sense that it rises when AY falls and falls when AY rises. This observation also explains the pattern exhibited by the trajectory of $SC(x; y)$ in the bottom-right

---

[19] We choose $n = 5$ because while it is small enough to be computationally tractable, it turns out to be sufficiently large to generate cyclical behavior in the population game π under the parameters we fix.

[20] We do not discuss the limit cycle of x in the state space Δ^n in this paper. Chakrabarti and Lahkar (2017) discuss such limit cycles in detail. Here, our interest is the fluctuations in aggregate income and inequality.

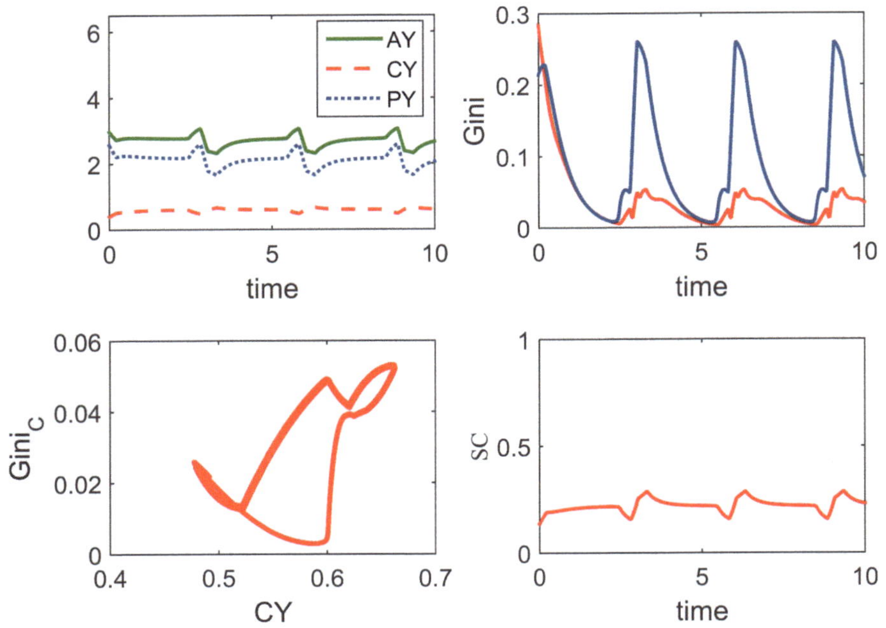

Fig. 1 Cyclical behavior of income and inequality aggregates when $n = 5$ and $y = \{0.5, 0.5\}$. The production function is $q(t) = 1 + 5\sqrt{t}$ and cost function is $c(t) = t$. The initial state is $x(0) = (\frac{1}{6}, \ldots, \frac{1}{6})$ and the perturbation parameter in the logit dynamic is $\eta = 0.001$. In the top-right panel, $Gini_C$ is in red and $Gini_P$ is in blue

panel of Fig. 1. This, as defined in (6), is the share of aggregate income accruing to cognitive workers. Thus, $SC(x; y)$ is countercyclical which implies that the share of GDP received by physical workers ($SP(x; y)$) is procyclical. Our simulations, therefore, suggest that the effect of business cycles is not the same for both types of workers. While physical workers suffer during a recession, cognitive workers actually benefit.

We now consider the trajectory of inequality within the two groups, which is measured by two Gini coefficients, $Gini_C$ and $Gini_P$, defined as in (7). We present these trajectories in the top-right panel of Fig. 1. The trajectory of $Gini_C$ is in red and that of $Gini_P$ is in blue. Clearly, the two Gini coefficients exhibit a cyclical pattern of evolution.

It is evident from Fig. 1 that aggregate income and the two Gini coefficients are fluctuating together. Thus, inequality in both groups of workers fluctuates along with GDP over the business cycle. The reason is obvious. Aggregate income and the two Gini coefficients are functions of the state variable x. Hence, the common origin of fluctuations in GDP and inequality in our model is the limit cycle of the state variable. This explains one of the empirical observations we sought to explain–GDP and inequality appear to share a mutual causal relationship which suggests they

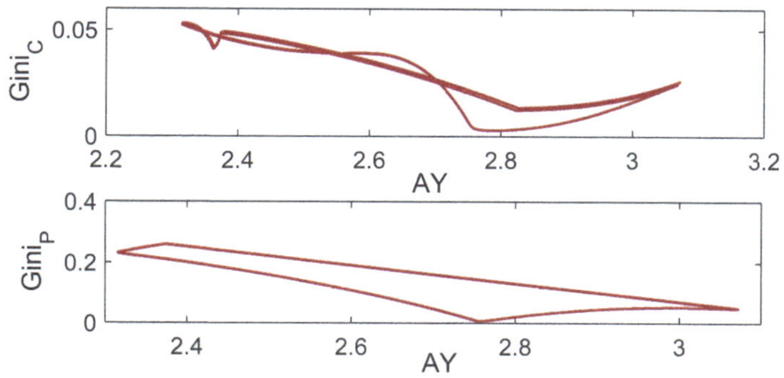

Fig. 2 Relationship between AY and $Gini_C$ (top panel) and AY and $Gini_P$ (bottom panel). The values of AY, $Gini_C$, and $Gini_P$ are as originally simulated in Fig. 1

arise from the same underlying economic process. In our model, this process is the dynamics of the state variable.

The second empirical feature we sought to explain is the countercyclical relationship between GDP and inequality over a business cycle. For this, we consider the relationship between AY and the two Gini coefficients. We have noted that it is not sensible in our model to calculate a single Gini coefficient for both populations combined. (see footnote 13). However, if it is the case that the two Gini coefficients exhibit a similar relationship with AY, then we can arrive at some conclusion about the relationship between AY and overall inequality.

To arrive at any such relationship, we note two features of interest in the cyclical behavior of the Gini coefficients.[21] First, as can be seen from the top right panel of Fig. 1, both $Gini_C$ and $Gini_P$ move synchronously. When inequality is high in one population, it is also high in the other population. Second, there is a positive relationship between CY and $Gini_C$. This is shown in the bottom-left panel of Fig. 1. Inequality is high in the population of cognitive workers when aggregate income in that population is high.

These two patterns shown by the Gini coefficients allow us to deduce the relationship between aggregate income AY and the overall pattern of inequality in the economy. We show these relationships in Fig. 2, where the three series of AY, $Gini_C$, and $Gini_P$ as originally simulated in Fig. 1. Clearly, both $Gini_C$ and $Gini_P$ has an inverse relationship with AY.

Both the inverse relationships in Fig. 2, arise from the observations we have made from Fig. 1. The negative relationship between AY and $Gini_C$ arises because AY and CY are themselves inversely related while CY and $Gini_C$ are directly related. The synchronous behavior of $Gini_C$ and $Gini_P$ then implies a similar inverse relationship between AY and $Gini_P$. We, therefore, conclude that under the conditions in which

[21] We also observe from Fig. 1 that the fluctuation in inequality among physical workers is much higher than that among cognitive workers. But as in the case of the income aggregates, the absolute values of the Gini coefficients are not of much significance as they arise from various exogenous factors in our model.

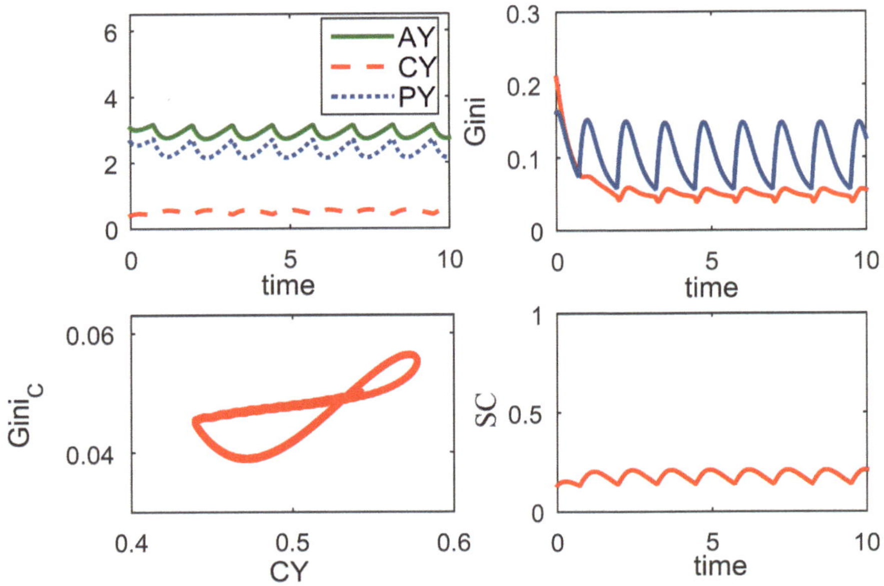

Fig. 3 Cyclical behavior of income and inequality aggregates when $n = 100$ and $y = \{0.5, 0.5\}$. The production function is $q(t) = 1 + 5\sqrt{t}$ and cost function is $c(t) = t$. The initial state is $x(0) = (\frac{1}{101}, \ldots, \frac{1}{101})$ and the perturbation parameter in the logit dynamic is $\eta = 0.001$. In the top-right panel, $Gini_C$ is in red and $Gini_P$ is in blue

we have simulated Figs. 1 and 2, inequality is a countercyclical variable. When GDP is high, inequality is low among both types of workers and, hence, in the economy as a whole. In our view, this inverse relationship between aggregate income and overall inequality is perhaps the most important insight of our analysis in this paper. One major implication of this insight is that the ill-effects of a recession includes not just lower GDP but also higher inequality. We provide a more detailed discussion of this relationship between GDP and inequality in Sect. 4.1.

To check the robustness of the patterns we observe in Fig. 1, we extend our simulations in two ways. First, we enlarge the strategy set of π to $S^{100} = \{0, 0.01, \ldots, 0.99, 1\}$ while keeping y, the production function, and the cost function unchanged. We apply the logit (0.001) dynamic to π from the initial state $\left(\frac{1}{101}, \ldots, \frac{1}{101}\right)$. We present the results of this simulation in Fig. 3. Clearly, the overall patterns of aggregate income and inequality are the same as in Fig. 1. We observe that CY is procyclical (in relation to AY) while PY is countercyclical. Both the Gini coefficients move synchronously while CY and $Gini_C$ share a positive relationship. This figure also provides an insight into the asymptotic behavior of aggregate income and inequality measures as the strategy set S^n becomes large. Figure 4 plots the relationships between AY and the two Gini coefficients that arise from Fig. 3.

For our second extension, we retain the same strategy set S^5 as in Fig. 1. We also keep y, the initial state, and the cost function unchanged from Fig. 1. However, we

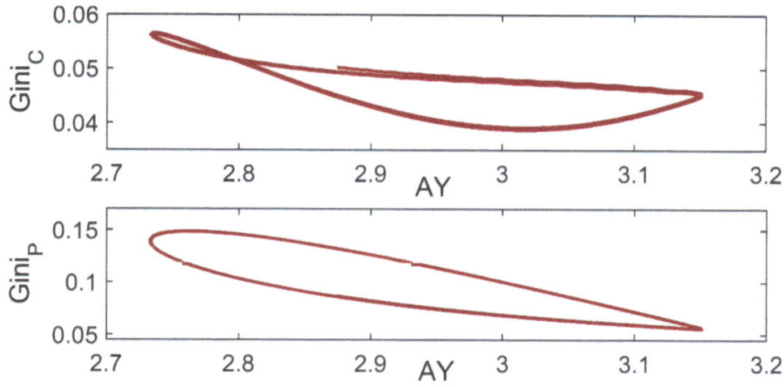

Fig. 4 Relationship between AY and $Gini_C$ (top panel) and AY and $Gini_P$ (bottom panel). The values of AY, $Gini_C$, and $Gini_P$ are as originally simulated in Fig. 3

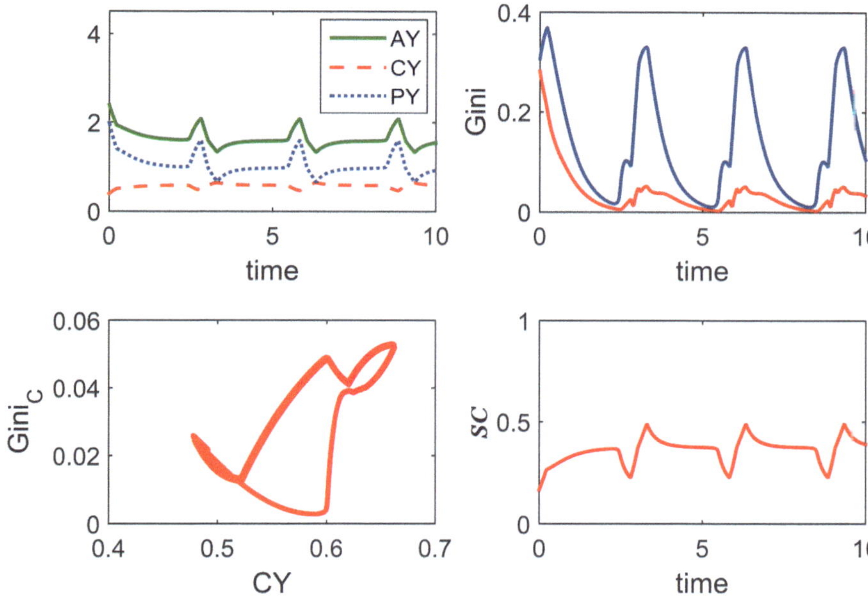

Fig. 5 Cyclical behavior of income and inequality aggregates when $n = 5$ and $y = \{0.5, 0.5\}$. The production function is $q(t) = 1 + 5t$ and cost function is $c(t) = t$. The initial state is $x(0) = (\frac{1}{6}, \ldots, \frac{1}{6})$ and the perturbation parameter in the logit dynamic is $\eta = 0.001$. In the top-right panel, $Gini_C$ is in red and $Gini_P$ is in blue

change the production function to $q(t) = 1 + 5t$. We show the resulting simulations under the logit (0.001) dynamic in Fig. 5. Once again, the behavior of the income and inequality measures is broadly similar to that in Figs. 1 and 3.

Our simulations, therefore, account for the two empirical observations we sought to explain—the common source of fluctuations in GDP and inequality and the countercyclical nature of inequality. We now provide a more detailed discussion of these cyclical fluctuations to have a better understanding of their origins. Since the patterns we observe in Figs. 1, 3, and 5 are broadly similar, it suffices to focus on Fig. 1 for the purpose of this discussion.

4.1 Discussion

Recall that the fundamental reason behind why cycles arise in our model is the characteristic of positive definiteness (Proposition 3.1). Positive definiteness causes a limit cycle of the state variable $x(t)$ which, as noted earlier, is the common origin of the cyclical fluctuations in income and inequality in our model.

To understand the relationships between different cycles in the income and inequality measures, we first focus on the three measures of aggregate income in our model. Recall from Fig. 1 that aggregate income (AY) and the income of physical workers (PY) exhibit a positive relationship while AY and the income of cognitive workers (CY) are negatively related. To understand these relationships, note from (4) that PY increases when the aggregate output $Q(x, y) = \sum_{i \in S^n} h_i(x; y)q(t_i)$ increases. Aggregate output, in turn, will increase if $h_i(x; y)$ is higher for relatively larger values of t_i, i.e., relatively more physical workers are using capital of higher quality. Hence, PY increases when a larger share of physical workers use better quality capital. Notice also from (5) that AY depends upon the difference $q(t) - c(t)$, which, for our specification of production and cost functions, is increasing in t. Therefore, when the share of physical workers using higher quality capital increases, then the aggregate value of this difference $\sum_{i \in S^n} h_i(x, y)(q(t_i) - c(t_i))$, which, by (5), is AY, will also increase. Hence, both PY and AY will increase when $h_i(x, y)$ is higher for higher values of t_i. On the other hand, it is evident from (3) that CY will decline when $h_i(x, y)$ increases for higher values of t_i because $c(t)$ is increasing in t. Thus, when larger shares of physical workers use higher quality capital, PY and AY increases while CY declines.

The behavior of $h(x, y)$ is, of course, dependent upon the behavior of x, which is the only state variable in our model. Consider the third panel of Fig. 6, which depicts the relationship between $E_x(t) = \sum_{i \in S^n} x_i t_i$ and $E_h(t) = \sum_{i \in S^n} h_i(x, y)t_i$ for the specific case $n = 5$. We interpret these values as, respectively, the average quality of capital and the aggregate quality of capital in the economy. There is clearly a positive relationship between these values. This is to be expected since $E_h(t)$ will be high when $h_i(x; y)$ is high for higher levels of t_i and $E_x(t)$ will be high when x_i is high for higher levels of t_i. But the share of physical workers who use higher quality levels of capital $(h_i(x, y))$ can be high only if the share of cognitive workers who produce higher quality capital (x_i) is itself high. The discussion in the previous paragraph then implies that when $E_x(t)$ and, hence, $E_h(t)$ is high, PY and AY are also high

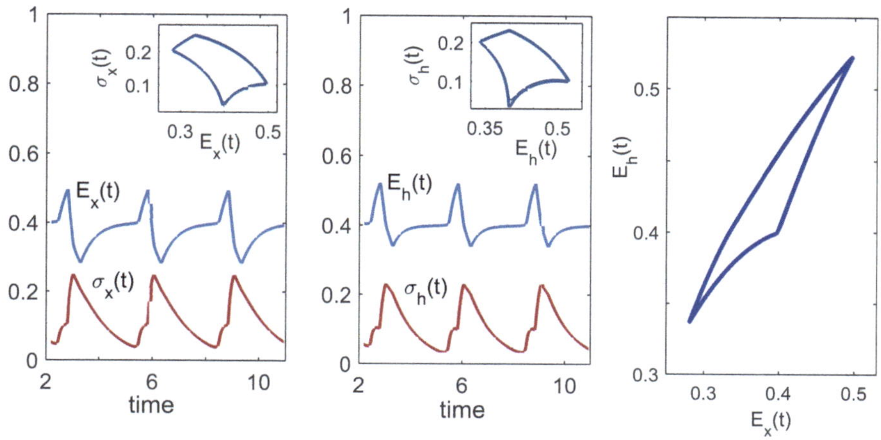

Fig. 6 Evolution of expected values and standard deviations of the quality of capital t with respect to the probability distributions x and $h(x; y)$. The strategy set is $S^5 = \{0, 0.2, 0.4, 0.6, 0.8, 1\}$, $y = \{0.5, 0.5\}$, the initial state is $x(0) = \{\frac{1}{6}, \dots, \frac{1}{6}\}$ and the dynamic is the logit (0.001) dynamic. *Left panel*: $E_x(t) = \sum_{i \in S^5} x_i t_i$, $\sigma_x(t) = \sqrt{\sum_{i \in S^5} (t_i - E_x(t))^2 x_i}$. *Middle panel*: Analogous measures with respect to the probability distribution h. *Right panel*: Relationship between $E_x(t)$ and $E_h(t)$ over time

while CY is low.[22] It then immediately follows that when $E_x(t)$ is high, $SC(x; y)$ is low while $SP(x; y)$ is high. Thus, $SC(x; y)$ is countercyclical while $SP(x; y)$ is procyclical. We, therefore, conclude that the most important determinant of income aggregates and the share of GDP accruing to the two groups in our model is $E_x(t)$, the average quality of capital produced by cognitive workers.

Next, we seek to understand the inverse relationship between inequality and AY as shown in Fig. 2. For this purpose, let us consider the first two panels in Fig. 6. The left panel plots $E_x(t)$ and the standard deviation of t with respect to x, which we denote as $\sigma_x(t)$. The middle panel plots $E_h(t)$ and the standard deviation of t with respect to $h(x; y)$, which we denote as $\sigma_h(t)$. It is clear that there is a negative relationship between the expected value and the standard deviation in both cases.[23]

Hence, when $E_h(t)$ is high, most physical workers are likely to be using high t. Therefore, not only will PY be high, the difference in income levels of physical workers is also likely to be low. We then have a situation of high PY and low inequality among physical workers. But since PY and AY move together, this implies an inverse relationship between AY and $Gini_P$, as depicted in Fig. 2. Similarly, when $E_x(t)$ is

[22]The fact that CY is low when $E_x(t)$ is high is quite evident. High $E_x(t)$ means a significant proportion of cognitive workers produce high-quality capital. But (1) then implies than the payoff of such agents will be low, which will cause CY to be low.

[23]We have already noted the positive relationship between $E_x(t)$ and $E_h(t)$ as shown in the right panel of Fig. 6. The two standard deviations, $\sigma_x(t)$ and $\sigma_h(t)$, will also be share a positive relationship. If the dispersion in the quality of capital being produced by cognitive workers is high, it is natural that the dispersion in the quality of capital being used by physical workers will also be high.

high, the low $\sigma_x(t)$ implies that most cognitive workers are producing high quality of t. Therefore, there will be low inequality in income levels among cognitive workers. But high $E_x(t)$ implies low CY. Thus, a high $E_x(t)$ causes both CY and $Gini_C$ to be low, which explains the direct relationship observed between these two variables in the bottom left panel of Fig. 1. But since there is an inverse relationship between CY and AY, the relationship between AY and $Gini_C$ is itself negative, as is seen in Fig. 2.

The inverse relationship between GDP and inequality in our model, therefore, ultimately arises from the inverse relationship between $E_x(t)$ and $\sigma_x(t)$. Recall our interpretation of the quality of capital as productivity of physical workers. A high $E_x(t)$ implies a high average productivity in our model which causes GDP to be high. But a high $E_x(t)$ is associated with a low $\sigma_x(t)$ which causes low inequality among cognitive workers. A high $E_x(t)$–low $\sigma_x(t)$ combination will be reflected in a similar high $E_h(t)$–low $\sigma_h(t)$ combination. As discussed in the previous paragraph, this leads to low inequality among physical workers. Therefore, when the average productivity in the economy is high, the dispersion in productivity will be low. The result will be high GDP and low inequality.

This discussion also provides us with a more nuanced understanding of the aggregate income and inequality within each group. Within the population of physical workers, the fact that high $E_h(t)$ and low $\sigma_h(t)$ occur together imply that low inequality is associated with high PY. On the other hand, for cognitive workers, low inequality is associated with low CY. Again, this is because a high $E_x(t)$ occurs together with a low $\sigma_x(t)$. Thus, even as inequality is low at the same time for both types of workers, the implications are different. Low inequality is associated with high aggregate income for physical workers but with low aggregate income for cognitive workers.

To complete our intuition, we should also explain why the expected values and standard deviations in Fig. 6 have an inverse relationship. It is sufficient to consider $E_x(t)$ and $\sigma_x(t)$ for this purpose since the corresponding values generated by h depend upon the behavior of the state variable x. Consider a situation where $E_x(t)$ is high. The fact that $E_x(t)$ then declines, as it must in a cycle, implies that for at least some cognitive workers using a high strategy, it becomes the best response to switch to a low strategy. The reason why this happens is that at the initial state of high $E_x(t)$, the mass of agents playing high strategies is sufficiently large that such agents are not able to distinguish themselves as producers of high-quality capital and, thereby, sell to the more informed physical workers (of mass y_2). Instead, they prefer to produce a lower quality of capital and just sell to the less informed physical workers (of mass y_1). But high $E_x(t)$ along with a high mass of agents producing better quality capital implies a low standard deviation.

5 Variations in the Search Distribution

In our simulations in Sect. 5, we fixed the value of the exogenous distribution y at $\{y_1, 1 - y_1\} = \{0.5, 0.5\}$. We then analyzed the cyclical behavior of aggregate income and inequality at this distribution. This naturally raises the question of whether such cyclical behavior will arise for other values of y_1.

In one sense, the answer to this question is in the affirmative. We know from Proposition 2.1 that for any y_1, if n is sufficiently large, all Nash equilibria will be mixed. Positive definiteness then implies that all logit equilibria will be unstable under the logit dynamic for η sufficiently small. Hence, for any y_1, if we make n sufficiently large, we will obtain a limit cycle in the state space and, therefore, cyclical fluctuations in aggregate income and inequality.

In this section, we approach the question from a different perspective. We fix n and then vary y_1 to see whether limit cycles emerge. In fact, we will fix n at a relatively small value. This is of interest because in reality, the range of qualities in which capital is available may be quite limited. If we can then show that the existence of a limit cycle depends upon the value of y_1, then that will suggest that the emergence of cyclical fluctuations in aggregate income and inequality depends upon the search behavior of the users of capital, at least for small values of n. Once again, the most feasible method of conducting such an analysis is through simulations.

For our simulations, we use the same parameters as in the simulation in Fig. 1. Thus, we fix n at the relatively small value of 5 so that the strategy set is $S^5 = \{0, 0.2, 0.4.0.6, 0.8, 1\}$. We use the production function $q(t) = 1 + 5\sqrt{t}$ and the cost function $c(t) = t$, and apply the logit (0.001) dynamic from the initial state $x(0) = \left(\frac{1}{6}, \ldots, \frac{1}{6}\right)$. Of course, as per our motivation, we will allow y_1 to vary from 0 and 1.

We focus on characterizing limit cycles in two variables of interest, viz., aggregate income (AY) and the Gini coefficient of cognitive workers ($Gini_C$).[24] Figure 1 already shows that for $y_1 = 0.5$, these variables exhibit limit cycles under the stated parameters. We note that from a time series perspective, such limit cycles can be understood as a scenario where these variables oscillate with finite ranges of fluctuations. Therefore, one way to determine computationally whether these variables do fluctuate cyclically for any y_1 is to impose a common threshold on some measure of fluctuations. In particular, we choose a simple measure of standard deviation. For computational purpose, we impose a positive but very small threshold of 10^{-3} on standard deviations. We then run a grid search algorithm over the space $y_1 \in [0, 1]$ to determine the standard deviations of two variables of our interest. If the evaluated standard deviation is greater than the threshold for a particular y_1, we conclude that the corresponding variable exhibits a limit cycle at that y_1. It is possible that there are limit cycles with standard deviation smaller than the imposed threshold, which our approach will not be able to detect. However, the chosen threshold is very small compared to the range of fluctuations found numerically. Also, we note that since

[24] We consider AY because it measures GDP in our model. The behavior of $Gini_P$ is similar to that of $Gini_C$.

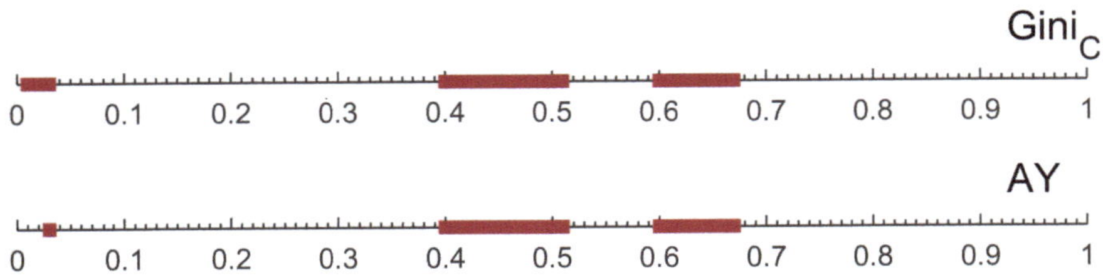

Fig. 7 Regions of instability with variation in y_1. The strategy set, production function, cost function, and initial state are as in Fig. 1. The dynamic is the logit (0.001) dynamic. Top panel shows the values of y_1, where $Gini_C$ shows limit cycles. Bottom panel shows the values of y_1, where AY shows limit cycles. In both cases, we have used a threshold of 10^{-3} for standard deviation, to extract the regions of instability. If simulations indicate the existence of a limit cycle, then we have denoted that y_1 with a filled square

our approach is computational, the possibility of missing some occurrences of limit cycles is inevitable. In any case, with such a small threshold as ours, it is unlikely that we will miss limit cycles of significance.

We present the results of our simulations in Fig. 7. We observe that for both variables, the values of y_1 for which these variables exhibit a limit cycle are broadly similar. Such values of y_1 are highlighted as filled squares. In this case, such consistency across variables arises because the underlying mechanism of these cycles is cyclical fluctuations in the state variable x. If for any particular value of y_1, x fluctuates, then both variables, being functions of x, will also fluctuate. From a computational point of view, however, presenting both variables in Fig. 7 is important for two reasons. One, it is possible that a variable may be an aggregation of multiple other variables and do not show any limit cycles due to counteracting movements of the underlying variables. Our numerical results suggest that this does not happen here. Second, it also validates the choice of the common threshold method to determine the existence of these limit cycles. The fact that both variables fluctuate at broadly similar values of y_1 suggest that our choice of threshold does not lead to inconsistent results.

The simulations in Fig. 7, therefore, suggest that the incidence of cyclical fluctuations in income and inequality in our model depends upon the search behavior of agents. There is an isolated region close to $y_1 = 0$, where our simulations show limit cycles. It is possible that limit cycles in this region are appearing due to reasons of numerical approximation. But barring this region, we do not observe limit cycles for y_1 close to 0. Similarly, our simulations fail to show limit cycles near $y_1 = 1$. Intuitively, this happens because as we know from Proposition 2.1, there exists pure Nash equilibria for any n if $y_1 = 0$ or $y_1 = 1$. Such pure equilibria (or the logit equilibria that approximate such Nash equilibria) attract solution trajectories of the logit (0.001) dynamic and, therefore, do not allow limit cycles to emerge. Another noteworthy feature of the simulations in Fig. 7 is that even for intermediate values of y_1, the range for which we obtain limit cycles is not continuous. This is evident from the fact that between $y_1 = 0.5$ and $y_1 = 0.6$, there is a significant interval for which there is no limit cycle. For such values of y_1, pure Nash equilibria exist which attract solution trajectories. We, therefore, conclude that the existence of limit cycles in our model does depend upon the distribution that governs the search behavior of physical workers.

One caveat that we note is that all our simulations in Fig. 7 are from the initial state $x(0) = \left(\frac{1}{6}, \ldots, \frac{1}{6}\right)$. Even if a limit cycle exists from this initial state for a particular value of y_1, it is quite possible that that limit cycle will no longer arise if we change the initial state. In fact, this caveat also applies to our simulations in Sect. 4, where we had used the initial state $x(0) = \left(\frac{1}{n+1}, \ldots, \frac{1}{n+1}\right)$.[25] Having said that, we do observe that robustness checks from a selection of other initial states failed to show significant differences in the behavior of the logit dynamic.

5.1 *Characterization of the Limit Cycles*

Figure 7 shows that limit cycles in our variables of interest arise only for specific subsets of the values of y_1 in [0, 1]. We now consider a related question of how far these limit cycles differ for different values of y_1. The question is of interest because if the limit cycles do differ, then it will show that the nature of fluctuations in aggregate income and inequality will depend upon the search behavior of agents.

To analyze this question, we transform the time domain information provided by these limit cycles into frequency domain so that we can characterize the limit cycles through their frequencies. Before discussing the results, we note that there is no independent scale of time here. That is to say, for the limit cycles, we cannot assign a physical meaning of time (1 year or 1 month). The choice of *unit time* is fixed outside the model and hence, we do not interpret the frequency of one limit cycle. Instead, we make comparison across limit cycles in terms of obtained frequencies. The power spectrum becomes useful here as such a spectrum will clearly delineate the portion of the power of the time series falling within the prespecified frequency bins.

We construct the power spectrum of the each of the two series via Fourier transformation. For a continuous series $z(t)$, we can find the power spectrum by computing $P(\omega) \propto |\tilde{z}(\omega)|^2$ where

$$\tilde{z}(w) = \int_{-\infty}^{\infty} z(t)exp(-i\omega t)dt. \tag{10}$$

For computational purpose, we need discrete Fourier transformation. For a finite time interval $[0,\ T]$ and a sampling interval Δ such that $T = N\Delta$, the transformation is

[25] Recall that $n = 5$ in Figs. 1 and 5, and $n = 100$ in Fig. 3.

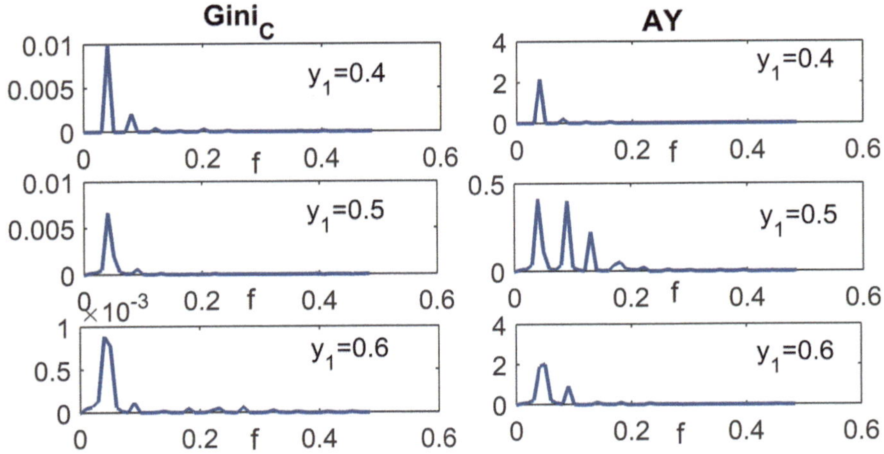

Fig. 8 Power spectra of the limit cycles of two macroeconomic variables: $Gini_c$ and AY. Horizontal axis denotes frequencies f. We have chosen three values of y_1 (0.4, 0.5, and 0.6) following Fig. 7. All other parameters and the dynamic are chosen as in Fig. 7

$$\tilde{z}_j = \sum_{0}^{N-1} z(t_k) exp(i\omega_j t_k), \tag{11}$$

where $\omega_j = \frac{2\pi j}{T}$ and $t_k = k\Delta$. Through computer simulations, we generate a discrete analogue of the continuous series. For the purpose of transformation, we sample data points from the series with uniform intervals. We compute the power spectrum of two series of exact same length ($Gini_c$ and AY) generated at three different values of y_1 (0.4, 0.5, and 0.6). The idea is that if the limit cycles do have similar characterization in the frequency domain, then we would get similar spectrum. As Fig. 8 shows the spectra are different for different limit cycles for the same macroeconomic variables, viz., Gini coefficient for the cognitive workers $Gini_c$ and aggregate income AY.

In order to analyze the spectrum in further detail, we have done the analysis for $y_1 \in [0.4,\ 0.65]$ in discrete intervals of 0.02. The interval choice is motivated by the findings of Fig. 7. Figure 9 clearly delineates the power spectra of the limit cycles in aggregate income for different values of y_1. To have a clear idea of the peaks of the spectra, we have also shown contour plots on the floor of the diagram. This analysis shows that what fraction of physical workers search only once (i.e., y_1) for a match, is the crucial factor behind occurrence of nontrivial dynamic behavior.

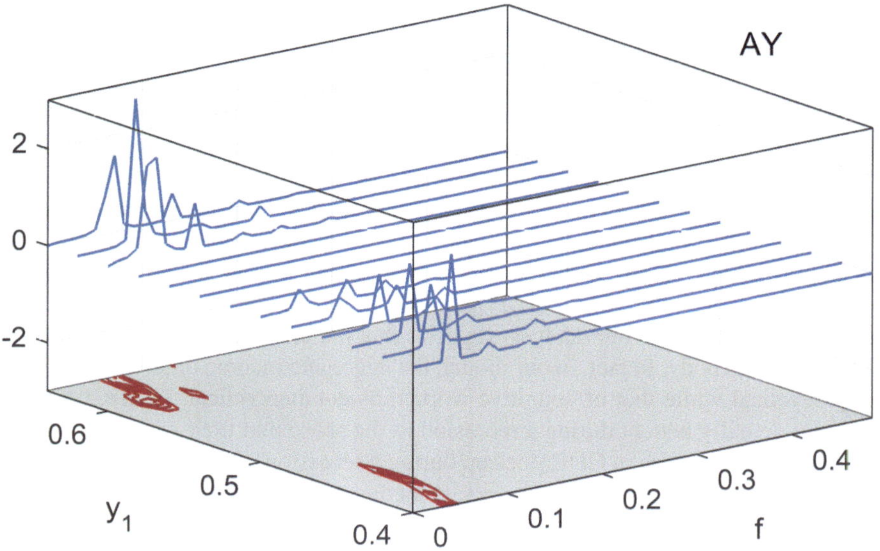

Fig. 9 Power spectra of the limit cycles of aggregate income AY as a function of y_1. Horizontal axis denotes frequencies f. In the floor of the figure, contour plot indicates the peaks of the spectra. The value of y_1 ranges from 0.4 to 0.65. All other parameters and the dynamic are chosen as in Fig. 7

6 Conclusion

In this paper, we have considered the relationship between inequality and the business cycle. Empirical evidence suggests a mutually causal relationship between GDP and inequality over the business cycle and the countercyclical nature of inequality. We have proposed an evolutionary game-theoretic model that recovers both these empirical observations. In our model, there are two types of workers; cognitive and physical. Cognitive workers produce capital of varying qualities. Physical workers search once or twice and use the highest quality of capital they obtain through their search process to produce output. The state variable of the model is the behavior of cognitive workers. We define the income levels for both groups of workers and interpret the aggregate income of both groups combined as GDP. We use the share of GDP received by each group as a measure of between group inequality and the Gini coefficients for each group as a measure of within-group inequality.

The model satisfies a technical condition called positive definiteness. This renders all mixed equilibria of the model as evolutionarily unstable. Instead, the state variable converges to a limit cycle causing cyclical fluctuations in functions of the state variable like aggregate income and Gini coefficients. We interpret the cyclical fluctuations in aggregate income as a business cycle and conclude that the limit cycle of the state variable is the common origin of both the business cycle and fluctuations in inequality, thereby appearing to mutually cause each other. Simulations then reveal an inverse relationship between aggregate income and Gini coefficients over the business cycle. This suggests that inequality is countercyclical. Our model, therefore, explains the two empirical observations which motivated this paper, namely, the mutual causality between GDP and inequality and the countercyclical nature of inequality. Finally, we present a spectral analysis to conclude that changes in the distribution that governs the search behavior of physical workers have implications on the nature of the limit cycles generated by aggregate income and Gini coefficients.

Our model has two major implications. First, it shows that the adverse consequences of a recession are not restricted to a fall in GDP. In addition, inequality also increases during a recession. This provides greater justification for policy interventions to combat recessions. Policy intervention during a recession is useful not just because it prevents the fall in GDP but also because it contributes toward a more equal society. Second, our analysis suggests that the effect of a recession is not the same for everybody. In fact, in our model, the aggregate income of physical workers is procyclical while that of cognitive workers is countercyclical. Hence, cognitive workers actually benefit during a recession in the sense that their aggregate income and, hence, their share in GDP, goes up during a recession. On the other hand, physical workers suffer a fall in their aggregate income and, hence, in their share in GDP. More generally, this finding suggests that different social may be affected differently during a recession. It may, therefore, be beneficial to target policy measures toward recessions in favor of such social groups who suffer the most from such recessions.

7 Appendix

7.1 Proofs in Sect. 2

Proof of Proposition 2.1: As in Chakrabarti and Lahkar (2017) (see Proof of Proposition 3.1 in that paper), we introduce the following change of variable. Let $t_{n-i} = 1 - t_i$, $n - i \in \{0, 1, \ldots, n\}$, and χ be the probability distribution over $\{t_{n-i}\}_{n-i=0}^{n}$. Clearly, $x_{n-i} = x_i$. Hence,

$$\sum_{j<i} x_j = \sum_{j=0}^{i-1} x_j = \sum_{j=n-(i-1)}^{n} x_j = \sum_{j>n-i} x_j.$$

We apply (1) to write the payoff of t_{n-i}, $n - i \in \{0, 1, \ldots, n\}$, as

$$\tilde{\pi}_{n-i}(x, y) = t_{n-i} \left[y_1 + 2y_2 \left(\frac{x_{n-i}}{2} + \sum_{j>n-i} x_j \right) \right]. \tag{12}$$

Apart from the difference in notation, (12) is of the same form as Eq. (1) of Lahkar (2011) when the distribution y is exogenously fixed at $\{y_1, y_2\}$.

The proof then follows from the proofs of Lemmas 2.1 and 2.2 in Lahkar (2011). Part 1 follows from Part 1 of Lemma 2.2 of Lahkar (2011). More directly, Part 1 of this proposition follows because if $y_1 = 1$, then $t_1 = 0$ is the dominant strategy. Part 2 of this proposition follows from Part 2 of Lemma 2.2 of Lahkar (2011). By that result, if $y_1 = 1$, then the game defined by (12) has three Nash equilibria; $x_0 = 1$, $x_1 = 1$ and $x_2 = 1$. The change of variable then implies that the game π defined by (1) has three Nash equilibria; $x_n = 1$, $x_{n-1} = 1$ and $x_{n-2} = 1$, the last only if $n \geq 2$. Part 3 follows from Lemma 2.1 of Lahkar (2011). ■

Compliance with Ethical Standards No funding was received for this paper. The authors declare that they have no conflict of interest.

References

Bonhomme S, Hospido L (2017) The cycle of earnings inequality: evidence from spanish social security data. Econ J 127(603):1244–1278

Braun RA, Nakajima T (2012) Uninsured countercyclical risk: an aggregation result and application to optimal monetary policy. J Eur Econ Assoc 10(6):1450–1474

Burdett K, Judd KL (1983) Equilibrium price dispersion. Econometrica 51(4):955–969

Chakrabarti AS, Lahkar R (2017) Productivity dispersion and output fluctuations: an evolutionary model. J Econ Behav Organ 137:339–360

Chen H, Michaux M, Roussanov N (2015) Houses as ATMs? mortgage refinancing and macroeconomic uncertainty

Edmond C, Veldkamp L (2009) Income dispersion and counter-cyclical markups. J Monetary Econ 56(6):791–804

Fudenberg D, Levine D (1998) Theory of learning in games. MIT press, Cambridge, MA

Hofbauer J, Sandholm W (2009) Stable games and their dynamics. J Econ Theory 144(4):1665–1693

Hoynes HW, Miller DL, Schaller J (2012) Who suffers during recessions? J Econ Perspect 26:27–47

Krebs T (2007) Job displacement risk and the cost of business cycles. Am Econ Rev 97(3):664–686

Krueger D, Perri F, Pistaferri L, Violante G (2010) Cross-sectional facts for macroeconomists. Rev Econ Dyn 13:1–14

Lahkar R (2011) The dynamic instability of dispersed price equilibria. J Econ Theory 146(5):1796–1827

Maestri V, Roventini A (2012) Stylized facts on business cycles and inequality. https://ideas.repec.org/p/aia/ginidp/30.html

Maynard Smith J, Price GR (1973) The logic of animal conflict. Nature 246:15–18

McKelvey RD, Palfrey TR (1995) Quantal response equilibria for normal form games. Games Econ Behav 10(1):6–38

Piketty T, Saez E (2014) Inequality in the long run. Science 344(6186):838–843

Sandholm W (2010) Population games and evolutionary dynamics. MIT Press, Cambridge, MA

Solon G, Barsky R, Parker JA (1994) Measuring the cyclicality of real wages: How important is composition bias. Q J Econ 109:1–25

Swanson ET (2004) Measuring the cyclicality of real wages: how important is the firm's point of view? Rev Econ Stat 86:362–377

Green A (2016) XVA: credit, funding and capital valuation adjustments. https://onlinelibrary.wiley.com/doi/book/10.1002/9781119161233

Simulated Maximum Likelihood Estimation of Agent-Based Models in Economics and Finance

Jiri Kukacka

Abstract This chapter presents a general simulation-based framework for estimation of agent-based models in economics and finance based on kernel methods. After discussing the distinguishing features between empirical estimation and calibration of economic models, the simulated maximum likelihood estimator is validated for utilization in agent-based econometrics. As the main advantage, the method allows for estimation of nonlinear models for which the analytical representation of the objective function does not exist. We test the properties and performance of the estimator in combination with the seminal Brock and Hommes (J Econ Dyn Control 22:1235–1274, 1998) asset pricing model, where the dynamics are governed by switching of agents between trading strategies based on the discrete choice approach. We also provide links to how the estimation method can be extended to multivariate macroeconomic optimization problems. Using simulation analysis, we show that the estimator consistently recovers the pseudo-true parameters with high estimation precision. We further study the impact of agents' memory on the estimation performance and show that while memory generally deteriorates the precision, the main properties of the estimator remain unaffected.

1 Introduction

Agent-based models (ABMs) are popular in many fields of social science, such as ecology and political science, because they enable the study of various complex systems. For economists, this methodology is particularly appealing in the construction of financial and macroeconomic models whose dynamics emerge from the interactions of many heterogeneous economic agents. ABMs have attracted considerable attention in reaction to the global financial crisis of 2007–2008. The mainstream

J. Kukacka (✉)
Faculty of Social Sciences, Institute of Economic Studies, Charles University, Opletalova 26, 110 00 Prague 1, Czechia
e-mail: jiri.kukacka@fsv.cuni.cz

Institute of Information Theory and Automation of the Czech Academy of Sciences, Pod Vodarenskou vezi 4, 182 00 Prague 8, Czechia

A. S. Chakrabarti et al. (eds.), *Network Theory and Agent-Based Modeling in Economics and Finance*, https://doi.org/10.1007/978-981-13-8319-9_10

economic approaches that assume perfect rationality and aggregation of all market participants to a single "representative" agent were not able to predict the looming collapse of the financial system. More importantly, these approaches also generally failed to provide policy recommendations to reduce the contagion effects and the devastating impact of the housing and stock market collapses on the real economy.

In terms of practical applications of agent-based analysis in economics and finance, recent research shows that various ABMs are easy to construct. However, the subsequent steps and actual utilization of the methodology in real-world decision-making are challenging because, in most cases, the models are overly complex for analytical solution and research. To address this issue, researchers can either calibrate a given model and analyze its qualitative properties via simulations or estimate the model parameters using empirical data. This traditional econometric approach of estimating the quantitative relationships between model variables remains, however, a major challenge in the field. Most ABMs cannot be estimated with standard econometric methods, such as the ordinary least squares (OLS) estimator or the maximum likelihood estimator (MLE), due to their nonlinearity stemming from interactions of agents or because the analytical representation of the objective function does not exist. This drawback greatly complicates the testing of statistical hypotheses and the selection among various models of the same economic phenomenon.

Fortunately, simulation-based inference provides a suitable toolkit for empirical validation of ABMs, taking advantage of the availability of powerful personal computers and computational clusters that enable extensive numerical exercises. Repeated simulations of a model under a large number of parameter combinations provide parameter estimates that most closely resemble the properties of the empirical data. The simulated method of moments (SMM), the most popular estimation approach, is simple to implement and avoids the restrictive theoretical assumptions that have been applied to many financial and, most recently, macroeconomic ABMs (Jang and Sacht 2016, 2019). The parameter estimates are obtained based on the minimization of the distance between a selected set of moments of simulations approximating the population and the empirical moments, such as the mean, variance, or correlations between variables. Utilization of other estimation methods in agent-based modeling remains scarce, but the pioneering attempts include Bayesian inference by Grazzini et al. (2017) and sequential Monte Carlo by Lux (2018).

This chapter presents an innovative computational framework for the estimation of agent-based models in economic and finance based on the simulated maximum likelihood estimator (SMLE), which was originally developed within the financial econometric literature (Altissimo and Mele 2009; Kristensen and Shin 2012; Lee and Song 2015). The estimation method can also be customized for multivariate economic estimation problems (Kukacka et al. 2018). As the main advantage, the method enables the estimation of nonlinear models for which the analytical representation of the objective function or its theoretical approximation does not exist. In a similar manner to the SMM, the method is based on a simulation-based approximation. Instead of population moments, however, the method obtains a simulated approximation of the entire conditional density function of the data-generating process by means of kernel density estimation. The SMLE is then derived similar to the infeasible MLE by maximization of the simulated likelihood function.

In a large Monte Carlo simulation study, we show that the estimation method can recover the pseudo-true parameters of the famous Brock and Hommes (1998) financial agent-based model consistently and with high estimation precision. We especially focus on the estimation performance for a switching parameter called the intensity of choice. This parameter governs the switching dynamics of the model, and its estimation is generally challenging (Bolt et al. 2014; Hommes and in 't Veld 2017). Moreover, its empirical estimates often remain insignificant in the recent agent-based econometric literature (Frijns et al. 2010; Bolt et al. 2011; ter Ellen et al. 2013; Verschoor and Zwinkels 2013; Chiarella et al. 2015; Lof 2015; Recchioni et al. 2015). As a novelty, we further study the impact of agents' memory on the estimation performance of the SMLE.

2 Estimation Versus Calibration in Economic Modeling

When a researcher tries to orientate themselves to contemporary econometric approaches within the field of agent-based modeling, several terms that are often used interchangeably may cause confusion. Windrum et al. (2007), Fagiolo et al. (2007), Guerini and Moneta (2017), Barde and van der Hoog (2017), Lamperti (2018b), Lamperti (2018a), Fagiolo et al. (2019) use the general term "validation", Barde (2016) introduces a "comparison" method, Recchioni et al. (2015); Lamperti et al. (2018); Platt (2019) provide and compare "calibration" methods, while many authors (Grazzini et al. 2013; Grazzini and Richiardi 2015; Franke and Westerhoff 2012, 2016; Ghonghadze and Lux 2016; Chen and Lux 2018; Chiarella et al. 2014; Cornea-Madeira et al. 2017; Goldbaum and Zwinkels 2014; Kouwenberg and Zwinkels 2015; Barunik and Kukacka 2015; Kukacka and Barunik 2017, to mention a few recent contributions) suggest and utilize various "estimation" methodologies. Lux and Zwinkels (2018) provide an excellent survey of these novel approaches in finance, and Franke and Westerhoff (2017) survey the approaches in behavioral macroeconomics.

In a recent contribution, Grazzini and Richiardi (2015) pointedly remark that the empirical grounding of ABMs is often limited to an ad hoc calibration of important parameters. They compare this practice to the state of the art in the dynamic stochastic general equilibrium (DSGE) modeling a decade ago, which now focuses more on formal estimation procedures. While recent advancements in the DSGE literature favor empirical estimation over calibration, the dominate method in the DSGE field in the past, the ABMs field remains immature in this respect. While we agree that empirical validation of ABMs via estimation is an essential step in the modeling cycle and an important premise for model comparison and resulting policy recommendations, we must also acknowledge the role and importance of model calibration not only as a starting point of ABM empirical analysis but as a complementary, rather than alternative, method of an optimal toolkit for ABMs validation procedures and sensitivity analyses. To quote from Hansen and Heckman (1996, p. 91),

> the distinction drawn between calibrating and estimating the parameters of a model is artificial at best,

and the justification of the two terms is often vague, confusing, and, in part, arbitrary, based on the scientific field and individual authors. For estimation of DSGE models, the Bayesian method is currently commonly used, and the priors are often set to very narrow ranges. According to Fernández-Villaverde (2010, p. 36):

> Some researchers prefer to select loose priors and let the likelihood dominate the posterior as much as feasible.... Other researchers favour tighter priors that sharpen our inference and guide the posterior to plausible regions.

The estimation then strongly resembles calibration, and there is not much difference between these otherwise different concepts. Furthermore, the evolution of the DSGE methodology in the past decade constitutes a crucial source of knowledge for the current development of validation methods for ABMs, as many challenges clearly overlap. Issues discussed in the following text are thus shared by the ABM field to a similar extent as the subject matter.

2.1 Calibration

The main benefit of calibration is that it allows a researcher to study a model consistently. Via calibration, a theoretical model is specified and can be repeatedly simulated. The calibration might be based on external knowledge, for instance, related literature comprising specific micro-studies regarding individual parameters or previous works focused on the same modeling framework. Another option is calibration founded on a relevant theory or some empirical criteria. The output of simulations can then be compared with empirical data; for example, various stylized facts of financial or macroeconomic data are compared with patterns and regularities in the simulation outputs. Another benefit of calibration is that it avoids difficulties related to imperfect estimates. This econometric issue may arise from many potential sources comprising problematic identification of parameters, possible estimation bias, lack of data, unmet theoretical assumptions, or misspecification of the theoretical model. On the basis of these benefits, one might consider simulation studies based on calibration more credible and scientifically valid.

On the other hand, calibration might be viewed as less trustworthy than estimation for various reasons. First, as aptly mentioned by Hansen and Heckman (1996), the process of calibration in economics is often different than that in natural or technical sciences, where precise data are available from experiments or measurements. This problem can lead to interdisciplinary misunderstandings. Originally, the term calibration referred to a graduation or a refinement of a measurement instrument, such as a Celsius thermometer (Kydland and Prescott 1996). In natural sciences, the term calibration is generally understood as a process of tuning a model. Researchers manipulate the independent variables to obtain an optimal match between the simulated and observed characteristics of the data, for instance, the distribution of the dependent variable or variables. This procedure can, of course, be practiced in economics, i.e., the model parameters can be manipulated until the model accurately mimics a carefully defined number of dimensions, for example, a set of stylized facts or moments of a distribution. However, calibration often denotes only a relatively simple process of setting the model parameters according to some best practice in the related literature before running simulations without subsequent iterations. This practice calls into question the standards for verification imposed by the calibrators.

For calibration procedures in natural sciences, available data are divided into two parts. The first set is used for calibrating the model, and the remaining data are used to verify the model. In econometrics, a comparable approach is (in-sample) estimation and (out-of-sample) testing. The process of calibration might be seen as vague and arbitrary as no coherent methodology usually exists for determining individual parameters from microeconomic data. For instance, Fernández-Villaverde and Rubio-Ramírez (2010) note that when utilizing the SMM, it is often unclear what set of moment conditions should be selected for the calibration procedure. The authors further argue that based on many years of econometric experience in the field with the methods of moments, choosing different sets of moments may well lead to different sets of resulting parameter estimates. Kydland and Prescott (1996) admit that in calibration, the model economy is sometimes made inconsistent with the empirical data on one dimension so that it resembles other important dimensions. Comparison with other differently calibrated versions of the same model is rarely discussed for a specific framework. Hansen and Heckman (1996) argue that little emphasis is devoted to assessing the quality of the resulting calibration. Therefore, formalization of the criteria for calibration and verification is suggested. This process is typically performed via optimization of a loss function, which increases the rigor and clarity behind the selection of a particular model.

Hansen and Heckman (1996, p. 97), further stress that it can be generally misleading and potentially incorrect to plug microeconometric parameter estimates into an economic model in cases where the environments of the two models are fundamentally different. The coefficients used for calibration might have been derived under assumptions incompatible with the modeling framework, resulting in a theoretical mismatch of the final model. Moreover, each individual micro-study is unique: the researcher may condition on different assumptions and thus implicate different economic interpretations. Micro-evidence synthesis is therefore a challenging task, as emphasized by Fernández-Villaverde (2010, p. 10):

> The researcher should be careful, though, translating this micro evidence into macro priors. Parameter values do not have an existence of their own, like a Platonic entity waiting to be discovered. They are only defined within the context of a model, and changes in the theory, even if minor, may have a considerable impact on the parameter values.

An extensive discussion of issues emerging from the application of microeconomics evidence and its incompatibility with economic models is provided by Browning et al. (1999). The authors highlight the heterogeneity of preferences, an essential finding of modern empirical microeconomics, that is, however, often abstracted from in DSGE models to gain computational tractability. Caution is recommended against weak the microempirical foundations of many DSGE models and possible improper use of synthesized microeconomic evidence within an internally consistent framework that may influence the structure of the model and, hence, alter its implications. Another area of problematic application is the notion of uncertainty. Aggregation may cause a disconnect between various individual sources because different economic agents may generally have fundamentally different risk preferences, thereby altering the predictability of the model. Calibrating the distributions of individual shocks thus becomes crucial. Similarly, ignoring real-world inefficiencies can lead to under- or overstating of the true risk faced by agents.

2.2 Estimation

In contrast to calibration, estimation is used to assess the approximate size of some phenomenon. Estimation is a traditional rigorous optimization approach for which a loss function is unambiguously defined. Compared to calibration, the estimated parameter values thus provide the best fit in some statistical sense. Various estimation methods, such as the OLS and related shrinkage methods (e.g., ridge regression, lasso), and the SMM, or different loss functions weight various features of the data differently; thus, the selection of a proper loss function may help to solve various modeling issues. According to Grazzini and Richiardi (2015, p. 148),

> ...estimation (as opposed to calibration) involves the attainment of clearly specified scientific standards in the way models are confronted with the data.

The acute need for such an approach for ABM is emphasized to gain support of the wide scientific community and confidence of the policy-makers in the use of a new economic decision-making tool. As a differentiating principle, the authors note the distinct goals of the two methods while admitting that a clear classification is problematic. Although calibration primarily aims for optimal tracking and describing of observed empirical data, the ultimate goal of estimation is recovering, via parameter values, the true data-generating process. Estimation is thus more concerned with the theoretical properties of the estimators and focuses to a large extent on the quantification of the uncertainty related to the point estimates, allowing for subsequent testing of statistical hypotheses.

Moreover, a crucial theoretical distinction between estimation and calibration is the notion of convergence. The selection of appropriate statistics for a stationary and ergodic series ensures estimates that are consistent in time. Conversely, when using cross-sectional statistics to describe a model, as in many calibration exercises, the consistency in population size cannot be assured due to possible autocorrelation between cross-sectional moments, which does not disappear with increasing sample size. Another important task is the final choice of variables used for the empirical estimation. On the one hand, an ABM may be too stylized to accommodate estimation of all the variables in the model. On the other hand, over-parametrization causes serious difficulties related to identification and practical aspects of estimation procedures, such as high computational burden. Additionally, in recent attempts to use DSGE models for forecasting, estimation appears to be favorable, as it often produces smaller mean squared errors than calibration; we refer to Del Negro and Schorfheide (2012), who provide an extensive study of the forecasting performance of DSGE models and a related literature review.

Compared to DSGE models and macroeconomic ABMs, a distinguishing feature of financial ABMs is the structure of the output, which is almost always a single time series. Thus, there are no cross-correlations observed, and usual autocorrelations are employed in a standard set together with other financial stylized facts. The modelers are thus likely to avoid possibly arbitrary decisions about what information should be used for calibration and what should be reserved for model testing. On the other hand, financial ABMs and DSGE models often share problems related to a flat likelihood function. The likelihood often encompasses little information in a direction of a particular parameter. In such cases, according to Fagiolo and Roventini (2012), informal calibration might be a more honest strategy to provide an internally consistent setup for a model. In terms of estimation, in contrast to DSGE models and macroeconomic ABMs with a relatively large number of parameters, where empirical estimation might not be feasible or advisable, simple financial ABMs typically contain only a few parameters that do not have any obvious empirically measurable counterparts. Moreover, some researchers might deliberately design simple stylized or highly aggregated models to facilitate the computation and estimation of a small set of crucial parameters, but such an approach is fully scientifically legitimate and follows the so-called KISS[1] modeling principle.

This chapter focuses on estimation of the Brock and Hommes (1998) model. A crucial component of this agent-based modeling framework is the switching parameter, called the intensity of choice, which is defined based on discrete choice theory. Since the economic literature lacks a general consensus about the existence or intensity of behavioral switching in various markets, the calibration approach would not be of much help in this situation. The switching parameter thus needs to be retrieved from data via some optimization technique. A similar uncertainty hinders the calibration of another important parameter of the model, which is the intensity of the market noise. Conversely, some of the model parameters allow for calibration based on micro-studies or literature surveys, e.g., the overall market risk aversion or specifications of dominant trading strategies (Allen and Taylor 1990; Frankel and Froot 1991).

3 Agent-Based Models in Economics and Finance

A general theoretical framework for ABMs in economic and finance follows Grazzini and Richiardi (2015), in which an agent $h \in \{1, \dots H\}$, $H \in \mathbb{N}$, at time $t \in \{1, \dots T\}$,

[1] An acronym for "Keep it Short and Simple", "Keep it Simple and Straightforward", or an impolite original meaning by the U.S. Navy "Keep It Simple, Stupid".

$T \in \mathbb{N}$, is fully characterized by agent-specific state variables $z_{h,t} \in \mathbb{R}^j$, $j \in \mathbb{N}$. The evolution of $z_{h,t}$ follows:

$$z_{h,t} = \mathfrak{f}_h(z_{h,t-1}, Z_{\neg h,t-1}, \epsilon_{h,t}; \theta_h), \tag{1}$$

where $Z_{\neg h,t-1} = (z_{1,t-1}, \dots z_{h-1,t-1}, z_{h+1,t-1}, \dots z_{H,t-1})$ comprises the state variables of all other agents at time $t-1$, $\epsilon_{h,t}$ is an independent and identically distributed (i.i.d.) stochastic term, and θ_h represents vectors of unknown parameters. $\mathfrak{f}_h$ then completely describes the agent-specific behavioral heuristics.

A system of H structural equations specifies the data-generating process according to Eq. (1), for which a compact state representation is given by

$$Z_t = F(Z_{t-1}, \epsilon_t; \theta), \tag{2}$$

where $Z_{t-1} = (z_{1,t-1}, \dots z_{H,t-1})$, $\epsilon_t = (\epsilon_{1,t}, \dots \epsilon_{H,t})$, $\theta = (\theta_1, \dots \theta_H)$, and F encompasses the behavioral rules of all the agents.

In terms of empirical data, one can observe only a vector of aggregate variables:

$$Y_t = Q(Z_t, \xi_t; \theta), \tag{3}$$

where ξ_t stands for potential additional stochastic shocks and Q is a projection function from the underlying state variables Z to observables Y.

In most of the financial ABMs, the state representation given by F does not have a closed form, as it is only implicitly defined by a system of H structural equations according to Eq. (1). Moreover, we also lack an analytical formulation for the projection function Q. Due to potential nonlinearities of $\mathfrak{f}_h$ and Q, the effect of ξ_t cannot be eliminated by taking advantage of the expectations. The only feasible tool for analyzing the mapping of the initial state (Z_0, θ) to observables Y_t is Monte Carlo simulation, in which the model is simulated for different combinations of initial states and vectors of parameters and each combination is repeated many times to obtain the distribution of Y_t.

Multiple candidates exist for the specific ABMs design to employ. We refer to LeBaron (2006), Hommes (2006), Chen et al. (2012), who provide excellent surveys of the current state of the art of the ABMs in finance, and to Franke and Westerhoff (2017), Dilaver et al. (2018), who survey the field of macroeconomic ABMs.

4 Simulated Maximum Likelihood Estimator

The general complex economic agent-based system presented in Sect. 3 often cannot be estimated via standard econometric methods due to the inherent nonlinearity stemming from the adaptive processes in $\mathfrak{f}_h$ and the nonexistence of the closed-form analytical solution or theoretical approximation of the objective function. Thus, only simulation-based inference remains in the toolkit of an economist for analyzing the properties or estimating the parameters.

The SMLE was originally developed within the financial econometric literature (Altissimo and Mele 2009; Kristensen and Shin 2012; Lee and Song 2015). Let us assume a generic time series process (y_t, x_t), $y_t : t \mapsto \mathbb{R}^k$, $k \in \mathbb{N}$; $x_t : t \mapsto \mathcal{X}_t$, $t = \{1, \ldots, \infty\}$. We observe T realizations $\{(y_t, x_t)\}_{t=1}^T$. Assume $\{y_t\}_{t=1}^T$ is generated by a fully parametric model function q:

$$y_t = q_t(x_t, \varepsilon_t, \theta), \quad t \in \{1, \ldots, T\}, \tag{4}$$

where $q : \{x_t, \varepsilon_t, \theta\} \mapsto \mathbb{R}^k$, $\theta \in \Theta \subseteq \mathbb{R}^u$ is an unknown parameter vector, and $\varepsilon_t \in \mathbb{R}^k$ is an i.i.d. sequence with known distribution $\mathcal{F}_\varepsilon$ (independent of t and θ). For a simple stylized univariate financial ABM, $k = 1$, while for the behavioral macroeconomic three-equation new Keynesian model suggested by De Grauwe (2010, 2011, 2012), $k = 3$. (y_t, x_t) can be nonstationary, y_t can directly represent empirical observables or latent variables, and x_t contains exogenous explanatory and lagged dependent variables y_t.

Assume further an associated conditional density $p_t(y|x; \theta)$:

$$P(y_t \in \mathcal{A}|x_t = x) = \int_{\mathcal{A}} p_t(y|x; \theta)\mathrm{d}y, \quad t \in \{1, \ldots, T\} \tag{5}$$

for any Borel set $\mathcal{A} \subseteq \mathbb{R}^k$. However, as the probability $p_t(m|x; \theta)$ does not have a closed-form representation, the standard MLE is infeasible.

To obtain a simulated approximation of the conditional density $p_t(y_t|x_t; \theta)$ $\forall t \in \{1, \ldots, T\}$, $y_t \in \mathbb{R}^k$, $x_t \in \mathcal{X}_t$, and $\theta \in \Theta$, we first generate $N \in \mathbb{N}$ i.i.d. draws from the k-dimensional distribution $\mathcal{F}_\varepsilon$, $\{\varepsilon_i\}_{i=1}^N$, to compute:

$$y_{t,i}^\theta = q_t(x_t, \varepsilon_i, \theta), \quad i \in \{1, \ldots, N\}. \tag{6}$$

These N simulated i.i.d. random variables (k-multiples for $k \geq 2$), $\{y_{t,i}^\theta\}_{i=1}^N$ follow the target distribution by construction, $y_{t,i}^\theta \sim p_t(\cdot|x_t; \theta)$, and thus allow for estimation of $p_t(y|x; \theta)$ via a standard kernel method:

$$\widehat{p_t}(y_t|x_t; \theta) = \frac{1}{N} \sum_{i=1}^{N} K_\eta(y_{t,i}^\theta - y_t), \tag{7}$$

where $K_\eta(\psi) = K(\psi/\sqrt{\eta})/\sqrt{\eta}$, $K : \mathbb{R}^k \mapsto \mathbb{R}$ is a generic kernel function (a symmetric density, multivariate for $k \geq 2$) and $\eta > 0$ is a bandwidth (a symmetric positive definite $k \times k$ matrix for $k \geq 2$). From $\widehat{p_t}(y_t|x_t; \theta)$, we derive the SMLE of θ:

$$\widehat{\theta} = arg \underbrace{max}_{\theta \in \Theta} \widehat{L_T}(\theta), \tag{8}$$

where $\widehat{L_T}\theta) = \sum_{t=1}^{T} log\ \widehat{p_t}(y_t|x_t;\theta)$. Under the regularity conditions on p_t and K (Kristensen and Shin 2012), conditions A.1-4, K.1-2, p. 80–81), $\widehat{p_t}(y_t|x_t;\theta) \xrightarrow{P} p_t(y_t|x_t;\theta)$, which implies that $\widehat{L_T}(\theta) \xrightarrow{P} L_T(\theta)$ as $N \longrightarrow \infty$ for a given $T \geq 1$. Thus, the SMLE, $\widehat{\theta}$, retains the properties of the infeasible MLE, $\tilde{\theta}$, as $T, N \longrightarrow \infty$. Moreover, the SMLE $\widehat{\theta}$ is first-order asymptotically equivalent to the infeasible MLE under suitable conditions satisfied by most models. Finally, the curse of dimensionality remains only of order $k \equiv dim(y_t)$, and the SMLE therefore behaves similar to other estimation techniques, including the MLE. For a detailed discussion of the advantages, disadvantages, and asymptotic properties of the SMLE, we refer to Kukacka and Barunik (2017), and for related theoretical derivations, to Kristensen and Shin (2012).

5 Agent-Based Model with Behavioral Switching

The seminal Brock and Hommes (1998) model constitutes one of the most widespread financial ABMs applied and analyzed in the literature. The model introduces a stylized financial market application, i.e., the endogenous evolutionary selection of heterogeneous expectation rules (Lucas 1978). In essence, the model is an expectation feedback system in which the variables partly depend on the known values and partly depend on the expectations of the future. The specification of the model utilized in this paper, comprehensively described in Hommes (2006, p. 1169), contains three interdependent equations:

$$Ry_t = \sum_{h=1}^{H} n_{h,t-1} f_{h,t} + \varepsilon_t \equiv \sum_{h=1}^{H} n_{h,t-1}(g_h y_{t-1} + b_h) + \varepsilon_t, \tag{9}$$

$$n_{h,t-1} = \frac{\exp(\beta U_{m,h,t-1})}{\sum_{h=1}^{H} \exp(\beta U_{m,h,t-1})}, \tag{10}$$

$$U_{m,h,t-1} = \frac{1}{m_h} \sum_{j=0}^{m_h-1} \left[(y_{t-1-j} - Ry_{t-2-j}) \frac{f_{h,t-2-j} - Ry_{t-2-j}}{a\sigma^2} \right] - C_h. \tag{11}$$

The pricing equation We start the description of the system with Eq. (9) where a fixed gross rate, $R = 1 + r$, represents a constant risk-free interest rate. Eq. (9) introduces a pricing formula for a heterogeneous market equilibrium price level $\mathfrak{p}_t$ of a risky asset stated in terms of deviation from the latent fundamental price $\mathfrak{p}_t^*$:

$$y_t = \mathfrak{p}_t - \mathfrak{p}_t^*. \tag{12}$$

Assuming i.i.d. stochastic risky asset dividend process d_t, expectation $E_t\{d_{t+1}\} = \bar{d}$ is a constant; thus, $\mathfrak{p}_t^*$ is derived as perpetuity: $\mathfrak{p}_t^* = \sum_{i=1}^{\infty} \frac{\bar{d}}{(1+r)^i} = \frac{\bar{d}}{r}$. The terms $n_{h,t-1}$ denote the market fractions of agents of classes $h \in \{1, \dots H\}$ at time $t-1$

satisfying $\sum_{h=1}^{H} n_{h,t-1} = 1$. $f_{h,t} = g_h y_{t-1} + b_h$ is a deterministic function defining a "h-type" trading strategy at time t, where g_h and b_h are the trend and bias parameters. Fundamentalists with $g_h = b_h = f_{h,t} = 0$ believe that an asset price always converges to its fundamental values. Chartists, on the other hand, use simple technical trading rules and an extrapolation toolkit. If $b_h = 0$, strategy h is pure trend-following if $0 < g_h \leq R$ and strong trend-following if $g_h > R$. Furthermore, strategy h is contrarian if $-R \leq g_h < 0$ and strong contrarian if $g_h < -R$. If $g_h = 0$, strategy h is purely upward-biased/optimistic if $b_h > 0$ and purely downward-biased/pessimistic if $b_h < 0$. Trading strategies combining $g_h \neq 0$ and $b_h \neq 0$ are of course possible. Finally, ε_t is an i.i.d. sequence accounting for the stochastic nature of the market and unpredictable events.

Market fractions Terms $n_{h,t-1}$, derived under the discrete choice probability framework, are defined by Eq. (10). $U_{m,h,t-1}$ is a profitability measure for strategy h, and $\beta \geq 0$ is the intensity of choice governing the willingness of agents to switch between alternative trading strategies.

Profitability measures with memory Finally, $U_{m,h,t-1}$, defined by Eq. (11), are derived as weighted averages of past realized values based on past realized profits, risk-aversion coefficient $a > 0$, and conditional variance σ^2. This approach to agents' memory is suggested by Barunik et al. (2009), who consider equal weights for a finite number of past realizations. The specific memory lengths m_h are randomly generated integers from the uniform distribution $U(1, m)$. Parameters $C_h \geq 0$ are the costs to obtain the necessary market information for application of the trading strategies. A different approach to agents' memory is an infinite gradual dilution of the weights of past realized values governed by dilution parameter $0 \leq \delta \leq 1$: $U_{m,h,t-1} = U_{h,t-1} + \delta U_{m,h,t-2} - C_h$, as suggested by Brock and Hommes (1998). Due to the computational burden of this memory specification, however, a simplification to $\rho = 0$ is often suggested in applied works, essentially neglecting the concept of memory in the model. The approach defined by Eq. (11) allows us to study the impact of memory while maintaining a manageable computational burden.

6 Monte Carlo Simulation Study

In the simulation study, we analyze the behavior of the estimation framework and the properties of the SMLE. We simulate the model under a pseudo-true parameter setup to generate a pseudo-true output on which we apply the simulation-based estimator. We are especially interested in to what extent the estimator is able to recover the pseudo-true parameter values of the model, what the properties of the simulated likelihood are, and what the numerical simulations can reveal about the consistency of the SMLE. By repeating this analysis multiple times under random initial conditions, we achieve statistical validity of the findings and markedly reduce the impact of randomness and uncertainty inherently related to empirical estimation based on a single realization of the data-generating process.

Because of its specific conceptual importance, in Sect. 6.2, we first focus on the estimation performance of the switching parameter, the intensity of choice β in Eq. (10), i.e., $u = 1$ and $\theta = \beta$. The coefficient β can also be called the "rationality" parameter in economic agent-based models because it expresses the willingness of traders to update their trading strategy in light of past profitability. After the ability of the estimator to recover the value of this crucial parameter is verified, we extend the unknown pseudo-true parameter vector to three parameters by means of the trend and bias belief parameters in the so-called two-type model (Chen et al. 2012) in Sect. 6.4, i.e., $u = 3$ and $\theta = \{\beta, g_2, b_2\}$. As a novel approach of this chapter, we analyze the impact of memory in the Brock and Hommes (1998) model setup on the estimation performance of the SMLE.

6.1 Setup

The model The computational design of the Brock and Hommes (1998) model follows Kukacka and Barunik (2013, 2017); Polach and Kukacka (2019). Starting with Eq. (9), the gross interest rate $R = 1 + r = 1.0001$ and the number of trading strategies $H = 4$. A strict fundamental strategy appears by default as $g_1 = b_1 = 0$, and the trend and bias parameters g_h and b_h, $h \in \{2, \dots H\}$, are randomly drawn from normal distributions $N(0, 0.4^2)$ and $N(0, 0.3^2)$, respectively. The stochastic noise term is randomly drawn from normal distributions $N(0, \sigma_\varepsilon^2)$, $\sigma_\varepsilon = \{10^{-8}, 0.1, 1, 2\}$, introducing miniscule, small, intermediate, and strong stochastic noise to the system.

In Eq. (10), the intensity of choice is set to $\beta = \{0, 0.1, 0.5, 1, 5, 10\}$ with a specific focus on $\beta = \{0.1, 1, 10\}$, representing a negligible, intermediate, and strong switching of agents in the model, during the subsequent analysis.

The linear scale factor for the profitability measure in Eq. (11) is set to $1/a\sigma^2 = 1$. This simplification does not affect the relative proportions of $U_{m,h,t}$, it only linearly scales the profitability measure of all trading strategies. Importantly, however, while the conditional variance σ^2 is kept constant, the model output is characterized by dynamic time-varying variance. The memory in Eq. (11) is represented by a maximal length $m = 5$, i.e., the agents consider the history of market prices and realized profits for up to one trading week. Finally, following Hommes (2006), $C_h = 0$ for all H potential trading strategies to keep the impact of agents' memory simple to analyze without an interference about the cost of market information.

The estimation method Reported results are mostly based on the time series length of the model output $t = 10000$ and the kernel estimation precision $N = 1000$. To analyze the consistency of the SMLE, we compare two levels of estimation precision $N = \{100, 1000\}$ and three time series lengths $t = \{100, 1000, 10000\}$ in Fig. 1. As a burn-in period to establish the model dynamics, 100 additional initial observations are discarded. The statistical validity of the analysis is ensured by 1000 random runs, if not stated otherwise, which is only the case for computationally demanding simulations in Figs. 3 and 4. The parameter space is restricted to $\langle -\beta, 3\beta \rangle$ for $\beta > 0$

and to $\langle -0.5, 0.5\rangle$ for $\beta = 0$. As an exception, we apply less narrow restrictions $\langle -8\beta, 10\beta\rangle$ for $\beta > 0$ for Fig. 1 to capture the impact of shorter time series lengths $t = \{100, 1000\}$. The conditional density $p_t(y|x;\theta)$ is estimated via the Gaussian kernel, and an automated rule of thumb from Silverman (1986) is applied to select the optimal bandwidth size:

$$\eta = [4/(3N)]^{1/5}\widehat{\sigma_\varepsilon}, \tag{13}$$

where $\widehat{\sigma_\varepsilon}$ denotes the estimated standard deviation of $\{y^{\theta}_{t,i}\}^{N}_{i=1}$.

Computations All simulations are conducted in `MATLAB`. To reduce the computational burden, we utilize the `MATLAB` Parallel Computing Toolbox in combination with the parfor command to execute parallel for-loop iterations. The fmincon procedure with the "`interior-point`" algorithm is used for optimization.

6.2 Estimation Performance of the Switching Parameter

We first focus on the estimation performance of the SMLE for the switching parameter, the intensity of choice β, which is not only conceptually important for capturing the behavioral switching of market agents among trading strategies but is also crucial for the dynamics of the model.

Aggregated results for various combinations of the three values of interest, namely, $\beta = \{0.1, 1, 10\}$, kernel estimation precision N, time series length t, and the three stochastic term specifications using different intensities σ_ε are presented in Fig. 1. The estimation method is able to recover the pseudo-true values of β, and the estimation accuracy rapidly increases with increasing time series length t. By contrast, the kernel estimation precision N has only a small effect on the accuracy of the estimates, as shown by the similar light gray and dark gray dashed lines marking the 95% sample confidence bands.

Another important finding is that although we allow for economically irrelevant negative values of β within the optimization search, for reasonable time series lengths t, the 95% sample confidence bands is completely located in the positive part of the parameter space. Figure 1 reveals the importance of the finite sample properties for practical utilization of the estimator. With large datasets typical in financial markets analysis, the method is able to recover even very weak signs of behavioral switching [panels (a) and (d)], while stronger manifestations of behavioral switching are detectable with only hundreds of observation [panels (c) and (f)], as is typical for macroeconomic datasets. The impact of memory, which is observable in the bottom part of Fig. 1, naturally decreases the estimation precision with increasing intensity of choice β. For $\beta \in \{0.1, 1\}$, the effect is small, but for the large $\beta = 10$, representing strong switching, memory markedly deteriorates the estimation accuracy.

From now on, all reported results are based on the time series length $t = 10000$ and kernel estimation precision $N = 1000$. Figure 2 and Table 1 depict different aspects of the estimation performance of the SMLE for the switching parameter

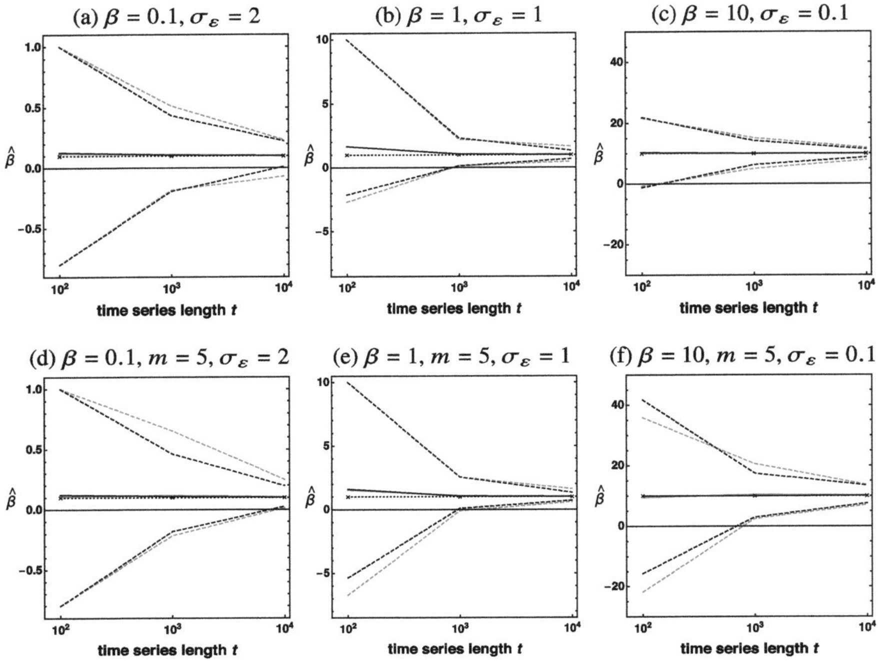

Fig. 1 Estimation of the pseudo-true $\beta \in \{0.1, 1, 10\}$. The upper part depicts the model without memory, and the bottom part represents the model with memory ($m = 5$). Stochastic term ε_t and $\{\varepsilon_i\}_{i=1}^{N}$ drawn from the normal distribution $N(0, \sigma_\varepsilon^2)$. The black dotted line with $\times$ symbol represents the pseudo-true β. The solid gray line depicts the sample means of the estimated $\widehat{\beta}$. The gray dashed lines represent 2.5 and 97.5% quantiles (light gray for $N = 100$ and dark gray for $N = 1000$)

β and further confirm the ability of the SMLE to produce precise estimates with reasonably narrow sample confidence intervals. The impact of memory is clearly observable only for the large $\beta = 10$. The shape of the estimated density of $\widehat{\beta}$ is markedly affected [panels (c) and (f) of Fig. 2], and the sample standard errors differ considerably for $\beta = 10$ [the last row in Table 1], where memory leads to a 50% increase compared to the model without memory.

Table 1 reports a selection of the most interesting combinations of β values and stochastic noise intensities. Complete results are summarized in Table 3 in the Appendix and enable study of the nontrivial interplay between the increasing model dynamics, represented by increasing β, and various stochastic noise intensities, represented by σ_ε, on the estimation performance of the SMLE. In the extreme case of miniscule noise with $\sigma_\varepsilon = 10^{-8}$, the behavior of the model is almost deterministic, resulting in an extremely precise estimation performance with negligible standard errors and a very narrow 95% confidence band between the 2.5 and 97.5% sample quantiles. However, for various realistic noise intensities, we observe that a small noise with $\sigma_\varepsilon = 0.1$ favors estimation of the largest $\beta = 10$, an intermediate standard normal noise provides model dynamics optimal for estimation of intermediate

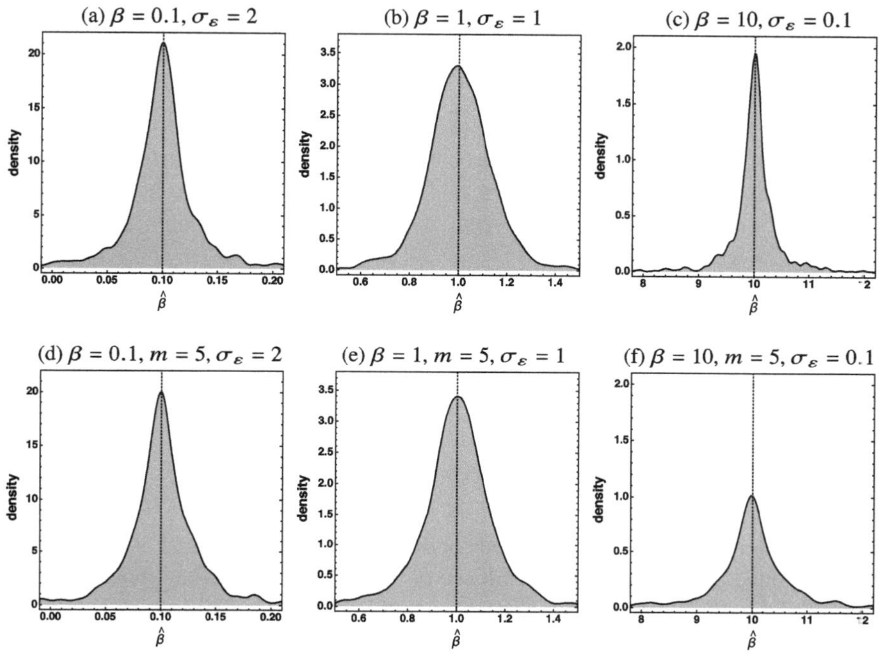

Fig. 2 Density of $\widehat{\beta}$. The upper part depicts the model without memory, and the bottom part depicts the model with memory ($m = 5$). Stochastic terms ε_t and $\{\varepsilon_i\}_{i=1}^N$ drawn from the normal distribution $N(0, \sigma_\varepsilon^2)$. The dotted vertical lines show the pseudo-true β

$\beta = \{1, 5\}$, and a strong stochastic noise $\sigma_\varepsilon = 2$ favors estimation of small to negligible $\beta = \{0, 0.1, 0.5\}$. Compared to the standard errors in Table 3, upper part, the agents' memory in Table 3, bottom part, generally deteriorates the estimation performance of the SMLE when the stochastic noise is small. However, for the intermediate standard normal noise, we do not observe any considerable impact, and for strong noise, we observe an impact only for intermediate and strong switching represented by $\beta = \{1, 5, 10\}$.

Various robustness checks are elaborated in Kukacka and Barunik (2017), who additionally study the impact of different distributions for the random generation of belief parameters g_h and b_h, the effects of inaccurate specification of the stochastic noise within the estimation procedure used to generate $\{y_{t,i}^\theta\}_{i=1}^N$, and the impact of a first-order autocorrelation of the stochastic noise entering the model but not considered within the estimation procedure that assumes i.i.d. noise.

Table 1 Selected results for β estimation

		(a) $\widehat{\beta}$, without memory					(b) $\widehat{\beta}$, with memory $m = 5$				
β	σ_ε	Median	Mean	SE	LQ	HQ	Median	Mean	SE	LQ	HQ
0	2	−0.00	−0.00	0.06	−0.13	0.12	−0.00	0.00	0.04	−0.08	0.08
0.1	2	0.10	0.10	0.05	−0.00	0.22	0.10	0.10	0.04	0.01	0.19
0.5	2	0.50	0.50	0.09	0.35	0.68	0.50	0.50	0.08	0.35	0.68
1	1	1.00	1.01	0.17	0.70	1.30	1.01	1.01	0.17	0.67	1.33
5	1	5.00	5.03	0.70	3.84	6.40	5.00	5.04	0.61	3.89	6.35
10	0.1	10.01	10.01	1.00	8.74	11.21	10.00	10.05	1.50	7.67	12.7

Stochastic terms ε_t and $\{\varepsilon_i\}_{i=1}^N$ drawn from the normal distribution $N(0, \sigma_\varepsilon^2)$. Sample medians, means, standard errors (SEs), and 2.5% (LQ) and 97.5% (HQ) quantiles reported. Results are rounded to two decimal places

6.3 *Shape of the Simulated Log-likelihood Function*

To ensure convergence of the estimated conditional density $\widehat{p_t}(y|x;\theta) \longrightarrow p_t(y|x;\theta)$ and guarantee the asymptotic equivalence of the SMLE, $\widehat{\theta}$, and the infeasible MLE, $\tilde{\theta}$, a set of regularity conditions (Kristensen and Shin 2012, conditions A.1-4, K.1-2, p. 80–81) must be verified for the model and its associated conditional density p_t and kernel K. K.1-2 are generally satisfied by the utilization of the Gaussian kernel; however, for the Brock and Hommes (1998) model, we are not able to verify A.1-4 analytically. We thus take advantage of numerical tools and assess the smoothness condition and the existence of a unique maximum computationally by exploring the numerically computed log-likelihood functions depicted in Fig. 3.

The simulated log-likelihood functions displayed for the pseudo-true $\beta = \{0.1, 1, 10\}$ have generally smooth shapes over the entire domain, which is a result of the Gaussian kernel approximation based on a large number $N = 1000$ of evaluated points. Moreover, a unique maximum is revealed in all 100 random runs. These results are in contrast with some of the findings of Hommes and in 't Veld (2017), who report a very flat shape of the likelihood function in the direction of the intensity of choice, and Grazzini and Richiardi (2015), who argue that together with the multitude of local optima, the flatness of the likelihood can lead to absurd parameter estimates.

The memory introduced in the bottom part of Fig. 3 results in flatter shapes, but both important characteristics remain unchanged. We can thus assume that whether the model is simulated with or without memory, the regularity conditions are satisfied for the model and its associated conditional density p_t, guaranteeing the identification of the parameters.

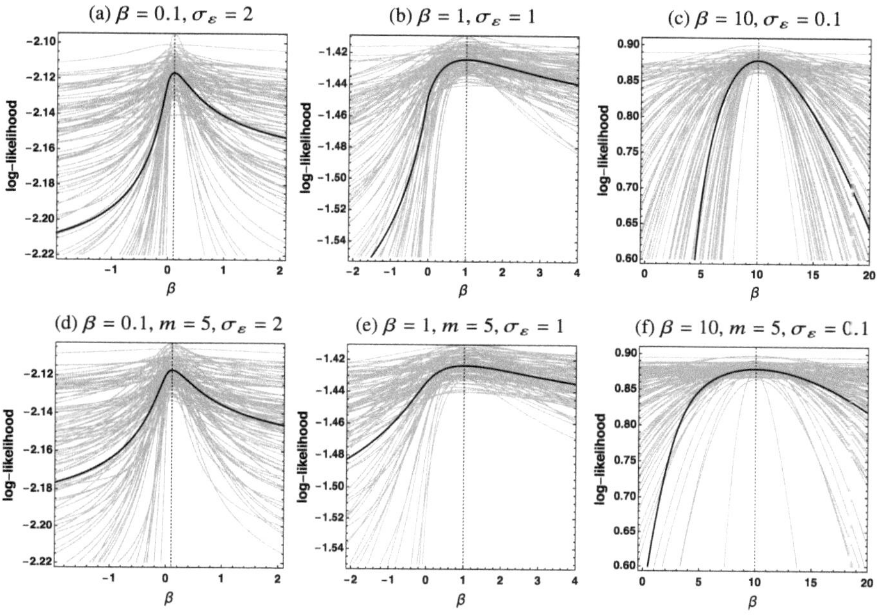

Fig. 3 Shape of the simulated log-likelihood functions. The upper part depicts the model without memory, and the bottom part depicts the model with memory ($m = 5$). Stochastic terms ε_t and $\{\varepsilon_i\}_{i=1}^{N}$ drawn from the normal distribution $N(0, \sigma_\varepsilon^2)$. The dotted vertical lines represent the pseudo-true β. Bold unbroken black curves depict the sample averages. Based on 100 random runs

6.4 Estimation Performance of the Two-Type Model

As a natural step after verification of the ability of the estimator to recover the value of the switching parameter β, we study the joint estimation performance of the SMLE for a triple of crucial parameters governing the dynamics of a two-type model. This simple version of the Brock and Hommes (1998) framework contains the fundamental and a chartist trading strategies. Empirical interactions between these two main types of investors drive the dynamics of financial markets (Allen and Taylor 1990; Frankel and Froot 1991). The belief coefficients of the trading strategies followed by chartists are thus added to the unknown parameter vector θ, i.e., $\theta = \{\beta, g_2, b_2\}$.

The computational setup must be slightly adjusted from that in Sect. 6.1. The number of available trading strategies is decreased to $H = 2$, the trend and bias belief coefficients g_h and b_h are fixed for all 1000 random runs, and the bounds for the parameter space of the two new parameters are set to $\langle -3|g_2|, 3|g_2|\rangle$ and $\langle -3|b_2|, 3|b_2|\rangle$, respectively. A grid of the pseudo-true switching coefficient β and belief coefficients of the chartist strategy g_2 and b_2 is defined in Table 2. A purposeful set of values is designed to cover different combinations of trend-following ($g_2 > 0$), contrarian ($g_2 < 0$), upward-biased ($b_2 > 0$), and downward-biased ($b_2 < 0$) strategies that may result in different market dynamics and estimation performance. Using

Table 2 Results of three-parameter estimation of a two-type model

		(a) $\widehat{\beta}$			(b) $\widehat{g_2}$		(c) $\widehat{b_2}$	
Upper part: without memory								
β, g_2, b_2	σ_ε	Median	Mean	SE	Mean	SE	Mean	SE
0.1, 0.5, 0.2	2	0.10	0.10	0.03	0.50	0.02	0.20	0.00
1, 0.5, 0.2	1	0.98	1.00	0.20	0.50	0.02	0.20	0.00
10, 0.5, 0.2	0.1	9.95	9.99	0.81	0.50	0.02	0.20	0.00
0.1, −0.5, −0.2	2	0.10	0.10	0.02	−0.50	0.02	−0.20	0.00
1, −0.5, −0.2	1	0.98	0.99	0.20	−0.50	0.02	−0.20	0.00
10, −0.5, −0.2	0.1	9.95	10.04	1.30	−0.50	0.02	−0.20	0.00
0.1, 0.5, −0.2	2	0.10	0.10	0.03	0.50	0.02	−0.20	0.00
1, 0.5, −0.2	1	1.00	1.01	0.19	0.50	0.02	−0.20	0.00
10, 0.5, −0.2	0.1	9.98	10.02	0.82	0.50	0.02	−0.20	0.00
0.1, −0.5, 0.2	2	0.10	0.10	0.02	−0.50	0.02	0.21	0.00
1, −0.5, 0.2	1	0.98	1.00	0.20	−0.50	0.02	0.20	0.00
10, −0.5, 0.2	0.1	9.97	9.99	1.32	−0.50	0.02	0.20	0.00
Bottom part: with memory $m = 5$								
0.1, 0.5, 0.2	2	0.10	0.10	0.04	0.50	0.02	0.19	0.00
1, 0.5, 0.2	1	0.99	1.01	0.20	0.50	0.02	0.20	0.00
10, 0.5, 0.2	0.1	10.01	10.05	1.52	0.50	0.03	0.20	0.00
0.1, −0.5, −0.2	2	0.10	0.10	0.03	−0.50	0.02	−0.20	0.00
1, −0.5, −0.2	1	0.99	1.00	0.17	−0.50	0.02	−0.20	0.00
10, −0.5, −0.2	0.1	9.94	10.00	2.24	−0.50	0.02	−0.20	0.00
0.1, 0.5, −0.2	2	0.10	0.10	0.03	0.50	0.02	−0.20	0.00
1, 0.5, −0.2	1	0.99	1.01	0.18	0.50	0.02	−0.20	0.00
10, 0.5, −0.2	0.1	9.94	9.97	1.62	0.50	0.03	−0.20	0.00
0.1, −0.5, 0.2	2	0.10	0.10	0.03	−0.50	0.02	0.20	0.00
1, −0.5, 0.2	1	0.99	1.00	0.17	−0.50	0.02	0.20	0.00
10, −0.5, 0.2	0.1	10.14	10.07	2.24	−0.50	0.02	0.20	0.00

Stochastic terms ε_t and $\{\varepsilon_i\}_{i=1}^N$ drawn from the normal distribution $N(0, \sigma_\varepsilon^2)$. Sample medians, means, and standard errors (SEs) are reported, rounded to two decimal places

a slightly different setup, Kukacka and Barunik (2017) further study a larger grid of combinations under different stochastic noise intensities and provide results of a joint estimation of five parameters in a three-type model.

Although the numerical point estimates of the switching parameter β and related sample standard errors are not directly comparable to the results in Tables 1 and 3 due to the different computational setup, we generally observe similarly favorable estimation performance of the SMLE. The method generally provides precise and unbiased estimates with comparable standard errors for $\beta = \{0.1, 1, 10\}$. Actually, for the negligible switching represented by $\beta = 0.1$, we gain overall statistical significance for all parameter combinations, which was not the case in Tables 1 and 3.

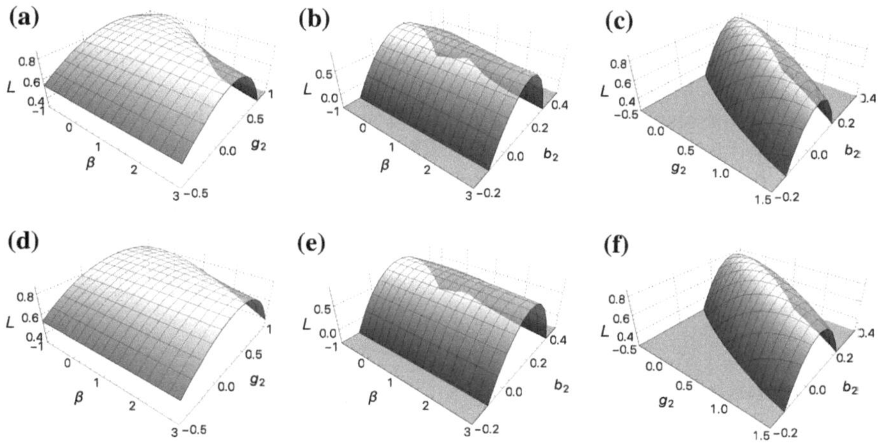

Fig. 4 Shape of the simulated sub-log-likelihood functions in 3D. The pseudo-true parameters: $\beta = 1, g_2 = 0.5, b_2 = 0.2$. The upper part depicts the model without memory, and the bottom part depicts the model with memory ($m = 5$). Stochastic terms ε_t and $\{\varepsilon_i\}_{i=1}^{N}$ drawn from the normal distribution $N(0, 0.1^2)$. L denotes the log-likelihood. Based on 30 random runs

Interestingly, the setup of the chartist belief coefficients does not appear to impact the estimation performance for models with negligible or intermediate switching with $\beta = \{0.1, 1\}$. However, the setup affects the estimation of $\beta = 10$, for which the standard error is markedly higher for the cases of contrarian ($g_2 = -0.5$) strategies than trend-following ($g_2 = 0.5$) strategies. The trend and bias belief coefficients g_h and b_h are estimated with a substantive precision. This finding clearly shows the importance of the belief coefficients for the behavior of the model, as well the ability of the SMLE to recover the pseudo-true values efficiently from all possible dynamics of the output implied by different setups of g_h and b_h.

The impact of memory, shown in the bottom part of Table 2, on the estimation of the belief coefficients appears to be negligible, but memory markedly decreases the precision of the estimation of the switching parameter β. For some combinations of pseudo-true parameters, the standard error of $\widehat{\beta}$ is almost double compared to that of the case without memory.

6.5 *Shape of the Simulated 4D Log-likelihood*

We also verify the smoothness condition and the existence of a unique maximum guaranteeing the identification of parameters in the case of joint estimation. For this purpose, Fig. 4 visualizes partial sub-shapes of the simulated 4D log-likelihood function as transversal 3D profiles of a 4D log-likelihood object in the directions of the displayed parameters. The remaining parameter is always fixed at its pseudo-true value: $\beta = 1, g_2 = 0.5, b_2 = 0.2$. The surface of the profiles numerically verifies the smoothness condition, and areas around global maxima are reasonably detectable.

These results are in contrast with some findings of Chen and Lux (2018), who report multiple local optima, flat shape, and roughness of the surface of the objective function for the SMM estimation.

Following the results in Table 2, a graphical visualization of the 4D shapes of the log-likelihood functions confirms the excellent identification of the bias coefficient b_2 [panels (b) and (c)] and trend coefficient g_2 [panels (a) and (c)] and the potential flatness of the log-likelihood function in the direction of the switching parameter β [panels (a) and (b)], leading to larger standard errors. The memory introduced in the bottom part of Fig. 4 results in a flatter shape in the direction of β, but the directions of g_2 and b_2 remain unaffected.

7 Conclusion

This chapter presents an innovative computational framework for estimation of agent-based models in economic and finance based on the SMLE. As the main advantage, the method allows for estimation of nonlinear models for which the analytical representation of the objective function or its theoretical approximation does not exist. In a large Monte Carlo simulation study, we test the properties and performance of the estimator in combination with the seminal Brock and Hommes (1998) asset pricing model, whose dynamics are governed by the switching of agents between trading strategies based on the discrete choice approach. We also provide links to how the method can be extended to multivariate macroeconomic optimization problems.

We show that the estimation method consistently recovers the pseudo-true parameters with high estimation precision. The simulated likelihood reveals a generally smooth shape over the entire domain, which numerically verifies the smoothness condition important for optimization. A unique global maximum is detectable, guaranteeing the identification of the parameters.

We focus on the estimation performance for the switching parameter called the intensity of choice. Estimation of this parameter is generally challenging in the recent agent-based econometric literature, and empirical estimates often remain insignificant. The crucial finding of our analysis is the ability of the method to recover this parameter; however, the estimation precision strongly depends on the size of the dataset. With a large dataset, which is typical for financial markets analysis, the method can recover even very weak signs of behavioral switching, while stronger manifestations of behavioral switching are detectable with only hundreds of observations, as is typical for macroeconomic datasets.

As a novelty, we further study the impact of agents' memory on the estimation performance of the SMLE. While memory generally deteriorates the precision, the main properties of the estimator remain unaffected.

Acknowledgements Jiri Kukacka gratefully acknowledges financial support from the Czech Science Foundation under the 17-12386Y "Multifractality analysis in finance" project, from the Charles University UNCE project (UNCE/HUM/035), and from the Charles University PRIMUS program (project PRIMUS/19/HUM/17).

8 Appendix

Table 3 Complete results for β estimation

β	Median	Mean	SE	LQ	HQ	Median	Mean	SE	LQ	HQ
Upper part: without memory										
	(a) $\widehat{\beta}$, $\sigma_\varepsilon = 10^{-8}$					(b) $\widehat{\beta}$, $\sigma_\varepsilon = 0.1$				
0	−0.00	0.00	0.04	−0.02	0.03	−0.00	0.01	0.24	−0.50	0.49
0.1	0.10	0.10	0.02	0.08	0.12	0.10	0.10	0.13	−0.10	0.30
0.5	0.50	0.50	0.04	0.47	0.54	0.49	0.49	0.33	−0.41	1.43
1	1.00	1.00	0.07	0.98	1.02	1.01	1.04	0.46	0.16	2.32
5	5.00	5.00	0.11	4.97	5.02	5.00	5.03	0.74	3.86	6.40
10	10.00	10.00	0.12	9.97	10.02	10.01	10.01	1.00	8.74	11.21
	(c) $\widehat{\beta}$, $\sigma_\varepsilon = 1$					(d) $\widehat{\beta}$, $\sigma_\varepsilon = 2$				
0	0.00	0.00	0.10	−0.24	0.22	−0.00	−0.00	0.06	−0.13	0.12
0.1	0.10	0.10	0.08	−0.10	0.30	0.10	0.10	0.05	−0.00	0.22
0.5	0.50	0.51	0.13	0.25	0.77	0.50	0.50	0.09	0.35	0.68
1	1.00	1.01	0.17	0.70	1.30	0.99	1.00	0.14	0.72	1.29
5	5.00	5.03	0.70	3.84	6.40	4.95	5.05	1.00	3.22	7.30
10	10.06	10.13	1.42	7.45	12.90	9.93	9.31	1.63	5.45	11.65
Bottom part: with memory $m = 5$										
	(e) $\widehat{\beta}$, $\sigma_\varepsilon = 10^{-8}$					(f) $\widehat{\beta}$, $\sigma_\varepsilon = 0.1$				
0	0.00	0.00	0.03	−0.02	0.02	−0.02	−0.01	0.34	−0.50	0.50
0.1	0.10	0.10	0.02	0.07	0.13	0.10	0.10	0.16	−0.10	0.30
0.5	0.50	0.50	0.06	0.48	0.52	0.49	0.49	0.54	−0.49	1.49
1	1.00	1.00	0.10	0.98	1.03	1.01	1.03	0.80	−0.92	2.98
5	5.00	5.00	0.13	4.97	5.02	5.00	5.00	1.19	2.65	7.55
10	10.00	9.99	0.19	9.96	10.02	10.00	10.05	1.50	7.67	12.76
	(g) $\widehat{\beta}$, $\sigma_\varepsilon = 1$					(h) $\widehat{\beta}$, $\sigma_\varepsilon = 2$				
0	0.00	0.00	0.11	−0.24	0.22	−0.00	0.00	0.04	−0.08	0.08
0.1	0.10	0.10	0.09	−0.10	0.30	0.10	0.10	0.04	0.01	0.19
0.5	0.50	0.50	0.14	0.25	0.82	0.50	0.50	0.08	0.35	0.68
1	1.01	1.01	0.17	0.67	1.33	1.00	1.02	0.19	0.72	1.39
5	5.00	5.04	0.61	3.89	6.35	4.93	5.04	1.17	2.92	7.82
10	9.98	10.13	1.50	7.41	13.55	10.07	9.87	2.30	5.21	14.68

Stochastic terms ε_t and $\{\varepsilon_i\}_{i=1}^N$ drawn from the normal distribution $N(0, \sigma_\varepsilon^2)$. Sample medians, means, standard errors (SEs), and 2.5% (LQ) and 97.5% (HQ) quantiles reported. Results are rounded to two decimal places

References

Allen H, Taylor MP (1990) Charts, noise and fundamentals in the London foreign exchange market. Econ J 100(400):49–59

Altissimo F, Mele A (2009) Simulated non-parametric estimation of dynamic models. Rev Econ Stud 76(2):413–450

Barde S (2016) Direct comparison of agent-based models of herding in financial markets. J Econ Dyn Control 73:329–353

Barde S, van der Hoog S (2017) An empirical validation protocol for large-scale agent-based models. Technical report, Bielefeld Working Papers in Economics and Management No. 04-2017

Barunik J, Kukacka J (2015) Realizing stock market crashes: stochastic cusp catastrophe model of returns under time-varying volatility. Quant Financ 15(6):959–973

Barunik J, Vacha L, Vosvrda M (2009) Smart predictors in the heterogeneous agent model. J Econ Interact Coord 4:163–172

Bolt W, Demertzis M, Diks C, van der Leij M (2011) Complex methods in economics: an example of behavioral heterogeneity in house prices. Technical report, De Nederlandsche Bank

Bolt W, Demertzis M, Diks CGH, Hommes CH, van der Leij M (2014) Identifying booms and busts in house prices under heterogeneous expectations. Technical report, De Nederlandsche Bank Working Paper No. 450

Brock WA, Hommes CH (1998) Heterogeneous beliefs and routes to chaos in a simple asset pricing model. J Econ Dyn Control 22:1235–1274

Browning M, Hansen LP, Heckman JJ (1999) Chapter 8: micro data and general equilibrium models. In: Handbook of Macroeconomics, Volume 1, Part A of Handbook of Macroeconomics, pp 543–633. Elsevier

Chen S-H, Chang C-L, Du Y-R (2012) Agent-based economic models and econometrics. Knowl Eng Rev 27:187–219

Chen Z, Lux T (2018) Estimation of sentiment effects in financial markets: a simulated method of moments approach. Comput Econ 52(3):711–744

Chiarella C, He X-Z, Zwinkels RC (2014) Heterogeneous expectations in asset pricing: empirical evidence from the SP500. J Econ Behav Organ 105:1–16

Chiarella C, ter Ellen S, He X-Z, Wu E (2015) Fear or fundamentals? heterogeneous beliefs in the european sovereign CDS markets. J Empir Financ 32:19–34

Cornea-Madeira A, Hommes C, Massaro D (2017) Behavioral heterogeneity in U.S. inflation dynamics. J Bus Econ Stat, 1–13

De Grauwe P (2010) Top-down versus bottom-up macroeconomics. CESifo Econ Stud 56(4):465–497

De Grauwe P (2011) Animal spirits and monetary policy. Econ Theory 47(2):423–457

De Grauwe P (2012) Booms and busts in economic activity: a behavioral explanation. J Econ Behav Organ 83(3):484–501. The Great Recession: motivation for re-thinking paradigms in macroeconomic modeling

Del Negro M, Schorfheide F (2012) DSGE model-based forecasting. Technical report, FRB of New York Staff Report No, p 554

Dilaver Ã, Calvert Jump R, Levine P (2018) Agent-based macroeconomics and dynamic stochastic general equilibrium models: Where do we go from here? J Econ Surv 32(4):1134–1159

Fagiolo G, Guerini M, Lamperti F, Moneta A, Roventini A (2019) Validation of agent-based models in economics and finance, pp 763–787. Springer International Publishing

Fagiolo G, Moneta A, Windrum P (2007) A critical guide to empirical validation of agent-based models in economics: methodologies, procedures, and open problems. Comput Econ 30(3):195–226

Fagiolo G, Roventini A (2012) Macroeconomic policy in dsge and agent-based models. Revue de l'OFCE 5(124):67–116

Fernández-Villaverde J (2010) The econometrics of DSGE models. SERIEs 1(1–2):3–49

Fernández-Villaverde J, Rubio-Ramírez JF (2010) Macroeconomics and volatility: data, models, and estimation. Working Paper 16618, National Bureau of Economic Research
Franke R, Westerhoff F (2012) Structural stochastic volatility in asset pricing dynamics: estimation and model contest. J Econ Dyn Control 36(8):1193–1211
Franke R, Westerhoff F (2016) Why a simple herding model may generate the stylized facts of daily returns: explanation and estimation. J Econ Interact Coord 11(1):1–34
Franke R, Westerhoff F (2017) Taking stock: a rigorous modelling of animal spirits in macroeconomics. J Econ Surv 31(5):1152–1182
Frankel JA, Froot KA (1991) Chartists, fundamentalists and the demand for dollars. Technical report, NBER Working Paper No. R1655
Frijns B, Lehnert T, Zwinkels RCJ (2010) Behavioral heterogeneity in the option market. J Econ Dyn Control 34:2273–2287
Ghonghadze J, Lux T (2016) Bringing an elementary agent-based model to the data: estimation via GMM and an application to forecasting of asset price volatility. J Empir Financ 37:1–19
Goldbaum D, Zwinkels RC (2014) An empirical examination of heterogeneity and switching in foreign exchange markets. J Econ Behav Organ 107, Part B, 667–684. Empirical Behavioral Finance
Grazzini J, Richiardi M (2015) Estimation of ergodic agent-based models by simulated minimum distance. J Econ Dyn Control 51:148–165
Grazzini J, Richiardi M, Sella L (2013) Indirect estimation of agent-based models. an application to a simple diffusion model. Compexity Econ 1(2):25–40
Grazzini J, Richiardi MG, Tsionas M (2017) Bayesian estimation of agent-based models. J Econ Dyn Control 77:26–47
Guerini M, Moneta A (2017) A method for agent-based models validation. J Econ Dyn Control 82:125–141
Hansen LP, Heckman JJ (1996) The empirical foundations of calibration. J Econ Perspect 10(1):87–104
Hommes C, 't Veld D (2017) Booms, busts and behavioural heterogeneity in stock prices. J Econ Dyn Control 80:101–124
Hommes CH (2006) Chapter 23: heterogeneous agent models in economics and finance. In: Tesfatsion L, Judd K (Eds) Handbook of computational economics, Vol 2 of Handbook of Computational Economics, pp 1109–1186. Elsevier
Jang T-S, Sacht S (2016) Animal spirits and the business cycle: empirical evidence from moment matching. Metroeconomica 67(1):76–113
Jang T-S, Sacht S (2019) Forecast heuristics, consumer expectations, and new-keynesian macroeconomics: a horse race. J Econ Behav Organ, forthcoming
Kouwenberg R, Zwinkels RCJ (2015) Endogenous price bubbles in a multi-agent system of the housing market. PLoS ONE 10(6):e0129070
Kristensen D, Shin Y (2012) Estimation of dynamic models with nonparametric simulated maximum likelihood. J Econ 167(1):76–94
Kukacka J, Barunik J (2013) Behavioural breaks in the heterogeneous agent model: the impact of herding, overconfidence, and market sentiment. Phys A: Stat Mech Appl 392(23):5920–5938
Kukacka J, Barunik J (2017) Estimation of financial agent-based models with simulated maximum likelihood. J Econ Dyn Control 85:21–45
Kukacka J, Jang T-S, Sacht S (2018) On the estimation of behavioral macroeconomic models via simulated maximum likelihood. Technical Report Economics Working Paper No 2018-11, Kiel University
Kydland FE, Prescott EC (1996) The computational experiment: an econometric tool. J Econ Perspect 10(1):69–85
Lamperti F (2018a) Empirical validation of simulated models through the GSL-div: an illustrative application. J Econ Interact Coord 13(1):143–171
Lamperti F (2018b) An information theoretic criterion for empirical validation of simulation models. Econ Stat 5:83–106

Lamperti F, Roventini A, Sani A (2018) Agent-based model calibration using machine learning surrogates. J Econ Dyn Control 90:366–389
LeBaron B (2006) Chapter 24: agent-based computational finance. In: Tesfatsion L, Judd K (Eds), Handbook of computational economics, Vol 2 of Handbook of Computational Economics, pp 1187–1233. Elsevier
Lee D, Song K (2015) Simulated maximum likelihood estimation for discrete choices using transformed simulated frequencies. J Econ 187(1):131–153
Lof M (2015) Rational speculators, contrarians, and excess volatility. Manag Sci 61(8):1889–1901
Lucas REJ (1978) Asset prices in an exchange economy. Econometrica 46:1429–1445
Lux T (2018) Estimation of agent-based models using sequential Monte Carlo methods. J Econ Dyn Control 91:391–408
Lux T, Zwinkels RC (2018) Chapter 8: empirical validation of agent-based models. In: Hommes C, LeBaron B (Eds), Handbook of computational economics, Vol 4, pp 437–488. Elsevier
Platt D (2019) A comparison of economic agent-based model calibration methods. Technical report, Mathematical Institute & INET, University of Oxford
Polach J, Kukacka J (2019) Prospect theory in the heterogeneous agent model. J Econ Interact Coord 14(1):147–174
Recchioni M, Tedeschi G, Gallegati M (2015) A calibration procedure for analyzing stock price dynamics in an agent-based framework. J Econ Dyn Control 60:1–25
Silverman BW (1986) Density estimation for statistics and data analysis. Chapman and Hall, London
ter Ellen S, Verschoor WF, Zwinkels RC (2013) Dynamic expectation formation in the foreign exchange market. J Int Money Financ 37:75–97
Verschoor WF, Zwinkels RC (2013) Do foreign exchange fund managers behave like heterogeneous agents? Quant Financ 13(7):1125–1134
Windrum P, Fagiolo G, Moneta A (2007) Empirical validation of agent-based models: alternatives and prospects. J Artif Societies Soc Simul 10(2):8

Emergent Urban Morphologies in an Agent-Based Model of Housing

Tsung-Wen Yen, Mikhail Filippov and Siew Ann Cheong

Abstract How do the cities form and emerge? How sensitive is this process to initial conditions of their locations, sizes and other parameters? To answer these questions many theories were developed over the recent years by policy makers, economists, urban planners and mathematicians. Most of these theories use top-down approaches. However, since housing markets consist of large number of heterogeneous agents with non-linear interactions, many have recognized the importance of studying them as complex systems. Drawing inspiration from the Alonso-Muth-Mills model, we developed an agent-based model within which household agents maximize their utilities subject to rent and transportation costs. Through simulating this agent-based model, we show that city-like dense urban agglomerates emerge automatically, and depending on the distribution of incomes, transportation modes, absence and presence of highways and secondary centers of employment, the model is capable of producing a diverse array of urban morphologies.

1 Introduction

Cities are morphologically diverse (Moosavi 2017). Besides having central business districts (CBDs) that are distinct from the surrounding urban regions (Brueckner et al. 1999; Caruso et al. 2007; Wu and Plantinga 2003), we also have monocentric cities as well as polycentric cities (Hartwick and Hartwick 1974; White 1976, 1988; Wieand 1987; Lemoy et al. 2017). While planning may be important in determining the overall structure and functions of urban agglomerates, most cities evolve according to rules constrained by economics (O'Sullivan 2007). Therefore, we can understand urban morphologies as emergent entities from the bottom up due to interactions between city dwellers and the built environment (Xie et al. 2006, 2007; Batty 2007).

T.-W. Yen · M. Filippov · S. Cheong (✉)
Division of Physics and Applied Physics, School of Physical and Mathematical Sciences, Nanyang Technological University, 21 Nanyang Link, Singapore 637371, Singapore
e-mail: cheongsa@ntu.edu.sg

Complexity Institute, Nanyang Technological University, 61 Nanyang Drive, Singapore 637335, Singapore

A. S. Chakrabarti et al. (eds.), *Network Theory and Agent-Based Modeling in Economics and Finance*, https://doi.org/10.1007/978-981-13-8319-9_11

Are there thus a small number of general principles that can generate these different urban morphologies, or do we need specific rules for specific patterns?

To answer a question like this, we turn first to the urban economics model developed independently by Alonso (1964), Muth (1969), and Mills (1967), to study factors affecting the location choices of households. In some sense, where the residents live determine to a great extent where the jobs will be, where the transport hubs will be, where the commercial activities will be, and where other amenities like schools, hospitals, and parks will be. The model was later formalized by Wheaton (1974), who called this the AMM model. Conceptually, the AMM model assumes a city with a fixed population and a given income level living around a central business district (CBD). In this framework, commuting cost becomes progressively more expensive with the distance away from the CBD. As a result, households will attempt to first locate near the CBD. However, this will cause the population pressure to increase and push some of the households outwards to more distant locations. After an equilibrium is reached, cities with different morphologies will emerge. Later analytical works by Brueckner et al. (1999) and Gofette-Nagot et al. (2000) use this model to explain the location choices of households, while Wu and Plantinga modified the AMM model to incorporate the influence of exogeneous amenities (Wu and Plantinga 2003). In 1989, Fujita wrote a comprehensive book on Urban Economic Theory (Fujita 1989) giving extensive coverage of the AMM model.

In the AMM model, we consider a utility function

$$u = \alpha \ln z + \beta \ln s \tag{1}$$

that has the Cobbs-Douglas form, where

$$Y = z + tx + ps \tag{2}$$

is the *total income* of the household, x is the distance between the household and the CBD, t is the transportation cost per unit distance, s is the total floor area, and p is the home price per unit area. When written this way, z can be thought of as disposable income, whereas α and β are parameters, satisfying $\alpha + \beta = 1$. The $\ln z$ term in the utility function makes the household happier if they have more disposable income, while the $\ln s$ term in the utility function makes the household happier if they have a bigger house, but both with diminishing return.

This model is analytically tractable if the CBD is a single point. Following Fujita (1989) and Lemoy (2011), we understand that under equilibrium conditions, the utility is all the same everywhere, and no agents will have the incentive to move further. The bid rent $\psi(x, u)$ would then be maximum with respect to floor area, and thus

$$\psi(x, u) = \max_{s} \frac{Y - tx - z(s, u)}{s}, \tag{3}$$

where $z(s, u) = s^{-\frac{\beta}{\alpha}} \exp\left(-\frac{u}{\alpha}\right)$ is the disposable income. Substituting $z(s, u)$ into Eq. (3) and setting the derivative with respect to s equals to zero, we obtain

$$s(x, u) = \alpha^{-\frac{\alpha}{\beta}}(Y - tx)^{-\frac{\alpha}{\beta}} e^{\frac{u}{\beta}} \quad (4)$$

and

$$\rho(x, u) = s_{\text{tot}} \alpha^{\frac{\alpha}{\beta}}(Y - tx)^{\frac{\alpha}{\beta}} e^{-\frac{u}{\beta}}. \quad (5)$$

Substituting $z(s, u)$ and $s(x, u)$ back into Eq. (3), we arrive at an expression for bid rents $\psi(x, u) = \alpha^{\frac{\alpha}{\beta}}\beta(Y - tx)^{\frac{1}{\beta}} e^{-\frac{u}{\beta}}$. Finally, the density of households can be evaluated using the relation $\rho(x, u) = \frac{s_{\text{tot}}}{s(x,u)}$ and Eq. (4),

$$\rho(x, u) = s_{\text{tot}} \cdot \alpha^{\frac{\alpha}{\beta}}(Y - tx)^{\frac{\alpha}{\beta}} e^{-\frac{u}{\beta}}. \quad (6)$$

In this model, there are two unknowns a priori, i.e. the equilibrium utility u and the radius of the city x_f. We assume that at the boundary of the city, the bid rent will equal to the agricultural rent R_a,

$$\psi(x_f, u) = \alpha^{\frac{\alpha}{\beta}}\beta(Y - tx_f)^{\frac{1}{\beta}} e^{-\frac{u}{\beta}} = R_a \quad (7)$$

and thus if we integrate the density up to x_f, it will equal to the total population N,

$$\int_0^{x_f} 2\pi x \rho(x) dx = N. \quad (8)$$

This is a simplistic conclusion. In general, we expect there to be a distribution of floor areas. Mathematically, this does not pose a problem, because the product $ps = \beta(Y - tx)$ is a function only of x, hence at a fixed x, a larger apartment will have a lower price per unit area, while a smaller apartment will have a higher price per unit area. The edge of the city will be smeared out when there is a distribution $f(s)$ of floor areas. Similarly, the edge of the city will be smeared out if there is a distribution $f(Y)$ for the income. For example, if there are two income levels, then the solution of the AMM model would be the superposition of a rich city on top of a poor city, with the rich city going further out from the CBD. In reality, there can be a second or third center in addition to the CBD, and many researchers have tried to add these into the original AMM model. These additional centers can function as an employment center or a secondary city center. Beside centers, city developers and policymakers are also interested to know how amenities such as public schools, green spaces, swimming pools, ...affect the population distribution. Unfortunately, after adding these factors into the AMM model, it is no longer tractable, and researchers like Raux et al. (2014), Lemoy et al. (2010, 2017), Lemoy (2011), and Kulish et al. (2012) have to turn to agent-based models to gain insights.

The term agent-based models (ABMs) was first coined by Henry Holland (1995), to refer to models where we simulate individual decision makers in a problem. When applied to social sciences, agents can therefore be individual persons, groups, corporations, or even nation states (Gilbert 2008). This form of modeling and simulation also became popular in the biological sciences, where agents can even be individual cells (González et al. 2003; Khan et al. 2003) or animals (Tang and Bennett 2010). It is synonymous with individual-based models (Grimm et al. 2005) and particle-based models (Helbing et al. 2000). In computational terms, an agent is an autonomous software object capable of: (1) setting goals (proactivity); (2) adjusting its course of action in response to changes in the environment (reactivity); and (3) exchanging information with other agents (interactivity). ABM is touted as bottom-up social science, in that we assume agents follow simple rules, but as we simulate interactions according to these rules, we observe complex collective dynamics emerging at the system level. In addition, the simulations and the model are dynamic, in contrast to the AMM model and most analytical models, which are equilibrium models. This time dependence allows us to study out-of-equilibrium dynamics. In general, with an ABM we can explore phenomena which are difficult or impossible to treat analytically.

The first social science to benefit from using ABM is economics, when Thomas Schelling used coins and graph papers to simulate household patterns in his 1971 paper on segregation (Schelling 1971). The first economics ABM to be simulated on a computer is the Santa Fe Artificial Market model (Palmer et al. 1994). This spurred the growth of the field of computational economics (LeBaron 2006). In his 2012 review paper, Chen et al. (2012) named this field *Agent-Based Computational Economics (ACE)*, and devised a taxonomy for various ACE models. In addition, they surveyed many ACE models in the literature and evaluated their performances, before discussing some of the stylized facts in asset markets that cannot be explained by traditional economic models. In other social sciences, Axtell et al. demonstrated with ABMs why the Kayenta Anasazi civilization of Long House Valley in the Black Mesa area of northeastern Arizona declined before the European colonists arrive (Axtell et al. 2002). In the early 1980s, Axelrod organized a competition for computer programs with different strategies to play the iterated Prisoner's Dilemma (Axelrod 1997a), to see which strategy wins in the long run. Later on, he used ABMs to studied political science from ethnocentrism (Hammond and Axelrod 2006; Axelrod and Hammond 2003) to the dissemination of culture (Axelrod 1997b).

In 1990s, Epstein and Axtell developed a large-scale ABM platform called Sugarscape to simulate and explore the role of social process such as death, disease, trade, wealth, sex and reproduction, culture, conflict and war, and externalities such as pollution. This culminated in the book titled *Growing Artificial Societies* to explain how we can use ABMs as a tool to understand the emergent functions of societies from the bottom up (Epstein and Axtell 1996). In 2005, Gilbert and Troitzsch published the book *Simulation for the Social Scientist* (Gilbert and Troitzsch 2005), addressing many social simulation topics. The use of ABMs in social science was then in full swing. From Carley (1999) at CMU, who explored the co-evolution of social networks and culture, to Hegselmann and Krause (2002) and Deffuant et al. (2000) who created formal models of belief dynamics and used ABMs to investigate how interacting agents form consensus or have polarizations in their beliefs according to the threshold levels in the differences of their opinions. They found that for high thresholds settings, they resulted in several opinion clusters whereas high thresholds yields a convergence of opinions toward a consensus. Later in 2006, Sun et al. (2006) developed ABMs to investigate the topics of human cognition, and called it cognitive social simulation. Finally, Reynolds' famous work (Reynolds 1987) on flocking models inspired some of the first biological ABMs that manifested social characteristics.

In this paper, we will start with a comprehensive survey on the use of ABMs to simulate housing markets in Sect. 2. In this section, we review ABMs that explore different factors affecting home choice, and leading to such phenomena as segregation, gentrification, and urban sprawl. We also look at the classification of ABMs depending on which classical theories they are based on, or those that are focused on the urbanization process, or those employing microsimulation. Then in Sect. 3 we describe the ABM we used, before describing our results in Sect. 4. The broad story we can tell is: (1) competition between rental cost and transportation cost lead to home price distribution in space, if the source of income/utility is a single point in the city in Sect. 4.1, (2) the structure of the city if there are two income groups in Sect. 4.2, (3) competition between two transportation modes (walking vs driving) results in the emergence of a central business district (CBD) centered about the point in the city, surrounded by a circular shell that forms the suburbs in Sect. 4.3, (4) distortion of the shape of the city by a highway that passes through the city in Sect. 4.4, when there is (4a) no exclusion region in Sect. 4.4.1, and when (4b) the CBD is an exclusion region devoid of housing in Sect. 4.4.2, and (5) the transition from a distorted household density to a separate satellite town caused by a secondary center of employment in Sect. 4.4.3. We then conclude in Sect. 5.

2 Background

Over the past two decades, we find in the urban economics literature equilibrium sorting models to study heterogeneous location preferences, including the quality of the public education, the quality of neighborhoods, pollutions, green spaces, etc. (Spencer Banzhaf and Walsh 2008; Bayer et al. 2007; Kuminoff et al. 2013; Calabrese et al. 2006). Housing market ABMs are similar to equilibrium models in many aspects, in that they take agents heterogeneity into consideration and study the dynamic processes that individual households sort themselves into neighborhoods based on their preferences and socio-demographic characteristics. However, there are important differences. One of them is that equilibrium sorting models impose market equilibrium conditions to settle the sorting process, whereas in ABMs, sorting emerges due to interactions between all types of market participants. Over the past years, researchers have been developing a variety of ABMs to study emergent spatial phenomena, such as segregation (Crooks et al. 2010; Fossett and Dietrich 2009; Feitosa et al. 2011) urban sprawl (Brown and Robinson 2006) and gentrification (Torrens and Nara 2007). In a series of works, Raux et al. (2014), Lemoy et al. (2010, 2017), Lemoy (2011) built an ABM based on the AMM model to demonstrate how an ABMs can be used to investigate various housing issues. Specifically, he used his ABM to investigate (1) the emergence of developments in a city, (2) polycentrism in cities and how it impacts the economy and society, (3) setting up two income groups and vary their preferences toward amenities and observe how it led to different social-demographical effects. Inspired by the AMM model as well, Kulish et al. (2012) constructed an ABM using Mathematica and calibrated the parameters using statistics collected from Australian cities. He then used his ABM to answer questions like how the structure of the city, including densities and the price of land and houses changes? How the provision of transportation, land use policies, and net population affect the city morphology and cause the housing price to redistribute? Their work shed light on how the authorities could come up with better urban plans for the development of cities.

In 2014, Huang et al. (2014) wrote a review article in which they did a comprehensive survey of the ABM works on household location choices. They classified three categories for these works and the model they proposed, i.e. (1) urban land-use models based upon classical theories such as Schelling's segregation model and the AMM model, (2) different stages of the urbanization process, and (3) integrated ABM and microsimulation models. In addition, they proposed to use some of the new attributes in ABM that have not been used for model classification in the past, i.e. agent heterogeneity (AH), the representations of the land-market processes, and the method of measuring the extensive model outputs. Regarding AH, they showed that Filatova's work (Filatova et al. 2011) on heterogeneity toward risk perceptions will indeed result in different outcomes in spatial and economic metrics in hazard-prone areas. Another example is from Chen et al. (2011) who found that heterogeneity in income can lead to leapfrog development in an exurban area. For land-market representation, they proposed to use *preferences*, *resource constraints*, *competitive bidding*, and *relocation* for classifying different ABM models. Finally, some of the metrics evaluating the performance of the ABM models are mentioned, including *aggregated-level outcomes*, *individual-level outcomes*, and *stochasticity and repetitive runs*. Using these new features, one can analyze differences in models that are grouped in the same category.

Other groups have used ABMs as a computational framework and to understand how house prices fluctuate as various factors change. One of them is McMahon et al. (2009) who set up an ABM environment using NetLogo to study the US housing market crash. In their ABM, we find Buyers, Houses, Banks, and Mortgages, and in spite of the simple settings, they succeeded in simulating a housing bubble and concluded that the interest rate was the reason behind it [a conclusion we also arrived at using a very different approach (Tan and Cheong 2016)]. In contrast, Gilbert et al. (2009) used NetLogo to simulate the housing market in the UK and concluded that the loan-to-value ratio has larger impact than the interest rate on the housing market. Through ABM simulations, Geanakoplos et al. (2012) concluded that leverage instead of interest rates seemed to be a more important factor driving the 1997–2009 US house market boom and burst. More recently in 2017, Ge (2017) developed an ABM to study the US housing market crash. She incorporated many attributes and options for her agents, and concluded that lenient lending and speculation instead of the interest rate are the major cause that increased the housing market volatility.

3 Model

Our model is similar in spirit to the AMM model, in that the main factors affecting location choice is the rental cost and the transportation cost. We do not assume any specific spatial dependence for the rental cost, and want this to emerge from simulation, but we do assume that transportation cost is directly proportional to the distance the agent has to travel from its home to the center of the city, which is the only source of utility in the simulation.

At the start of the simulation, we randomly scatter N agents on a two-dimensional grid with $L_x \times L_y$ cells. Unless otherwise mentioned, we use $L_x = L_y = 80$ for most of our simulations. Very roughly, each cell can be treated as a city block, and each agent treated as a household. Therefore, we assume that each cell has a total surface area of 12 units. This total surface area is to be partitioned amongst the agents residing in the cell. A cell can also be left empty, in which case it would have value per unit area equal to the agricultural rent R_a. In all our simulations, we always have a point-like CBD at the center of the grid. Land prices are initially set equal to the agricultural rent $p_0 = R_a$.

During our simulations, the population N is kept fixed. At each time step n, all agents can move to cells nearby without incurring relocation costs (naturally, we would need to make this nonzero for the ABM to be more realistic). Suppose the price of a cell a distance x from the CBD is p_n at time step n. Assuming the AMM model correctly describes the behavior of households, an agent moving in to this cell would buy a house with optimal area $s = \alpha\frac{Y-tx}{p_n}$, if the total land area utilized by agents within the cell does not exceed 12, and retain a disposable income after rent of $k = \beta(Y - tx)$, if its utility u_f after the move is higher than its utility u_i before the move. If $\Delta U = u_f - u_i < 0$, the agent stays at its original location. If the cell's total floor area is not fully utilized, we will decrease the price a little, simulating a competitive bidding effect in our model.

The first agent to move in an empty cell does not raise its price. However, when a cell is close to its maximum capacity, and many agents wish to move into it, they must draw random bids for the remaining vacancies and the highest bidders get to move in. When an agent bids higher, the price is changed immediately for all other agents in the cell. Their consumption of land is also changed according to $s = \alpha\frac{Y-tx}{p_{n+1}}$ to accommodate the new rent. Furthermore, their utility functions will be re-computed. This feature of the model defines a competition for lands between agents that is similar in spirit to the AMM model. However, we can also allow situations that do not occur in the theoretical model, like allowing an agent who wish to move into a cell that is already full, and thus make offers to agents already in the cell to move out.

For those cells that lies close to the CBD, we expect them to be more crowded than others. Therefore, we set an upper-limit of 14 to the number of agents that can reside in a cell. When cells become fully occupied, we will relocate excess agents to a specific cell at the upper right corner of the simulation grid, to ensure that the number of agents is less than 14. We call this temporary residence a *hotel*. Agents in

Table 1 Parameters used in the simulation

Parameters	Description	Default value
α, β	Preferences for the disposable wealth and floor areas in the utility function	0.9; 0.1
Y_r, Y_p	Income for the rich and poor group	30,000; 48,000
t_r	Transportation cost for the rich group	650–700
N	Population	700
R_a	Agricultural rent	1000
s	Total floor area in a cell	12
γ	Preference for the highway entry ramps, and the employment center	0.1–0.2
a_h, a_e	Attractiveness of the highway entry ramps, and the employment center	4; 4
b_h, b_e	Range of attraction of the highway entry ramps, and the employment center	2; 2

the hotel will be offered one chance per time step to relocate to a new cell that is not yet fully occupied, or have available floor area for agents to buy. If they fail to find a new place to relocate, their utility will be decreased by 0.2, so after one to three time steps, all of them will be eager to find a new place to move into due to their lowered utilities. This is to ensure that agents do not stay in the hotel for too long.

Besides this standard ABM, we also simulated several scenarios, including (1) two groups with different income levels; (2) two groups choosing between different modes of transportation; (3) the effect of a nearby highway; (4) the effects of a nearby second employment center. In cases (1) and (2), the income levels and transportation cost can be varied by assigning different values to each of the groups. Since each group will calculate their utilities based upon these parameters, the equilibrium properties can be modulated by these parameters to reveal changes in the household distribution. For case (3) we treat each entry ramp as an amenity, which affect the agents through an utility term $\gamma \ln\left[1 + a \exp\left(-\frac{r}{b}\right)\right]$, where r is the distance an agent is from an entry ramp. For these simulations we set $\gamma = 0.1$, $a = 4$, and $b = 2$, as shown in Table 1. The locations of entry ramps as well as their attractiveness will determine their influences on the agents and the household distributions. Case (4) is conceptually similar to case (3), except that now the amenity changes from an entry ramp to an employment center.

4 Results

4.1 Population Density, Home Prices, and Floor Areas

When we simulate the ABM, starting from an initial condition where the agents are uniformly distributed across the region, we find the evolution of home locations shown in Fig. 1. We find homes starting to concentrate around the CBD as early as 25 time steps, and a compact circular 'city' starting to take shape after 100 time steps. This evolution towards equilibrium can be seen clearly from the house price, density, and floor area as functions of time in Fig. 2. One important feature to note is that for agents living farther from the CBD, there are larger fluctuations in their purchased floor area. We believe this is because at the boundary of the city, the density value tend to become close to zero, and since the purchased floor area is inversely proportional to the density, it will have larger fluctuations caused by the variations in density.

In this first part of our simulations, we assume all the residents have the same income, i.e. 30,000 per month. After averaging over time and over 10 simulations, we show the house price, resident density, and floor area as functions of the distance

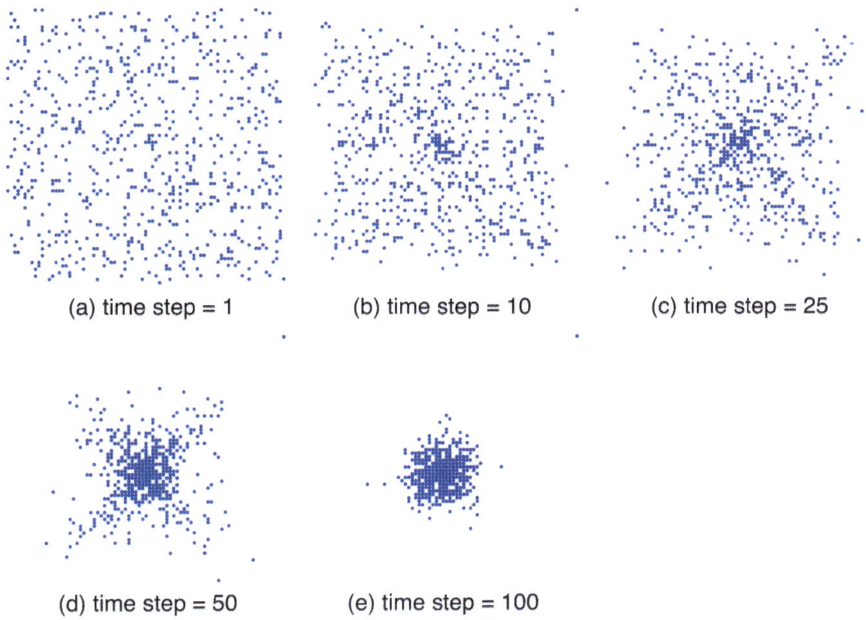

Fig. 1 Simulation of the Netlogo program with 700 agents in the 80 × 80 region. All agents collect the same income of 30,000 by traveling to the center of the region, and spend part of this income to pay for rent and transportation cost. An agent can increase its utility by moving closer to the center of the region, and if we start from an initial state where the agents are uniformly distributed across the region, after 100 time steps we find a city emerging from the simulation

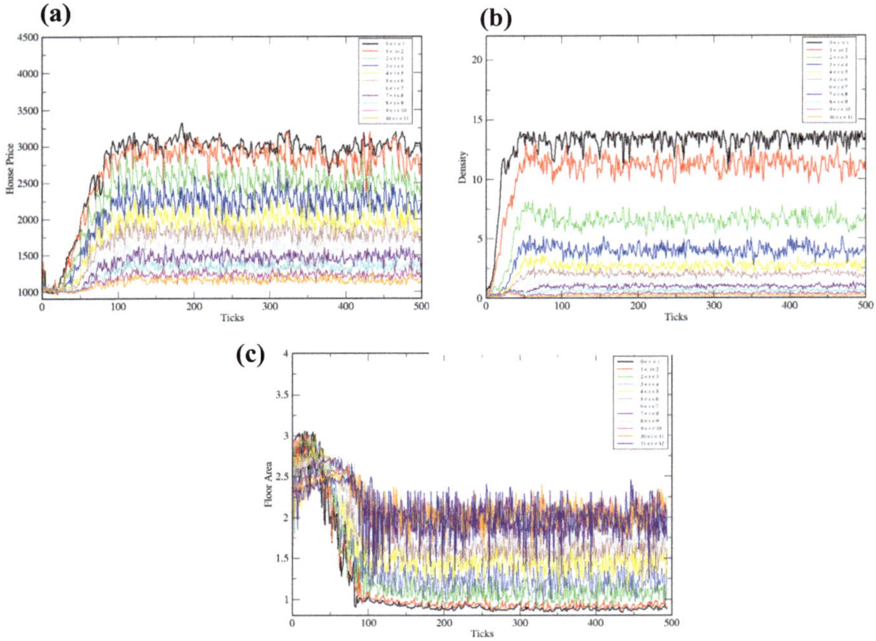

Fig. 2 The **a** house price, **b** resident density, and **c** floor area as functions of time steps

to the CBD in Fig. 3. These results are very close to Lemoy's ABM, Lemoy (2011) and the analytical AMM model. In particular, the house price is highest at the origin of the CBD, amounting to 3000 per unit area, and decreases gradually with respect to the distance away from the origin and finally reach the agricultural rent value of 1000 when the distance becomes $d \geq 11$. We observe a similar trend for the resident density, which is the highest in the CBD at 13 residents per cell, and decreases to zero when $d \geq 11$. The floor area per resident is around 1 at the center of CBD, and increases up to 2.1 at $d = 11$. This is understandable, as the houses near the CBD are more expensive, and therefore the floor area each resident can rent become smaller, very much like the case in Hong Kong.

4.2 *Competition Between Two Income Groups*

In the second part of our simulations, we create two income groups where one group of agents (the red, poor group) always derive the benchmark income of 30,000 from the CBD, while the other group (the blue, rich group) extracts an income that varies from 30,000 to 80,000 from the CBD. Other parameters remained the same as the first part of our simulations. In particular, for both groups of agents, the transportation

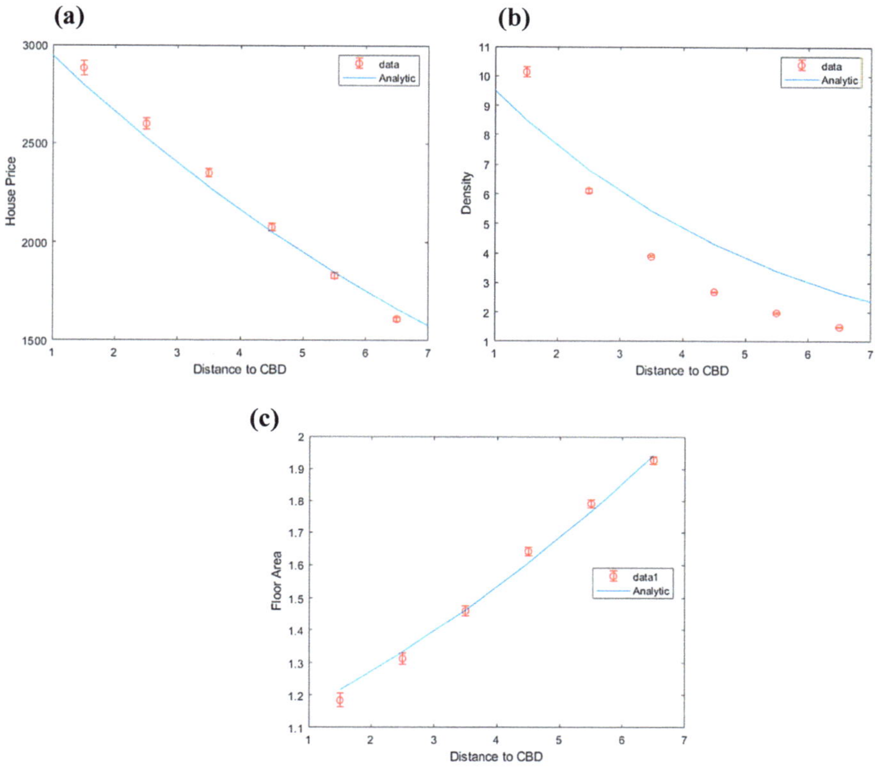

Fig. 3 The time averaged **a** house price, **b** resident density, and **c** floor area with respect to the distance to the CBD. The error bars are calculated from ten separate runs

cost is fixed at 600 per unit distance throughout the simulation. Our results are shown in Fig. 4. When the rich group's income is in excess of 50,000 (Fig. 4a–d), we observe a clear red spot emerging in the middle of the CBD. This indicates the poor agents are accumulating at the center of the CBD, a pattern seen in some North-American cities. This segregation by income level disappears when the rich agents' income level drops below 40,000. In this more equitable situation, the color of the CBD is purple, suggesting well mixing between the two groups of agents.

4.3 Competition Between Two Transportation Modes

In this part of the simulation, we study of the effects of assigning agents with an income level of 30,000 to two different transportation modes. The transportation cost per unit distance for the first group of 700 blue agents taking transportation 1 is $t_1 = 600$, while that for the second group of 700 red agents taking transportation

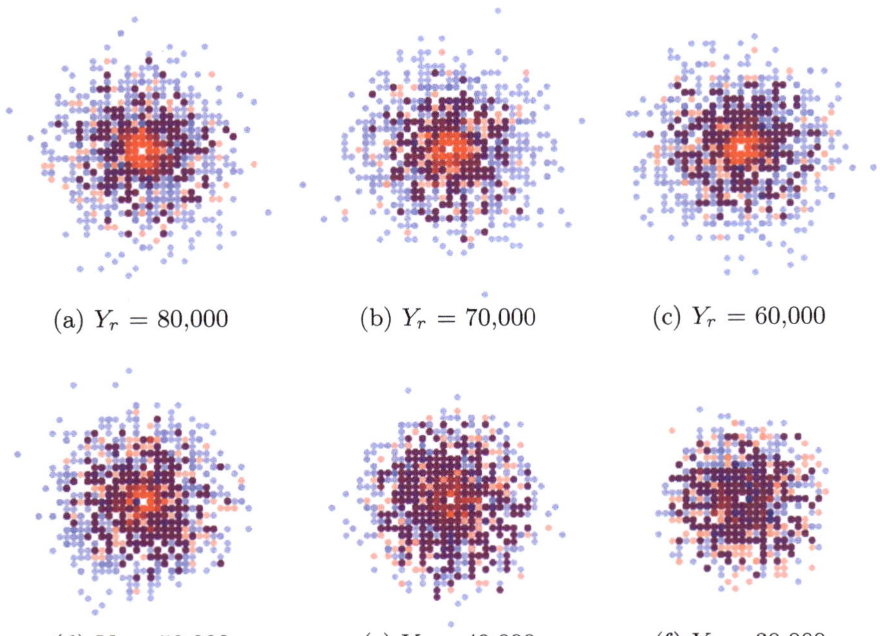

(a) $Y_r = 80{,}000$ (b) $Y_r = 70{,}000$ (c) $Y_r = 60{,}000$

(d) $Y_r = 50{,}000$ (e) $Y_r = 40{,}000$ (f) $Y_r = 30{,}000$

Fig. 4 We set two groups of agents, rich agents in blue, and poor agents in red. The income level of the poor agents are set at 30,000, while that of the rich agents are varied from **a** 80,000 to **f** 30,000 in steps of 10,000

2 is $t_2 = 100, 200, 300, 400, 500, 600$. When $t_2 = t_1$, the two groups of agents are well mixed. When $t_2 < t_1$, in particular when $t_2 = 100$, the agents having the higher transportation cost form a blue disk centered around the CBD, while the agents having the lower transportation cost moved further away from the CBD to form a red suburb. The main result we obtain here is thus the emergence of a larger CBD (for agents with larger transportation cost) and a suburb (for agents with smaller transportation cost) that surrounds it (Fig. 5).

4.4 Effects of Highways

In North-American cities, highways usually go through the center of the cities, whereas in European cities, highways tend to go through the suburb areas instead. We would like to learn how these two different conditions cause the household distribution to change in our simulation. To this end, we simulated several scenarios. In the first scenario, a single highway goes through the center of the CBD, as shown in Fig. 6a. The overall distribution of households becomes deformed into an elliptical shape with the longer axis lying parallel with the highway. The on/off entry ramp

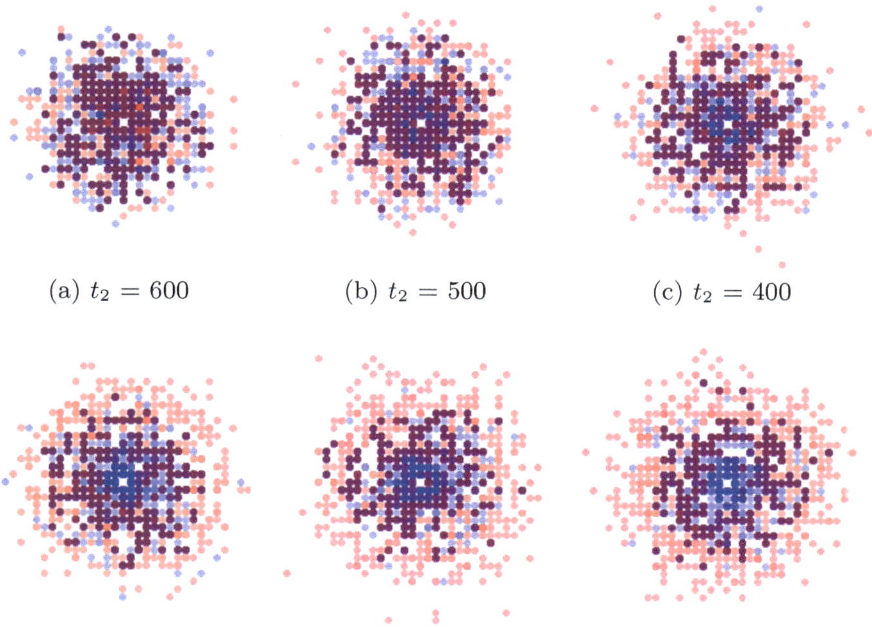

(a) $t_2 = 600$ (b) $t_2 = 500$ (c) $t_2 = 400$

(d) $t_2 = 300$ (e) $t_2 = 200$ (f) $t_2 = 100$

Fig. 5 We set two agents taking different transportation (blue and red). We fix the transportation cost for blue agents as 600 per unit distance, and varies the transportation cost of the red agents from 600 till 100, at a decrease of 100 each time. The results are shown in sub-figure (**a**)–(**f**). If a cell contains only blue or red agents, the color of the cell will be in color gradients of the white (no agents) to blue (red) when the cell is fully-occupied. If a cell contains both red and blue agents, then the color of the cell will be a mix of red and blue colors

at the center of the CBD also attracted more people to live near it, resulting in a denser CBD. In Fig. 6b we shifted the highway to the left by a distance of $d_v = 5$. We observed that the entry ramp closest to the CBD 'pulled' the high-density region away from the CBD. At $d_v = 10$ (Fig. 6c), the nearest entry ramp and the CBD form two separate high-density regions. For $d_v = 15$ (Fig. 6d) and $d_v = 20$ (Fig. 6e), a simple highway has little effect on the distribution of households in the city. Therefore, we added a service highway that connects the nearest entry ramp of the highway to midway between the CBD and the highway. In Fig. 6d, we see that this additional entry ramp cause the two high-density regions in Fig. 6c to merge together again. In contrast, in Fig. 6e, the two high-density regions remain separate.

4.4.1 No Exclusion Region

In addition to the case where there is only one highway, we simulate two highways perpendicular to each other intersecting at the CBD and show them in Fig. 7a. We

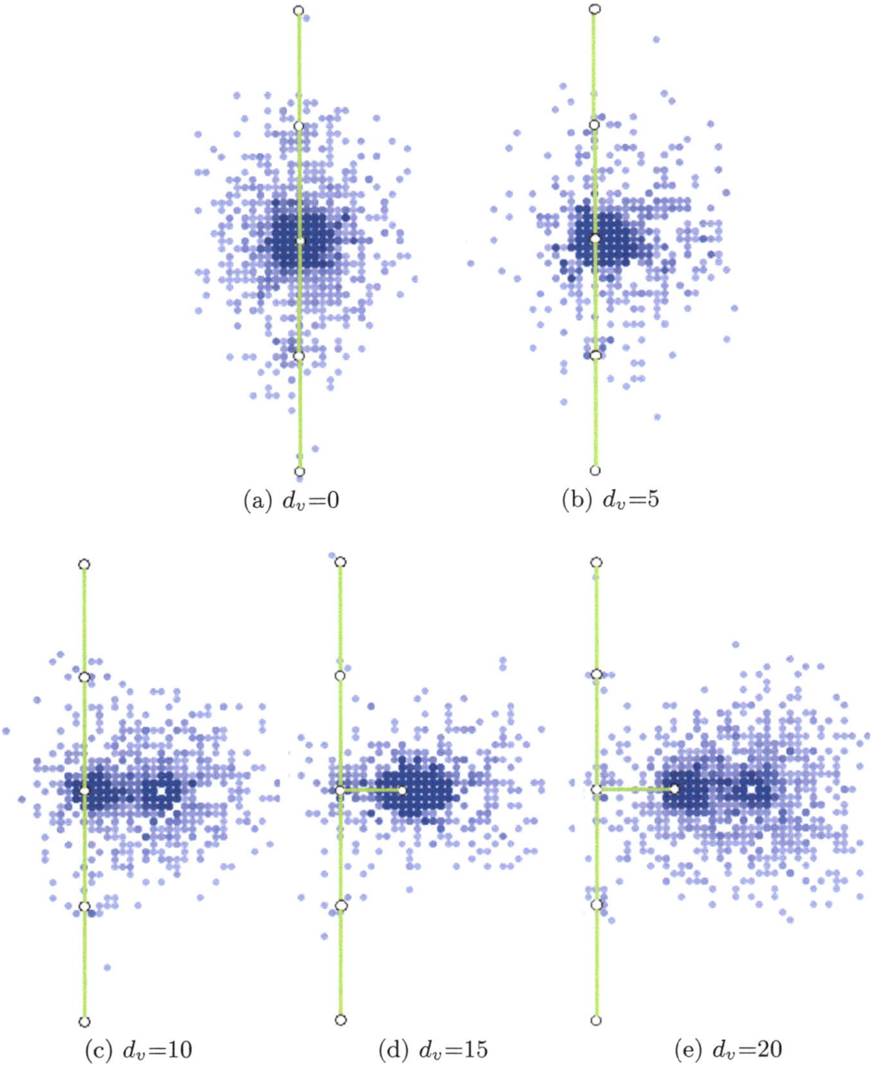

Fig. 6 We assume a highway that has five entry ramps (green dots) separating 20 unit distances away from each other. The vertical distance d_v to the CBD is set at 0, 5, 10, 15, and 20 showed in sub-figures (**a**)–(**e**)

believe the crossing-highways infrastructures can be discovered a lot in major US traffic hubs, e.g. Philadelphia, Atlanta city, Indianapolis, etc. From Fig. 7a, we see that the two perpendicular highways indeed cause the distribution to generate a flower-like shape, which is quite similar in shape to the population distribution maps of major US traffic hubs (Fig. 8).

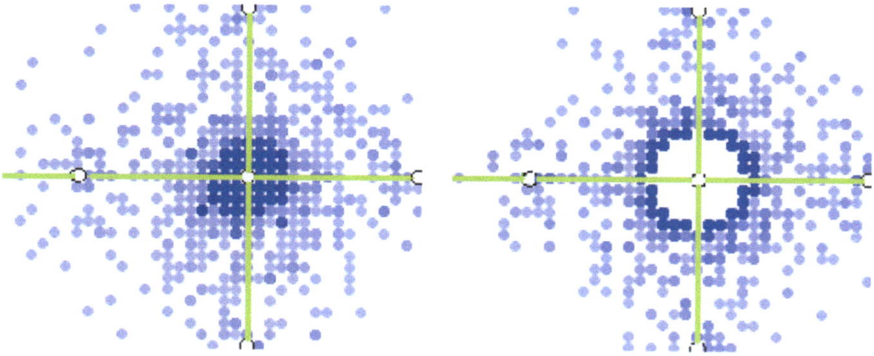

(a) No exclusion zone. (b) An exclusion zone for $d < 4$

Fig. 7 In **a** and **b** we simulated two vertical highways crossing over at the center of CBD, with **a** having no exclusion zone, and **b** having an exclusion zone for $d < 4$. **b** Resembles the case in Singapore

4.4.2 Presence of Exclusion Region

To better understand the household distribution of Singapore where the general scenario is for commuters to travel to the CBD to work during the day and return to their homes outside the CBD at night, we assume an exclusion zone within a distance $d < 4$ of the CBD, and show the simulation results in Fig. 7b. We see that at the edge of the exclusion zone, the density is quite high and agents generally prefer to reside in the concentric area between the inner radius $d = 4$ and the outer radius formed by the four nearest entry ramps. This finding has implications for Singapore policymakers who aim to design better transportation regulations and policies to improve the traffic congestion, but at the same time provide affordable residential locations for the commuters to rent or purchase.

4.4.3 Secondary Center of Employment

Finally, we studied the effect of a nearby secondary center of employment to the distribution of households in the city. Our results are shown in Fig. 9. As we varied the distance d between the secondary employment center and the CBD from $d = 2$ to $d = 25$, we find for $d < 7$ an elliptical distortion to the high-density region surrounding the CBD. Beyond $d = 10$, this high-density region breaks into two smaller parts, one surrounding the CBD, and the other surrounding the secondary employment center. As we increase d further up to 15, the attractiveness of the employment center to the households become lesser, and the household distribution returns to that of a monocentric city. We believe this is because we treated the secondary employment center as an amenity (Table 1). If we were to implement it as a true secondary source of utility, it should at least be able to support a satellite town of its own. Indeed, this

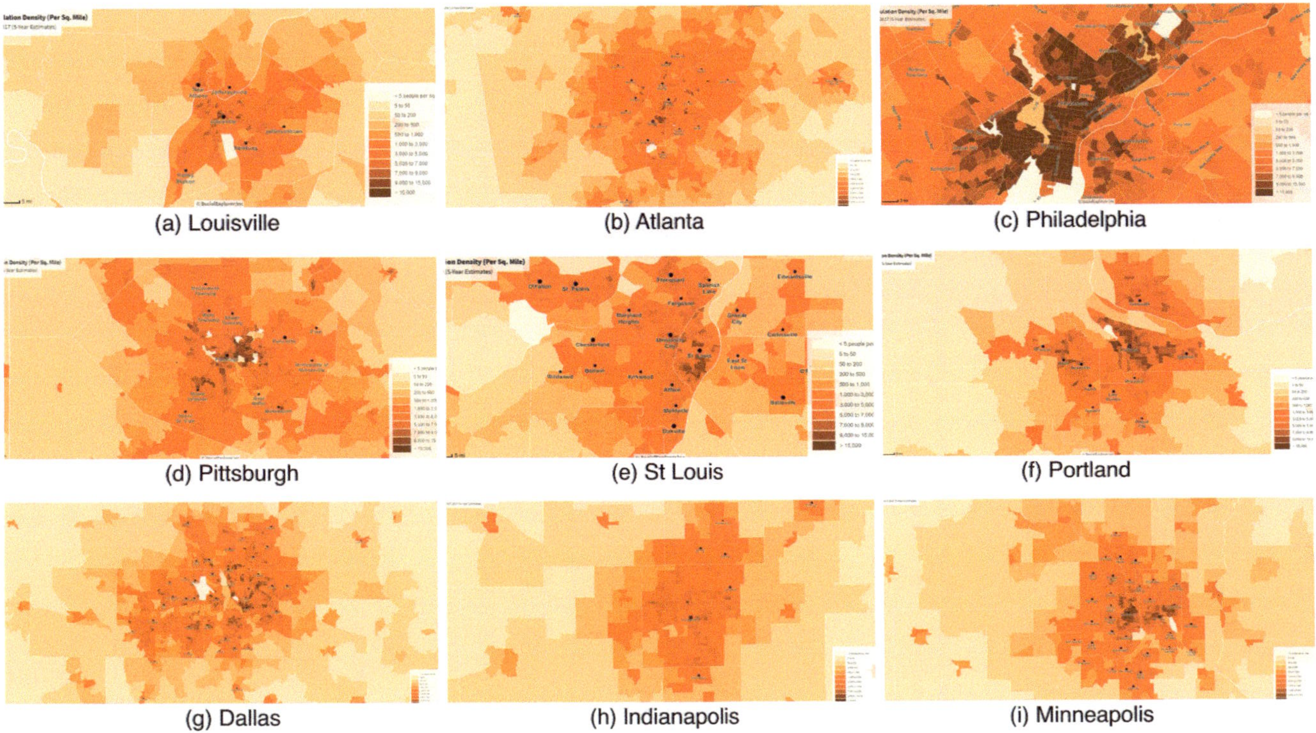

Fig. 8 Population density maps of nine major US traffic hubs, which manifest a flower-like shape. Pictures are retrieved from Social Explorer website. https://www.socialexplorer.com/explore-maps

is what we see in many real-world cities that have secondary employment centers developed far away enough that they become satellite towns.

5 Conclusions

To conclude, in this paper we developed an ABM inspired by the AMM model to explore macroscopic urban morphologies that emerge as a result of microscopic interactions between household agents that choose their housing locations to maximize the utility gained from employment at a point-like CBD, balancing the costs of rent and transportation. We first demonstrated that the combination of a center of utility

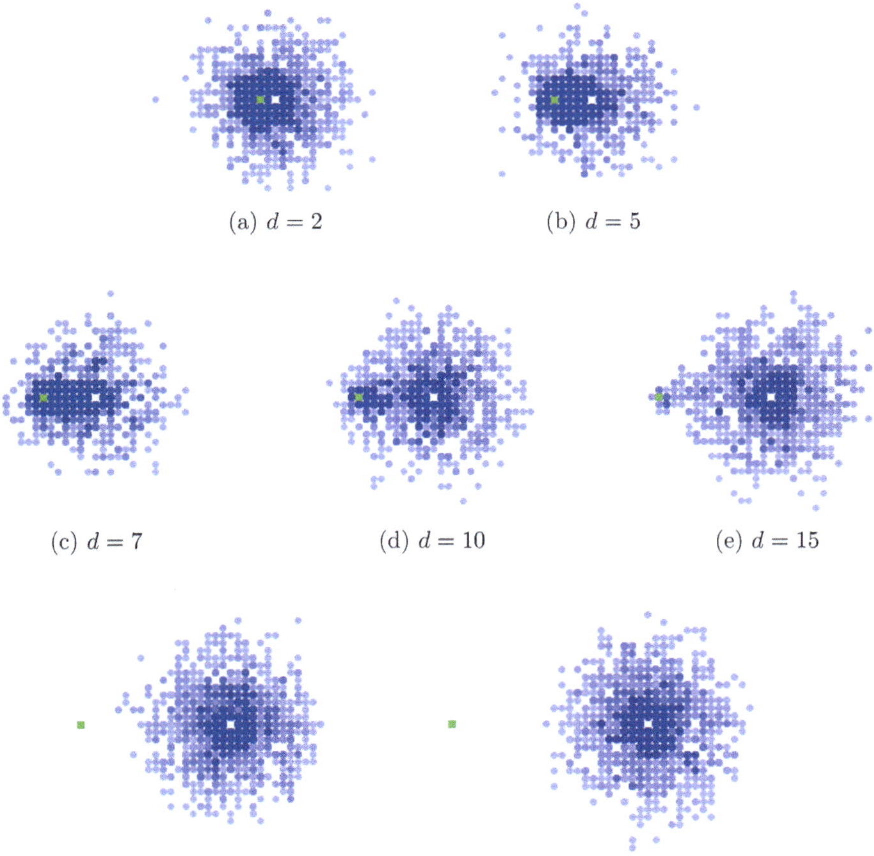

(a) $d = 2$ (b) $d = 5$

(c) $d = 7$ (d) $d = 10$ (e) $d = 15$

(f) $d = 20$ (g) $d = 25$

Fig. 9 We assume an employment center that locates at a distance d from the CBD. d varies from 2 to 25, which corresponds to sub-figures (**a**)–(**g**) production, and balancing the costs of rent and transportation is sufficient to produce dense urban agglomerates that are city-like. We then explored how the housing distribution will change when there are two income groups, versus two transportation modes, and found the emergence of a North-American-like poor inner city when the gap between the two income groups is large, and the emergence of an European-like rich inner city when there is great disparity between the costs of the two modes of transportation.

We also investigated the effects highways have on urban morphologies, and found that a single highway passing through the CBD (North-American-like) has the effect of producing an elongated city, which does not happen when the single highway skirts the suburbs of the city (European-like). When we have a city criss-crossed by two perpendicular highways, we find density enhancements shaped like a four-petal flower centered around the entry ramps. In the presence of an exclusion region, i.e. a region around the CBD where no homes exist (because of competition with commercial use), a high-density ring forms just outside the exclusion region. Finally, we investigated the effects of a secondary center of employment, and found that once it is situated too far from the CBD, the city fragments into a city proper, and a satellite town. These findings have implications on how to develop urban and transportation plans that are compatible with the emergent properties of cities.

References

Alonso W (1964) Location and land use: toward a general theory of land rent. Harvard University Press

Axelrod R (1997a) The complexity of cooperation: agent-based models of competition and collaboration, vol 1. Princeton University Press

Axelrod R (1997b) The dissemination of culture: a model with local convergence and global polarization. J Confl Resolut 41(2):203–226

Axelrod R, Hammond RA (2003) The evolution of ethnocentric behavior. In: Midwest political science convention, vol 2

Axtell RL, Epstein JM, Dean JS, Gumerman GJ, Swedlund AC, Harburger J, Chakravarty S, Hammond R, Parker J, Parker M (2002) Population growth and collapse in a multiagent model of the kayenta anasazi in long house valley. Proc Natl Acad Sci 99(Suppl 3):7275–7279

Batty M (2007) Cities and complexity: understanding cities with cellular automata, agent-based models, and fractals. The MIT Press

Bayer P, Ferreira F, McMillan R (2007) A unified framework for measuring preferences for schools and neighborhoods. J Polit Econ 115(4):588–638

Brown D, Robinson D (2006) Effects of heterogeneity in residential preferences on an agent-based model of urban sprawl. Ecol Soc 11(1)

Brueckner JK, Thisse J-F, Zenou Y (1999) Why is central Paris rich and downtown detroit poor?: an amenity-based theory. Eur Econ Rev 43(1):91–107

Calabrese S, Epple D, Romer T, Sieg H (2006) Local public good provision: voting, peer effects, and mobility. J Public Econ 90(6–7):959–981

Carley KM (1999) On the evolution of social and organizational networks. Res Sociol Organ 16

Caruso G, Peeters D, Cavailhès J, Rounsevell M (2007) Spatial configurations in a Periurban city. A cellular automata-based microeconomic model. Reg Sci Urban Econ 37(5):542–567

Chen S-H, Chang C-L, Du Y-R (2012) Agent-based economic models and econometrics. Knowl Eng Rev 27(2):187–219
Chen Y, Irwin EG, Jayaprakash C (2011) An agent-based model of exurban land development. In: Agricultural and Applied Economics Association (AAEA), 2011 Annual Meeting, 24–26 July 2011
Crooks AT (2010) Constructing and implementing an agent-based model of residential segregation through vector GIS. Int J Geogr Inf Sci 24(5):661–675
Deffuant G, Neau D, Amblard F, Weisbuch G (2000) Mixing beliefs among interacting agents. Adv Complex Syst 3(01n04):87–98
Epstein JM, Axtell R (1996) Growing artificial societies: social science from the bottom up. Brookings Institution Press
Feitosa FF, Le QB, Vlek PLG (2011) Multi-agent simulator for urban segregation (MASUS): a tool to explore alternatives for promoting inclusive cities. Comput Environ Urban Syst 35(2):104–115
Filatova T, Parker DC, van der Veen A (2011) The implications of skewed risk perception for a dutch coastal land market: insights from an agent-based computational economics model. Agric Resour Econ Rev 40(3):405–423
Fossett M, Dietrich DR (2009) Effects of city size, shape, and form, and neighborhood size and shape in agent-based models of residential segregation: are schelling-style preference effects robust? Environ Plan B: Plan Des 36(1):149–169
Fujita M (1989) Urban economic theory: land use and city size. Cambridge University Press
Ge J (2017) Endogenous rise and collapse of housing price: an agent-based model of the housing market. Comput Environ Urban Syst 62:182–198
Geanakoplos J, Axtell R, Doyne Farmer J, Howitt P, Conlee B, Goldstein J, Hendrey M, Palmer NM, Yang C-Y (2012) Getting at systemic risk via an agent-based model of the housing market. Am Econ Rev 102(3):53–58
Gilbert N (2008) Agent-based models, no 153. Sage
Gilbert N, Troitzsch K (2005) Simulation for the social scientist. McGraw-Hill Education, UK
Gilbert N, Hawksworth JC, Swinney PA (2009) An agent-based model of the english housing market. In: AAAI spring symposium: technosocial predictive analytics, pp 30–35
Gofette-Nagot F, Thomas I, Zenou Y (2000) Structure urbaine et revenu des ménages, chapter 10. economie géographique: approches théoriques et empiriques. Economica, Paris
González PP, Cárdenas M, Camacho D, Franyuti A, Rosas O, Lagúnez-Otero J (2003) Cellulat: an agent-based intracellular signalling model. Biosystems 68(2–3):171–185
Grimm V, Revilla E, Berger U, Jeltsch F, Mooij WM, Railsback SF, Thulke H-H, Weiner J, Wiegand T, DeAngelis DL (2005) Pattern-oriented modeling of agent-based complex systems: lessons from ecology. Science 310(5750):987–991
Hammond RA, Axelrod R (2006) The evolution of ethnocentrism. J Confl Resolut 50(6):926–936
Hartwick PG, Hartwick JM (1974) Efficient resource allocation in a multinucleated city with intermediate goods. Q J Econ 88(2):340–352
Hegselmann R, Krause U (2002) Opinion dynamics and bounded confidence models, analysis, and simulation. J Artif Soc Soc Simul 5(3)
Helbing D, Farkas I, Vicsek T (2000) Simulating dynamical features of escape panic. Nature 407(6803):487
Henry Holland J (1995) Hidden order: how adaptation builds complexity. Helix Books
Huang Q, Parker DC, Filatova T, Sun S (2014) A review of urban residential choice models using agent-based modeling. Environ Plan B: Plan Des 41(4):661–689
Khan S, Makkena R, McGeary F, Decker K, Gillis W, Schmidt C (2003) A multi-agent system for the quantitative simulation of biological networks. In: Proceedings of the second international joint conference on Autonomous agents and multiagent systems. ACM, pp 385–392
Kulish M, Richards A, Gillitzer C (2012) Urban structure and housing prices: some evidence from Australian cities. Econ Rec 88(282):303–322
Kuminoff NV, Kerry Smith V, Timmins C (2013) The new economics of equilibrium sorting and policy evaluation using housing markets. J Econ Lit 51(4):1007–1062

LeBaron B (2006) Agent-based computational finance. Handb Comput Econ 2:1187–1233
Lemoy R (2011) The city as a complex system. Statistical physics and agent-based simulations on urban models. Ph.D. thesis, Université Lumière-Lyon II
Lemoy R, Raux C, Jensen P (2010) An agent-based model of residential patterns and social structure in urban areas. Cybergeo Eur J Geogr
Lemoy R, Raux C, Jensen P (2017) Exploring the polycentric city with multi-worker households: an agent-based microeconomic model. Comput Environ Urban Syst 62:64–73
McMahon M, Berea A, Osman H (2009) An agent-based model of the housing market. https://www.semanticscholar.org/paper/An-Agent-Based-Model-of-the-Housing-Market%3A-Mcmahon-Berea/ae1fe95de30c501b523fa1fa32cfcef5b5253460
Mills ES (1967) An aggregative model of resource allocation in a metropolitan area. Am Econ Rev 57(2):197–210
Moosavi V (2017) Urban morphology meets deep learning: exploring urban forms in one million cities, town and villages across the planet. CoRR abs/1709.02939
Muth RF (1969) Cities and housing: the spatial pattern of urban residential land use. University of Chicago Press
O'sullivan A (2007) Urban economics. McGraw-Hill/Irwin
Palmer RG, Brian Arthur W, Holland JH, LeBaron B, Tayler P (1994) Artificial economic life: a simple model of a stockmarket. Phys D: Nonlinear Phenom 75(1–3):264–274
Raux C, Ma T-Y, Lemoy R, Ovtracht N (2014) A LUTI agent-based model of Lyon area: welfare analysis of some scenarios. Transp Res Arena (TRA)
Reynolds CW (1987) Flocks, herds and schools: a distributed behavioral model. ACM SIGGRAPH computer graphics, vol 21. ACM, pp 25–34
Schelling TC (1971) Dynamic models of segregation. J Math Sociol 1(2):143–186
Spencer Banzhaf H, Walsh RP (2008) Do people vote with their feet? An empirical test of Tiebout. Am Econ Rev 98(3):843–863
Sun R et al (2006) Cognition and multi-agent interaction: from cognitive modeling to social simulation. Cambridge University Press
Tan J, Cheong SA (2016) The regime shift associated with the 2004–2008 us housing market bubble. PLoS ONE 11(9):0162140
Tang W, Bennett DA (2010) Agent-based modeling of animal movement: a review. Geogr Compass 4(7):682–700
Torrens PM, Nara A (2007) Modeling gentrification dynamics: A hybrid approach. Comput Environ Urban Syst 31(3):337–361
Wheaton WC (1974) A comparative static analysis of urban spatial structure. J Econ Theory 9(2):223–237
White MJ (1976) Firm suburbanization and urban subcenters. J Urban Econ 3(4):323–343
White MJ (1988) Location choice and commuting behavior in cities with decentralized employment. J Urban Econ 24(2):129–152
Wieand KF (1987) An extension of the monocentric urban spatial equilibrium model to a multicenter setting: the case of the two-center city. J Urban Econ 21(3):259–271
Wu J, Plantinga AJ (2003) The influence of public open space on urban spatial structure. J Environ Econ Manag 46(2):288–309
Xie Y, Batty M, Zhao K (2007) Simulating emergent urban form using agent-based modeling: Desakota in the Suzhou-Wuxian region in China. Ann Assoc Am Geogr 97(3):477–495
Xie Y, Mei Y, Bai Y, Xing X (2006) Ecological analysis of an emerging urban landscape pattern-desakota: a case study in Suzhou. China. Landsc Ecol 21(8):1297–1309

The Transferability of Human Capital, the Brain Drain and the Brain Gain

Akira Shimada

Abstract I investigate the effects of transferability of human capital on the labour-sending country's total human capital, i.e. which of the brain drain and the brain gain happens in a steady state and in the short run under the different degrees of human capital transferability. I find that if transferability of human capital is low, education demand is small and no individuals migrate. Conversely, if it is high, education demand is large and individuals with high innate ability attempt to migrate. I also find that even with high transferability, the labour-sending country experiences the brain drain in both the short run and steady state if the wage disparity is small between labour-sending and labour-receiving countries. However, if it is large, high transferability brings about the brain gain in a steady state whereas the brain drain might occur in the short run. Accordingly, highly transferable human capital does not necessarily contribute to raising the labour-sending country's total human capital. The labour-sending country might be faced with a trade-off in raising its total human capital in the short run and steady state.

1 Introduction

In this chapter, I deal with the problem related to the transferability of human capital and the possibilities of the brain drain or the brain gain in the labour-sending country. In particular, I investigate how labour migration is related to the transferability of human capital among individuals with different innate abilities. I also investigate which of the brain drain and the brain gain happens under different wage disparities when transferability of human capital is high.

Individuals prefer to work in a country with higher wages instead of one with lower pay. The country with lower wages may experience an outflow of workers. In that case, human capital that remains in the country may be diminished. However, at the same time, individuals in the country with lower wages may receive more education and build larger human capital when migration to the country with

A. Shimada (✉)
Faculty of Economics, Nagasaki University, Nagasaki, Japan
e-mail: shimada@nagasaki-u.ac.jp

A. S. Chakrabarti et al. (eds.), *Network Theory and Agent-Based Modeling in Economics and Finance*, https://doi.org/10.1007/978-981-13-8319-9_12

higher wages is possible because the return of education is raised by migration possibilities. Consequently, the country's total human capital may increase. Given these two opposing effects, it cannot be determined a prior whether the wage disparity between labour-sending and labour-receiving countries causes a brain drain, i.e. smaller human capital under migration possibilities than otherwise or a brain gain, i.e. larger human capital under migration possibilities than otherwise.

However, wage disparities are not the only factors causing the brain drain or the brain gain. Human capital is not perfectly transferable to the foreign country. Migrants cannot utilise human capital built in the home country fully in the host country. Wages that migrants actually receive in the host country will be much lower than the wage disparity indicates if human capital is less transferable.

This limitation suggests that which of the brain drain or the brain gain happens depends not only on the wage disparity but also on human capital transferability. Previous studies on the brain drain did not pay sufficient attention to the effects of the imperfect transferability of human capital on its formation in the labour-sending country.

In addition, because human capital evolves over time, the wage disparity and transferability of human capital will likely affect short- and long-run human capital formation differently. However, migration's effects on human capital formation in the short run and long run have not been distinguished clearly until now although there are some exceptions.

As individuals are not homogenous, how much education they receive and how much human capital they accumulate differ between individuals. This suggests that the accumulation of the country's total human capital should be considered from the perspective of the maximising behaviour of individuals having such heterogeneous qualities.

In this chapter, I assume an overlapping-generations economy with heterogeneous individuals and investigate the effects of the international transferability of human capital on an individual's decisions with regard to education and migration. I further investigate the effects of human capital transferability on short-run human capital formation, as well as those on steady-state human capital formation.

I find that if transferability of human capital is low, education demand is small and no individuals migrate. In contrast, if it is high, education demand is high and individuals with high innate ability have an incentive to migrate, whereas those who do not prefer not to migrate.

I also find that even if human capital is highly transferable, the labour-sending country experiences the brain drain both in the short run and in the steady state if the wage disparity is small. If the wage disparity is large, the brain gain occurs in steady state due to high transferability. However, even in such a situation, it cannot be determined a prior whether the labour-sending country experiences the brain drain or the brain gain in the short run. Accordingly, the trade-off relationship might exist between short-run and steady-state human capital formation.

This chapter has the following structure: Sect. 2 presents a literature review. Section 3 models a small open economy. Section 4 solves the utility maximisation problems. Section 5 examines the migration decision of individuals. Section 6 derives the labour-sending country's human capital under different degrees of human capital transferability. Section 7 examines which of the brain drain and the brain gain occurs in the short run and in steady state when human capital is highly transferable. Section 8 provides concluding remarks.

2 Literature Review

According to the traditional view of the international migration of labour, migration causes the outflow of skilled workers and the labour-sending country suffers from reduced human capital. In other words, the labour-sending country experiences the brain drain. Bhagwati and Hamada (1974), Hamada and Bhagwati (1975), Miyagiwa (1991) and Haque and Kim (1995) emphasised this negative effect of migration.

After the mid-1990s, the positive effect of migration was offered by Mountford (1997), Stark et al. (1997, 1998) and others. Migration possibilities encourage individuals in the labour-sending country to receive education and build human capital. As a result, total human capital may be larger than otherwise under conditions where migration is possible. Beine et al. (2001, 2008) and Docquier and Rapoport (2009) found that the labour-sending country might raise human capital due to migration possibilities.

The net resultant of these two opposing effects of migration on a labour-sending country's total human capital, i.e. whether they experience the brain drain or the brain gain is not uniform across countries. Beine et al. (2011) found the positive effect in the low-income countries with the low skilled emigration rate, but they also found a negative effect in the middle- and high-income countries. The effect was comprehensively investigated by Boeri et al. (2012). Numerous studies including Batista et al. (2012) and Dinkelman and Mariotti (2016) have been conducted on the effect on human capital of the specific labour-sending countries down to the present.

If human capital is not perfectly transferable to the foreign country, part or all of the skills built in the home country may not be available in the host country after migration (Docquier and Rapoport, 2012, p. 700). This scenario happens due to differences in language or the way of doing certain jobs. Chiswick and Miller (1992, 2009), Özden (2006), Mattoo et al. (2008), Sanromá et al. (2015) and Tzanakou and Behle (2017) presented evidence pertaining to the low degree of international transferability of human capital.

Docquier and Rapoport (2012) suggested that the transferability of human capital changes the likelihood of either a brain drain or a brain gain. Higher transferability may encourage human capital formation in the labour-sending country because the returns to education are higher and individuals are inspired to study more. It may also lower total human capital because individuals become more likely to leave the country.

Thum and Uebelmesser (2000) and Pires (2015) are a few studies that incorporated transferability of human capital into the discussion of the brain drain and the brain gain. According to Thum and Uebelmesser (2000), transferability raises human capital in the labour-sending country.

Based on Docquier and Rapoport's (2009) stylised model on self-selection and the brain drain, Pires (2015) showed that low transferability causes the brain waste and makes the brain gain less likely because the brain waste reduces the incentives of individuals to receive education.

His analysis is interesting. However, there are only two kinds of workers in his model: skilled and unskilled. Because decisions pertaining to education and migration are usually made by each individual (or family), individuals demand education and accumulate human capital differently, depending on their innate ability which usually differs between individuals. It is thus best to assume that individuals are *heterogeneous*, rather than focusing on the representative individual. In addition, in his model, there is only one generation and the economy begins in the first period and ends in the second. However, clearly, human capital has a long-lasting effect. Human capital accumulated by the past generations tends to have a positive effect on the present generation's human capital formation. To incorporate such an intergenerational effect (Galor and Stark, 1994; Vidal, 1998) into the analysis, the analysis should be conducted within a *dynamic* context. Doing so will enable us to examine both the short- and long-run effects of migration possibilities, as in Byra (2013), even though imperfect transferability is not taken into account.

3 Model

This section models a small open economy with overlapping-generations and heterogeneous individuals.

The small open economy is connected to the rest of the world through labour emigration. In this chapter, the small open economy is called the *home country* and the rest of the world is called the *foreign country*. Individuals in the home country live for two periods and there are two generations in each period. They may migrate to—and work in—the foreign country.

Individuals are heterogeneous in their innate ability. In particular, in each period an individual i is born with innate ability a_i. Individuals' innate abilities are distributed uniformly between $\underline{a}(> 0)$ and $\bar{a}(> \underline{a} + 1)$ where $\underline{a}$ is sufficiently close to 0 and $\bar{a}$ is sufficiently large.

In the first period of life, i.e. in the youth-age, an individual receives education. An individual i receives education by paying the fee privately and accumulates the human capital that is utilised to work during the old-age. Human capital accumulated by the previous (old) generation and remains in the home country also helps the present (young) generation build human capital. Accordingly, an intergenerational externality of human capital is operative. For simplicity, I assume that an individual does not consume during his youth and that receiving an education itself does provide an individual with utility.

In the second period of life, i.e. in the old-age, an individual works in the home country or migrates to the foreign country to work. The migration probability is given exogenously. If he works in the home country, he can fully utilise his human capital accumulated during youth. However, he cannot fully utilise it if he works in the foreign country because human capital built in the home country cannot be perfectly transferred to the foreign country. An individual consumes as well as works during the old-age.

An individual attempts to maximise utility derived from net earnings that are spent on consumption in the old-age by manipulating the amount of education. In particular, he borrows money to finance education when he is young and pays it back from wages when he is old. He also borrows money to finance migration cost during his youth if he attempts to migrate to the foreign country in the old-age. He determines how much education to receive at the beginning of his life. An individual decides whether to migrate or not by comparing the utility derived from non-migration with that from migration.

Some individuals in the home country may opt to work in the foreign country In such cases, part of the human capital accumulated in the home country flows to that country, and the home country may suffer from the brain drain, i.e. the total human capital that remains in the home country is *smaller* when emigration occurs than when it does not. However, even if emigration occurs, it is possible that the home country experiences the brain gain, too, i.e. the total human capital that remains in the home country is *larger* when emigration occurs than when it does not. The brain gain happens when the higher return of education encourages human capital formation and this positive effect dominates the negative one arising from the outflow of human capital.

4 Utility Maximisation

In this section, I solve the individual's utility maximisation problem in the case where he does not attempt to migrate and he definitely stays in the home country to work, and in another case where he attempts to migrate to the foreign country, in order to work there, with a certain probability.

4.1 The Case Where an Individual Does not Attempt to Migrate

This subsection deals with the case where an individual does not attempt to migrate to the foreign country.

An individual born at the beginning of period t with innate ability a_i receives education by $e_{i,t}$ in period t and accumulates human capital by $h^{nm}_{i,t+1}$

$$h^{nm}_{i,t+1} = \beta(a_i + e_{i,t}) + (1 - \beta)h_t, 0 < \beta < 1, \tag{1}$$

where h_t is the sum of human capital of individuals existing in the home country in period t. Equation (1) represents human capital measured in efficiency units of labour that is available to an individual with innate ability a_i to work in the home country in period $t+1$. The coefficients β and $1-\beta$ represent the relative importance of innate ability and education in building human capital and the relative importance of intergenerational externality, respectively. An individual with innate ability a_i utilises $h_{i,t+1}^{nm}$ when providing his labour in period $t+1$ in the home country.

Undertaking one's education incurs a cost $(c/2a_i)e_{i,t}^2$, where c is a positive constant. An individual with higher innate ability can receive the same amount of education at a lower cost. Its marginal cost is increasing because a larger amount of money is needed to receive the additional amount of education as he proceeds to a higher level. He borrows money by $(c/2a_i)e_{i,t}^2$ when young and pays back when old.

Accordingly, his net earnings in period $t+1$ are $h_{i,t+1}^{nm}w'_H - (1+r)(c/2a_i)e_{i,t}^2$, where $w'_H > 0$ is wages per efficiency in the home country and $r > 0$ is the world interest rate, which is given exogenously and is a constant.

His lifetime utility is higher as his net earnings are larger. I represent the utility of such an individual by

$$u_{i,t,t+1}^{nm} = h_{i,t+1}^{nm}w_H - (c/2a_i)e_{i,t}^2, \tag{2}$$

where $w_H \equiv w'_H/(1+r)$. He maximises lifetime utility at the beginning of his life, i.e. at the beginning of period t by manipulating education.

An individual with innate ability a_i demands education by

$$e_{i,t}^{nm} = \beta\frac{w_H}{c}a_i. \tag{3}$$

Equation (3) suggests that education demand increases with wages per efficiency and innate ability and decreases with c. By substituting Eq. (3) into Eq. (1), it can be found that his human capital is larger as he is more innately able.

Substituting Eq. (3) into Eq. (2), his utility is determined as

$$u_{i,t,t+1}^{nm\ *} = \left(\beta w_H + \beta^2\frac{w_H^2}{2c}\right)a_i + (1-\beta)w_H h_t. \tag{4}$$

As Eq. (4) states, his utility increases with his innate ability and the human capital built in the previous period and available in the current period.

4.2 The Case Where an Individual Attempts to Migrate

This subsection deals with the case where an individual attempts to migrate to the foreign country and actually migrates with a certain probability.

An individual who attempts to migrate is faced with two possibilities. One possibility is to migrate to—and work in—the foreign country with a probability $0 < p < 1$, where p is a constant. Another is not to migrate and instead stay and work in the home country with a probability $1 - p$. As human capital possessed by individuals cannot be easily seen and distinguished by the foreign country's firms, it is likely that individuals, even with different amounts of human capital, migrate and get the job with almost the same probability. For these reasons, I assume that the migration probability is identical to all individuals.

A wage disparity exists between the home and foreign countries. In particular, wages per efficiency in the foreign country are $\alpha > 1$ times as high as those in the home country.

In cases where an individual with innate ability a_i migrates, human capital available to him in the foreign country in period $t + 1$ is

$$h_{i,t+1}^{m,p} = \beta(a_i + \theta e_{i,t}) + (1/\alpha)(1 - \beta)h_t, 0 < \theta < 1, \tag{5}$$

where θ measures the degree of transferability of human capital built by education. As mentioned already, human capital formed in the home country cannot be fully exerted in the foreign one. How much human capital built by education can be transferred to the foreign country depends on the kind of education (e.g. so-called *global* education or *country-specific* education) so that its transferability can assume various values. Accordingly, θ falls between 0 and 1. In contrast, the transferability of human capital built via intergenerational externality is less likely to vary across individuals because the intergenerational externality indirectly affects emigrants' human capital formation. For this reason, I assume that its transferability is a constant and equal to $1/\alpha$. This assumption suggests that even with higher wages in the foreign country, labour provided in the foreign country by utilising human capital built via intergenerational externality generates earnings equal only to those generated by working in the home country. This assumption reflects the fact that the larger the wage disparity, the larger the technological differences between the home and foreign countries and the less important are the effects of intergenerational externality. It also enables us to solve the model easily.

The cost of migration is $MC > 0$, which is a constant, being independent of an individual's human capital.

He borrows money by $(c/2a_i)e_{i,t}^2 + MC$ when young and pays it back from his earnings $h_{i,t+1}^{m,p}\alpha w_H$ when old.

Accordingly, his net earnings in period $t + 1$ evaluated in period t are

$$h_{i,t+1}^{m,p}\alpha w_H - (c/2a_i)e_{i,t}^2 - MC.$$

In another case where he does not migrate and stays in the home country, human capital available to him in that country in period $t + 1$ is

$$h_{i,t+1}^{m,1-p} = \beta(a_i + e_{i,t}) + (1 - \beta)h_t. \tag{6}$$

This is the same as $h^{nm}_{i,t+1}$ [Eq. (1)]. In this case, his net earnings in period $t+1$ evaluated in period t are

$$h^{m,1-p}_{i,t+1} w_H - (c/2a_i)e^2_{i,t}.$$

Summing up net earnings in two cases with weights, I represent lifetime utility at the beginning of period t by

$$u^m_{i,t,t+1} = ph^{m,p}_{i,t+1}\alpha w_H + (1-p)h^{m,1-p}_{i,t+1} w_H - \frac{c}{2a_i}e^2_{i,t} - pMC. \tag{7}$$

By utility maximisation, his demand for education is

$$e^m_{i,t} = \beta(p\theta\alpha + 1 - p)\frac{w_H}{c}a_i. \tag{8}$$

Similar to Eq. (3), education demand increases with wages per efficiency and innate ability and decreases with c. Equation (8) also suggests that higher migration possibilities raise demand for education if $\theta\alpha > 1$ due to the higher return of education. It should be noted that education demand is not necessarily higher with greater migration probabilities. It is also related with the transferability of human capital.

His utility is obtained by substituting Eq. (8) into Eq. (7).

$$\begin{aligned} u^{m\ *}_{i,t,t+1} &= \left\{\beta(p\alpha + 1 - p)w_H + \beta^2(p\theta\alpha + 1 - p)^2\frac{w_H^2}{2c}\right\}a_i \\ &\quad + (1-\beta)w_H h_t - pMC. \end{aligned} \tag{9}$$

According to Eq. (9), his utility increases along with his innate ability and human capital built by the previous generation, which is available when he receives his education in the youth-age.

It is possible to take that utility under no migration attempts [Eq. (4)] is a special case of utility under migration attempts [Eq. (9)] because $\lim_{p\to 0} u^{m\ *}_{i,t,t+1} = u^{nm\ *}_{i,t,t+1}$.

5 The Migration Decision

In this section, I examine how individuals with heterogeneous innate abilities make the decision to migrate under the different degrees of human capital transferability. Not wanting to complicate the analysis, I focus on the situations that will most likely happen.

By subtracting Eq. (4) from Eq. (9), I derive the difference of utility

$$u^{m\ *}_{i,t,t+1} - u^{nm\ *}_{i,t,t+1}$$

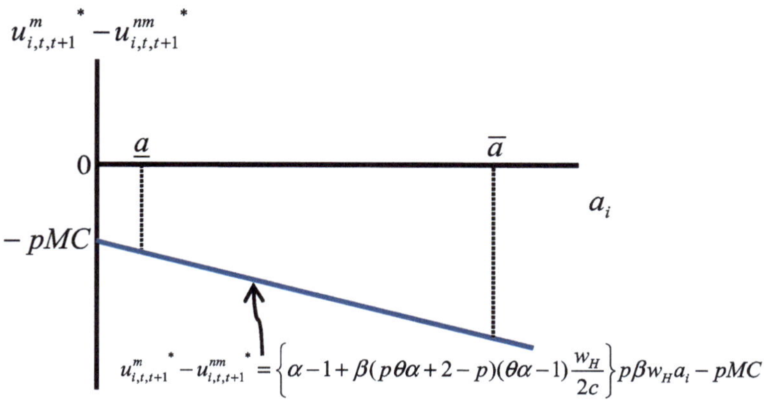

Fig. 1 The difference of utility when the transferability of human capital is low

$$= \left\{\alpha - 1 + \beta(p\theta\alpha + 2 - p)(\theta\alpha - 1)\frac{w_H}{2c}\right\} p\beta w_H a_i - pMC. \tag{10}$$

If the right-hand side of Eq. (10) is positive, an individual with a_i attempts to migrate. If it is negative, he decides not to.

When the transferability of human capital is low and $\theta\alpha < 1$, $u^{m}_{i,t,t+1}{}^{*} - u^{nm}_{i,t,t+1}{}^{*}$ is likely to be negative for any a_i.[1] If this is the case, no individuals decide to migrate to the foreign country and all of them stay in the home country (see Fig. 1). Accordingly, migration does not occur if the transferability of human capital is low.

In contrast, when the transferability of human capital is high and $\theta\alpha > 1$, $u^{m}_{i,t,t+1}{}^{*} - u^{nm}_{i,t,t+1}{}^{*}$ would be positive for any a_i and all individuals would attempt to migrate *if the migration cost were very small*. Also, $u^{m}_{i,t,t+1}{}^{*} - u^{nm}_{i,t,t+1}{}^{*}$ would be negative for any a_i and all individuals would decide not to migrate *if the migration cost were very large*. However, these two situations are unlikely to happen.

Rather, it is more likely that $u^{m}_{i,t,t+1}{}^{*} - u^{nm}_{i,t,t+1}{}^{*}$ can be both positive and negative for $a_i \in [\underline{a}, \bar{a}]$ when the transferability of human capital is high and $\theta\alpha > 1$. In this situation $\alpha - 1 + \beta(p\theta\alpha + 2 - p)(\theta\alpha - 1)(w_H/2c)$ is definitely positive; thus, $u^{m}_{i,t,t+1}{}^{*} - u^{nm}_{i,t,t+1}{}^{*}$ increases with a_i. Accordingly, $u^{m}_{i,t,t+1}{}^{*} - u^{nm}_{i,t,t+1}{}^{*} \leq 0$ for individuals with $\underline{a} \leq a_i \leq \tilde{a}$ and $u^{m}_{i,t,t+1}{}^{*} - u^{nm}_{i,t,t+1}{}^{*} > 0$ for individuals with $\bar{a} \geq a_i > \tilde{a}$, where

$$\tilde{a} \equiv \frac{MC}{\{\alpha - 1 + \beta(p\theta\alpha + 2 - p)(\theta\alpha - 1)(w_H/2c)\}(\beta w_H)}. \tag{11}$$

In other words, individuals with low innate ability do not migrate and those with high innate ability attempt to do so (see Fig. 2). Accordingly, even if migration occurs, all

[1] Even in a case where the transferability is very low, i.e. θ is very close to 0 so that $\theta\alpha$ is nearly equal to 0, the parenthesis is positive if α or c is sufficiently large. In such a case, $u^{m}_{i,t,t+1}{}^{*} - u^{nm}_{i,t,t+1}{}^{*}$ can be positive if MC is sufficiently small.

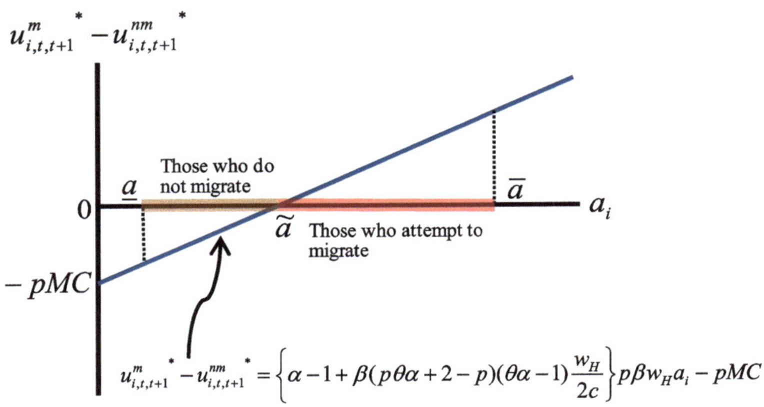

Fig. 2 The difference of utility when the transferability of human capital is high

individuals do not necessarily migrate. Individuals with high innate ability attempt to migrate. In other words, migrants are positively self-selected.

According to Eq. (11), $\tilde{a}$ increases with decreases in θ. This suggests that the low transferability of human capital enhances positive self-selection. On the other hand, lower migration costs and higher migration probabilities weaken positive self-selection because they make $\tilde{a}$ smaller.

To summarise this section, the migration decision is associated with the transferability of human capital and one's innate ability. In particular, lower transferability discourages individuals from migrating, while higher transferability encourages individuals to do so. However, not all of them attempt to migrate; a positive self-selection of migrants occurs.

6 Home Country's Human Capital

In this section, I derive the home country's human capital under both low and high transferability of human capital. First, I calculate the sum of individuals' human capital when there is no migration. I then calculate the sum of individuals' human capital that remains in the home country when some individuals attempt to migrate.

In the previous section, I found that no individuals attempt to migrate when the transferability of human capital is low, i.e. $\theta\alpha < 1$. I call this situation *the non-migration regime*. In contrast, when transferability is high, i.e. $\theta\alpha > 1$, individuals with innate ability $\underline{a} \leq a_i \leq \tilde{a}$ do not attempt to migrate or are indifferent to migration and non-migration whereas individuals with innate ability $\bar{a} \geq a_i > \tilde{a}$ do attempt to do so. I call this situation *the partial migration regime*.

First, I deal with the non-migration regime. Because no individuals migrate, all individuals' human capital surely remains in the home country in this regime. Accordingly, I substitute Eq. (3) into Eq. (1) to derive

$$h_{i,t+1}^{nm} = \beta\{a_i + \beta(w_H/c)a_i\} + (1-\beta)h_t (\equiv h_{i,t+1}^{NMR}).$$

Given human capital in the home country in period t in the non-migration regime h_t^{NMR}, human capital in period $t+1$ in the non-migration regime h_{t+1}^{NMR} is

$$h_{t+1}^{NMR} = \{1/(\bar{a} - \underline{a})\} \int_{\underline{a}}^{\bar{a}} h_{i,t+1}^{NMR} da_i = \beta\left(1 + \beta\frac{w_H}{c}\right)\frac{\bar{a} + \underline{a}}{2} + (1-\beta)h_t^{NMR}. \tag{12}$$

The evolution of human capital is illustrated in Fig. 3.

Assuming that $h_t^{NMR} = h_{t+1}^{NMR} \equiv h^{NMR*}$, steady-state human capital under no migration h^{NMR*} is

$$h^{NMR*} = \left(1 + \beta\frac{w_H}{c}\right)\frac{\bar{a} + \underline{a}}{2}. \tag{13}$$

The steady state is stable.

Secondly, I deal with the partial migration regime. Because individuals with $\underline{a} \leq a_i \leq \tilde{a}$ decide not to migrate, all of their human capital surely remains in the home country. Their individual human capital is determined by Eqs. (1) and (3). The sum of their human capital is $\{1/(\bar{a} - \underline{a})\} \int_{\underline{a}}^{\tilde{a}} h_{i,t+1}^{nm} da_i$. Given the home country's human capital in period t in the partial migration regime h_t^{PMR}, it is calculated as

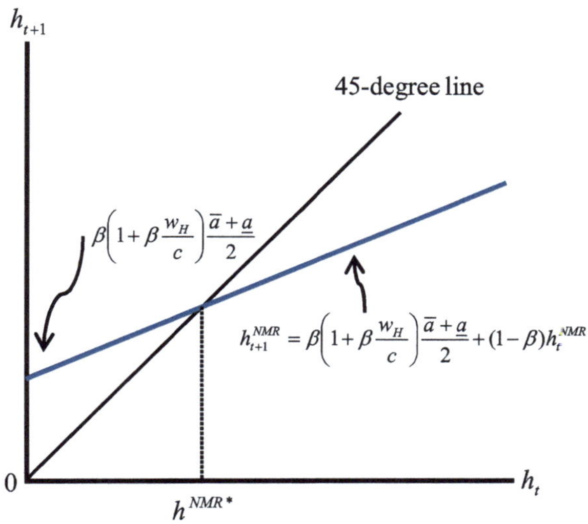

Fig. 3 The evolution of human capital under the non-migration regime

$$
\begin{aligned}
&\{1/(\bar{a} - \underline{a})\} \int_{\underline{a}}^{\tilde{a}} h_{i,t+1}^{nm} da_i \\
&= \{1/(\bar{a} - \underline{a})\} \int_{\underline{a}}^{\tilde{a}} [\beta\{a_i + \beta(w_H/c)a_i\} + (1-\beta)h_t^{PMR}]da_i \\
&= \frac{1}{\bar{a} - \underline{a}} \left\{ \beta\left(1 + \beta\frac{w_H}{c}\right)\frac{\tilde{a}^2 - \underline{a}^2}{2} + (1-\beta)h_t^{PMR}(\tilde{a} - \underline{a}) \right\}. \qquad (14)
\end{aligned}
$$

On the other hand, as individuals with $\tilde{a} < a_i \leq \bar{a}$ attempt to migrate and do so with a probability p, they demand education according to Eq. (8) and the individual human capital of those who failed to migrate and remain in the home country is determined by Eq. (6). They remain in the home country with a probability $1 - p$. Accordingly, the sum of their human capital that remains in the home country is

$$
\begin{aligned}
&\{(1-p)/(\bar{a} - \underline{a})\} \int_{\tilde{a}}^{\bar{a}} h_{i,t+1}^{m,1-p} da_i \\
&= \{(1-p)/(\bar{a} - \underline{a})\} \int_{\tilde{a}}^{\bar{a}} [\beta\{1 + \beta(p\theta\alpha + 1 - p)(w_H/c)\}a_i + (1-\beta)h_t^{PMR}]da_i \\
&= \frac{1-p}{\bar{a} - \underline{a}} \left[\beta\left\{1 + \beta(p\theta\alpha + 1 - p)\frac{w_H}{c}\right\}\frac{\bar{a}^2 - \tilde{a}^2}{2} + (1-\beta)h_t^{PMR}(\bar{a} - \tilde{a}) \right]. \qquad (15)
\end{aligned}
$$

Therefore, total human capital that remains in the home country in period $t + 1$ in the partial migration regime h_{t+1}^{PMR} is calculated by summing up Eqs. (14) and (15):

$$
\begin{aligned}
h_{t+1}^{PMR} = &\frac{1}{\bar{a} - \underline{a}} \left[\beta\left(1 + \beta\frac{w_H}{c}\right)\frac{\tilde{a}^2 - \underline{a}^2}{2} + (1-p)\beta\left\{1 + \beta(p\theta\alpha + 1 - p)\frac{w_H}{c}\right\}\frac{\bar{a}^2 - \tilde{a}^2}{2} \right] \\
&+ \frac{1}{\bar{a} - \underline{a}}(1-\beta)\{\tilde{a} - \underline{a} + (1-p)(\bar{a} - \tilde{a})\}h_t^{PMR}. \qquad (16)
\end{aligned}
$$

The evolution of human capital is illustrated in Fig. 4.

Assuming that $h_t^{PMR} = h_{t+1}^{PMR} \equiv h^{PMR*}$, steady-state human capital under partial migration h^{NMR*} is

$$
h^{PMR*} = \frac{\frac{1}{\bar{a}-\underline{a}}\left[\beta\left(1 + \beta\frac{w_H}{c}\right)\frac{\tilde{a}^2 - \underline{a}^2}{2} + (1-p)\beta\left\{1 + \beta(p\theta\alpha + 1 - p)\frac{w_H}{c}\right\}\frac{\bar{a}^2 - \tilde{a}^2}{2}\right]}{1 - \frac{1}{\bar{a}-\underline{a}}(1-\beta)\{\tilde{a} - \underline{a} + (1-p)(\bar{a} - \tilde{a})\}}. \qquad (17)
$$

The steady state is stable.

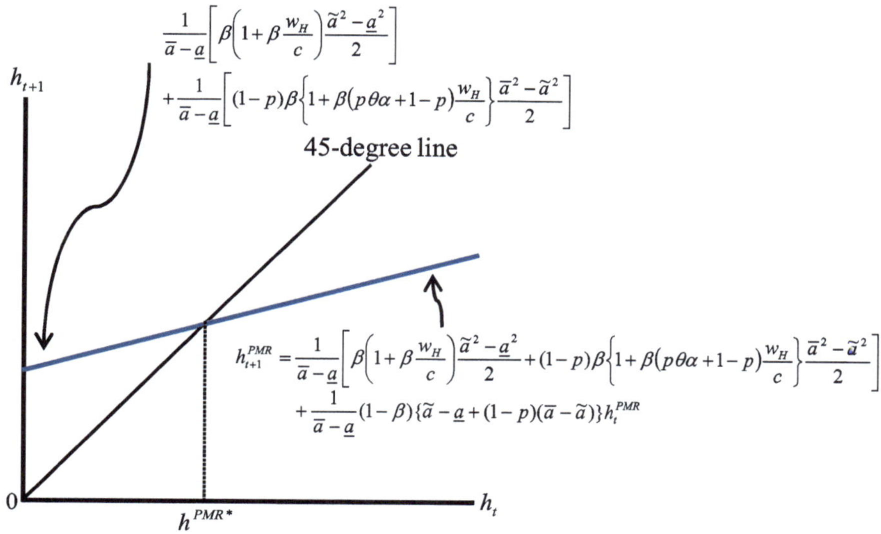

Fig. 4 The evolution of human capital under the non-migration regime

7 The Brain Drain or the Brain Gain in Steady State and the Short Run

In this section, I examine whether the home country experiences a brain gain or a brain drain when the transferability of human capital is high. In particular, I attempt to find whether the home country's total human capital is larger when individuals with high innate ability attempt to migrate due to the higher transferability of human capital than when no individuals migrate due to lower transferability. Unlike previous analysis, I make this comparison not only in steady state but also in the short run.

Firstly, I deal with steady state. A comparison of Eqs. (17) and (13) reveals that

$$\begin{aligned} & h^{PMR*} - h^{NMR*} \gtreqless 0 \\ \Leftrightarrow\ & \frac{1}{\bar{a}-\underline{a}}\frac{\bar{a}^2-\tilde{a}^2}{2}\left[-\left(1+\beta\frac{w_H}{c}\right)\beta + (1-p)\beta^2(\theta\alpha-1)\frac{w_H}{c}\right] \\ & -\left(1+\beta\frac{w_H}{c}\right)\frac{\bar{a}+\underline{a}}{2}\frac{1-\beta}{\bar{a}-\underline{a}}(\bar{a}-\tilde{a}) \gtreqless 0. \end{aligned} \tag{18}$$

Equation (18) suggests that if $\theta\alpha(>1)$ is close to 1, that is, if human capital is highly transferable but the wage disparity is *small*, both terms on the left-hand side are negative and thereby $h^{PMR*} - h^{NMR*} < 0$. In other words, the brain drain occurs. On the other hand, if $\theta\alpha(>1)$ is much larger than 1, that is, if human capital is highly transferable and the wage disparity is *large*, the first term can be positive and larger than the absolute value of the second term and thereby $h^{PMR*} - h^{NMR*} > 0$. In other words, the brain gain occurs.

Accordingly, in steady state when the transferability of human capital is high, the small wage disparity causes the brain drain and the large wage disparity causes the brain gain.

Secondly, I deal with the short run. To compare h_{t+1}^{PMR} with h_{t+1}^{NMR}, I calculate the difference of the constant terms in Eqs. (16) and (12) and the difference of the coefficient of h_t^{PMR} in Eq. (16) and the coefficient of h_t^{NMR} in Eq. (12).

The constant term in Eq. (16) − the constant term in Eq. (12) $\gtreqless 0$

$$\Leftrightarrow -\left(1 + \beta\frac{w_H}{c}\right) + (1-p)\beta(\theta\alpha - 1)\frac{w_H}{c} \gtreqless 0. \tag{19}$$

The coefficient of h_t^{PMR} in Eq. (16) − the coefficient of h_t^{NMR} in Eq. (12)

$$= -\frac{1}{\bar{a} - \underline{a}}(1-\beta)p(\bar{a} - \tilde{a}) < 0. \tag{20}$$

The sign of Eq. (19) is affected by the wage disparity. The sign of Eq. (20) suggests that the effect of intergenerational externality on human capital formation is weaker when the transferability of human capital is high than when it is low because more individuals attempt to migrate when it is more transferable (see Eq. 11).

Equations (19) and (20) reveal that if human capital is highly transferable but the wage disparity is sufficiently small, Eq. (19) will also be negative. In such cases, $h_{t+1}^{PMR} - h_{t+1}^{NMR}$ is definitely negative. This suggests that the small wage disparity causes the brain drain in the short run.

However, if the wage disparity is large, it cannot be determined a priori whether the home country experiences the brain drain or the brain gain in the short run. This outcome occurs because Eq. (19) can be positive whereas Eq. (20) is negative.

When human capital in period t, $h_t^{PMR} = h_t^{NMR} (\equiv h_t)$ is small, in particular, when

$$h_t < \frac{1}{1-\beta}\frac{\bar{a}+\tilde{a}}{2}\beta\left[-\left(1+\beta\frac{w_H}{c}\right) + (1-p)\beta(\theta\alpha - 1)\frac{w_H}{c}\right] (\equiv \hat{h}_t)$$

$h_{t+1}^{PMR} - h_{t+1}^{NMR} > 0$ and the home country experiences the brain gain in the short run. When human capital in period t is small, the negative effect on human capital due to higher transferability is small.

When human capital in period t is large, in particular, when

$$h_t > \hat{h}_t$$

$h_{t+1}^{PMR} - h_{t+1}^{NMR} < 0$ and the home country experiences the brain drain in the short run. When human capital in period t is large, the negative effect on human capital due to higher transferability is large.

Therefore, if the wage disparity is small, high transferability of human capital likely causes brain drain in both the steady state and the short run. However, if wage disparity is large, high transferability might generate different effects on human capital accumulation in a steady state and in the short run, depending on the level of human capital accumulated and left over from the previous period. In particular, a brain gain occurs in steady state but a brain drain occurs in the short run if human capital in the previous period is large.

These results have the following implications: In terms of human capital accumulation, it is not necessarily effective for the labour-sending country to provide education that helps raise the transferability of human capital. In particular, such education should not be given when the wage disparity is small because it makes some individuals attempt to migrate and doing so might cause the brain drain in both a steady state and the short run.

Also, when the wage disparity is large, the home country might be faced with a trade-off between steady-state and short-run human capital accumulation. In particular, by increasing the transferability of human capital, steady-state human capital increases and we may experience the brain gain but we might also end up with lower short-run human capital due to the brain drain. In such cases, the conflict of interest might arise between different generations.

Summarising this section, high transferability of human capital does not necessarily cause the brain gain. Whether the home country experiences the brain gain or the brain drain depends on the degree of the wage disparity. Moreover, whether the brain gain or the brain drain happens in steady state does not necessarily coincide with which one happens in the short run. We might be faced with the trade-off between steady state and short-run human capital formation.

8 Concluding Remarks

In this chapter, I sought to find how labour-sending country's human capital accumulation is associated with international transferability of human capital as well as the wage disparity. Unlike previous analyses, I assumed the heterogeneous individuals on the ground that education and migration decisions are usually made on an individual basis. I also employed the dynamic framework to shed light on the effects of transferability both in short-run and steady-state human capital accumulation.

The results revealed that the high transferability encourages individuals to migrate, and some individuals actually do so, but the brain drain occurs both in the short run and a steady state if the wage disparity is small. This fact suggests that providing education that helps raise the transferability of human capital is not always beneficial to the labour-sending country in terms of human capital formation. They also revealed that the labour-sending country cannot raise human capital in the short run and a steady state simultaneously if the wage disparity is large, suggesting that the trade-off between the short-run and steady-state human capital formation may happen. Accordingly, the labour-sending country may be faced with the conflict among different generations.

This study contributes to the brain drain and brain gain literatures by showing the non-unique effects of the human capital transferability on human capital formation and the possible trade-off between short-run and steady-state human capital formation.

This study can be extended in the following directions: First, I assumed that the degree of the transferability of human capital built by intergenerational externality takes a specific value. Doing so simplified the analysis. However, to make the results more generalised, it should be assumed otherwise. Second, I assumed that the migration probability does not differ across individuals due to the difficulty in distinguishing migrants' human capital on the part of the firms in the labour-receiving country. Even if this is true, migration probability may take a different value when the home country's average human capital changes. The migration probability can be related to the average human capital. In particular, migrants will be more likely employed in the foreign country as average human capital of their home country increases. Third, I assumed that education is financed privately. Alternatively, it could be assumed that education is financed at least partially by taxes. With increases in the mobility of workers, it has become difficult for the government to reap the benefits of spending taxes on education. Even so, the government will not lose an incentive for the public finance due to the nature of education.

Acknowledgements The preliminary version of this chapter was presented at the 16th Annual Conference of the European Economics and Finance Society, University of Ljubljana, Slovenia, June 2017; the 9th International Symposium on Human Capital and Labor Markets, Central University of Finance and Economics, China, December 2017; the China Meeting of the Econometric Society, Fudan University, China, June 2018; the 23rd Annual Workshop on Economic Science with Heterogeneous Interacting Agents, International Christian University, Japan, July 2018. I wish to thank participants at these conferences for valuable comments. This study was supported by JSPS KAKENHI Grant Number JP15K03356.

References

Batista C, Lacuesta A, Vicente PC (2012) Testing the 'brain gain' hypothesis: micro evidence from Cape Verde. J Dev Econ 97(1):32–45

Beine M, Docquier F, Rapoport H (2001) Brain drain and economic growth: theory and evidence. J Dev Econ 64(1):275–289

Beine M, Docquier F, Rapoport H (2008) Brain drain and human capital formation in developing countries: winners and losers. Econ J 118(528):631–652

Beine M, Docquier F, Oden-Defoort C (2011) A panel data analysis of the brain gain. World Dev 39(4):523–532

Bhagwati J, Hamada K (1974) The brain drain, international integration of markets for professionals and unemployment: a theoretical analysis. J Dev Econ 1(1):19–42

Boeri T, Brücker H, Docquier F, Rapoport H (eds) (2012) Brain drain and brain gain: the global competition to attract high-skilled migrants. Oxford University Press, Oxford
Byra L (2013) Rethinking the brain drain: dynamics and transition. Int Rev Econ Financ 28(Special Issue):19–25
Chiswick BR, Miller PW (1992) Language in the immigrant labour market. In: Chiswick BR (ed) Immigration, language, and ethnicity: Canada and the United States. American Enterprise Institute, Washington DC, pp 229–296, 471–476
Chiswick BR, Miller PW (2009) The International transferability of immigrants' human capital. Econ Educ Rev 28(2):162–169
Dinkelman T, Mariotti M (2016) The long-run effects of labor migration on human capital formation in communities of origin. Am Econ J: Appl Econ 8(4):1–35
Docquier F, Rapoport H (2009) Skilled immigration: the perspective of developing countries. In: Bhagwati J, Hanson G (eds) Skilled immigration today: prospects, problems, and policies. Oxford University Press, Oxford, pp 247–284
Docquier F, Rapoport H (2012) Globalization, brain drain, and development. J Econ Lit 50(3):681–730
Galor O, Stark O (1994) Migration, human capital formation, and long-run output. In: Siebert H (ed) Migration: a challenge for Europe: symposium 1993. J. C. B, Mohr (Paul Siebeck), Tübingen, pp 59–68
Hamada K, Bhagwati J (1975) Domestic distortions, imperfect information and the brain drain. J Dev Econ 2(3):265–279
Haque NU, Kim S-J (1995) "Human capital flight": impact of migration on income and growth. IMF Staff Papers 42(3):577–607
Mattoo A, Neagu IC, Özden Ç (2008) Brain waste? educated immigrants in the US labor market. J Dev Econ 87(2):255–269
Miyagiwa K (1991) Scale economies in education and the brain drain problem. Int Econ Rev 32(3):743–759
Mountford A (1997) Can a brain drain be good for growth in the source economy? J Dev Econ 53(2):287–303
Özden Ç (2006) Educated migrants: is there brain waste? In: Özden Ç, Schiff M (eds) International migration, remittances, and the brain drain. The World Bank and Palgrave Macmillan, Washington DC, pp 227–244
Pires AJG (2015) Brain drain and brain waste. J Econ Dev 40(1):1–34
Sanromá E, Ramos R, Simón H (2015) Portability of human capital and immigrant overeducation in Spain. Popul Res Policy Rev 34(2):223–241
Stark O, Helmenstein C, Prskawetz A (1997) A Brain gain with a brain rain. Econ Lett 55(2):227–234
Stark O, Helmenstein C, Prskawetz A (1998) Human capital depletion, human capital formation, and migration: a blessing or a "curse"? Econ Lett 60(3):363–367
Thum C, Uebelmesser S (2000) Brain gain and the transferability of skills. University of Munich, Mimeo
Tzanakou C, Behle H (2017) The intra-European transferability of graduates' skills gained in the U.K. J Ethn Migr Stud 43(8):1379–1397
Vidal J-P (1998) The effect of emigration on human capital formation. J Popul Econ 11(4):589–600

Is Life (or at Least Socioeconomic Aspects of It) Just Spin and Games?

Shakti N. Menon, V. Sasidevan and Sitabhra Sinha

Abstract The enterprise of trying to explain different social and economic phenomena using concepts and ideas drawn from physics has a long history. Statistical mechanics, in particular, has often been seen as most likely to provide the means to achieve this, because it provides a lucid and concrete framework for describing the collective behavior of systems comprising large numbers of interacting entities. Several physicists have, in recent years, attempted to use such tools to throw light on the mechanisms underlying a plethora of socioeconomic phenomena. These endeavors have led them to develop a community identity—with their academic enterprise being dubbed as "econophysics" by some. However, the emergence of this field has also exposed several academic fault lines. Social scientists often regard physics-inspired models, such as those involving spins coupled to each other, as over-simplifications of empirical phenomena. At the same time, while models of rational agents who strategically make choices based on complete information so as to maximize their utility are commonly used in economics, many physicists consider them to be caricature of reality. We show here that while these contrasting approaches may seem irreconcilable, there are in fact many parallels and analogies between them. In addition, we suggest that a new formulation of statistical mechanics may be necessary to permit a complete mapping of the game-theoretic formalism to a statistical physics framework. This may indeed turn out to be the most significant contribution of econophysics.

1 Introduction

The physicist Ernest Rutherford is believed to have once distinguished physics from the other sciences, referring to the latter as merely "stamp collecting" (Bernal 1939). While Rutherford may have been exceptional in explicitly voicing the traditional

Shakti N. Menon · Sitabhra Sinha (✉)
The Institute of Mathematical Sciences, CIT Campus, Taramani, Chennai 600113, India
e-mail: sitabhra@imsc.res.in

V. Sasidevan
Department of Physics, Cochin University of Science and Technology, Cochin 682022, India

A. S. Chakrabarti et al. (eds.), *Network Theory and Agent-Based Modeling in Economics and Finance*, https://doi.org/10.1007/978-981-13-8319-9_13

arrogance of physicists towards other branches of knowledge, it is true that the spectacular success of physics in explaining the natural world has led many physicists to believe that progress has not happened in other sciences because those working in these fields are not trained to examine observed phenomena from the perspective of physics. Intriguingly, practitioners in several branches of knowledge have also occasionally looked at physics as a model to aspire to, a phenomenon sometimes referred to as "Physics-envy". For instance, the science of economics has undergone such a phase, particularly in the late nineteenth century, and concepts from classical physics, such as equilibria and their stability, were central to the development of the field during this period (Mirowski 1989). However, this situation gradually changed starting at the beginning of the twentieth century, curiously just around the time when physics was about to be transformed by the "quantum revolution", and economics took a more formal, mathematical turn. The development of game theory in the 1920s and 1930s eventually provided a new *de facto* language for theorizing about economic and social phenomena. However, despite this apparent "parting of ways" between economics and physics, there have been several attempts, if somewhat isolated, throughout the previous century to build bridges between these two fields. In the 1990s, these efforts achieved sufficient traction and a subdiscipline sometimes referred to as "econophysics" emerged with the stated aim of explaining economic phenomena using tools from different branches of physics (Sinha et al. 2010).

In earlier times, the branch of physics now known as dynamical systems theory had been a rich source of ideas for economists developing their field. More recently, however, it has been the field of statistical mechanics, which tries to explain the emergence of system-level properties at the macroscale as a result of interactions between its components at the microscale, that has become a key source of concepts and techniques used to quantitatively model various social and economic phenomena. The central idea underlying this enterprise of developing statistical mechanics-inspired models is that, while the behavior of individuals may be essentially unpredictable, the collective behavior of a large population comprising many such individuals interacting with each other may exhibit characteristic patterns that are amenable to quantitative analysis and explanation, and could possibly even be predicted. This may bring to one's mind the fictional discipline of "psychohistory", said to have been devised by Hari Seldon of Isaac Asimov's *Foundation* series fame (Asimov 1951), that aimed to predict the large-scale features of future developments by discerning statistical patterns inherent in large populations. Asimov, who was trained in chemistry (and was a Professor of Biochemistry at Boston University), in fact used the analogy of a gas, where the trajectory of any individual molecule is almost impossible to predict, although the behavior of a macroscopic volume is strictly constrained by well-understood laws.

A large number of statistical mechanics-inspired models for explaining economic phenomena appear to use the framework of interacting spins. This is perhaps not surprising given that spin models provide possibly the simplest descriptions of how order can emerge spontaneously out of disorder. An everyday instance of such a self-organized order–disorder transition is exemplified by the so-called effect of a staring crowd (Kikoin and Kikoin 1978). Consider a usual city street where pedestrians walking along the sidewalk are each looking in different arbitrarily chosen directions. This corresponds to a "disordered" situation, where each component is essentially acting independently and no coordination is observed globally. If however a pedestrian at some point persistently keeps looking at a particular object in her field of view (which corresponds to a fluctuation event arising through chance), this action may induce other pedestrians to also do likewise—even though there may actually be nothing remarkable to look at. Eventually, it may be that the gaze of almost all pedestrians will be aligned with each other and each of them will be staring into the same point in space that is devoid of any intrinsic interest. This situation will correspond to the spontaneous emergence of "order" through interactions between the components, i.e., as a result of the pedestrians responding almost unconsciously to each other's actions. It is of course also possible to have everyone stare toward the same point by having an out-of-the-ordinary event (a "stimulus") happen there. In this case, it will be the stimulus extrinsic to the pedestrians—rather than interactions between the individuals—that causes the transition from the disordered to ordered state.

The simplest of the spin models, the *Ising model*, was originally proposed to understand spontaneous magnetization in ferromagnetic materials below a critical temperature. It assumed the existence of a large number of elementary spins, each of which could orient in any one of two possible directions ("up" or "down", say). Each spin is coupled to neighboring spins through exchange interactions, which makes it energetically favorable for neighboring spin pairs to be both oriented in the same direction. However, when the system is immersed in a finite temperature environment, thermal fluctuations can provide spins with sufficient energy to override the cost associated with neighboring spins being oppositely aligned. The spins could also be subject to the influence of an external field that will break the symmetry between the two orientations and will make one of the directions preferable to the spins. By associating temperature to the degree of noise or uncertainty among agents, field to any external influence on the agents, and exchange coupling between spins to interaction between individuals in their social milieu, it is easy to see that the Ising model can be employed to quantitatively model a variety of social and economic situations involving a large number of interacting individuals. Such modeling is particularly relevant when the question of interest involves qualitative changes that occur in the collective behavior as different system parameters are varied. The nature of the transition may also be of much interest as external field-driven ordering typically manifests as a first-order or discontinuous transition, while transitions orchestrated entirely through interactions between the components has the characteristics of a second-order or continuous transition. As the latter is often associated with so-called power laws, it is not unusual that these are often much sought after by physicists modeling social or economic phenomena (sometimes to the puzzlement of economists).

The popularity of spin models in the econophysics community has however not percolated to mainstream social scientists, who, probably justifiably, find such models to be overly simplified descriptions of reality. Many economic and social phenomena are therefore quantitatively described in terms of game-theoretic models, where the strategic considerations of individuals, who rationally choose between alternatives in order to maximize their utilities or payoffs, come to the fore. However, such approaches have also been criticized as being based upon an idealized view of the capabilities of individual agents and of the information that they have access to for making decisions. A complete description of aspects of economic life is possibly neither provided by spin models nor by game-theoretic ones—but being two very different types of caricature of reality, an attempt to integrate them may provide us with a more nuanced understanding of the underlying phenomena. With this aim in view, in the following two sections, we describe in brief the essential framework of these two approaches that are used to understand collective behavior in a population of agents. We show that despite their differences, there are in fact many parallels and analogies between spin model-based and game-theoretic approaches to describing social phenomena. We conclude with the suggestion that the statistical mechanics approach used at present may not be completely adequate for describing strategic interactions between many rational agents, which is the domain of game theory. This calls for the development of a new formalism that will allow for a seamless integration of statistical mechanics with game theory—which will possibly be the most enduring contribution of econophysics to the scientific enterprise.

2 Collective Decision-Making by Agents: Spins…

We can motivate a series of models of the dynamics of collective decision-making by agents that differ in terms of the level of details or information resolution that one is willing to consider. We begin by considering a group of N agents, each of whom is faced with the problem of having to choose between a finite number of possible options at each time step t, where the temporal evolution of the system is assumed to occur over discrete intervals. To simplify matters, we consider the special case of binary decisions in which the agents, for instance, simply choose between "yes" or "no". Thus, in the framework of statistical physics, the state of each agent (representing the choice made by it) can be mapped to an Ising spin variable $S_i = \pm 1$. Just as spin orientations are influenced by the exchange interaction coupling with their neighbors in the Ising model, agents take decisions that can, in principle, be based on the information regarding the choices made by other agents (with whom they are directly connected over a social network) in the past—as well as the memory of its own previous choices. If an agent needs to explicitly identify the specific choice made by each neighbor in order to take a decision, then this constitutes the most detailed input information scenario. Here, each agent i considers the choices made by its k_i neighbors in the social network of which it is a part (if its own choices also need to be taken into account we may assume that it includes itself in its set of neighbors). Furthermore, each agent i has a memory of the choices made by its neighbors in the preceding m_i time steps. Thus, the agent, upon being presented with a history represented as a $m_i \times k_i$ binary matrix, has to choose between -1 and $+1$. As there are $2^{m_i k_i}$ possible histories that the agent may need to confront, this calls for formulating an input–output function f_i for the agent that, given a string of $m_i k_i$ bits, can generate the probability that the agent will make a particular choice, viz., $\Pr(S_i = +1) = f_i(\{\pm 1, \pm 1, \ldots \pm 1\}_{m_i k_i})$ and with $\Pr(S_i = -1) = 1 - \Pr(S_i = +1)$. In other words, the choice of each agent i will be determined by a function whose domain is a $m_i k_i$-dimensional hypercube and whose range is the unit interval $[0, 1]$.

The previous situation is simplified by assuming that agents do not know the exact identity of the choices made by each of its neighbors but only have access to the aggregate information as to how many chose a particular option, e.g., -1. A natural extension of this is the scenario where, instead of an explicit network, agents are considered to essentially interact with the entire group. Such an effectively "mean-field" like situation (where pairwise interactions between spins are replaced by a self-consistent field representing the averaged effect of interactions of a spin with the collective) will arise when, in particular, an agent's choice is made on the basis of a global observable that is the record of the outcome of choices made by all agents. For instance, one can model financial markets in this manner, with agents deciding whether or not to trade in an asset based entirely on its price, a variable that is accessible to all agents and which changes depending on the aggregate choice behavior of agents—with price rising if there is a net demand (more agents choose to buy than sell) and falling if the opposite is true (more agents choose to sell than to buy). Thus, if N_+ and N_- are the number of agents choosing $+1$ and -1, respectively, then agents base their decision on their knowledge of the net number of agents who choose one option rather than the other, i.e., $N_+ - N_- = \sum_i S_i = NM$, with M being the magnetization or average value of spin state in the Ising model. In this setting, the choice of the ith agent having memory (as stated previously) is made using information about the value of M in the preceding m_i time steps. Therefore, the input–output function specifying the choice behavior of the agents maps a string of m continuous variables[1] lying in the interval $[-1, 1]$ to a probability for choosing a particular option, viz., $\Pr(S_i = +1) = f_i(M_1, M_2, \ldots, M_m)$ where M_j is the value of magnetization j time steps earlier. One can view several agent-based models that seek to reproduce the stylized features of price movements in financial markets as special cases of this framework, including the model proposed by Vikram and Sinha (2011) that exhibits heavy-tailed distributions for price fluctuations and trading volume which are quantitatively similar to that observed empirically, as well as volatility clustering and multifractality.

A further simplification can be achieved upon constraining the function f_i to output binary values, so that $\Pr(S_i = +1)$ can only be either 0 or 1. The set of functional values realized for all possible values of the argument (i.e., all possible histories that an agent can confront) which defines the *strategy* of the agent can, in this case, be written as a binary string of length $2^{m \log_2(N+1)} = (N+1)^m$. It is easy to see that the total number of possible distinct strategies is $2^{(N+1)^m}$. In reality, of course, many of these possible strategies may not make much sense and one would be focusing on the subset for which f_i has some well-behaved properties

[1] We however note that as there are only N agents whose choices need to be summed, the relevant information can be expressed in $\log_2(N+1)$ bits. As N diverges, m becomes continuous.

such as monotonicity. To simplify the situation even more, the granularity of the information on choices made in the past can be reduced (Sasidevan et al. 2018). In the most extreme case, the information about the aggregate or net choice of agents at a particular instant can be reduced to a single bit, viz., sign(M_j) instead of M_j. This will be the case, for instance, when one only knows whether a particular option was chosen by the majority or not, and not how many opted for that choice. The number of possible different histories that an agent may confront is only 2^m in this situation, and thus, the total number of possible strategies is 2^{2^m}. The well-known *Minority Game* (Moro 2004) can be seen as a special case of this simplified formalism. It is the very antithesis of a coordination game, with each agent trying to be contrary to the majority. In other words, each agent is aiming to use those f_i that would ensure $S_i \times \text{sign}(M) = -1$ in each round of the game.

In the detailed input information scenario described previously, a Minority Game (MG) like setting will translate into an Ising model defined over a network, where connected spin pairs have antiferromagnetic interactions with each other. Such a situation will correspond to a highly frustrated system, where the large number of energy minima would correspond to the various possible efficient solutions of the game. However, if the system remains at any particular equilibrium for all time, this will not be a fair solution as certain individuals will always form the minority and thus get benefits at the expense of others. A possible resolution that may make it both efficient and fair is to allow for fluctuations that will force the collective state to move continuously from one minima to another, without settling down into any single one for a very long time (see, e.g., Dhar et al. 2011).

An important feature of the MG is the ability of agents to adapt their strategies, i.e., by evaluating at each time step the performance or payoff obtained by using each of the strategies, the agent can switch between strategies in order to maximize payoff. One can ask how the introduction of "learning" into the detailed input information scenario will affect the collective dynamics of the system. In the classical MG setting, each agent begins by randomly sampling a small number of fs (typically 2) from the set of all possible input–output functions and then scores each of them based on their performance against the input at each time step, thereafter choosing the one with the highest score for the next round. In the detailed information setting, we need to take into account that an agent will need to consider the interaction strength it has with each of its neighbors in the social network it is part of. Thus, agents could adapt based on their performance not just by altering strategy but also by varying the importance that they associate with information arriving from their different neighbors (quantified in terms of weighted links). Hence, link weight update dynamics could supplement (or even replace) the standard strategy scoring mechanism used by agents to improve their payoffs in this case. For example, an agent may strengthen links with those neighbors whose past choices have been successful (i.e., they were part of the minority) while weakening links with those who were unsuccessful. Alternatively, if agent i happened to choose S_i correctly, i.e., so as to have a sign opposite to that of sign(M), while its neighbor agent j chose wrongly, learning may lead to the link from j to i becoming positive (inducing j to copy the choice made by i in the future) while the link from i to j becomes negative (suggesting that i will choose the opposite of what j has chosen).

It may be worth noting in this context that the role of a link weight update rule on collective dynamics has been investigated in the context of spin models earlier, although in the different context of coordination where agents prefer to make similar choices as their neighbors (Singh et al. 2014). Using a learning rule that is motivated by the Hebbian weight update dynamics that is often used to train artificial recurrent neural network models, it has been seen that, depending on the rate at which link weights adapt (relative to the spin state update timescale) and the degree of noise in the system, one could have an extremely high diversity in the time required to converge to *structural balance* (corresponding to spins spontaneously segregating into two clusters, such that within each cluster all interactions are ferromagnetic and all interactions between spins belonging to different clusters are antiferromagnetic) from an initially frustrated system. It is intriguing to speculate as to what will be observed if instead the learning dynamics tries to make the spins misalign with their neighbors, which would be closer to the situation of MG.

3 Collective Decision-Making by Agents: ... and Games

We now shift our focus from the relatively simpler spin model-inspired descriptions of collective behavior of agents to those that explicitly incorporate strategic considerations in the decision-making of agents. Not surprisingly, this often involves using ideas from game theory. Developed by John von Neumann in the early part of the twentieth century, the mathematical theory of games provides a rigorous framework to describe decision-making by "rational" agents.

It appears intuitive that the states of binary Ising-like spins can be mapped to the different choices of agents when they are only allowed to opt between two possible actions. We will call these two options available to each agent as action A and action B, respectively (e.g., in the case of the game Prisoner's Dilemma, these will correspond to "cooperation" and "defection", respectively). However, unlike in spin models, in the case of games, it is difficult to see in general that the choices of actions by agents are somehow reducing an energy function describing the global state of the system. This is because instead of trying to maximize the total payoff for the entire population of agents, each agent (corresponding to a "spin") is only trying to maximize its own expected payoff—sometimes at the cost of others. Possibly the only exception is the class of the Potential Games wherein one can, in principle, express the desire of every agent to alter their action using a global function, viz., the "potential" function for the entire system.

Let us take a somewhat more detailed look into the analogy. For a spin model, one can write down the effective time-evolution behavior for each spin from the energy function as the laws of physics dictate that at each time step the spins will try to adopt the orientation that will allow the system as a whole to travel "downhill" along the landscape defined by the energy function

$$E = -\sum_{ij} J_{ij} S_i S_j + h \sum_i S_i.$$

Here, J_{ij} refers to the strength of interaction between spins i and j, the summation $\sum_{ij}$ is performed over neighboring spin pairs and h refers to an external field. In the absence of any thermal fluctuations (i.e., at zero temperature), it is easy to see that the state of each spin will be updated according to

$$S_i(t+1) = \text{sign}(\sum_j J_{ij} S_j + h).$$

For the case of a symmetric two-person game, the total utility resulting from the choice of actions made by a group of agents whose collective behavior can be decomposed into independent dyadic interactions will be given by

$$U = R f_{AA} + P f_{BB} + (S+T) f_{AB}.$$

Here R and P refer to the payoffs obtained by two agents when both choose A or both choose B, respectively, while if one chooses A and the other chooses B, the former will receive S while the latter will receive T. The variables f_{AA}, f_{BB} and f_{AB} refer to the fraction of agent pairs who both choose A, or both choose B, or where one chooses A while the other chooses B, respectively. On the other hand, for an individual agent, the payoff is expressed as

$$U_i = \sum_j p_i p_j R + p_i(1-p_j)S + (1-p_i)p_j T + (1-p_i)(1-p_j)P,$$

where p_i, p_j refer to the probabilities of agents i and j, respectively, to choose action A. As an agent i can only alter its own strategy by varying p_i, it will evaluate $\partial U_i / \partial p_i$ and increment or decrement p_i so as to maximize U_i, eventually reaching an equilibrium.

Different solution concepts will be manifested according to the different ways an agent can model the possible strategy p_j used by its opponent j (which of course is unknown to the agent i). Thus, in order to solve the previous equation, the agent i actually replaces the variable p_j by its assumption $\hat{p}_j$ about that strategy. In the conventional Nash solution framework, the agent is agnostic about its opponent's strategy so that $\hat{p}_j$ is an unknown. To physicists, this approach may sound similar to that of a maximum entropy formalism, where the solution is obtained with the least amount of prior knowledge about the situation at hand. However, advances in cognitive science and attempts to develop artificial intelligence capable of semi-human performance in various tasks have made us aware that human subjects rarely approach a situation where they have to anticipate their opponent's move with a complete "blank slate" (so to say). Even if the opponent is an individual who the subject is encountering for the first time, she is likely to employ a *theory of mind* to try to guess the strategy of the opponent. Thus, for example, a goalie facing a penalty kick will make a decision as to whether to jump to the left or the right as soon as the kick is taken (human response time is too slow for it to make sense for the goalie to wait until she actually sees which direction the ball is kicked) by trying to simulate within her mind the thought process of the player taking the kick. In turn, the player taking the penalty kick is also attempting to guess whether the goalie is more likely to jump toward the left or the right, and will, so to say, try to "get inside the mind" of the goalie. Each player is, of course, aware that the other player is trying to figure out what she is thinking and will take this into account in their theory of mind of the opponent. A little reflection will make it apparent that this process will ultimately lead to an infinite regress where each individual is modeling the thought process of the opponent simulating her own thought process, to figure out what the opponent might be thinking, and so on and so forth (Fig. 1).

The coaction solution framework (Sasidevan and Sinha 2015, 2016) solves the problem of how agents decide their strategy while taking into account the strategic considerations of their opponent by assuming that if both agents are rational, then regardless of what exact steps are used by each to arrive at the solution, they will eventually converge to the same strategy. Thus, in this framework, $\hat{p}_j = p_i$. This results in solutions that often differ drastically from those obtained in the Nash framework. For example, let us consider the case of the two-person *Prisoner's Dilemma* (PD), a well-known instance of a *social dilemma*. Here, the action chosen by each of the agents in order to maximize their individual payoffs paradoxically results in both of them ending up with a much inferior outcome than that would have been obtained with an alternative set of choices. In PD, each agent has the choice of either cooperation (C: action A) or defection (D: action B) and the payoffs for each possible pair of actions chosen by the two (viz., DC, CC, DD or CD) have the hierarchical relation $T > R > P > S$. The value of the payoff T is said to quantify the temptation of an

Fig. 1 A schematic diagram illustrating the infinite regress of theories of mind (viz., "she thinks that I think that she thinks that I think that...") that two opponents use to guess the action that the other will choose. Figure adapted from a drawing of the cover of *The Division Bell*, a music album by Pink Floyd, which was designed by Storm Thorgerson based on illustrations by Keith Breeden

agent for unilateral defection, while R is the reward for mutual cooperation, P is the penalty paid when both agents choose defection and S is the so-called sucker's payoff obtained by the agent whose decision to cooperate has been met with defection by its opponent. Other symmetric two-person games can be defined upon altering the hierarchy among the values of the different payoffs. Thus, $T > R > S > P$ characterizes a game referred to as *Chicken* (alternatively referred to as *Hawk-Dove* or *Snowdrift*) that has been used extensively to model phenomena ranging from nuclear saber-rattling between nations (with the prospect of mutually assured destruction) to evolutionary biology. Another frequently studied game called *Stag Hunt*, which is used to analyze social situations that require agents to coordinate their actions in order to achieve maximum payoff, is obtained when $R > T \geq P > S$.

In the Nash framework, the only solution to a one-shot PD (i.e., when the game is played only once) is for both agents to choose defection. As is easily seen, they therefore end up with P, whereas if they had both cooperated they would have received R which is a higher payoff. This represents the dilemma illustrated by the game, namely, that choosing to act in a way which appears to be optimal for the individual may actually yield a suboptimal result for both players. Indeed, when human subjects are asked to play this game with each other, they are often seen to instinctively choose cooperation over defection. While this may be explained by assuming irrationality on the part of the human players, it is worth noting that the apparently naive behavior on the part of the players actually allows them to obtain a higher payoff than they would have received had they been strictly "rational" in the Nash sense. In fact, the rather myopic interpretation of rationality in the Nash perspective may be indicative of more fundamental issues. As has been pointed out in Sasidevan and Sinha (2015), there is a contradiction between the two assumptions underlying the Nash solution, viz., (i) the players are aware that they are both equally rational and (ii) that each agent is capable of *unilateral deviation*, i.e., to choose an action that is independent of what its opponent does. The coaction framework resolves this by noting that if a player knows that the other is just as rational as her, she will take this into account and thus realize that both will eventually use the same strategy (if not the same action, as in the case of a mixed strategy). Therefore, cooperation is much more likely in the solution of PD in the coaction framework, which is in line with empirical observations.

A much richer set of possibilities emerges when one allows the game to be played repeatedly between the same set of agents. In this iterative version of PD (IPD), mutual defection is no longer the only solution even in the Nash framework, because agents need to now take into account the history of prior interactions with their opponents. Thus, direct reciprocity between agents where, for example, an act of cooperation by an agent in a particular round is matched by a reciprocating act of cooperation by its opponent in the next round, can help in maintaining cooperation in the face of the ever-present temptation toward unilateral defection. Indeed, folk theorems indicate that mutual cooperation is a possible equilibrium solution of the infinitely repeated IPD. Multiple reciprocal strategies, such as "tit-for-tat" and "win-stay, lose-shift" have been devised and their performance tested in computer tournaments for PD. Intriguingly, it has been shown that when repeated interactions are allowed between rational agents, the coaction solution is for agents to adopt a Pavlov strategy. In this, an agent sticks to its previous choice if it has been able to achieve a sufficiently high payoff but alters the choice if it receives a low payoff, which allows robust cooperation to emerge and maintain itself (Sasidevan and Sinha 2016).

Moving beyond dyadic interactions to general N-person games, the analysis of situations where an agent simultaneously interacts with multiple neighbors can become a formidable task, especially with increasing number of agents. Thus, one may need to simplify the problem considerably in order to investigate collective dynamics of a group of rational agents having strategic interactions with each other. One possible approach—which deviates from assuming a strictly rational nature of the agents—invokes the concept of *bounded rationality*. Here, the ability of an agent to find the optimal strategy that will maximize its payoff is constrained by its cognitive capabilities and/or the nature of the information it has access to. A notable example of such an approach is the model proposed by Nowak and May (1992), where a large number of agents, arranged on a lattice, simultaneously engage in PD with all their neighbors in an iterative fashion. As in the conventional two-player iterated PD, each agent may choose to either cooperate or defect at each round, but with the difference that the agents nominate a single action that it uses in its interactions with each of its neighbors. At the end of each round, agents accumulate the total payoff received from each interaction and compare it with those of its neighbors. It then copies the action of the neighbor having the highest payoff to use in the next round. Note that each agent only has access to information regarding the decisions of agents in a local region, viz., its topological neighborhood, and hence, the nature of the collective dynamics is intrinsically dependent on the structure of the underlying connection network. Nowak and May demonstrated that the model can sustain a nonzero fraction of cooperating agents, even after a very large number of rounds. In other words, limiting interactions to an agent's network neighborhood may allow cooperation to remain a viable outcome—a concept that has been referred to as *network reciprocity*.

This model has been extremely influential, particularly in the physics community, where it has motivated a large number of studies that have built upon the basic framework provided by Nowak and May. Beyond the implications for how cooperation can be sustained in a population of selfish individuals, these studies have revealed tantalizing links between game theory and statistical physics. For instance, by considering the distinct collective dynamical regimes as phases, one may describe the switching between these regimes in terms of nonequilibrium phase transitions. The nonequilibrium nature is manifest from the breakdown of detailed balance (where the transition rate from one state to another is exactly matched by that of the reverse process) because of the existence of absorbing states. These states, once reached by the system, are defined by the cessation of further evolution and correspond to either all agents being cooperators or all being defectors. The system cannot escape these states as agents can only copy actions that are still extant in the population.

While Nowak and May had considered a deterministic updating procedure (viz., the "imitate the best" rule described previously), there have been several variants that have incorporated the effect of uncertainty into an agent's decision-making process.

One of the most commonly used approaches is to allow each agent i to choose a neighbor j at random and copy its action with a probability given by the Fermi distribution function:

$$\Pi_{i \to j} = \frac{1}{1 + \exp(-(\pi_j - \pi_i)/K)},$$

where π_i and π_j are, respectively, the total payoffs received by agents i and j in the previous round, and K is the effective temperature or noise in the decision-making process (Szabó and Tőke 1998). The utility of this function is that it allows one to smoothly interpolate between a deterministic situation in the limit $K \to 0$ (viz., agent i will copy agent j if $\pi_j > \pi_i$) and a completely random situation in the limit $K \to \infty$ (viz., agent i will effectively toss a coin to decide whether to copy agent j). Implementing this scheme in a population of agents whose interactions are governed by different connection topologies allows us to investigate the spectrum of collective dynamical states that arise, and the transitions between them that take place upon varying system parameters (Menon et al. 2018).

Figure 2 shows the different collective states of the system that occur at various regions of the (K, T) parameter space. It is tempting to compare this with the phase diagrams obtained by varying the temperature and external field in spin systems. First, the state of an agent, i.e., the action chosen by it at a particular time instant, can be mapped to a spin orientation—e.g., if the ith agent chooses cooperation, then the corresponding spin state can be designated $S_i = +1$, whereas $S_i = -1$ implies that the agent has chosen defection. Typically, there is symmetry between the two orientations $\{-1, +1\}$ that a spin can adopt. However, in games such as PD one of the actions may be preferable to another under all circumstances (e.g., unconditional defection or $p = 0$ is the dominant strategy in PD). This implies the existence of an effective external field, whose magnitude is linearly related to the ratio of the temptation for defection and reward for cooperation payoffs, viz., $1 - (T/R)$, that results in one of the action choices being more likely to be adopted by an agent than another. We also have noise in the state update dynamics of the agents as, for a finite value of K, an agent stochastically decides whether to adopt the action of a randomly selected neighbor who has a higher total payoff than it. This is not unlike the situation where spins can sometimes switch to energetically unfavorable orientations because of thermal fluctuations, when the system is in a finite temperature environment.

Analogous to ordered states in spin systems (corresponding to the spins being aligned), we have the collective states all C (all agents choose to cooperate) or all D (all agents have chosen defection), and similar to a disordered state we observe that the collective dynamics of agents can converge to a fluctuating state F, in which agents keep switching between cooperation and defection. Just as in spin systems, the phases are distinguished by using an order parameter, namely, magnetization per spin $m = \sum_i S_i/N \in [-1, 1]$, we can define an analogous quantity $2f_C - 1$, which is a function of the key observable for the system of agents, viz., the fraction of agents who are cooperating at any given time f_C. As for m, the value of this quantity is bounded between -1 (all D) and $+1$ (all C), with the F state yielding values close to 0 provided sufficient averaging is done over time.

Note that despite this analogy between the parameters (viz., temperature/noise and field/payoff bias) governing the collective dynamics of spin systems and that of a population of agents that exhibit strategic interactions with each other, there are in fact significant differences between the two. As is manifest from Fig. 2, an increase in the noise K does not quite have the same meaning as raising the temperature in spin systems. Unlike the latter situation, agents do not flip from cooperation to defection with equal probability as the temperature/noise increases. Instead, with equal probability agents either adopt the action chosen by a randomly selected neighbor or stick to their current action state. Not surprisingly, this implies that all C and all D states will be stable (for different values of the field T, the payoff value corresponding to

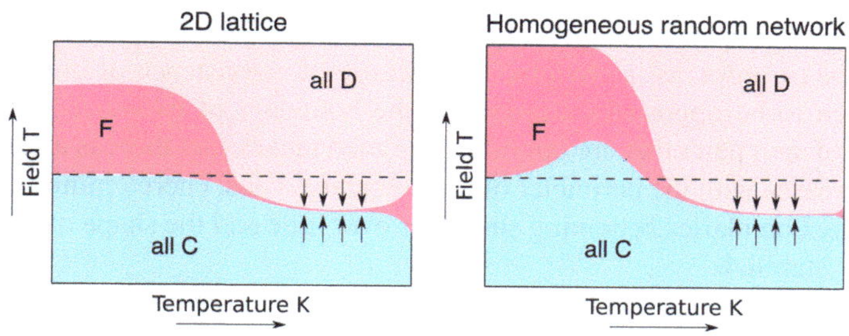

Fig. 2 Schematic parameter space diagrams illustrating the dependence on the contact network structure of the collective dynamics of a system of agents that synchronously evolve their states (representing actions) through strategic interactions with their neighbors. Each agent in the system adopts one of two possible actions at each round, viz., cooperate or defect, and receives an accumulated payoff based on each of their neighbors choice of action. The agents update their action at every round by choosing a neighbor at random and copying their action with a probability that is given by a Fermi function, where the level of temperature (noise) is controlled by the parameter K. The broken horizontal line in both panels corresponds to the case where the temptation T (payoff for choosing defection when other agent has chosen cooperation) is equal to the reward R for mutual cooperation. Hence, the region above the line corresponds to the case where agents play the Prisoner's Dilemma game, while that below corresponds to the case where they play the Stag Hunt game. Note that the temptation T can be viewed as a field, in analogy to spin systems, as its value biases an agent's preference for which action to choose. The three regimes displayed in each case correspond to situations where the system converges to a state where all the agents cooperate ("all C"), all agents choose defection ("all D") or the states of the agents fluctuate over time ("F"). We note that the region corresponding to fluctuations appears to comprise two large segments connected by a narrow strip. However, the nature of the collective behavior is qualitatively different in the two segments, as the dynamics observed for large K can be understood as arising due to extremely long transience as a result of noise. The left panel displays the regimes obtained when agents are placed on a two-dimensional lattice, where each agent has eight neighbors, while the right panel displays the situation where agents are placed on a homogeneous random network where all nodes have eight neighbors. The difference in the collective dynamics between the two scenarios is most noticeable at intermediate values of K, where the system can converge to an all C state even in the Prisoner's Dilemma regime in the right panel

temptation for unilateral defection, relative to the reward for mutual cooperation) even when K diverges.

In addition, even in the absence of noise (i.e., at $K = 0$), we observe that agents can keep switching between different actions. In other words, unlike the situation in spin systems at zero temperature, the system will keep evolving dynamically. When an agent determines that a randomly selected neighbor has higher total payoff than it, the agent will switch to the action chosen by its neighbor deterministically. Therefore, if there is a coexistence of cooperation and defection states there will be switching between these two actions—thereby ensuring the existence of the fluctuating state at $K = 0$.

Spin systems are also characterized by coarsening dynamics, wherein spins of similar orientation coalesce over time to form domains. Existence of such domains in a spin system, whereby spins of opposite orientations can coexist even in the ordered phase, means that even at low temperatures, the global magnetization of a sufficiently large system can yield quite small values. This happens not because of the absence of order, as is obvious, but because of coexistence of ordered regions that happen to be oppositely aligned. At the boundary of two such domains, the existence of spin pairs that are oppositely aligned means that there is an energy cost which increases with the perimeter of the boundary. Thus, energy minimization will result in the boundaries becoming smoother over time and the shape of the domains eventually stabilize.

Agents on lattices or networks will also exhibit the spontaneous formation of domains or clusters of interacting agents who have chosen the same action. Indeed, in order to maintain cooperation in the system for any length of time (in the presence of defectors), the cooperators will have to form clusters. Within these clusters, agents receive a sufficiently high payoff from cooperating neighbors to prevent them from switching to defection, despite the potential for being exploited by any neighbor that chooses to defect. However, the collective dynamics leads to a form of "anti-coarsening". This is because agents choosing defection would like to be surrounded by as many cooperating agents as possible in order to maximize their payoff, so that the boundary between groups of cooperators and defectors will tend to develop kinks and corners over time, instead of becoming smoother as in the case of spins. Furthermore, as the cooperators would tend to prefer as few defectors as possible at neighboring positions, we would observe ceaseless flux in the shape of the domain boundaries unless the system eventually converges to any one of the two absorbing states, all C or all D.

As already mentioned earlier, the mechanism of agents copying the action of neighbors who are more successful than them—although helping to simplify the dynamics—is somewhat dissatisfactory as the agents are now no longer strictly rational. For instance, if the collective dynamics results in the system converging to the all C absorbing state, all agents will always cooperate with each other from that time onwards, as there is no agent left to copy the defection action from. Yet, in a one-shot PD game, defection is always the dominant strategy as will be realized by any agent who is being "rational" and works out the implications of its action in light of the payoff matrix (instead of blindly copying its neighbor). Of course, in the iterated PD, it is no longer true that unconditional defection is the best strategy (Axelrod 1984). Nevertheless, an all C state is highly unstable as it provides a lucrative target for agents who choose to defect, knowing that they will reap an extremely high payoff at the expense of the cooperators. One possible way to prevent global cooperation from being an absorbing state in the modeling framework described previously is to introduce a mutation probability. This will allow agents to spontaneously switch to a particular action with a low probability, independent of whether any of their more successful neighbors are using it or not. This will ensure that even if a population has reached an all C state, it need not remain there always.

A more innovative approach that reintroduces the essential rationality of agents in the context of studying the collective dynamics of a large number of agents interacting over a social network has been introduced in Sharma et al. (2019). Although formulated in the specific context of agents making rational decisions as to whether to get vaccinated (based on information about the incidence of a disease and knowledge of how many neighbors have already gotten vaccinated), the framework can be generally applied to understand many possible situations in which a large number of agents make strategic decisions through interactions with other agents. In this approach, each agent plays a symmetric two-person game with its "virtual self", rather than with any of its neighbors, in order to decide its action. The interaction with neighbors is introduced by making specific entries in the payoff matrix that an agent uses for its decision process into functions of the number of its neighbors who have chosen a particular action. Thus, in the context of vaccination, if all its neighbors have already chosen to vaccinate themselves, an agent is already protected from disease and is most likely to choose not to get vaccinated (thereby avoiding any real or imagined cost associated with vaccination, e.g., perceived side effects). As the neighborhood of each agent is different (in general) when considering either a lattice or a network, this means that each agent is playing a distinct game. Not only will the games played by each other differ quantitatively (i.e., in terms of the payoffs of the game) but also qualitatively. Thus, for instance, one agent may be playing what is in effect PD while another may be playing Chicken. Initial explorations suggest that such spatiotemporal variation of strategies may give rise to a rich variety of collective dynamical phenomena, which have implications for problems as diverse as designing voluntary vaccination programs so as to have maximum penetration in a population and predicting voter turnout in elections.

4 In Lieu of a Conclusion

The brief presentation in this chapter of several approaches toward understanding the collective dynamics of a population of interacting agents, by using both physics-inspired spin models and game-theoretic models of rational individuals making strategic decisions, has hopefully made it clear that there are clear parallels and analogies between the two frameworks. Although both are at best caricatures of reality, albeit of different types, comparing and contrasting between the results generated by both of these approaches should help us understand better how and why large groups or crowds behave in certain ways. While physicists may harbor the hope of revolutionizing the understanding of society through the use of simple models of self-organizing phenomena, it may also be that the contribution may be the other way around. In general, for a group of rational agents, unlike the case in spin models, there appears to be no single global function (such as energy) whose minimization leads to the collective states. Thus, it appears that the traditional tools of statistical mechanics may be inadequate for describing situations where the same collective state may have different utilities for each agent. For instance, in PD, agent 1 choosing C while agent 2 choosing D may be the best of all possible outcomes for 2—but it is the worst of all possible outcomes for agent 1. Therefore, while agent 2 may be desirous of nudging the system to such an outcome, agent 1 maybe as vehemently trying to push the system away from such a state. How then would one proceed to model the collective activity of such systems using the present tools of statistical mechanics? It does appear that we may need to have a new formulation of statistical mechanics that applies to the situation outlined previously. Thus, it may well turn out that the lasting significance of econophysics will be in not what it does for economics, but rather in the new, innovative types of physical theories, particularly in statistical physics, that it may spawn.

Acknowledgements We thank our collaborator Anupama Sharma, whose joint work with us forms the basis of several ideas discussed in this research, and Deepak Dhar, whose insightful comments had first gotten us intrigued about the relation between strategic games and statistical physics. The research reported here has been funded in part by the Department of Atomic Energy, Government of India through the grant for Center of Excellence in Complex Systems and Data Science at IMSc Chennai.

References

Asimov I (1951) Foundation. Gnome Press, New York

Axelrod R (1984) The evolution of cooperation. Basic Books, New York

Bernal JD (1939) The social function of science. G Routledge & Sons, London, p 9

Dhar D, Sasidevan V, Chakrabarti BK (2011) Emergent cooperation amongst competing agents in minority games. Physica A 390:3477

Kikoin AK, Kikoin IK (1978) Molecular physics. Mir Publishers, Moscow, pp 436–437 (English trans, Leib G)

Menon SN, Sasidevan V, Sinha S (2018) Emergence of cooperation as a non-equilibrium transition in noisy spatial games. Front Phys 6:34

Mirowski P (1989) More heat than light: economics as social physics, physics as nature's economics. Cambridge University Press, Cambridge

Moro E (2004) The minority game: an introductory guide. In: Korutcheva E, Cuerno R (eds) Advances in condensed matter and statistical mechanics. Nova Science Publishers, New York

Nowak MA, May RM (1992) Evolutionary games and spatial chaos. Nature 359:826

Sasidevan V, Sinha S (2015) Symmetry warrants rational cooperation by co-action in social dilemmas. Sci Rep 5:13071

Sasidevan V, Sinha S (2016) Co-action provides rational basis for the evolutionary success of Pavlovian strategies. Sci Rep 6:30831

Sasidevan V, Kushal A, Sinha S (2018) When big data fails: adaptive agents using coarse-grained information have competitive advantage. Phys Rev E 98:020301
Sharma A, Menon SN, Sasidevan V, Sinha S (2019) Epidemic prevalence information on social networks can mediate emergent collective outcomes in voluntary vaccine schemes. PLOS Comput Biol 15:e1006977
Singh R, Dasgupta S, Sinha S (2014) Extreme variability in convergence to structural balance in frustrated dynamical systems. EPL 105:10003
Sinha S, Chatterjee A, Chakraborti A, Chakrabarti BK (2010) Econophysics: an introduction. Wiley-VCH, Weinheim
Szabó G, Tőke C (1998) Evolutionary prisoner's dilemma game on a square lattice. Phys Rev E 58:69
Vikram SV, Sinha S (2011) Emergence of universal scaling in financial markets from mean-field dynamics. Phys Rev E 83:016101

Financial and Economic Linkages

An Agent-Based Model of BCVA and Systemic Risk

D. Ladley and C. T. M. Tran

Abstract An agent-based model is used to determine whether the stability of a financial system can be improved by incorporating BCVA into the pricing of OTC derivatives contracts. The results illustrate that the adjustments of financial institutions' credit cannot only improve the stability of financial counterparties in credit events but also reduce systemic risk of the entire network. The equity cushion provided by BCVA shields financial intuitions against unanticipated losses. This reduces the frequency and scale of systemic events and the spread of losses. The scale of the benefit is dependent upon the leverage of institutions and is significantly affected by connectivity and the premiums of derivatives contracts.

Keywords Systemic risk · Derivatives markets · BCVA · Agent-based model

1 Introduction

The 2007–8 financial crisis highlighted significant weaknesses in Basel II's treatment of derivative contracts. Basel II regulated the capital requirements for all loans from banks to corporations, but it did not consider the default correlations between corporations and banks. Lehman Brothers was regarded as a risk-free financial institution by most of its counterparties, but it was allowed to fail on the 15 September 2008. This financial institution's collapse impacted not only the US market but also the economy of the EU. The global credit crisis forced investors to carefully evaluate the default possibility of all financial institutions.

This paper aims to understand the response to this issue; if financial institutions consider creditworthiness of their counterparties when pricing their derivatives contracts, is there an improvement in their stability and does the entire network become

D. Ladley · C. T. M. Tran (✉)
School of Business, University of Leicester, Leicester, UK
e-mail: tmct3@leicester.ac.uk

D. Ladley
e-mail: dl110@leicester.ac.uk

A. S. Chakrabarti et al. (eds.), *Network Theory and Agent-Based Modeling in Economics and Finance*, https://doi.org/10.1007/978-981-13-8319-9_14

more stable? Whilst there is a significant amount of research focusing on systemic risk in lending markets, there is little that examines the impact of the deterioration of counterparties' credit rating on systemic risk in derivatives markets. The rapid development of Basel standards after the global credit crisis poses the question of whether Credit/Debit Value Adjustment play a role in improving the stability of financial systems and to what extent. This paper, thus, examines the relationship between the interconnected architecture and systemic risk in a financial derivatives network.

To study this question, we construct an agent-based model of the financial system in which financial institutions are connected by a network of derivative contracts. Changes in the values of these contracts result in payments between institutions. If these changes in value are larger than the capital of the associated institutions they may force defaults. The failure of institutions potentially results in the spread of losses as institutions are no longer able to make their required payments pushing losses onto their counterparties. We consider this system with and without CVA and DVA price adjustments in order to understand how they effect the market stability.

2 Credit Risk and Value Adjustments

Counter-party credit risk is the risk that each party in a transaction could suffer in the form of a loss from its counterparties' failures before the expiration of their contracts. In an Over-The-Counter (OTC) derivatives market, the issue of counter-party credit risk is potentially serious. For example, in an Interest Rate Swaps (IRS) transaction, one counter-party agrees to pay a fixed rate per annum on a notional principal, and in return, the other counter-party agrees to pay a floating rate which can be based on, for example, the London Inter-bank Offered Rate (LIBOR). The time intervals at which payments are exchanged are initially specified by both parties. At the points of exchange, one counter-party receives a payment from the other depending on the value of the contract. Exposure can be to either the fixed rate payer or the floating rate payer (bilateral exposure).

Since 1999, large banks have started to incorporate Credit Value Adjustments (CVA) in order to evaluate the cost of counter-party risk. If the default probability of an entity increases, its counter-party faces an equivalent increase in credit risk if the entity is unable to pay its contractual obligations. The counter-party, therefore, needs to evaluate its credit risk and factor this into the pricing of contracts. This adjustment is referred to a CVA and is calculated form the downgrade of the entity's credit rank. At the same time counterparties will be evaluating the institutions rating and applying similar changes in value to reflect changes in default probabilities. These too must be reflected in the price as Debt Value Adjustment (DVA). Thus, the dual component of CVA—DVA—should both be considered when calculating an asset value. Beyond this, there are further components that increase or decrease the market values of contracts such as collateral agreements, initial margins, required capital. These elements should also be included, e.g. Collateral Value Adjustment (ColVA), Margin Value Adjustment (MVA) and Capital Value Adjustment (KVA). The set of these adjustments are referred to as xVA, however, in this paper, we focus just on the first two CVA and DVA. Below we set out the mechanism by which CVA and DVA are calculated.

2.1 CVA/DVA

CVA is the price of counter-party risk, i.e. the expected loss due to counter-party default(s) in the future (Gregory 2010; Crepey 2014). The market value of a derivative should therefore be

$$\text{Market price} = \text{Fair price} - \text{CVA}$$

Gregory (2010) derived the efficient formula to compute CVA in the assumption of no wrong way risk as follows:

$$CVA \approx (1 - \overline{\delta}) \sum_{j=1}^{m} B(t_j) EE(t_j) q(t_{j-1}, t_j) \tag{1}$$

where
Loss Given Default, $(1 - \overline{\delta})$, gives the proportion amount of expected loss in default event.
Discount Factor, $B(t_j)$, denotes the risk-free discount factor at time t_j.
Expected Exposure, $EE(t_j)$, is calculated for the relevant dates in the future given by (t_j) for $j = 0, n \to m$.
Default Probability, $q(t_{j-1}, t_j)$, is the marginal default probability in the interval between date t_{j-1} and t_j.

However, unlike lending markets where the exposure is usually one-way or unilateral, i.e. in where lenders face the risk that borrowers could default but not the opposite situation, transactions in derivatives markets are two-way or bilateral so both counterparties have credit risk from the default possibilities of others. This is why DVA needs to be considered as the cost of institutions' own risk when financial firms fail before maturity.

DVA can be understood as the contrasting component of CVA and *vice versa* (Green 2016). Thus, in a bilateral derivatives contract with two counterparties A and B, the below equation can be reached:

$$CVA_A = DVA_B; \; CVA_B = DVA_A$$

Market value of a counter-party in a derivatives transaction is now equal to:

$$\text{Market price} = \text{Fair price} - \text{CVA} + \text{DVA}$$

2.2 BCVA

The combination of CVA and DVA in pricing a derivatives contract is known as Bilateral Credit Value Adjustment (BCVA). A simplified approach by Pallavicini et al. (2011) combines both components into one concept of BCVA equal to:

$$BCVA = CVA - DVA$$

Gregory (2010) argues, however, that the counter-party credit risk of one institution is only a concern if the institution survives after its counter-party default. Therefore, Gregory's adjusted BCVA is derived under one further assumption of no simultaneous defaults. Brigo et al. (2011) also details this as

$$\begin{aligned} BCVA \approx &(1-\overline{\delta})\sum_{j=1} B(t_j)EE(t_j)S_I(t_{i-1})q(t_{j-1},t_j) \\ &-(1-\overline{\delta}_I)\sum_{j=1} B(t_j)NEE(t_j)S(t_{i-1})q_I(t_{j-1},t_j) \end{aligned} \tag{2}$$

where
$S(.)$ and $S_I(.)$ are the survival probabilities of institution and counter-party, respectively.
q_I and $\overline{\delta}_I$ represent the default probability and recovery of the institution.
$NEE(t_j)$ denotes the negative expected exposure. That is the EE from the perspective of the counter-party.

3 Literature Review

Systemic risk in derivative markets are received less attention than inter-bank lending markets. The majority of the work that has gone on has focused on the role of the credit derivatives market in financial crisis and the benefits of Central Counter-party clearing house (CCP) in reducing financial distress.

Bliss and Kaufman (2004) examined whether protective characteristics of derivatives transactions, such as netting, collateral and close-out, can reduce systemic risk. They found that the positive impacts are unclear because netting and collateral can increase systemic risk by permitting the concentration of risk in dealers. However, they may also decrease distress by giving dealers an effective tool to manage their counter-party risk and reduce unanticipated defaults. Meanwhile, close-out can be a potential source of systemic risk since applying close-out increases the difficulty of managing insolvent major dealers. Conversely, Ali and Vause (2016), Singh (2010) and Hull (2010) agreed that the OTC derivatives market is a potential source of systemic risk, but financial institutions can effectively manage counter-party risk by using bilateral netting or collateralisation agreements.

Kiff et al. (2009) and Russo (2010) argued that the credit derivatives market can increase systemic risk and the inter-connectedness of large financial institutions. A CCP can help to reduce systemic risk, however, the effectiveness of multiple CCPs is arguable. This proposal is supported by Zigrand (2010), Wellink (2010), Yavorsky (2010), as well as Cont and Minca (2014) who used network-based measures of systemic risk and demonstrated that the default of an entity in the CDS market exposes losses for both its counterparties and protection sellers. If financial institutions do not have enough reserves to fulfil their CDS liabilities, the credit event also causes the bankruptcy of protection sellers and widens the default contagion. Nonetheless, if all major dealers join multilateral clearing with a CCP, the stability of the market can increase. Whilst, Borovkova and Mouttalibi (2013) concludes that a CCP can reduce the network's fragility but only for a homogeneous financial system; a CCP's presence can exacerbate contagion defaults for non-homogeneous financial networks, especially small financial firms.

4 Model

We construct a model of a financial system that is statistically similar to the real network that was in place before the global credit crisis in 2008. There were at least 8,000 counterparties working in the American derivatives market at this time. They were connected to one another by approximately 1 million derivatives transactions including interest rate swaps, foreign exchange and credit default swaps. The total notional outstanding was approximately \$800 billion. It is not feasible to engage in computational simulation with such a significant number of repetitions on a network of 8,000 counterparties, involving over 1 million derivatives transaction. Hence, this paper attempts to develop a similar financial network on a smaller scale. Similarly, we focus on a single type of deviate contract—interest rate swaps (IRS). This is because these contracts account for approximately 80% of the derivatives market (BISb 2015).

4.1 *Financial Network*

Consider a closed-economy which comprising n risk-neutral financial institutions. Each financial institution has a stylised balance sheet comprising derivative contracts at their fair value (Y), capital (E) and a balancing term Residual Assets (A). The value of the derivative position can be positive, negative or zero depending upon the value of each derivative contract being an asset or liability for the institution. If Y is negative the derivatives are a net liability—the bank owes more than it is owed. The residual assets term represent all other assets and liabilities, including customer deposits, cash and loan position through. Note, like the derivative position, A can be positive or

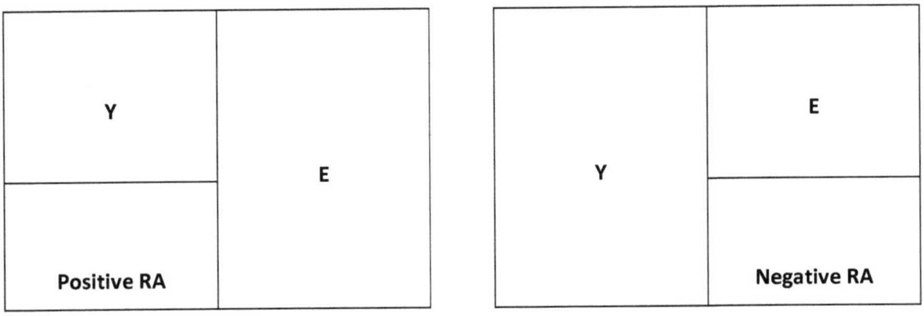

Fig. 1 Stylised balance sheet

negative depending on the balance or the non-equity and derivative asset and liability terms. The expression of the stylised balance sheet is illustrated in Fig. 1.

These figures are related through the following accounting equation:

$$E_i = Y_i + RA_i$$

The derivatives position of financial institution i, Y_i is comprised of the sum of its contracts with other financial institutions. The value of a contract between institution i and institution j is denoted by $y^t_{ij} = -y^t_{ji}$. If there is no contract between bank i and bank j then $y^t_{ij} = y^t_{ji} = 0$. The fair value of bank i's derivative position at time t is equal to $Y^t_i = \sum^n_{j=1, j \neq i} y^t_{ij}$. Note we do not specify that a given pair of banks have at most one derivative transaction connecting them, however, the sum of transactions between a pair of banks may be netted together to be considered as one single aggregate transaction. The set of derivatives contracts across all institutions forms a network in which nodes are financial institutions and edges are derivative positions.

4.2 Contracts and Payments

Each derivative contract has a floating rate and a fixed rate payer. The fixed rate is specified at the start of the simulation whilst the floating rate is determined in each time step. The net payment is the difference between these interest rates multiplied by the notional value of the contract. This value is transferred from the payer of the higher rate to that of the lower rate.

If one counter-party has a total amount of payables across all contracts which is larger than its sum of receivables, it has an excess obligation RA_i. If a financial institution has insufficient equity to cover the excess obligation, bank i is insolvent according to the default rule:

$$\text{Bank } i \text{ defaults when } E_i < RA_i \text{ where } RA_i = Payment - Receive$$

The loss from an insolvency can negatively impact other financial institutions through the edges of the network leading to further insolvencies. Such insolvencies result in second round defaults. The chain of defaults from the first to the final failure is the financial contagion which is measured by the total number of insolvent counterparties.

4.3 Derivative Network

Define L_{ij}^t as the nominal liability of node i to node j at time t. Liabilities are non-negative and no counter-party has a claim against itself. Let p_i^t represent the total payment by counter-party i to other counterparties at time t and define

$$p_i^t = \sum_{j=1}^{n} L_{ij}^t$$

The vector $p^t = \{p_1^t, p_2^t, \ldots, p_n^t\}$ represents the liabilities at time t of all financial institutions, respectively.

Let Π^t denote the relative liability matrix which captures the proportion of the nominal liabilities by counter-party i to be paid to counter-party j, therefore:

$$\Pi_{ij}^t = \begin{cases} \dfrac{L_{ij}^t}{p_i^t} & \text{if } p_i^t > 0 \\ 0 & \text{otherwise} \end{cases}$$

Under equal priority of payments, the payment by counter-party i to counter-party j is $p_i^t \Pi_{ij}^t$. Thus, the total payments received by node i are equal to $\sum_{j=1}^{n} \Pi_{ij}^t p_j^t$.

The fair value, or residual capital, of node i is the difference between the total payments from creditors and total payments from creditors which is equivalent to:

$$Y_i^t = \sum_{j=1}^{n} \Pi_{ij}^t p_j - p_i^t \tag{3}$$

Given the above, the fair value of each node is used to evaluate the stability of each counter-party and the system as a whole via the algorithm of Eisenberg and Noe (2001).

4.4 BCVA

We analyse the model with and without BCVA. In the presence of BCVA the market value of the derivatives contract is equal to the price of a risk-free derivative minus BCVA.

$$\text{Market price} = \text{Fair price} - \text{BCVA}$$

$BCVA_i$ of financial institution i is the sum of the $bcva_i$ for all its transactions:

$$BCVA_i^t = \sum_{j=1, j \neq i}^{n} bcva_{ij}^t$$

In a derivative transaction between bank i and bank j, if bank j defaults, bank i suffers a loss which is equivalent to the payment from bank j to bank i. Conversely, if bank i is insolvent, bank j could not gain from bank i as its initial commitment. Therefore, the $bcva_{ij}$ is estimated by

$$bcva_{ij}^t = \overline{\delta}_I \times receive_{ij}^t \times PD_j^t \times PS_i^t - \overline{\delta}_I \times payment_{ij}^t \times PD_i^t \times PS_j^t \quad (4)$$

where $\overline{\delta}_I$ is the recovery rate; $receive_{ij}$ is the receivable which bank i gains from bank j; $payment_{ij}$ denotes the payable which bank i pays to bank j. PD and PS are the respective probability of default and of survival. The discounted interest rate in the original formula (2) is ignored because we focus on the determination of insolvency at a single point in time.

4.5 Parameters

In order to understand the effect of BCVA on market stability we simulate the model. Parameters are assigned based on data prior to the last financial crisis in the period 2003–2007.

4.5.1 Exit Price and Duration

The notional value of the IRS contracts are randomly drawn from a uniform distribution with range 10,000–15 million currency units. Similarly, transaction duration are randomly drawn with uniform probability from a discrete distribution with intervals at 3 months ranging from 6 months to 5 years.

4.5.2 Fixed Rate and Floating Rate

Fixed rates are randomly assigned from a continuous uniform distribution spanning 3.734–6.007% which is the range of average 3-month LIBOR in the data window. Meanwhile, floating rates are drawn randomly from the uniform distribution with range 3.391 and 6.904%, reflecting the minimum and maximum LIBOR in the same period.

4.5.3 Credit Spread and Probability of Default

Credit spreads of financial institutions are assumed to be positive values which are normally distributed with a mean of 330 and a standard deviation of 500. The conditional probability of default in calculating CVA/DVA is then:

$$\bar{\lambda} = \frac{s(t)}{1 - \bar{\delta}_I} \tag{5}$$

where:
$s(T)$ is the credit spread for a maturity of T
$\bar{\delta}_I$ denotes the recovery rate which is assumed to be 40% for all financial institutions
$\bar{\lambda}$ is the probability of first default between time 0 and time t of one counter-party. Later probabilities of default are calculated based on the first default probability.

4.5.4 Equity

Equity is randomly drawn from a continuous distribution with range 10,000–1 million currency units.

5 Results

In order to estimate the impact of BCVA on systemic risk, we simulate systemic risk on the identical networks with and without BCVA. The difference between numbers of defaults in the two scenarios is the effect of BCVA on systemic risk.

We use a Monte Carlo approach to evaluate how systemic risk happens in the without-BCVA condition. Contagion is triggered by a shock of interest rates. We draw a value of the floating rate interest rate from the distribution described above. Institutions face difficulties in fulfilling their obligations if their payables are larger than their own capital. Consequently, one or more insolvencies may occur. These are considered first-round insolvencies. After that, any financial institutions which are connected to the first defaulting institution are negatively influenced due to credit losses. These losses are equal in value to the payments not received from the defaulting institutions. Systemic risk is evaluated as the number of defaults from the second round until no more defaults occur.

The experiment is then repeated for the same financial network with the same financial institutions but with the addition of BCVA. In this scenario financial institutions have to have higher capital to cover their credit risk, which is equivalent to BCVA value; i.e. the obligation of each financial entity includes payables plus BCVA. The processes of applying the default rule and determining systemic risk are the same as with the non-BCVA simulation. The effect of accounting for BCVA in pricing derivatives on systemic risk is illustrated by the difference in default numbers between two Monte Carlo simulations.

5.1 Systemic Risk Versus Leverage

The effect of the introduction of BCVA on systemic risk is demonstrated in Fig. 2 across three levels of derivatives contract sizes: 15 million (Panel a), 30 million (Panel b) and 40 million (Panel c). Each case has the same level of equity. This configuration allows the effect of BCVA to be estimated across different levels of potential risk.

The results clearly indicate that the stability of the financial network is improved by accounting for counter-party credit risk in pricing derivatives. The level of improvement is dependent upon the differences between the total notional principals of derivative contracts and the equity levels of financial institutions. The increased equity available within the system protects institutions against the failure of their counterparties. Larger exposures reduce the scale of this protection. Formally we test the following hypothesis:

H_{0a}: Higher leverage results in a more fragile network
H_{1a}: Higher leverage does not result in a more fragile network
H_{0b}: Higher leverage reduces improvement of systemic risk caused by BCVA
H_{1b}: Higher leverage does not reduce improvement of systemic risk caused by BCVA.

In both cases, the alternative hypotheses of H_{1a} and H_{1b} are rejected at the 1% significance level. Larger leverage leads to a more fragile system. Additionally, higher leverage reduces the role of BCVA in enhancing the stability of the system.

Whilst the effect of leverage was perhaps intuitive, we further investigate the impact on systemic risk caused by network structure (also known as connectivity) and premium (understood as the additional rate which floating rate payers agree to commit with fixed rate payers).

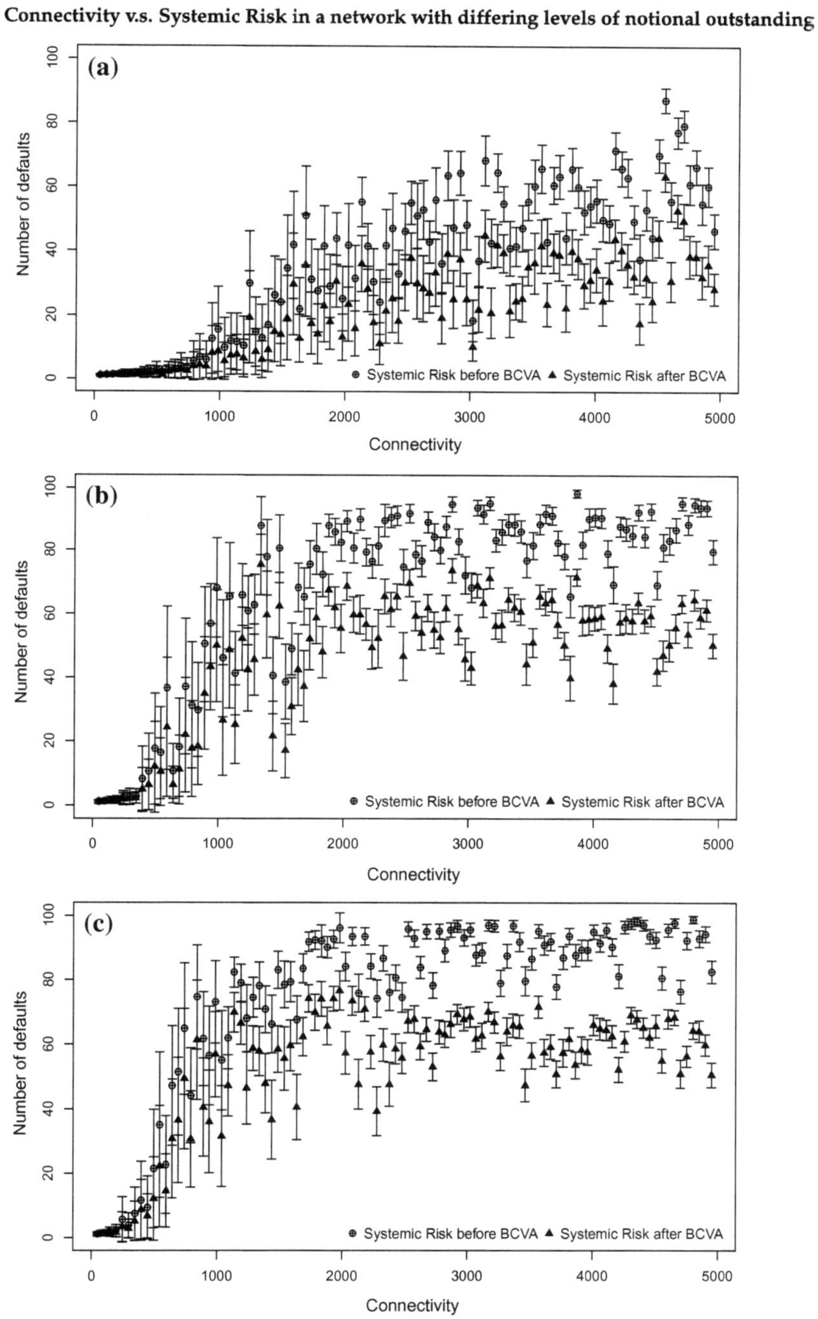

Fig. 2 **a** Contract size from 10 thousand to 15 million currency unit, **b** Contract size from 10 thousand to 30 million currency unit, **c** Contract size from 10 thousand to 40 million currency unit

5.2 Systemic Risk Versus Connectivity

Connectivity has been shown in numerous papers to effect systemic risk, e.g. (Ladley 2013). We, therefore, examine the impact of connectivity associated with notional principals of derivatives contracts. Figures 2 and 3 display the effect of connectivity on systemic risk with two conditions for notional outstanding. The first, Fig. 2 increases notional outstanding as the number of connections increases, in essence adding more contracts to the market. The second fixes the notional outstanding as connections increase, in effect dispersing or reducing the per connection contract size (Fig. 3).

Figure 2a–c indicate that more connections leads to higher the systemic risk. This finding supports work done by Blume et al. (2011) and Blume (2013). The shape of the relationship between connectivity and systemic risk changes, however, with the strength of connections between institutions. For relatively small contract sizes, systemic risk increases for the full range of connectivity's all the way up to a fully connected network. Each additional connection increases the risk of the spread of failures by bringing more capacity to spread shocks into the system. Increasing the levels of the contract size, however, changes the shape of this relationship. Figure 2b shows that the network saturates at approximately 2,500 connections, whereas for Fig. 3 is saturated at a lower level. In these cases, larger contract sizes suggest that banks are more likely to fail and therefore spread further failures. As contract size increases the point at which all banks that can fail occurs with fewer connections due to the larger losses associated with each connection.

Although the above result is intuitive and speaks to the growth of these markets, it is also interesting to consider the effect of changing the structure of the market whilst holding the value of contracts constant. Such a change has often been discussed by regulators and commentators. These results are presented in Fig. 3 and have a markedly different shape. The positive monotonic relationship between connectivity and systemic risk observed previously is no longer uniformly present. Whilst it is still the case that higher value contracts increase the number of failures observed connectivity also plays a role. For small contracts, failures are maximised for relatively small numbers of contracts, however, as contract size increases the maxima also increases. Notably for intermediate size contracts, the relationship between connectivity and defaults is n-shaped. This pattern somewhat mirrors the results observed in Ladley (2013) and Acemoglu et al. (2015). Effectively, as shock size increases the network changes from being risk-spreading to failure-spreading. For small contract sizes, and therefore small shocks, the network spreads risk—as the network becomes better connected fewer institutions fail. As the contracts get larger, this relationship is replaced by the network acting to spread failures resulting in the peak in the distribution moving rightwards.

Both Figs. 2 and 3 indicate that as the size of constricts increases, BCVA has a greater positive effect. In the case of higher notional principals of derivative contracts, a financial institution has to reserve more capital for its counter-party credit risk

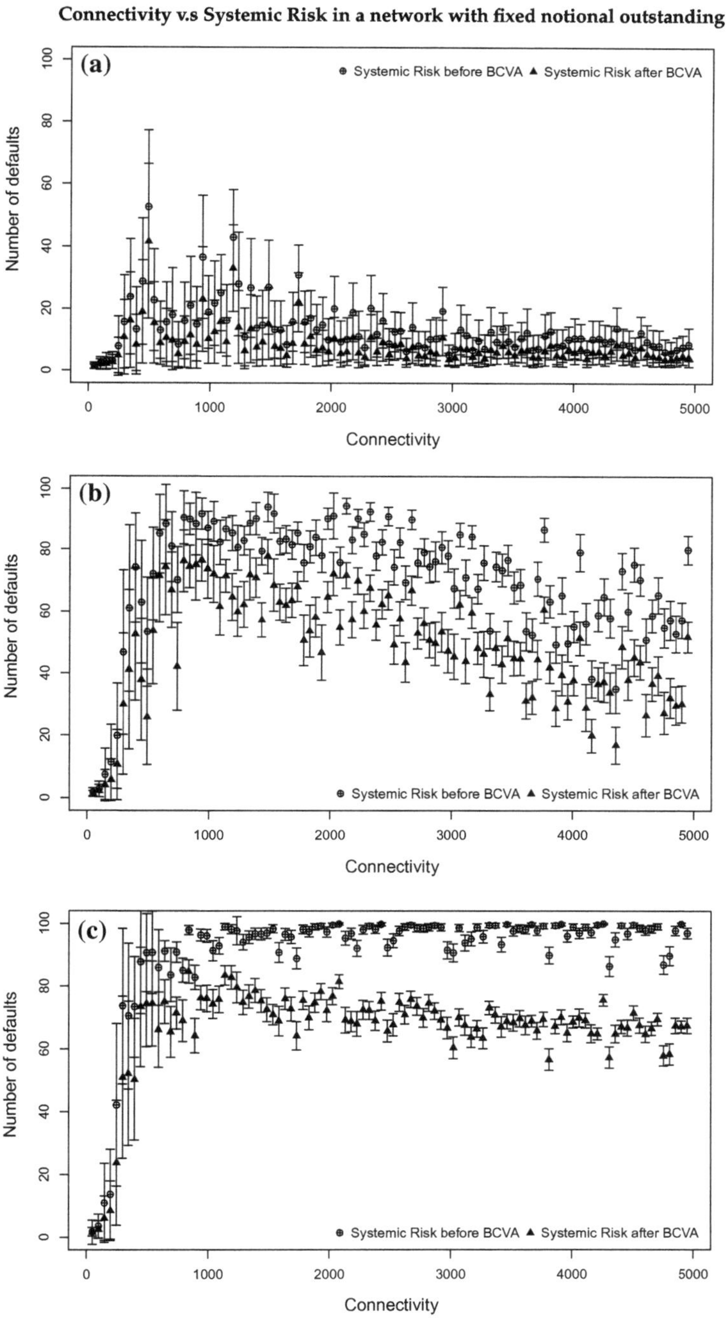

Fig. 3 **a** Total notional outstanding of 10 billion for entire network, **b** Total notional outstanding of 70 billion for entire network, **c** Total notional outstanding of 500 billion for entire network

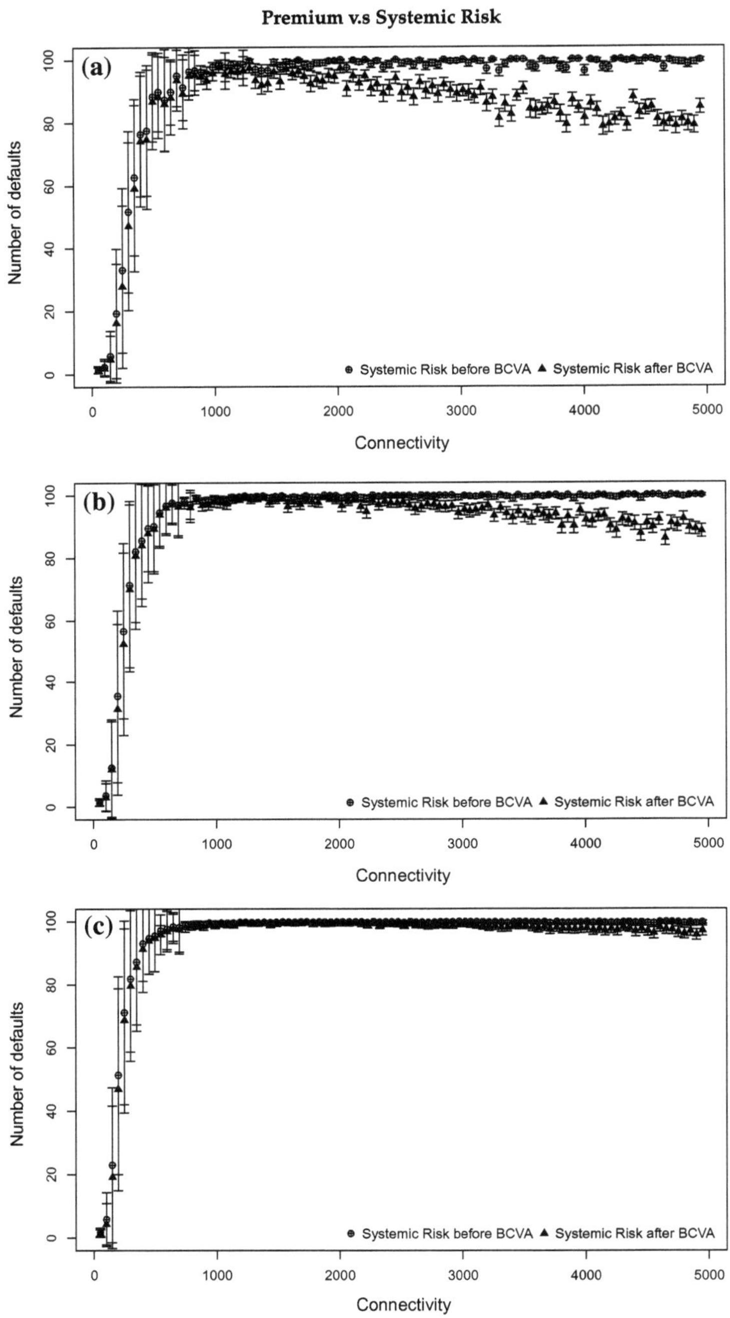

Fig. 4 **a** 0.5% premium, **b** 1% premium, **c** 3% premium

(BCVA). As a result, the default probability of each financial entity decreases and the stability of the network is enhanced as it has greater loss-absorbing capacity.

5.3 *Systemic Risk Versus Premiums*

In addition to the effect of leverage and connectivity, premiums also have a substantial impact on systemic risk. We consider and evaluate three levels of premium: 0.5, 1 and 3%. The level of premium dictates the size of the payment owed from one counter-party to the other. Larger premiums equate to a larger gap—effectively a larger payment shock. Figure 4 illustrates the changes in systemic due to BCVA for different levels of premiums.

The results suggest that BCVA is more effective for smaller premiums. For markets with low connectivity, BCVA has little effect. It only starts to have an impact on greater numbers of connections. The higher connectivity results in shocks from failures being better spread, therefore allowing the BCVA to more effectively absorb the shock. More connections effectively lead to more BCVA reserves being available to absorb the failure. This benefit of BCVA, however, decreases as the strength of connections increases. Stronger connections limit the ability of BCVA to absorb the failures.

6 Conclusion

Our paper is the first to analyse systemic risk in derivative markets under the effect of xVA. This paper numerically analyses the fragility of a financial network before and after accounting for bilateral credit valued adjustments. The results illustrate that BCVA indeed has a positive influence on systemic risk of financial networks although its effect is dependent upon the leverage ratio in the network. In particular, the higher the leverage, the more effective the BCVA's impact. This paper provides evidence supporting the perspective that a more complete system, which can propagate negative effects from one or more defaulting counterparties to other entities reduces systemic failures. If shocks exceed a certain value, however, more interconnections do not reduce contagion but instead act to effectively transfer negative shocks from financial institution(s) to others and widen systemic risk. Our results also highlight the possible improvement of systemic risk by BCVA if the variation of the premium is small enough.

References

Acemoglu D, Ozdaglar A, Tahbaz-Salehi A (2015) Systemic risk and stability in financial networks. Am Econ Rev 105(2):564–608

Ali R, Vause N (2016) Systemic risk in derivatives markets: a pilot study using CDS data. Financ Stab Paper 38:1–21

BISb (2015) Statistical release: OTC derivatives statistics at end-June 2015. Discussion paper, Bank for International Settlements

Bliss RR, Kaufman GG (2004) Derivatives and systemic risk: netting, stay, and closeout, pp 1–27

Blume L (2013) Network formation in the presence of contagious risk. In: 12th ACM conference on electronic commerce, pp 1–23

Blume L, Easley D, Kleinberg J, Kleinberg R, Tardos V (2011) Which networks are least susceptible to cascading failure? In: IEEE 52nd annual symposium on foundations of computer science, pp 303–402

Borovkova S, Mouttalibi HLE (2013) Systemic risk and centralized clearing of OTC derivatives: a network approach, Working paper, pp 1–54

Brigo D, Buescu C, Morini M (2011) Impact of the first to default time on bilateral CVA, pp 1–14

Cont R, Minca A (2014) Credit default swaps and systemic risk, pp 1–24

Crepey S (2014) XVA: about CVA, DVA, FVA and other market adjustments, pp 1–30

Eisenberg L, Noe TH (2001) Systemic risk in financial systems. Manag Sci 47:236–249

Green A (2016) XVA: credit, funding and capital valuation adjustments. https://onlinelibrary.wiley.com/doi/book/10.1002/9781119161233

Gregory J (2010) Counterparty credit risk: the new challenge for global financial markets, 3rd edn. Wiley

Hull J (2010) OTC derivatives and central clearing: can all transactions be cleared? Financ Stab Rev Deriv Financ Innov Stab 14:71–80

Kiff J, Elliott J, Kazarian E, Scarlata J, Spackman C (2009) Credit derivatives: systemic risks and policy options, IMF working paper, Working paper 09/254, pp 1–35

Ladley D (2013) Contagion and risk-sharing on the inter-bank market, Discussion papers in economics, Department of Economics, University of Leicester, 11/10, pp 1–45

Pallavicini A, Perini D, Brigo D (2011) Funding valuation adjustment: a consistent framework including CVA, DVA, collateral, netting rules and re-hypothecation, pp 1–23

Russo D (2010) OTC derivatives: financial stability challenges and responses from authorities. Financ Stab Rev Deriv Financ Innov Stab (14), 101–112

Singh M (2010) Collateral, netting and systemic risk in the OTC derivatives market, IMF working paper, pp 1–15

Wellink N (2010) Mitigating systemic risk in OTC derivatives markets. Financ Stab Rev Deriv Financ Innov Stab (14), 131–135

Yavorsky A (2010) OTC derivatives market structure and the credit profiles of wholesale investment banks. Financ Stab Rev Deriv Financ Innov Stab (14), 143–151

Zigrand J-P (2010) What do network theory and endogenous risk theory have to say about the effects of central counterparties on systemic stability? Financ Stab Rev Deriv Financ Innov Stab (14), 153–160

Additional Default Probability in Consideration of Firm's Network

Takuya Kaneko and Masato Hisakado

Abstract In this paper, we propose a methodology to effectively capture credit risk from firms' network. In short, our target is to numerically obtain additional credit risk from connected firms on network. Recently, commercial networks are available for investing and managing risk on professional information terminals like Bloomberg and Reuters. They enable us to check commercial connection of firms. We utilize them to assess positive and negative effect on observing firms from neighbor firms, especially, when the neighbor firms have any credit events. We propose a methodology to analyze/measure credit impact, which observing firms potentially receive from their neighbors. We applied Merton model (Merton and Robert in J Financ 29(2):449–470, 1974) which is generally utilized for credit risk management to calculate additional risk and simplified the formula for practicability/usability. Also, it enables us to escape from having any difficulties in computation time. We introduce our approach with over-viewing simple model guidance and explaining a few samples of numerical experiments.

Keywords Default probability · Credit risk management · Network · CB pricing

1 Introduction

In this paper, we propose a methodology to logically capture additional credit risk of firms from neighbor firms on networks. When firms go bankruptcy, there are two reasons. They are general and special factors. When firms default due to their

We would like to thank Dr. Toshinao Yoshiba (Bank of Japan) for useful discussion. We thank anonymous reviewers for their helpful comments and suggestions. The first author was supported by a grant-in-aid from Zengin Foundation for Studies on Economics and Finance.

T. Kaneko (✉)
ICU and KDDI Research, 3–10–2 Osawa, Mitaka-shi, Tokyo 181-0015, Japan
e-mail: tkaneko@icu.ac.jp

M. Hisakado
FSA, 1–6–5, Kasumigaseki,, Chiyoda-ku, Tokyo 100-8967, Japan
e-mail: masato.hisakado@fsa.go.jp

A. S. Chakrabarti et al. (eds.), *Network Theory and Agent-Based Modeling in Economics and Finance*, https://doi.org/10.1007/978-981-13-8319-9_15

own problems (idiosyncratic risk) or economical depression (systematic risk), these cases are classified into the first category. When firms default due to affection from neighbor firms' default (chain reaction bankruptcies), these cases are classified into the second category. We already have many kinds of default probability estimating methodology for general factors. On the other hand, there are a few methodologies for special factors. As regulators request to effectively manage credit risk by supposing historical stressed scenarios (Basel Committee on Banking Supervision 2009; Board of Governors of the Federal Reserve System 2009), we focus on proposing practicable methodology to capture the credit risk from special factors for improving credit risk management.

Merton model (Merton and Robert 1974) is well known as the standard methodology in measuring default probability. It is frequently utilized in many financial institutions for their credit risk management. It defines firm's default as observing firm's total asset's being smaller than its total liability (debt amount) at some fixed future time horizon T. It provides default probability clearly by supposing fixed debt amount and the dynamics of total assets as being log-normal model. If you cautiously check the detail of total asset in its balance sheet of financial reports, you can easily find many types of assets ordered from high liquidity assets toward low. Also, you can easily recognize the firm's trading partners from its notes receivables, account receivables, note payables, account payables, and so on. Namely, firms are basically not independent but belonging to some commercial networks. When a firm goes bankruptcy, this credit event does not remain within the firm but spreads out to other firms especially toward closely trading partners as negative impacts. And this is rarely propagated. Watt's model (Watts and Duncan 2002) is frequently used for these phenomena. In this paper, our main aim is to propose a framework for estimating each firm's possible negative impacts receiving from their trading partners by combining well known standard models: Merton model and Watt's model. There are many articles which propose stress testing by supposing the same shock to all firms uniformly and not consistent with the individual corporate bond pricing under the stressed scenario (ex. Bonti 2005; Jakubik et al. 2008; Stein and Roger 2014). In these papers, the authors mainly target to observe extended tail part of portfolio loss distribution for VaR (Value at Risk) and Expected Shortfall. While our approach targets to estimate micro first by analyzing individual stress level based on each network: bespoke analytic settings. Then this paper would be the first trial to quantify each additional default probability in consideration of the commercial network and this could be utilized for pricing individual bond under the stressed scenario consistently with stressed portfolio risk estimation. Also, we visually indicate correctness of our model by comparison with market index: J-TED.[1]

To consider default event propagation in credit risk management, intensity-based approaches like top-down approach initially introduced by Azizpour and Giesecke (2008) are frequently utilized for modeling. Researchers tried expanding many applications to adopt to the actual financial data (ex. Kaneko and Takuya

[1]This index is difference between Yen LIBOR and risk-free interest rate: Japanese government bond yield.

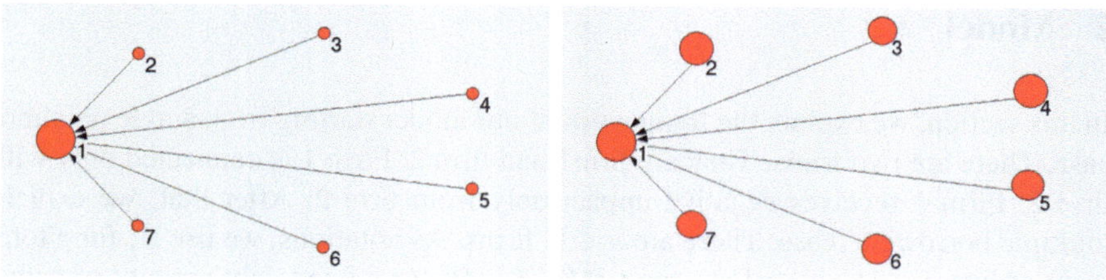

Fig. 1 In above figures, circles indicate firms and arrows indicate commercial connections between firms. Size of circles indicates default probabilities of the firms. The direction of arrows indicates risk flow and firm 1 is a lender to other six firms. Firm 1 in the left figure should be regarded as lower credit risk than firm 1 in the right figure because of the default probabilities of their trading partners: borrowers

2016; Kaneko et al. 2010). One of the characteristics of these approaches is their directly generating default events based on their default intensity functions. While, we use structural approach for its tractability to figure out the mechanism of these phenomena in this paper. As simply explained in above, firms, which trade with good counterparties, would be evaluated as high credit quality. While firms, which trade with bad ones, would be evaluated as low quality through this model.

Circles in the Fig. 1 indicate firms and their sizes are default risk. The arrows means a commercial relation between lender and borrower. Arrowheads are lenders and starting points of arrows are borrowers. The firm 1 in both figures lend to other firms and the borrower's credit qualities are the same. Even though the firm 1 in both figures (left figure and right figure) have the same credit quality, they should be evaluated differently by considering the credit qualities of their trading partners: borrowers. We target this point and our model enables us to distinguish them. Also, we need to consider the set of firm's trading partners and their volumes are continuously changing in the real commercial environment. Firms trading with other firms sometimes receive negative impacts from their trading partner's default because we need to depreciate the amount of assets issued by defaulted firms. And if the depreciated total asset of event receiving firm becomes smaller than its debt amount, the receiver also goes bankruptcy. Namely, the default infection happens. In Watt's model, agents in a network transfer impacts to each other with small, fixed, and same thresholds to propagate the event rarely. In this paper, we set the debt amount as default threshold. Then high debt ratio firms are easy to transfer and receive default events, while low debt ratio firms are not easy to transfer and receive the events under our setting. We explain the framework of our model in the next section, and show experimental result by using actual financial data.

2 Model

In this section, we explain the framework of our model starting from single neighbor case. There are two firms. They are firm1 and firm 2. Firm 1 is connected only with firm 2. Firm 1 receives negative impact only from firm 2. After that, we explain multiple borrowers' case. There are $n \in \mathbf{N}$ firms. As notations, we use A_t^i for a total asset of firm $i \in \{1, \dots, n\}$ at time $t \in [0, T]$, D^i for a fixed debt amount of firm i during time $\forall t \in [0, T]$. $\mathbf{1}_{\{C\}}$ is an indicator function outputs one when the condition C is true, otherwise outputs zero. We define a firm's default as its total asset becomes smaller than its debt amount based on Merton model. We use function $\mathbf{p}(i)$ to express the default probability of firm i at some time horizon T as of time zero. We observe total asset only at time horizon T.

$$\mathbf{p}(i) = \mathbf{E}\left[\mathbf{1}_{\{A_T^i < D^i\}}\right] = \mathbf{P}(A_T^i < D^i) \tag{1}$$

We suppose log-normal dynamics for total assets as follows by characterizing with μ_i for drift, σ_i for volatility, and Brownian motion W_T.

$$A_T^i = A_0^i \exp\left\{\left(\mu_i - \frac{1}{2}\sigma_i^2\right) T + \sigma_i W_T\right\} \tag{2}$$

We calculate default probability under risk-neutral probability sets based on risk-free interest rate r.

$$\mathbf{p}(i) = \mathbf{N}\left(\frac{\ln\left(\frac{D^i}{A_0^i}\right) - \left(r - \frac{1}{2}\sigma_i^2\right) T}{\sigma_i \sqrt{T}}\right) \tag{3}$$

where $\mathbf{N}(\cdot)$ indicates cumulative value of function $\phi(\cdot)$, $\mathbf{N}(a) = \int_{-\infty}^{a} \phi(\xi)\mathbf{d}\xi$, $\phi(\cdot)$ is standardized normal density function.

2.1 Single Neighbor Case

We estimate a negative impact from single neighbor. There are two firms: firm 1 and 2. We estimate firm 1's potentially receiving negative impact from firm 2. One important point is that this is not the estimation of simultaneous default but default infection phenomenon. Firm 2 is the neighbor of firm 1. Let ξ_1, ξ_2, ξ are random variables obtained from standardized normal distributions. First two are standardized assets of firm 1 and 2. They are correlated under correlation of ρ_{12}. Last one is independent from others. ξ_1 can be expressed as follows by using ξ_2 and ξ in consideration of the correlation of ρ_{12}. $\xi_1 = \rho_{12}\xi_2 + \sqrt{1 - \rho_{12}{}^2}\xi$ And d_1 and d_2 are default thresholds for firm 1 and 2. p_1 and p_2 are default probabilities for firm 1 and 2. There are four possible situations of two firms. They are $\xi_1 > d_1$ and $\xi_2 > d_2$ (case 1), $\xi_1 < d_1$ and $\xi_2 > d_2$ (case 2), $\xi_1 > d_1$ and $\xi_2 < d_2$ (case 3), and $\xi_1 < d_1$ and $\xi_2 < d_2$ (case 4). We focus only on the case 3. Firm 1 did not default. Namely, its total asset value index ξ_1 was greater than its default threshold d_1. And we observe that its neighbor firm 2 defaulted. We suppose that credit environment around firm 1 differs before and after its close business counter party's default. We check firm's situation with d_1 first. When firm 2 goes bankruptcy, we recheck firm 1 based on revised threshold $\overline{d_{12}}$ to assess firm 1's situation. Namely, we check whether firm 1's asset index which was greater than d_1 is greater than new threshold $\overline{d_{12}}$ or not. Relationship between parameters and default probability is as follows:

$$d_i := \mathbf{N}^{-1}(p_i) = \frac{\ln\left(\frac{D^i}{A_0^i}\right) - \left(r - \frac{1}{2}\sigma_i^2\right)T}{\sigma_i\sqrt{T}} \tag{4}$$

We utilize the idea proposed in Hisakado et al. (2016), which is the analytical solution of conditional default probability to escape from calculating double integral which requires long calculation time. When some firm defaults, it's impossible to observe its asset amount fluctuation like live firms because of its dynamics completely stopping. To reflect this, we intentionally stop updating neighbor's asset and fix it exactly at its default threshold d_2. The firm 1 defaults when its random variable ξ_1 becomes less than its default threshold d_1.

$$\mathbf{P}(\xi_1 < d_1|\xi_2 = d_2) = \mathbf{P}(\rho_{12}d_2 + \sqrt{1-\rho_{12}{}^2}\xi < d_1) = \mathbf{N}\left(\frac{d_1 - \rho_{12}d_2}{\sqrt{1-\rho_{12}{}^2}}\right) \tag{5}$$

Namely, default threshold of firm 1 is extended to $\overline{d_{12}}$ when firm 2 defaults. We put soft and reasonable assumption on parameters: d_2 and ρ_{12} such as d_2 is less than zero[2] and ρ_{12} is positive. New threshold $\overline{d_{12}}$ becomes larger than d_1. We recheck firm 1's situation based on more strict threshold in consideration of potential default infection from sole trading partner.

$$\overline{d_{12}} = \mathbf{N}^{-1}(\mathbf{P}(\xi_1 < d_1|\xi_2 = d_2)) = \frac{d_1 - \rho_{12}d_2}{\sqrt{1-\rho_{12}{}^2}} \tag{6}$$

Then the additional default probability for single neighbor case is as follows. This quantity is obtained from simulation.

$$Apd_{12} = \mathbf{E}[\mathbf{1}_{\{d_1 < \xi_1 < \overline{d_{12}}\}} \times \mathbf{1}_{\{\xi_2 < d_2\}}] \tag{7}$$

[2]This implies that firm 2's default probability is less that 50%.

2.2 Multiple Neighbors Case

We explain about two neighbors' case. We estimate the default impact which non defaulted firm 1 received from firm 2 and/or firm 3's default. We express this quantity as Apd_{123}. We can classify this case into three situations. They have impact only from firm 2, impact only from firm 3, and impact from firm 2 and 3. We can quantify the first two cases by referring to the threshold obtained from single neighbor case. Then we focus on the third case.

$$
\begin{aligned}
&Apd_{123} \\
&= \mathbf{E}[\mathbf{1}_{\{\xi_2<d_2\}}\mathbf{1}_{\{\xi_3>d_3\}}\mathbf{1}_{\{d_1<\xi_1<\overline{d_{12}}\}} + \mathbf{1}_{\{\xi_2>d_2\}}\mathbf{1}_{\{\xi_3<d_3\}}\mathbf{1}_{\{d_1<\xi_1<\overline{d_{13}}\}} + \mathbf{1}_{\{\xi_2<d_2\}}\mathbf{1}_{\{\xi_3<d_3\}}\mathbf{1}_{\{d_1<\xi_1<\overline{d_{123}}\}}] \quad (8)
\end{aligned}
$$

The threshold $\overline{d_{123}}$ indicates that firm 1 did not default due to its own reason but received simultaneous default impacts of firm 2 and 3. Let $\phi(\xi_1, \xi_2)$ be two-dimensional standardized normal density function and $\phi(\xi_1, \xi_2, \xi_3)$ be three-dimensional standardized normal density function.

$$
\mathbf{P}(\xi_1 < d_1|\xi_2 = d_2, \xi_3 = d_3) = \frac{\int_{-\infty}^{d_1} \phi(\xi, d_2, d_3)\mathbf{d}\xi_1}{\phi(d_2, d_3)} = \mathbf{N}(\overline{d_{123}}) \quad (9)
$$

Based on the above equation, the threshold $\overline{d_{123}}$ can be obtained from inverse function $\mathbf{N}^{-1}(\cdot)$ as follows:

$$
\overline{d_{123}} = \frac{d_1(1-\rho_{23}{}^2) - (\rho_{12} - \rho_{23}\rho_{13})d_2 - (\rho_{13} - \rho_{23}\rho_{12})d_3}{\sqrt{1-\rho_{23}{}^2}\sqrt{1+2\rho_{12}\rho_{13}\rho_{23} - \rho_{12}{}^2 - \rho_{13}{}^2 - \rho_{23}{}^2}} \quad (10)
$$

Figure 2 is the default threshold for two neighbors default case. We skip to explain the details of calculation here and wrote them in the following appendix. By utilizing LU-decomposition technique, we can increase the number of neighbors around target firm easily without having complicated calculation as we wrote in the appendix.

3 Numerical Experiment

We calculate the additional default probability for the largest financial institutions in Japan. There are the three largest banks. Their names are Mitsubishi, SMBC, and Mizuho. Their direct trading volume might not be large compared with total asset amount. But their lending exposures overlap and possible provide indirect effect. They are usually financing the same companies. For credit risk diversification, they do not dominantly finance to firms but oligopoly, especially for listed firms loan market. Then it is natural to suppose that one of the mega banks' default directly/indirectly

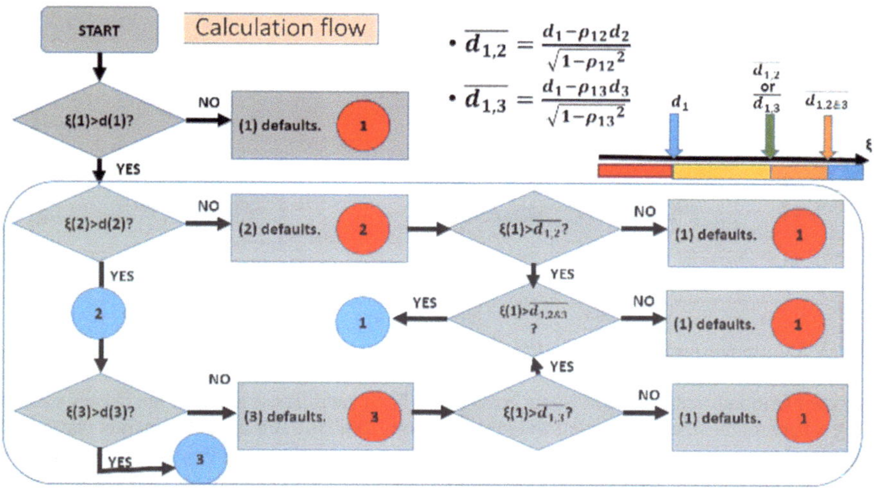

Fig. 2 The picture above indicates simulation flow to calculate the additional default probability for two neighbors' case. In this flowchart, $\xi(1)$, $\xi(2)$, and $\xi(3)$ are standardized asset random variables for firm 1, 2, and 3. $d(1)$, $d(2)$, and $d(3)$ are default thresholds, respectively. First, we check the target's situation and go checking only when the target is not default ($\xi(1) > d(1)$). When firm 2 defaults and firm 3 didn't default, we check $\xi(1)$ again by the threshold $\overline{d_{12}}$ explained in the Eq. (6). When firm 3 defaults and firm 2 didn't default, we check $\xi(1)$ again by the threshold $\overline{d_{13}}$. When firm 2 and 3 defaults, we check $\xi(1)$ again by the threshold $\overline{d_{123}}$ explained in the Eq. (10)

affects the others. When one bank defaults, it does not matter only for the bank but the risk would be infected to others. We set Mitsubishi as targeting firm to observe. We estimate additional default probability which Mitsubishi potentially receives from SMBC and Mizuho. There are mainly three reasons for this. First, Mitsubishi is the largest commercial bank in Japan,[3] so we can say that it is the most influential bank in Japan. Second, its credit quality is the best of all based on the historical 1-year default probability downloaded from Bloomberg terminal,[4] so its default probability due to internal reason should be small and its additional default probability would be relatively large, we thought. Third, it is contributing to calculating Yen LIBOR interest rate the longest among them, then its credit risk has been reflected to the spread between Yen LIBOR and risk-free interest rate. We calculate this Japanese TED (J-TED) spread by calculating difference between 12 months Yen LIBOR and 1-year Japanese government bond. We set J-TED spread as the benchmark for comparison with the additional default probability. If it becomes close values especially under the stressed economic situation (For example, subprime problem in 2007 and Lehman shock in 2008), we think that this methodology can

[3]Mitsubishi's total asset is 303 trillion yen. SMBC is 198 trillion yen, and Mizuho is 201 trillion yen as of 2017.

[4]For your information, Standard and Poor's is reporting as Mitsubishi's credit rating level is currently A (Positive), SMBC is also A (Positive), and Mizuho (Mizuho Financial Group) is A-(Stable).

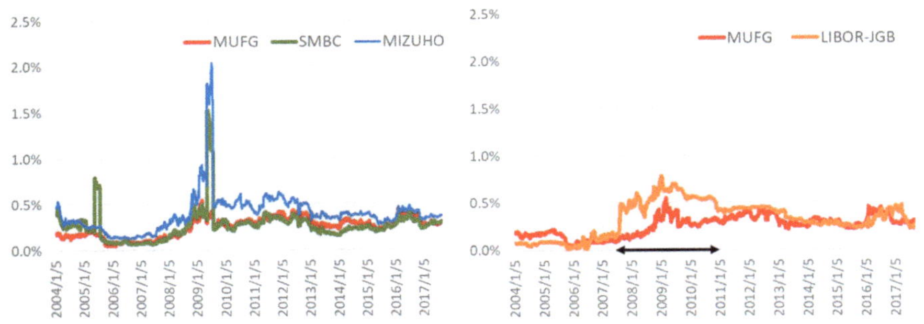

Fig. 3 Left figure indicates 1-year default probability for Mitsubishi (Red), SMBC (Green), and Mizuho (Blue). Right figure indicates 1-year default probability for Mitsubishi again and the gap between 12 month yen Libor and 1-year government bond yield. These two lines are closely overlapping except for the financial turmoil which started from 2007. In 2008 and 2009, market participants require additional risk premium. More conservative risk estimation methodology is needed to properly make up for the uncovered risk from neighbors especially under the stressed economic situation

be regarded as useful to systematically capture default probabilities under the stress scenario (stressed PD, this is Normal default probability plus our Additional PD) and could be utilized for improving credit risk management.

As shown in the Eqs. (5) and (10), we use default probability for each firm and asset correlation to calculate the additional default probability. And for benchmark, we use 1-year Yen LIBOR rate and 1-year Japanese government bond yield. We obtained above daily indexes from January 2004 to September 2017 by using Bloomberg terminal. For consistency, we checked that 1-year default probability on the Bloomberg terminal is calculated based on the Merton model. We calculated daily default thresholds and simulated 100,000 times for estimating daily additional default probability. We estimated correlation daily basis based on the previous 10 business days by utilizing log return of stock price.

The left-hand side of Fig. 3 shows historical 1-year default probability and the right-hand side of Fig. 3 shows historical J-TED spread and Mitsubishi's 1-year default probability take almost the same level. From the right figure, there are obvious gap between historical J-TED spread and Mitsubishi's PD starting from the year of 2007 and converged to the same level in 2011. We think that the professional market participant contributing to Yen LIBOR rate demanded additional credit risk premium to Mitsubishi in this period. It was not due to its internal reason because its default probability is stably low but due to the external reason as we can easily observe neighbors' PD are recording high on the left of Fig. 3. Figure 4 shows the additional default probability and same indexes on the right-hand side of Fig. 3. The additional default probability adding to normal PD is considerably matching with J-TED spread especially under the financial turmoil.

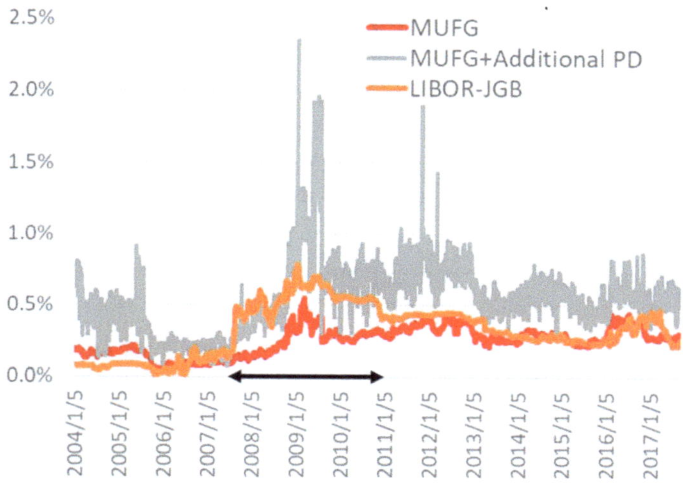

Fig. 4 The figure above indicates historical additional PD+normal PD (Gray line) and J-TED spread (Orange). They overlap, especially under the financial turmoil. This implies that professional market participants contribution to Yen LIBOR require additional risk premium when the default infection possibly happens

4 Conclusion

As shown in the above figures, it is not enough to manage credit risk for portfolio management and individual pricing just based only on normal default probabilities especially under the stressed economic situation. Additionally, for pricing individual financial instruments, we need to utilize consistent methodology with portfolio risk management. As we mentioned, there are no reasonable methodology proposed which fulfills consistent with market indexes and also total risk management (portfolio and individual loan/bond pricing) to consider neighbors' credit risk when we recognize downside abnormal business cycle. We strongly believe that it is useful to utilize our model to capture risk from others and reasonably consider it for appropriate pricing. Although it is complicated to calculate default threshold from the equation operations, we showed alternative methodology to obtain the threshold through LU decomposition of correlation matrix in the appendix. This enables us to obtain default threshold of multiple neighbors case easily. As our future work, we are continuously improving our methodology by considering recovery rate and testing tail part of loss distribution for value at risk and Expected Shortfall.

5 Appendix

We explain how to calculate default threshold for two neighbors' case at first. Namely, we calculate conditional default probability $\mathbf{P}(\xi_1 < d_1|\xi_2 = d_2, \xi_3 = d_3)$ and obtain the threshold from $\mathbf{N}^{-1}[\mathbf{P}(\xi_1 < d_1|\xi_2 = d_2, \xi_3 = d_3)]$. Three dimensional standardized probability density function for asset is as follows.

$$\phi(\xi_1, \xi_2, \xi_3) = \frac{1}{(2\pi)^{\frac{3}{2}}\sqrt{|\Sigma|}}\mathbf{exp}\left\{-\frac{1}{2}(\xi_1\xi_2\xi_3)\Sigma^{-1}(\xi_1\xi_2\xi_3)^\top\right\} \tag{11}$$

where ξ_1, ξ_2, and ξ_3 are standardized asset random variables for firm 1, 2, and 3 respectively. Σ is 3×3 correlation matrix.

$$\Sigma = \begin{pmatrix} 1 & \rho_{12} & \rho_{13} \\ \rho_{21} & 1 & \rho_{23} \\ \rho_{31} & \rho_{32} & 1 \end{pmatrix} \tag{12}$$

For later calculation, we prepare inverse matrix of Σ.

$$\Sigma^{-1} = \frac{1}{|\Sigma|}\begin{pmatrix} 1-{\rho_{23}}^2 & \rho_{13}\rho_{23}-\rho_{12} & \rho_{12}\rho_{23}-\rho_{13} \\ \rho_{13}\rho_{23}-\rho_{12} & 1-{\rho_{13}}^2 & \rho_{12}\rho_{13}-\rho_{23} \\ \rho_{12}\rho_{23}-\rho_{23} & \rho_{12}\rho_{13}-\rho_{23} & 1-{\rho_{12}}^2 \end{pmatrix} \tag{13}$$

where $|\Sigma| = 1 + 2\rho_{12}\rho_{13}\rho_{23} - {\rho_{12}}^2 - {\rho_{13}}^2 - {\rho_{23}}^2$

$$\begin{aligned}
(\xi_1\xi_2\xi_3)\Sigma^{-1}(\xi_1\xi_2\xi_3)^\top &= \frac{1}{|\Sigma|}(\xi_1\xi_2\xi_3)\begin{pmatrix} (1-{\rho_{23}}^2)\xi_1 + (\rho_{13}\rho_{23}-\rho_{12})\xi_2 + (\rho_{12}\rho_{23}-\rho_{13})\xi_3 \\ (\rho_{13}\rho_{23}-\rho_{12})\xi_1 + (1-{\rho_{13}}^2)\xi_2 + (\rho_{12}\rho_{13}-\rho_{23})\xi_3 \\ (\rho_{12}\rho_{23}-\rho_{23})\xi_1 + (\rho_{12}\rho_{13}-\rho_{23})\xi_2 + (1-{\rho_{12}}^2)\xi_3 \end{pmatrix} \\
&= \frac{1}{|\Sigma|}\{(1-{\rho_{23}}^2){\xi_1}^2 + (1-{\rho_{13}}^2){\xi_2}^2 + (1-{\rho_{12}}^2){\xi_3}^2 \\
&\quad +2(\rho_{13}\rho_{23}-\rho_{12})\xi_1\xi_2 + 2(\rho_{12}\rho_{23}-\rho_{13})\xi_1\xi_3 \\
&\quad +2(\rho_{12}\rho_{23}-\rho_{13})\xi_1\xi_3 + 2(\rho_{12}\rho_{13}-\rho_{23})\xi_2\xi_3\}
\end{aligned} \tag{14}$$

This result is part of coefficient of exponential in Eq. (11). We subtract coefficient of two dimensional normal distribution $\frac{1}{1-{\rho_{23}}^2}({\xi_2}^2 - 2\rho_{23}\xi_2\xi_3 + {\xi_3}^2)$ from Eq. (14). And we obtain below. We utilize this to calculate conditional default probability for two neighbors' case.

$$=> \frac{1}{|\Sigma|(1-{\rho_{23}}^2)}\left[\left\{(1-{\rho_{23}}^2)\xi_1 + (\rho_{13}\rho_{23}-\rho_{12})\xi_2 + (\rho_{12}\rho_{23}-\rho_{13})\xi_3\right\}^2\right] \tag{15}$$

$$\frac{\int_{-\infty}^{d_1} \phi(\xi, d_2, d_3)\mathbf{d}\xi_1}{\phi(d_2, d_3)}$$
$$= \int_{-\infty}^{d_1} \frac{1}{\sqrt{2\pi}\sqrt{\frac{|\Sigma|}{1-\rho_{23}^2}}} \mathbf{exp}\left\{-\frac{1}{2}\frac{(1-\rho_{23}^2)}{|\Sigma|}\left(\xi_1 + \frac{(\rho_{13}\rho_{23}-\rho_{12})d_2 + (\rho_{12}\rho_{23}-\rho_{13})d_3}{(1-\rho_{23}^2)}\right)^2\right\}\mathbf{d}\xi_1$$
$$= \mathbf{N}\left(\frac{d_1(1-\rho_{23}^2) - (\rho_{12}-\rho_{23}\rho_{13})d_2 - (\rho_{13}-\rho_{23}\rho_{12})d_3}{\sqrt{1-\rho_{23}^2}\sqrt{1+2\rho_{12}\rho_{13}\rho_{23}-\rho_{12}^2-\rho_{13}^2-\rho_{23}^2}}\right) \tag{16}$$

Next, we explain about LU decomposition method to escape from having above complicated calculations and consider multiple neighbor case. Through above operation, we saw two neighbors case and compare the result with threshold from Eq. (16). Let $\omega_1, \omega_2, \omega_3$ be independent random variables from standardized normal distribution. Let U be upper triangle correlation matrix and L be lower triangle correlation matrix. Temporary, we set firm 3 as target firm and calculate possible default impact from firm 1 and 2. After calculating the threshold based on LU decomposition method, we replace firm number to be consist with above methodology and compare the solutions.

$$\Sigma = LU = \begin{pmatrix} 1 & 0 & 0 \\ \rho_{12} & \sqrt{1-\rho_{12}^2} & 0 \\ \rho_{13} & \frac{\rho_{23}-\rho_{12}\rho_{13}}{\sqrt{1-\rho_{12}^2}} & \sqrt{1-\rho_{13}^2 - \frac{(\rho_{23}-\rho_{12}\rho_{13})^2}{1-\rho_{12}^2}} \end{pmatrix} \begin{pmatrix} 1 & \rho_{12} & \rho_{13} \\ 0 & \sqrt{1-\rho_{12}^2} & \frac{\rho_{23}-\rho_{12}\rho_{13}}{\sqrt{1-\rho_{12}^2}} \\ 0 & 0 & \sqrt{1-\rho_{13}^2 - \frac{(\rho_{23}-\rho_{12}\rho_{13})^2}{1-\rho_{12}^2}} \end{pmatrix} \tag{17}$$

By using above LU decomposition, we obtain correlated random variables ξ_1, ξ_2, ξ_3 as below.

$$(\xi_1 \quad \xi_2 \quad \xi_3) = (\omega_1 \quad \omega_2 \quad \omega_3)U$$
$$\xi_1 = \omega_1$$
$$\xi_2 = \rho_{12}\omega_1 + \sqrt{1-\rho_{12}^2}\omega_2$$
$$\xi_3 = \rho_{13}\omega_1 + \frac{\rho_{23}-\rho_{12}\rho_{13}}{\sqrt{1-\rho_{12}^2}}\omega_2 + \sqrt{1-\rho_{13}^2 - \frac{(\rho_{23}-\rho_{12}\rho_{13})^2}{1-\rho_{12}^2}}\omega_3 \tag{18}$$

When standardized asset of firm 1 and 2 are equal to threshold d_1 and d_2. Random variables ω_1 and ω_2 become as follows.

$$\omega_1 = d_1$$
$$\omega_2 = \frac{d_2 - \rho_{12}d_1}{\sqrt{1-\rho_{12}^2}}$$

We insert these to Eq. (18) and solve ω_3 of inequality by setting left-hand side of Eq. (18) being smaller than d_3.

$$d_3 > \rho_{13}d_1 + \frac{\rho_{23} - \rho_{12}\rho_{13}}{\sqrt{1-\rho_{12}{}^2}}\frac{d_2 - \rho_{12}d_1}{\sqrt{1-\rho_{12}{}^2}} + \sqrt{1-\rho_{13}{}^2 - \frac{(\rho_{23}-\rho_{12}\rho_{13})^2}{1-\rho_{12}{}^2}}\,\omega_3 \quad (19)$$

$$\Leftrightarrow \ \omega_3 < \frac{d_3(1-\rho_{12}{}^2) - (\rho_{31}-\rho_{12}\rho_{32})d_1 - (\rho_{32}-\rho_{12}\rho_{31})d_2}{\sqrt{1-\rho_{12}{}^2}\sqrt{1+2\rho_{31}\rho_{32}\rho_{12}-\rho_{31}{}^2-\rho_{32}{}^2-\rho_{12}{}^2}} \quad (20)$$

The right-hand side of Eq. (20) is new threshold for the setting of target firm 3 and neighbor firm 1 and 2. Also, this result is equal to Eq. (16) by replacing firm number accordingly. We showed two neighbors case here and LU-decomposition method enables us to increase the number of neighbors easily.

References

Azizpour S, Giesecke K (2008) Self-exciting corporate defaults: contagion vs. frailty, Working Paper, Stanford University

Basel Committee on Banking Supervision (2009) Principles for Sound Stress Testing Practices and Supervision, Basel Committee Publications No. 147. www.bis.org/publ/bcbs147.pdf

Board of Governors of the Federal Reserve System (2009) The Supervisory Capital Assessment Program: Design and Implementation, Board of Governors of the Federal Reserve System, www.federalreserve.gov/newsevents/press/bcreg/bcreg20090424al.pdf

Bonti G et al (2005) Credit risk concentrations under stress, Bank for International Settlement, October 2005, https://www.bis.org/bcbs/events/crcp05bonti.pdf

Hisakado M, Kaneko T (2016) Financial Risk Management methodology, No. 5953416, Japan Patent Office

Jakubik P, Schmieder C (2008) Stress testing credit risk: comparison of the Czech Republic and Germany. Financial Stability Institute, Bank for International Settlement, September 2008. https://www.bis.org/fsi/awp2008.pdf

Kaneko Takuya (2016) A new framework for credit risk management with the randomness of correlation matrix. J Soc Sci 81:17–28

Kaneko T, Nakagawa H (2010) Credit portfolio risk assessment based on top-down approach in consideration of yield curve forecast and individual firms credit situation. Kin'yu-kenkyu, Bank of Japan, vol 29(3), pp 19–44 (in Japanese)

Merton Robert C (1974) On the pricing of corporate debt: the risk structure of interest rates. J Financ 29(2):449–470

Stein Roger M (2014) The role of stress testing in credit risk management. J Invest Manag 10(4):64–90

Watts Duncan J (2002) A simple model of global cascades on random networks. Proced Natl Acad Sci 99(9):5766–5771

Systemic Risk: Fire-Walling Financial Systems Using Network-Based Approaches

V. Sasidevan and Nils Bertschinger

Abstract The latest financial crisis has painfully revealed the dangers arising from a globally interconnected financial system. Conventional approaches based on the notion of the existence of equilibrium and those which rely on statistical forecasting are seen to be inadequate to describe financial systems in any reasonable way. A more natural approach is to treat financial systems as complex networks of claims and obligations between various financial institutions present in an economy. The generic framework of complex networks has been successfully applied across several disciplines, e.g., explaining cascading failures in power transmission systems and epidemic spreading. Here we review various network models addressing financial contagion via direct inter-bank contracts and indirectly via overlapping portfolios of financial institutions. In particular, we discuss the implications of the "robust-yet-fragile" nature of financial networks for cost-effective regulation of systemic risk.

1 Introduction

Over the past few decades, the generic framework of complex networks has been used to address problems in a variety of scientific disciplines. The study of evolutionary dynamics of populations (Lieberman et al. 2005), spreading of diseases (Newman 2002), spreading of failures in power transmission systems (Dobson et al. 2007),

V. Sasidevan (✉)
Department of Physics, Cochin University of Science and Technology, Cochin 682022, India
e-mail: sasidevan@cusat.ac.in; sasidevan@gmail.com

N. Bertschinger
Frankfurt Institute for Advanced Studies, Frankfurt am Main, Germany

Department of Computer Science, Goethe University, Frankfurt am Main, Germany
e-mail: bertschinger@fias.uni-frankfurt.de

A. S. Chakrabarti et al. (eds.), *Network Theory and Agent-Based Modeling in Economics and Finance*, https://doi.org/10.1007/978-981-13-8319-9_16

dynamics of opinion formation (Castellano et al. 2009), and working of genetic regulation (Hasty et al. 2001) all heavily rely upon viewing and understanding the underlying systems as complex networks which represent the interaction between their various constituent elements. It is now well known that the emergent or collective behavior in these systems is greatly influenced by the topology of the underlying network of connections together with the particular kind of dynamic processes which run on it (Strogatz 2001; Amaral and Ottino 2004; Boccaletti et al. 2006).

One of the latest entries into the class problems to be analyzed using methods based on the theory of complex networks is the issue of spreading of failures in financial/economic systems. Often dubbed as systemic risk, this refers to the possibility of spread of initial distress at a single or a few financial institutions such as banks to a significant fraction of a financial network. Though earlier studies of such contagion exist, it is safe to say that it is only in the aftermath of the financial crisis of 2007–09 that viewing systems such as interbank markets from the standpoint of the theory of complex networks became mainstream (Schweitzer et al. 2009; Haldane 2015).

More and more direct and indirect connections are being made every day between financial institutions as the world increasingly becomes an economically globalized one. This together with modern-day financial products such as CDS (credit default swaps) or CDO (collateralized debt obligations) make today's financial systems a very complex one and hard to analyze. Conventional approaches based on the notion of the existence of equilibrium and those which rely on statistical forecasting have seen to be inadequate to describe financial systems in any reasonable way (Farmer and Foley 2009). Conventional theory often makes unrealistic assumptions and is capable of analyzing only simple situations. Most importantly, it cannot handle nonlinearities and feedback mechanisms ubiquitous in financial systems (Catanzaro and Buchanan 2013). Inadequacy of the conventional approaches together with the spread of concepts from complex network theory to the economic/financial community leads to the conclusion that a more natural way to describe financial systems is to treat them as complex networks of claims and obligations between various financial institutions present in an economy. It is to be noted that a significant push toward this direction was provided by people working on ecological and epidemiological systems (e.g., May et al. 2008). The result is that there has been an exponential increase in the studies of financial networks employing ideas and tools from the complex networks literature and the terminology of latter nowadays finds commonplace in the discussions related to contagion in financial systems.

In the simplest form, the nodes of a financial network are institutions involved in financial intermediation (called banks in short hereafter), and the links between the nodes represent lending/borrowing relationships. The links are therefore directed (usually represented as directed from borrower to lender indicating the anticipated direction of cash flow after the link has been established), and weight of a link represents the amount of money involved in the transaction between the nodes connected by that link. The banks may have other kinds of direct and indirect connections between them which are discussed in Sect. 2. Given such a financial network, the term systemic risk refers to the possibility of an initial financial distress at a single or a few banks (say their defaulting due to a loss of value of their asset holdings) to spread through the network causing the default of a significant number of other banks in the network. Such a system-wide collapse has serious repercussions for the financial system and global economy in general and could affect the lives of millions in a negative way as painfully revealed in the crisis of 2007–09.

We can easily see the parallels of the problem of the collapse of a financial network with that of an ecological network or the spread of an infectious disease. Although all these problems deal with the transmission of a "failure" at a node to its neighbors in a general sense, there are also marked differences between them (Haldane and May 2011). A feature which distinguishes financial networks from others is that unlike the failure of a node in an ecological network or an epidemiological network, the failure of a node in a financial network depends upon the "internal structure" of that node (i.e., balance sheet structure of the bank) and the "cumulative health" of its neighbors in the network. So unlike simple stochastic rules used to model for, e.g., the spread of an infectious disease (as in SIR model), here we need to consider the internal balance sheet dynamics of a node in detail.

Our aim in this review is to give a broad overview of the various developments in the field of systemic risk of financial networks and discuss some future directions. In this respect, we will mainly focus on the literature that concentrates on the use of tools from the study of complex networks. There is an extensive literature on banking systems based on more traditional point of views like the ones using equilibrium models. We will occasionally mention other strands of research but will refrain from discussing them in any detail.

This review is organized as follows. In Sect. 2, we give an account of the structure of a financial network and a detailed description of what constitutes its nodes and links. We describe the kind of questions one would like to answer about such networks. In Sect. 3, we review various past efforts in modeling contagion via interbank credit networks and in Sect. 4, we review contagion via fire sale of commonly held assets by banks. We discuss some of our findings related to asset contagion. In Sect. 5, we conclude and discuss some future possibilities.

2 Nodes and Links of a Financial Network

As mentioned in the introduction, the nodes of a financial network are banks, and the links represent various credit relationships. A sample financial network based on data from the European Banking Authority (EBA) is given in Fig. 1. To study the propagation of "failures" or "defaults" in a financial network, we may represent a bank by its balance sheet structure which lists its various assets and liabilities at any time. A schematic of a balance sheet is shown in Fig. 2. Broadly, we may classify the assets of a bank into two types, namely, interbank assets (the money that the bank lent to other banks) and external assets (investments like the money that the bank lent to other parties such as business firms often receiving a collateral, deposits with a central bank, etc). The external assets can be further classified into liquid assets

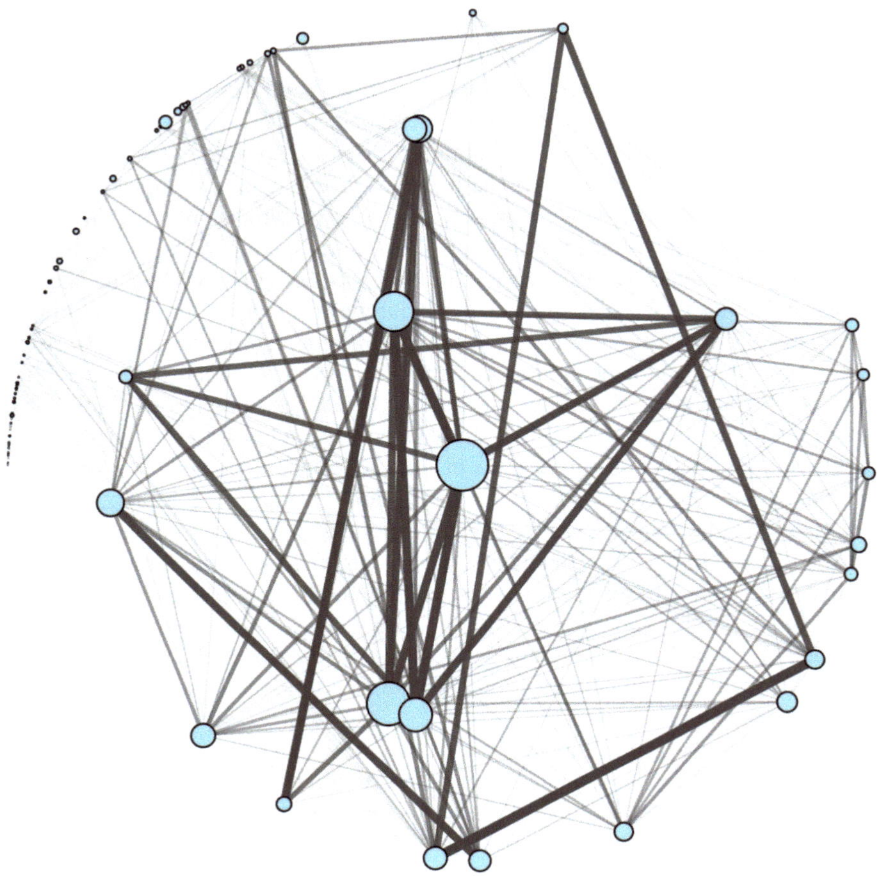

Fig. 1 A graphical representation of the banking network of major European banks created using data from the European Banking Authority (EBA). The size of the nodes represents the relative size of the balance sheets of the banks, and the thickness of the links represents the value of common asset exposures

(assets for which there is a market and hence can be easily converted into cash; examples include cash, central bank reserves, high-quality government bonds, etc.) and illiquid ones (like secured or unsecured mortgages which cannot be used to raise cash quickly). More detailed classification of assets is possible by further considering divisions such as "Collateral assets" (assets which may be used as collateral in repo transactions) and "Reverse repo assets" (i.e., collateralized lending), etc., (Gai et al. 2011). Considering such divisions is vital as the percentage of assets in various classes decides whether a bank in need of immediate cash for its promised payments can raise it in time or not. In general, the size of the balance sheet, i.e., the total value of assets (or equivalently liabilities) varies across banks. It is often found that the size distribution is heterogeneous having a heavy tail (Karimi and Raddant 2016).

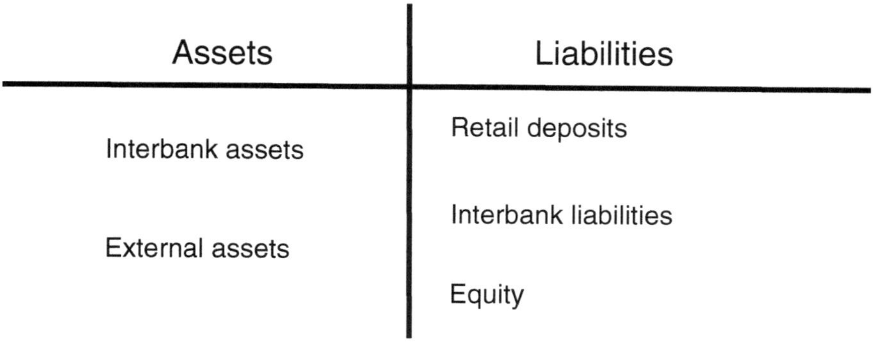

Fig. 2 A simplified schematic of a balance sheet of a bank indicating a minimal division of its assets and liabilities into various classes. Interbank assets represent the money that the bank lent to other banks and external assets represents the likes of money lent to industry. Interbank liabilities represent the money that is owed to other banks. Finer subdivisions of assets and liabilities can be considered depending upon the specific contexts (see text). The difference between the total assets and liabilities gives its equity. The bank is insolvent if its net liabilities exceed the value of its assets

The liability side of the balance sheet usually consists of interbank liabilities (money that is owed to other banks) and retail deposits. Finer divisions such as "repo liabilities" (i.e., borrowing secured with collateral) may also be considered (Gai et al. 2011). The difference between the total assets and liabilities of a bank gives its equity, also commonly called net worth or capital. The equity is written on the liability side of the balance sheet indicating that it is money owed to promoters and shareholders of the bank.

There are two important considerations regarding the "health" of a node.

- **Insolvency**: A bank is considered insolvent if it has a negative net worth. The net worth may become negative, say, e.g., due to a drop in the prices of its assets (in a mark-to-market framework, the value of assets are updated according to its current market price) or due to failed investments. We may say that the bank cannot operate anymore as it cannot honor its liabilities and should close down with appropriate insolvency proceedings, though in reality, a number of things may happen like possible government intervention, etc.
- **Illiquidity**: A bank is illiquid if its short-term liabilities exceed liquid assets, and hence, it cannot honor the former.

In both cases, the bank is in trouble, and this will affect other banks to which it owes money. Also, the assets of the bank that is in trouble will face devaluation, thus affecting other banks holding the same class of assets on their balance sheets. Due to such feedbacks, initial distress can spread to other banks in the network and may affect a large part of it.

The links between nodes represent lending/borrowing relations between banks. They are directed and weighted, the weight of a link representing the value owed (see Fig. 3). Note that banks may interact with others not only through the above

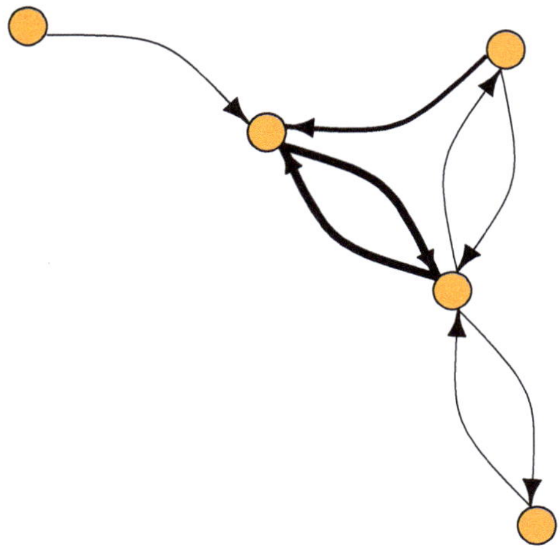

Fig. 3 A directed and weighted network consisting of five banks. The thickness of the links represent the value of credit exposures

lending/borrowing relationship. Their balance sheet may be affected by the fact that they are exposed to common assets. A sell-off or even a negative news about a bank might cause depreciation in the value of the assets it holds and other banks which have exposure to these assets will be affected. The sell-off of an asset by a distressed bank will cause excess supply of that asset and if the market depth of that asset is shallow, will lower its value significantly. Such fire-sale losses can quickly spread among the banks in the network. These two mechanisms of interbank lending and fire sale of assets form two important channels for spreading of crisis in a financial network. There are other channels possible and which ones are more relevant in reality is debatable. Upper (2011), e.g., argues that there is not much evidence for interbank channels propagating distress. In any case, these are possible channels of contagion under conceivable circumstances, and hence, it is important to understand them from a regulatory and individual point of view.

Being able to understand the "behavior" of a bank is crucial in the study of systemic risk. This is because the way a bank organizes its balance sheet and how they form connections will have a significant role in deciding whether or not an initial default will propagate in the network causing significant damage. An important factor which determines the actions of a bank is various regulatory constraints set by a central authority (like a central bank). For example, there could be a capital ratio constraint set by the authority stipulating that the activities of the bank should obey the constraint that its capital or net worth should be a minimum percentage of its illiquid assets (8% is now a typical figure under Basel III). Also, the bank itself may have policies to keep its balance sheet structure in a specific way (e.g., maintaining a constant leverage ratio (Adrian and Shin 2010)).

3 Contagion via Interbank Linkages

Lending and borrowing relationships with other banks serve the purpose of meeting the liquidity needs of a bank and profiting from the lending of idle liquid money. While lending, there is, of course, the risk of default. Presumably, forming more connections or dividing total exposure across several banks will reduce such risk. On the other hand, a bank with more number of connections will also have a higher probability of being hit by a defaulting counterparty. The interplay between these two effects determines whether the default of a single or a small set of banks will spread through the system.

3.1 Contagion via Default Spreading

One of the earlier works which explicitly incorporated the network aspects of interbank lending/borrowing is by Allen and Gale (2000). The interbank network they considered is rather simple compared to any real-world banking network and involved only four banks. Two different connectivity structures of the network are considered, and they showed that a complete network in which each bank is connected to every other bank readily absorbs any initial shock. However, an incomplete network structure makes the system prone to the spread of an initial default.

A pioneering work which considered the effects of varying the important features of interbank networks is by Nier et al. (2007). The network structures considered were Erdos–Renyi (ER) type and a tiered one. In the latter, a few large banks are connected to a large number of small banks where the latter are mostly connected to the large ones. The effect of level of capitalization of the banks, degree of the ER network, size of the interbank exposures, etc., on the number of defaults is determined using simulations. It is found that at smaller degrees of the network, a small increase in the degree of the network causes an increase in the number of defaults, but at relatively higher degrees, the effect of an initial default is contained. Predictably, higher capital levels found to reduce the risk of contagion, but the effect is found to be nonlinear. The paper also briefly considers the "fire sale" channel in which a distressed bank sells off its asset causing a depreciation in the market value of the assets.

A landmark work is that of Gai and Kapadia (2010) in which they formalized the previous setting by Nier et al. (2007) and distinguished probability of contagion from its extent. (A comparison between these models is given in May and Arinaminpathy (2010).) The former denotes the probability of a systemic event conditional on an initial shock while the latter denotes the percentage of banks affected conditional on a systemic event. One should note that what constitutes a systemic event is somewhat subjective. In this paper and several later works, a contagion is considered as systemic if more than 5% of the banks default. This might seem like a very low number, but reflects the fact that even the default of 5% of banks is a severe event in an economy affecting a large number of people and potentially wiping out millions of dollars. Their theoretical analysis closely follows the study of cluster size distribution in a complex network using generating functional formalism (see, e.g., Newman (2010)). The key finding was that the phenomena of contagion in a financial network have a "robust-yet-fragile" nature. That is, even if the probability of contagion is very low, whenever a contagion happens, it wipes out most of the system. The typical dependency of probability and the extent of contagion on the average degree of the ER network are as shown in Fig. 4. Note that when the average degree is below 1, there is no possibility of contagion as the network is a disconnected one, and there is no large connected component present. After that, the probability of contagion increases with an increase in average degree, reaches a maximum value, and then decreases. This clearly shows the interplay of the two effects of risk sharing and default spreading mentioned earlier. Higher degrees diffuse the initial impact among many neighbors, thus reducing the possibility of contagion. However, once a contagion starts to spread, higher degrees can act as an ideal setting for the propagation of defaults since many counterparties of a bank will be simultaneously affected. In a nutshell, this has the effect that the banking network may respond in entirely different ways to similar shocks at different nodes in the network.

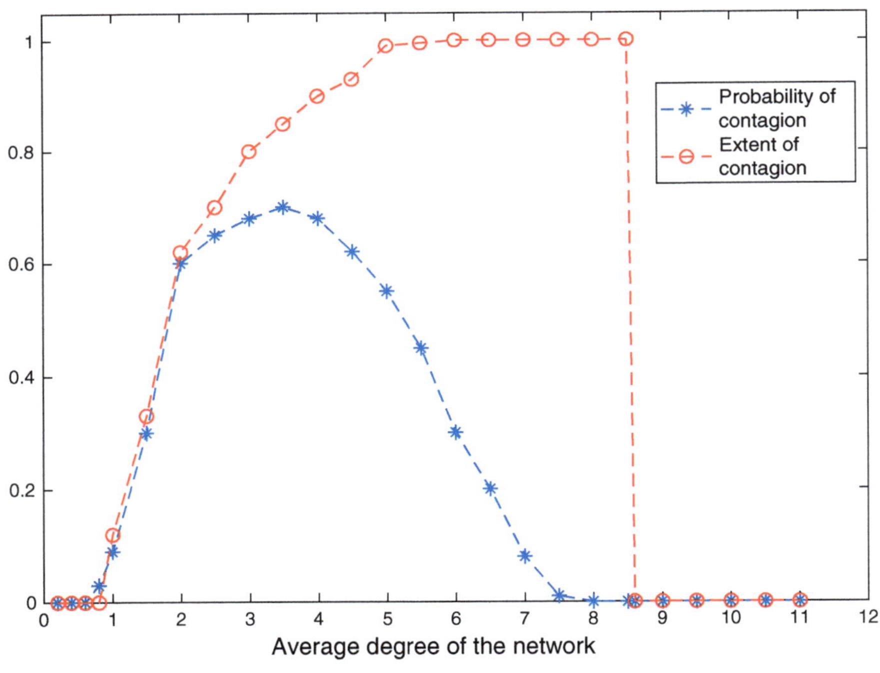

Fig. 4 A schematic representation of the dependence of probability of contagion and extent of contagion on the average degree of an ER network

The assumption made in Gai and Kapadia (2010) regarding the network structure as being ER type and that all banks are of the same size is not in line with empirical evidence about real-world banking networks which are found to have heavy-tailed distribution for both connectivity and size of banks (Huser 2015). Caccioli et al. (2012) extended the Gai and Kapadia model to incorporate these effects. It was found that a heterogeneous degree distribution makes the network more resilient toward random defaults but more vulnerable to targeted attacks on well-connected nodes. It was also found that a power-law distribution for the size of the banks makes the contagion events more frequent. The effects of policies such as increased capital requirements are also considered which are seen to improve the resilience of the system. A recent work discussing the role of assumptions made about the connectivity and size of banks is by Karimi and Raddant (2016) in which features of real-world empirical networks were considered in some detail.

3.2 Contagion via Valuation Adjustments

The seminal work of Eisenberg and Noe (2001) provides an alternative viewpoint on default contagion, namely, as a problem of contract valuation. While each bank has promised the repayment of a nominal amount, it might be unable to hold up to its promise when the payment is due. As each repaid credit—in full or in part—is an income to the lending counterparty, Eisenberg and Noe proposed that the payments of all banks have to be considered simultaneously and solved in a self-consistent fashion. The resulting *clearing vector* can then be regarded as the value of all interbank contracts. Thus, instead of modeling the spread of defaults directly, the model provides a self-consistent solution for the final state when all payments are settled. This point of view readily connects with more standard approaches for financial valuation and has since spurred a range of theoretical (Paul Glasserman and Young 2015) and computational research (Upper 2011).

In particular, the model of Suzuki (2002) is noteworthy in this respect. It not only generalizes the abovementioned model allowing for cross-holdings of equity as well as interbank debt but also clearly formulates it as an extension of the Merton model to multiple firms. In his groundbreaking work, Merton (1974) showed that liabilities and equity of a firm could be considered and valued as a short *put* and long *call* option on the firm's assets, respectively. Nowadays this insight forms the foundation of structural credit risk modeling and has been developed into established industry practice. Valuation in a network context is considerably more complicated as all option values are interdependent and have to be solved in a self-consistent fashion. Correspondingly, almost no analytic results are known, and only recently, some extensions including interbank liabilities of different seniorities (Fischer 2014) or maturities (Kusnetsov and Veraart 2019) have been proposed.

From the perspective of a single firm, an approximate valuation can be obtained by taking into account the solvency risk of its counterparties. Barucca et al. (2016) have shown that several network valuation models, including the Eisenberg and Noe model and many of its extensions, can be unified in this way. In their framework, each firm locally adjusts the value of interbank contracts according to its expected recovery value. Different valuation models are then obtained via different assumptions about when nominal contract values are adjusted—either after an actual default or before an expected default. In the first case, the valuations obtained by each firm agree with the values derived from self-consistent network considerations whereas they only approximately reflect market values of interbank contracts in the latter.

Within the framework of network valuation, financial contagion is naturally quantified in marginal terms, i.e., as the impact of infinitesimal asset price shocks (Ota 2014). Demange (2016) proposed a threat index considering a bank as systemically important to the extent that the values of all other banks are sensitive to devaluation of its assets. Taking this idea even further, Bertschinger and Stobbe (2018) have shown how to compute network Greeks, i.e., sensitivities with respect to several risk factors of interest besides asset prices. In principle, this allows to quantify the systemic impact of changes in interest rate, volatilities, or asset correlations and naturally extends standard risk management practices. Indeed, the Greeks of single asset options are routinely used in order to assess and hedge the risk of trading portfolios.

3.3 Strategic Reactions and Network Structure

An assumption that is made in the abovementioned studies is that the banks remain passive in the face of spreading of defaults. While there is some merit to this argument, it is doubtful that banks remain passive when aware of a possible default by a counterparty or news of trouble elsewhere in the financial system. There has been only minimal effort to model the effect of the response of banks on financial contagion. Anand et al. (2012) incorporate a strategic decision-making component to the network which makes rewiring of connections a possibility. They model liquidity hoarding behavior of banks in which a bank refuses to roll over an interbank loan in fear of default of the counterparty. The outcome of a coordination game among the neighbors of a bank decides whether they roll over or withdraw loans made to that bank. Another work is by Arinaminpathy et al. (2012) in which confidence effects are included in the response of banks. Among other defining features of the model, it is assumed that long-term loans can be made short-term and interbank loans may be withdrawn.

A natural question is what kind of network structure will increase the stability of a financial system. The question has been a much debated one regarding all complex systems from the time of the influential article by Simon (1962). Stability of many benchmark network structures of financial systems has been studied by Battiston et al. (2013) with the conclusion that network topology plays a prominent role only when the liquidity is low, and there is no single optimal connection structure.

4 Contagion via Fire Sale of Assets

As already mentioned, apart from the direct interbank claims and obligations, another way in which the balance sheets of banks influence each other is by carrying the same or similar asset classes. Portfolio overlaps thus produce another channel for initial defaults to spread, and evidence suggests that this one is more relevant than the direct channel in real-world systems. Though several earlier studies like that by Gai and Kapadia did consider minimally the effect of fire sale of a single asset held by all banks, it is in more recent literature that this channel of contagion is analyzed more thoroughly. A general message here is that too much diversification may not be desirable for the health of the system even though it might appear as a wise strategy from an individual bank's point of view. In fact, prior to the 2007–09 crisis, the prevailing view was that portfolio diversification by individual banks will make them safer and this will automatically translate into improved safety of the overall financial system. However, this ignored the crucial fact that diversification by firms will also create strong correlations across balance sheets, thus rendering the whole system susceptible to price changes of commonly held assets.

4.1 Contagion via Overlapping Portfolios

An early work which focuses on the network aspects of portfolio overlaps is by Caccioli et al. (2014) who considered a bipartite network of banks and assets. In

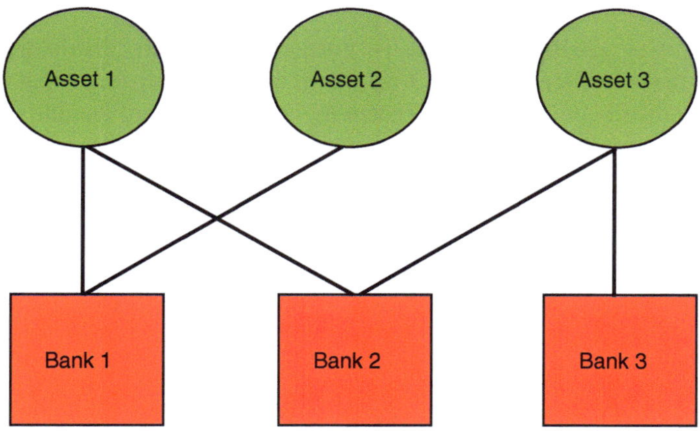

Fig. 5 A schematic representation of a bipartite network of banks and assets. If Bank 1 liquidates its assets in a fire sale, it will depreciate the prizes of Asset 1 and Asset 2. This will affect Bank 2 as it has exposure to Asset 1. In case Bank 2 also forced to liquidate its assets, Bank 3 will be affected as it has exposure to Asset 3. Thus, a contagion can spread due to correlated balance sheet structures

this network, banks have only connections to the assets they hold in their portfolio (see Fig. 5). Two kinds of initial shocks are considered, namely, (1) devaluation of the price of a randomly chosen asset to zero and (2) making a randomly selected bank default. The initial shock then causes defaulting banks to sell the assets in their portfolio en masse thereby depreciating their prizes. A specific functional form for the decrease in the price of an asset is assumed which in general could depend upon market depth and the volume of assets on sale. The depreciating prizes could cause further defaults thereby initiating a cascade. The effects of leverage (asset to equity ratio), market crowding (number of banks per assets), diversification (as measured by the average degree of banks), and market impact (price change of an asset as a function of its volume) on the cascading process are studied. A major finding is that there is a critical leverage below which financial networks are immune to systemic events, but above which there is a finite probability for them. A theory based on generalized branching process is also developed. A point to note here is that the probability and extent of contagion due to fire sales behaves more or less in a similar way to that in the case of direct counterparty contagion as depicted in Fig. 4.

Compared to interbank lending network, the asset contagion network can be more reliably constructed from the data available with regulatory authorities. Unlike interbank lending relations, the structure of the balance sheet of banks is partially publicly available and can be used to construct empirical banks-assets networks. Huang et al. (2013), ran an asset contagion model on US commercial banks asset network. The latter is constructed using data from US Commercial Banks Balance Sheet Data (CBBSD) which classifies assets into 13 different types. By running simulations on the empirical network, they could reproduce a large portion of actually failed banks during the 2007–09 crisis. Similar balance sheet structure of major European Banks is made available by the European Banking Authority (EBA) and has been used in several studies.

Banwo et al. (2016) considered the effects of power-law distributions for the degree of the network as well as for balance sheet size. They found that the power-law degree distribution makes the system less resilient toward the initial default of a randomly chose bank. Targeted initial defaults of biggest or less connected banks increase the probability of a default cascade. A power-law distribution for the assets, on the other hand, increases the resilience of the network with respect to random initial defaults but not with respect to targeted defaults. Caccioli et al. (2015) studied the effect of both interbank and asset contagion channels working together. It is found that systemic effects are amplified when both channels are present.

4.2 Fire-Walling Financial Networks

By now it should be clear that a crucial question in these contexts is to determine which of the banks are most important as far as the stability of the network is concerned. Identifying systemic institutions is crucial in developing strategies and policies to protect the financial system from contagion. A simple answer to this is the "too-big-to-fail" paradigm in which it is proposed that the largest banks are the most important ones (which are also presumably well connected). A complementing one to this is the "too-central-to-fail" paradigm introduced by Battiston et al. (2012) where they introduced a centrality measure called Debt rank. Another centrality measure is the Systemicness of a bank introduced by Greenwood et al. (2015) which is proportional to the product of the size of a bank, its leverage, and connectedness. Information regarding these variables may not be readily obtainable, and Gangi et al. (2015) discuss how to obtain Systemicness based on only partial data about the network. A recent addition to the list of various centrality measures is the one proposed by Cont and Schaanning (2016) which is based on liquidity weighted portfolio overlaps. Also, the threat index by Demange (2016) and the network Δ proposed by Bertschinger and Stobbe (2018) are noteworthy in this context.

Cost of Preventing Contagion

The proposed centrality measures should not just provide a better understanding of the systemic importance of financial institutions, but also should play a key role in designing cost-effective countermeasures to curtail the financial crisis. Here, we present a simulation study illustrating this idea. In particular, we study the model of Cont and Schaanning using portfolio holding data from the EBA (see Cont and Schaanning (2016) for additional details). The data set contains information about investments of the 90 largest European banks in about 140 asset classes, most finely disaggregated across European government bonds. We consider a shock scenario where a single asset is devalued by 30% and simulate ensuing banking defaults resulting from multiple rounds of fire-sale contagion. We investigate two ways of fire-walling the system against financial contagion: first via direct bail-out guarantees and second via asset price guarantees.

Figure 6 shows the effectiveness of different bail-out strategies. Here, each dot corresponds to a shock scenario and is slightly jittered to reduce overplotting. The all-or-nothing character of contagion is clearly visible with either almost no or all banks defaulting. Furthermore, providing bail-out guarantees to the most central banks first, i.e., 10% with the highest centrality, is more effective in preventing contagion than rescuing banks at random. Still, to fully avoid a systemic crisis, guarantees on more than 5 trillion euros have to be granted toward more than 30 major banks (30% of 90 banks).

Figure 7 shows the corresponding results of different buy-out strategies. In this case, instead of rescuing individual banks, prices of certain assets are guaranteed. Thus, even in case of a fire sale, no price impact occurs on the guaranteed assets preventing any contagion via them. Again, providing buy-out guarantees in the order of decreasing systemic importance is found to be more effective than randomly buying them out. Still, at comparable sizes, buy-out guarantees are found to be less effective in preventing contagion than direct bank bail-outs. Note that the nominal size of the guarantee—as stated here and often quoted by the media during the financial crisis—can be much larger than the actual implementation cost (Sasidevan and Bertschinger 2019) especially if contagion is prevented.

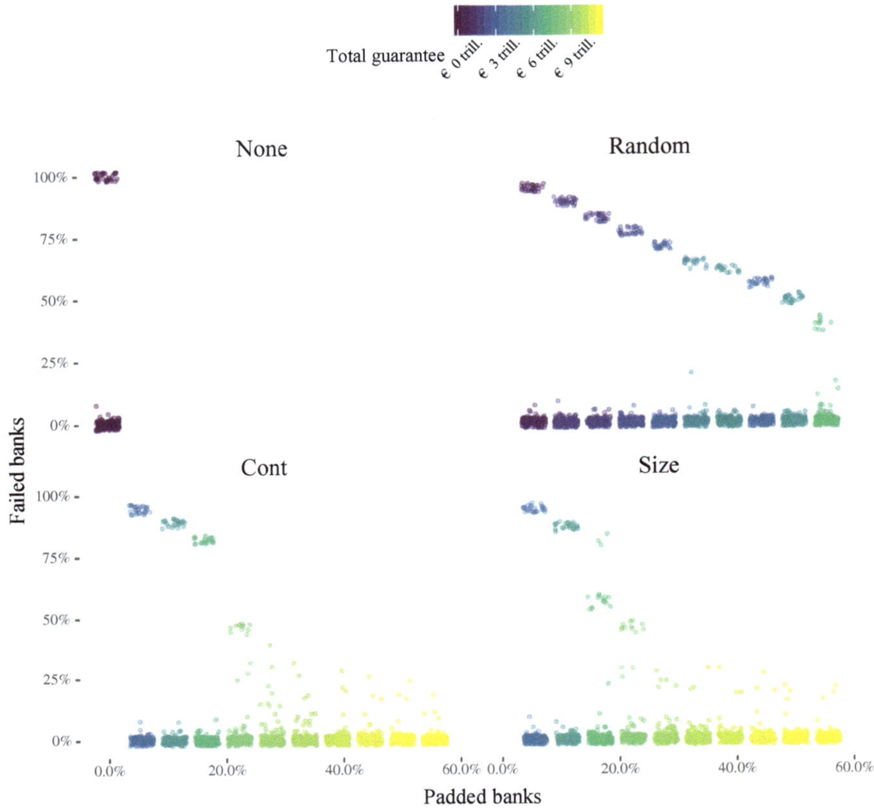

Fig. 6 Effectiveness of different bail-out strategies to prevent contagion from the fire sale of assets. Bail-out guarantees are provided to a fraction of all banks (padded banks) ranked either randomly (top-right) or according to two different centrality measures. The two centrality measures considered are the size of banks (bottom-right) and the one by Cont and Schaanning based on liquidity weighted portfolio overlaps of banks (bottom-left) (Cont and Schaanning 2016). The case with no bail-out guarantees is also shown (top-left)

5 Conclusion and Future Directions

The rapidly increasing connectivity and complexity of the global financial system is a major concern in today's world as it allows problems originating in one part to be quickly transmitted to seemingly distant and uncorrelated parts. Thus, it has become ever more critical to understand the collective behavior of financial systems so that we can prevent or minimize contagious events. Viewing financial systems as complex networks is a significant step in this direction. The importance of macroprudential regulation in which more consideration is given to the health of the entire system than of individual institutions has now been recognized and is increasingly becoming a part of regulatory action.

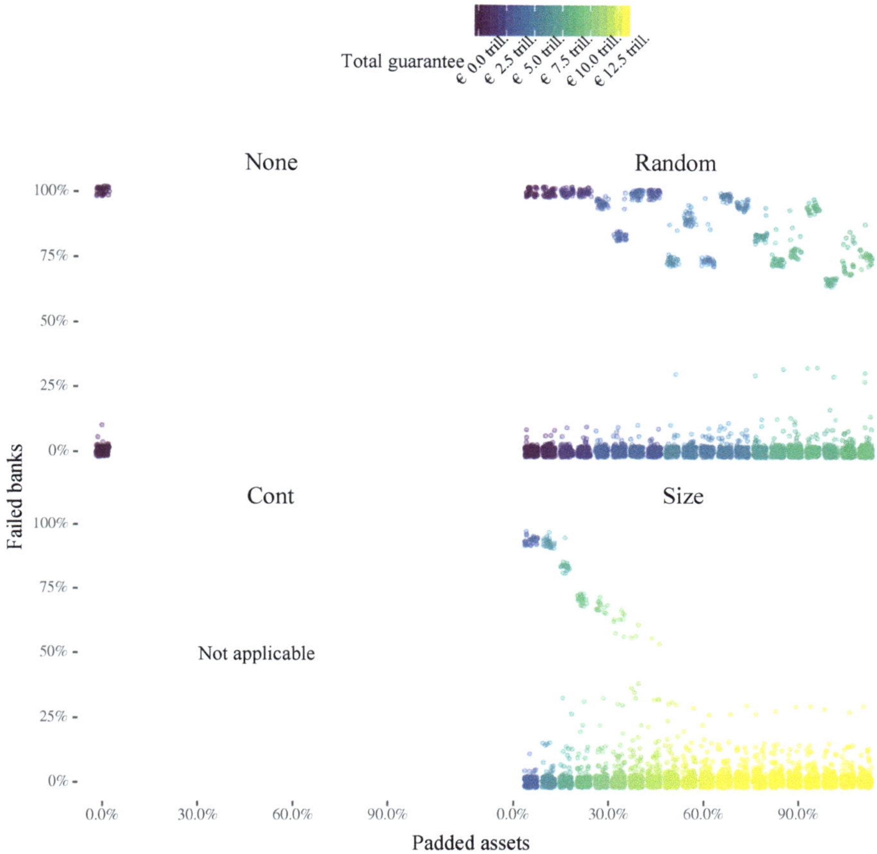

Fig. 7 Effectiveness of different buy-out strategies to prevent contagion from the fire sale of assets. Buy-out guarantees are provided to stabilize the prices of a fraction of all assets (padded assets) ranked either randomly (top-right) or according to the volume of assets (bottom-right). The centrality measure proposed by Cont and Schaanning based on liquidity weighted portfolio overlaps of banks (Cont and Schaanning 2016) is not applicable here. The case with no buy-out guarantees is also shown (top-left)

In this review, we have tried to give a broad overview of various attempts to understand and model financial networks. A major theme arising out of these studies is the "robust-yet-fragile" character of financial networks where contagious events are rare, but when they occur, they are widespread. Probability of contagion may be reduced by various regulatory measures such as requiring higher capital ratios and higher liquidity ratios. Identifying systemic institutions in a financial network is very important, and various centrality measures have been proposed for that. Protecting systemic institutions is crucial for the health of the entire system, but an issue here is how to set policies such that the institutions behave in a responsible way toward the health of the entire financial system. There is the problem of "moral hazard" where a systemically important institution, with the full knowledge that it will be rescued in a crisis by taxpayers money, engages in reckless risk-taking behavior.

Financial networks are ever evolving and have continuously been subjected to many internal and external influences like regulatory constraints and factors from the economy. The technology underlying the formation of financial networks itself could change in the coming years with the use of Artificial Intelligence and formalizing financial contracts (Brammertz and Mendelowitz 2018). Another area to watch out for is the increasing presence of cryptocurrency (Borri 2019). All these have the potential to change the global landscape of financial networks. Including the effects of these and other future possibilities is a major challenge going forward. As we mentioned earlier, including the behavior of the banks in the face of a potential crisis is essential to model banking systems realistically. An area in which the current models lack realism is in the exclusion of central banks from modeling as the presence and actions of such a bank can have a significant effect on systemic risk (Georg 2013).

Banks and regulatory authorities often do not have a complete picture of various direct and indirect exposures of institutions in a financial network (Battiston et al. 2012). Many recent studies try to integrate theoretical models with the available data to come up with conclusions about the possibility of systemic events and strategies or policies to contain them. With the availability of more and more real-time data and realistic models of bank behavior, it is expected that effective real-time monitoring of the health of financial systems will be a reality in the near future. Together with innovative taxation schemes, such as the one proposed by Poledna and Thurner (2016), this might pave the way toward an economic future without systemic financial crises.

Acknowledgements VS acknowledges support by University Grants Commission-BSR Start-up Grant No:F.30- 415/2018(BSR). NB thanks Dr. h.c. Maucher for funding his position.

References

Adrian T, Shin HS (2010) Liquidity and leverage. J Financ Intermed 19:418–437

Allen F, Gale D (2000) Financial contagion. J Polit Econ 108(1):1–33

Amaral LA, Ottino JM (2004) Complex networks. Eur Phys J B 38(2):147–162

Anand K, Gai P, Marsili M (2012) Rollover risk, network structure and systemic financial crises. J Econ Dyn Control 36:1088–1100

Arinaminpathy N, Kapadia S, May RM (2012) Size and complexity in model financial systems. Proc Natl Acad Sci 109(45):18338–18343

Banwo O, Caccioli F, Harrald P, Medda F (2016) The effect of heterogeneity on financial contagion due to overlapping portfolios. Adv Complex Syst 19:1650016-1–1650016-20

Barucca P, Bardoscia M, Caccioli F, D'Errico M, Visentin G, Battiston S, Caldarelli G (2016) Network valuation in financial systems. SSRN id 2795583:

Battiston S, Puliga M, Kaushik R, Tasca P, Caldarelli G (2012) DebtRank: too central to fail? Financial networks, the FED and systemic risk. Sci Rep 2:541

Battiston S, Gatti DD, Gallegati M, Greenwald B, Stiglitz JE (2012) Default cascades: when does risk diversification increase stability? J Financ Stab 8:138–149

Bertschinger N, Stobbe J (2018) Systemic greeks: measuring risk in financial networks. arXiv:1810.11849
Boccaletti S, Latora V, Moreno Y, Chavez M, Hwanga DU (2006) Complex networks: structure and dynamic. Phys Rep 424:175–308
Borri N (2019) Conditional tail-risk in cryptocurrency markets. J Empir Financ 50:1–19
Brammertz W, Mendelowitz AI (2018) From digital currencies to digital finance: the case for a smart financial contract standard. J Risk Financ 19(1):76–92
Caccioli F, Catanach TA, Farmer JD (2012) Heterogeneity, correlations and financial contagion. Adv Complex Syst 15(2):1250058
Caccioli F, Shrestha M, Moore C, Farmer JD (2014) tability analysis of financial contagion due to overlapping portfolios. J Bank Financ 46:233–245
Caccioli F, Farmer JD, Foti N, Rockmore D (2015) Overlapping portfolios, contagion, and financial stability. J Econ Dyn Control 51:50–63
Castellano C, Fortunato S, Loreto V (2009) Statistical physics of social dynamics. Rev Mod Phys 81(2):591
Catanzaro M, Buchanan M (2013) Network opportunity. Nature 9:121–123
Cont R, Schaanning E (2016) Fire sales, indirect contagion and systemic stress-testing. SSRN id 2541114
Demange G (2016) Contagion in financial networks: a threat index. Manag Sci 64(2):955–970
Dobson I, Carreras BA, Lynch VE, Newman DE (2007) Complex systems analysis of series of blackouts: cascading failure, critical points, and self-organization. Chaos Interdiscip J Nonlinear Sci 17(2):026103
Eisenberg L, Noe TH (2001) Systemic risk in financial systems. Manag Sci 47(2):236–249
Farmer JD, Foley D (2009) The economy needs agent-based modelling. Nature 460:685–686
Fischer T (2014) No-arbitrage pricing under systemic risk: accounting for cross-ownership. Math Financ 24(1):97–124
Gai P, Kapadia S (2010) Contagion in financial networks. Proc R Soc A 466:2401–2423
Gai P, Haldane A, Kapadia S (2011) Complexity, concentration and contagion. J Monet Econ 58:453–470
Gangi DD, Lillo F, Pirino D (2015) Assessing systemic risk due to fire sales spillover through maximum entropy network reconstruction. SSRN id 2639178
Georg C-P (2013) The effect of the interbank network structure on contagion and common shocks. J Bank Financ 37:2216–2228
Greenwood R, Landier A, Thesmar D (2015) Vulnerable banks. J Financ Econ 115:471–485
Haldane AG (2015) On microscopes and telescopes. Speech. www.bankofengland.co.uk/publications/Pages/speeches/default.aspx
Haldane AG, May RM (2011) Systemic risk in banking ecosystems. Nature 469:351–355
Hasty J, McMillen D, Isaacs F, Collins JJ (2001) Computational studies of gene regulatory networks: in numero molecular biology. Nat Rev Genet 2(4):268–279
Huang X, Vodenska I, Havlin S, Stanley HE (2013) Cascading failures in Bi-partite graphs: model for systemic risk propagation. Sci Rep 3:1219
Huser AC (2015) Too interconnected to fail: a survey of the interbank networks literature. SAFE Working Paper No. 91, SSRN id 2577241
Karimi F, Raddant M (2016) Cascades in real interbank markets. Comput Econ 47:49–66
Kusnetsov M, Veraart LAM (2019) Interbank clearing in financial networks with multiple maturities. SIAM J Financ Math 10(1):37–67
Lieberman E, Hauert C, Nowak MA (2005) Evolutionary dynamics on graphs. Nature 433(7023):312–316
May RM, Arinaminpathy N (2007) Systemic risk: the dynamics of model banking systems. J R Soc Interface 7:823–838
May RM, Levin SA, Sugihara G (2008) Complex systems: ecology for bankers. Nature 451:893–895
Merton RC (1974) On the pricing of corporate debt: the risk structure of interest rates. J Financ 29(2):449–470

Newman ME (2002) Spread of epidemic disease on networks. Phys Rev E 66(1):016128
Newman M (2010) Networks: an introduction. Oxford University Press, Oxford
Nier E, Yang J, Yorulmazer T, Alentorn A (2007) Network models and financial stability. J Econ Dyn Control 31:2033–2060
Ota T (2014) Marginal contagion: a new approach to systemic credit risk. Bank of England, working paper
Paul Glasserman P, Young HP (2015) How likely is contagion in financial networks? J Bank Financ 50:383–399
Poledna S, Thurner S (2016) Elimination of systemic risk in financial networks by means of a systemic risk transaction tax. Quant Financ 16(10):1599–1613
Roukny T, Bersini H, Pirotte H, Caldarelli G, Battiston S (2013) Default cascades in complex networks: topology and systemic risk. Sci Rep 3:2759
Sasidevan V, Bertschinger N (2019) Fire-walling banks and assets: a network based approach. forthcoming
Schweitzer F, Fagiolo G, Sornette D, Vega-Redondo F, Vespignani A, White DR (2009) Economic networks: the new challenges. Science 325:422–425
Simon HA (1962) The architecture of complexity. Proc Am Philos Soc 106:467–482
Strogatz SH (2001) Exploring complex networks. Nature 410(6825):268–276
Suzuki T (2002) Valuing corporate debt: the effect of cross-holdings of stock and debt. J Oper Res Soc Jpn 2
Upper C (2011) Simulation methods to assess the danger of contagion in interbank markets. J Financ Stab 7:111–125

Spectral and Network Method in Financial Time Series Analysis: A Study on Stock and Currency Market

Pradeep Bhadola and Nivedita Deo

Abstract In this work, we summarize some of the recent statistical, spectral, and network methods for the financial time series analysis. The main focus is first on spectral analysis of the eigenvalues and eigenvector of the cross-correlations between the different financial commodities, and second, the threshold networks derived from the cross-correlation matrix. We used an information theoretic measures called eigenvector localization defined as the "eigenvector entropy" to derive the community structure and interactions between the financial commodities in a system. Lastly, network analysis is performed on the system which shows how the interaction within the system changes with time. In this article, we present and compare the results of two different systems: (1) global financial indices of 31 countries and (2) foreign exchange rates of 21 different currencies using a rolling window method.

1 Introduction

Financial systems have been considered as highly complex evolving systems having many unknown agents and interactions that govern its dynamics. Understanding this complex nature of financial markets is emerged as a challenging task and in recent decades has been attracting considerable interest around the globe (Zhu et al. 2017; Kenett et al. 2015; Junior and Franca 2012; Wang et al. 2018; Ren and Zhou 2014). The complex interaction among the constituents of the system is sometimes highly stochastic in nature which is usually quantified by calculating the cross-correlations among the agents in the systems. In the current digital age, the financial data is growing at an enormous rate which is electronically accessible, thus making it possible to use the enormous financial data to predict the behavior of the financial

P. Bhadola
The Institute for Fundamental Study, Naresuan University, Phitsanulok, Thailand
e-mail: bhadola.pradeep@gmail.com

N. Deo (✉)
Department of Physics and Astrophysics, University of Delhi, Delhi 110007, India
e-mail: ndeo@physics.du.ac.in; ndeo007@gmail.com

A. S. Chakrabarti et al. (eds.), *Network Theory and Agent-Based Modeling in Economics and Finance*, https://doi.org/10.1007/978-981-13-8319-9_17

system. Most of the research in understanding the behavior of the financial time series over the last two decades is based mainly on the calculation of the cross-correlation between the price change of the financial commodities (stock, currency, goods, etc.) (Mantegna and Stanley 1999; Junior and Franca 2012; Wang et al. 2018; Ren and Zhou 2014). The is a large quantity of work on cross-correlation that includes random matrix theory (RMT) (Kumar and Deo 2012; Ren and Zhou 2014; Junior and Franca 2012), principal component analysis (PCA) (Billio et al. 2012), or the construction of the network (Nobi et al. 2014; Birch et al. 2016) not only in the field of econophysics but also in the field of biology (Bhadola and Deo 2016; Cocco et al. 2013), wireless communication (Tulino et al. 2004), medical science (Batushansky et al. 2016).

In the current article, to extract the information from the cross-correlation matrix, we first present a detailed spectral analysis of the eigenvalues and eigenvectors of the cross-correlation matrix; we introduce an entropic measure on the eigenvectors to quantify the information content and localization of the eigenvectors. This measure is used to identify the clusters of highly interacting agents within the system. A threshold method is used to analyze the structure and evolution of the interaction between the constituents element of the system.

2 System and Data

The current analysis is performed on the two diverse systems which are given below:

1. The first system comprises the daily adjusted close price of the 31 global financial indices from the European, Asian, and American markets downloaded from yahoo finance (2019) given in Table 1.
2. The second system consists of the daily FX rates for 21 different currencies from the Federal Reserve Bank of St. Louis (2019) and is shown in Table 2 denoted according to the ISO 4217 standards three-letter code. All the rates are expressed in terms of USD which is the base currency.

These two systems, namely, global financial indices (Bhadola and Deo 2017; Kumar and Deo 2012; Nobi et al. 2014) and foreign exchange rates (Bhadola and Deo 2019), are well studied in the literature. The common factor between the two systems is that in both the systems, the constituent elements (global indices and currencies) are represented as a time series. The data for both the systems are taken from the beginning of 2006 to the end of 2015 which spans a period of 10 years.

After a series of preprocessing steps on the data, the filtering is carried out to eliminate any numerical artifacts. The analysis is done on the daily logarithmic returns $R_i(t)$, which for a commodity i is calculated as

$$R_i(t) = ln(S_i(t + \Delta t)) - ln(S_i(t)) \tag{1}$$

Table 1 List of 31 financial indices symbol (SYM) used for the analysis

SYM	Country	SYM	Country	SYM	Country
^MERV	Argentina	^BVSP	Brazil	^BSESN	India
^KLSE	Malaysia	^MXX	Mexico	^KS11	South Korea
^AORD	Australia	^ATX	Austria	^FCHI	France
^HSI	Hongkong	^TA100	Israel	^N225	japan
^SSMI	Switzerland	^FTSE	UK	^GSPC	USA
^BUX	Hungary	^NZ50	New Zealand	^KSE	Pakistan
^IBEX	Spain	^CSE	Sri Lanka	^AEX	Netherland
^GSPTSE	Canada	^STOXX50E	Euro Zone	GD.AT	Greece
^TWII	Taiwan	^JKSE	Indonesia	^GDAXI	Germany
^STI	Singapore	^SSEC	China	RTS.RS	Russia
^BFX	Belgium				

Table 2 List of 21 currencies with the respective countries used for the analysis

Country	Currency	Country	Currency	Country	Currency
UK	GBP	Brazil	BRL	Switzerland	CHF
Mexico	MXN	China	CNY	Taiwan	TWD
Norway	NOK	Thailand	THB	Canada	CAD
Sweden	SEK	Hong Kong	HKD	Australia	AUD
Denmark	DKK	South Africa	ZAR	Euro	EUR
Japan	JPY	Singapore	SGD	New Zealand	NZD
India	INR	Sri Lanka	LKR	South Korea	KRW

where $S_i(t)$ is the price of commodity i (country index or currency) at time t and $\Delta t = 1$ day is the time lag. The normalized logarithmic returns are then given by

$$r_i(t) = \frac{R_i(t) - < R_i >}{\sigma_i} \tag{2}$$

where $< R_i >$ is the time average of the returns over the time period and σ_i is the standard deviation of $R_i(t)$ given by $\sigma_i =< R_i^2 > - < R_i >^2$.

3 Correlation Matrix

The study of the correlation between different commodities in a system is very crucial for risk management, stability analysis, portfolio management, etc. But the correlations are often plagued with the noise, and the quality of information depends on the ratio of the length of time series L and N number of commodities. This ratio should be greater than 1; otherwise, the resulting correlation matrix will be very noisy and is not good for analysis.

We use the Pearson cross-correlation coefficient which is calculated from the normalized log returns for any commodity pair i and j and defined as $C_{i,j} =< r_i(t)r_j(t) >$. The elements of the cross-correlation lie in the range $-1 \leq C_{i,j} \leq 1$, where $C_{i,j} = 0$ corresponds to no correlation between the commodity i and j, $C_{i,j} = 1$ is the perfect correlation, and $C_{i,j} = -1$ represents perfect anticorrelation. The correlation matrix for the period complete period 2006–2015 is calculated and shown in Fig. 1 for the 31 global financial indices and Fig. 2 for the foreign exchange rates which shows that the financial indices are positively correlated as compared to the foreign exchange rates. Foreign exchange rates show both high positive and negative correlation spanning the complete range of the possible correlations from [−1, 1] whereas for the global financial indices, this range is [−0.25, 1] whereas most of the weight (more than 90%) is concentrated on the positive side.

Next, we study the dynamics of both the systems on the yearly basis from 2006 to 2015 a total of 10 years. The correlation matrix is calculated for each year (window) and the properties are studied. First, we study the statistical properties of the cross-correlation for each window. First, we investigate the probability density function for each system (global indices and currencies) for the complete system and each window which is shown in Fig. 3. The PDF for both the systems global indices and foreign exchange rates is a nonsymmetrical distribution with a positive mean. The global financial indices for all window shows that the mass is concentrated on the positive side whereas for the foreign exchange rates the weight is evenly divided

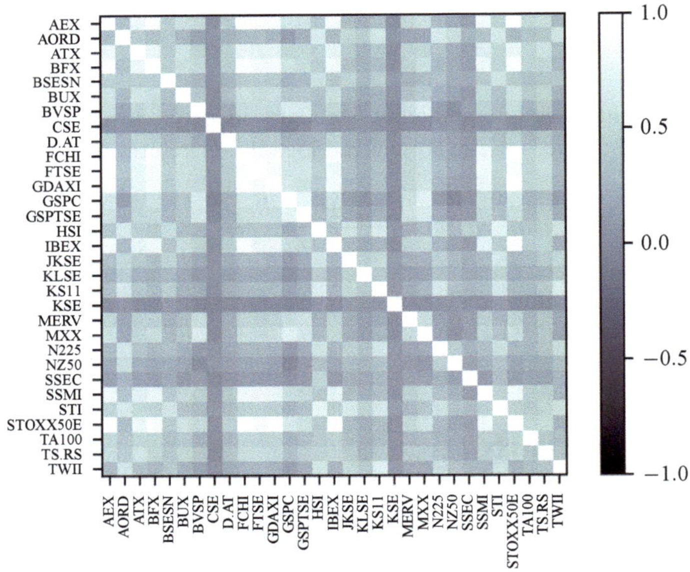

Fig. 1 Heatmap of C_{ij} for the 31 global financial indices from 2006 to 2015

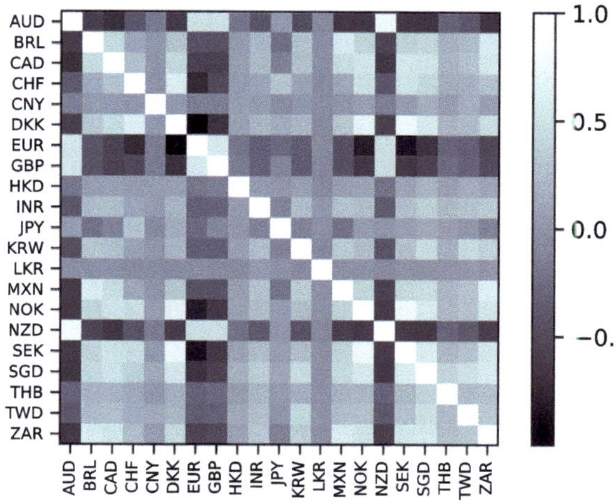

Fig. 2 Heatmap of C_{ij} for the foreign exchange rates from 2006 to 2015

among the complete range. The global indices show only high positive correlation with a very little anticorrelation whereas for the foreign exchange rates both high correlation and anticorrelation are possible.

Figure 4 shows the average of the off-diagonal elements of the cross-correlation matrix for global financial indices and foreign exchange rates. The average correlation between the global financial indices increases in the years 2007, 2008, and 2011. These three years have the highest average correlations and correspond to subprime mortgage crisis (2007), global financial crisis (2008), and European sovereign debt crisis (2011). For foreign exchange rate, $< C >$ shows a decrease average correlation in 2007 and 2008 but the highest average correlation among all the years considered. We also look into the average anticorrelation, i.e., negative off-diagonal elements of the correlation matrix which is shown in Fig. 5 for global financial indices (left) and foreign exchange rates (right). The Fig. 5 shows that anticorrelation are negligible for global financial markets and all the markets positively correlated. Whereas for the foreign exchange rates, there are significant anticorrelations which are lowest in 2011. For the complete period from 2006 to 2015, global financial markets are more correlated than the foreign exchange rates, whereas the foreign exchange rate how high degree of anticorrelation (10 times higher) than global financial markets.

4 Spectral Properties

The correlation structure can be studied by the eigenspectra of the correlation matrix C. For each year, the eigenvalues λ_i and eigenvectors υ_i are determined from the corresponding correlation matrix. The eigenvalues are then arranged in ascending order such that $\lambda_1 \leq \lambda_2 \cdots \leq \lambda_N$.

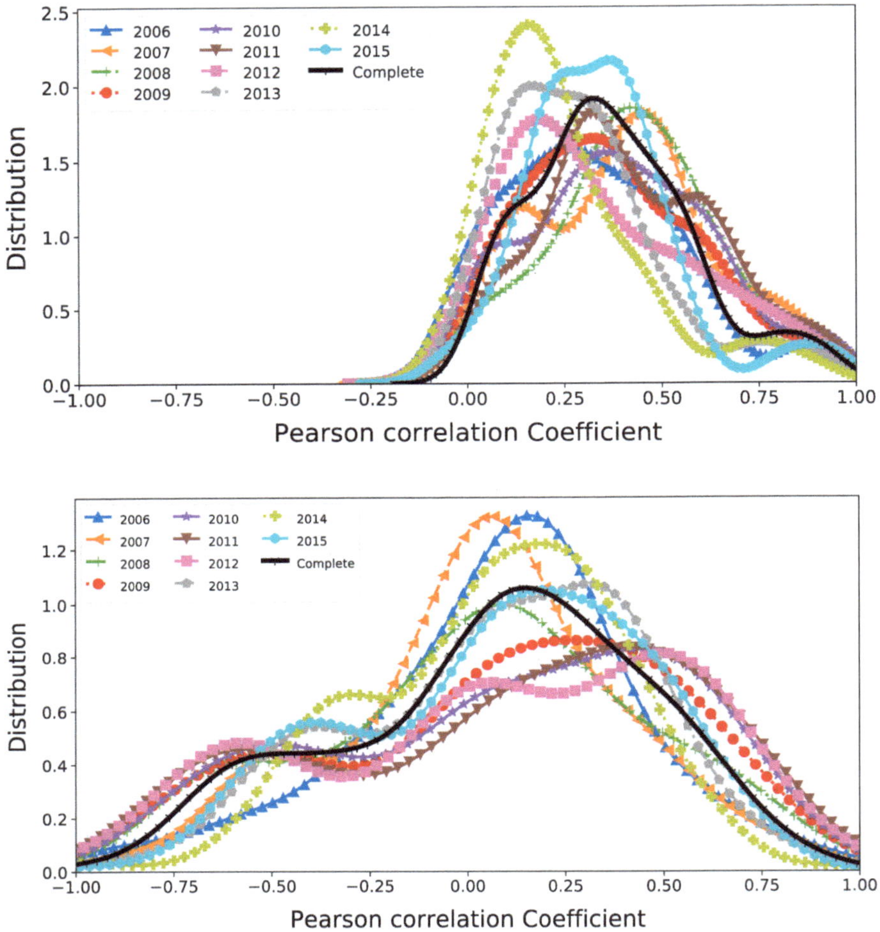

Fig. 3 First figure shows the PDF of C_{ij} for the global financial indices and second figure shows PDF of C_{ij} between the FX rates of different currencies for the complete interval from 2006 to 2015 and for the each window

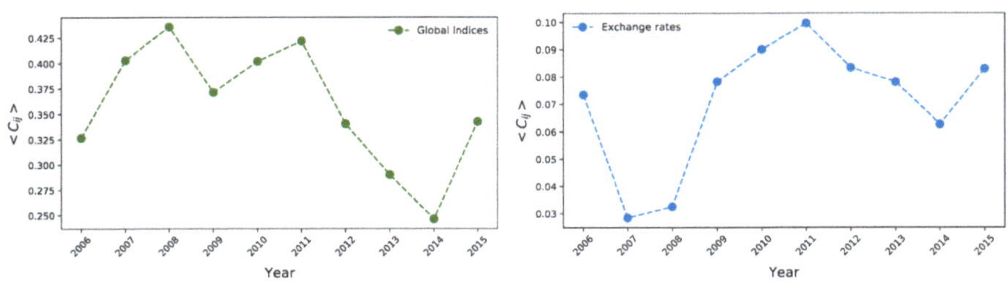

Fig. 4 Average off-diagonal elements of cross-correlation coefficients C_{ij} of the global financial indices (left) and foreign exchange rates (right)

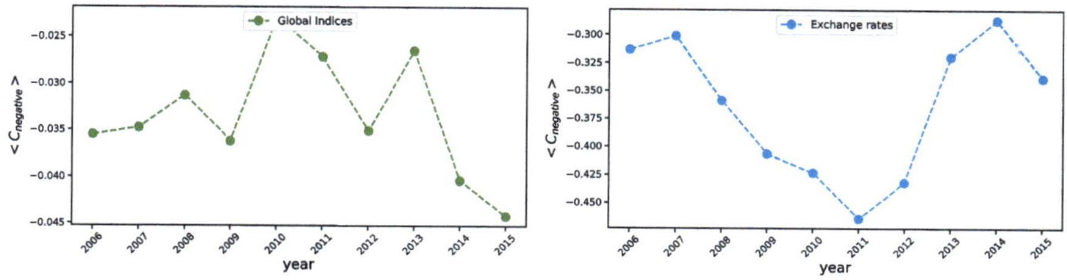

Fig. 5 Average negative off-diagonal elements of cross-correlation coefficients C_{ij} of the global financial indices (Left) and foreign exchange rates (right)

The cross-correlation matrix often contains random noise (Pandey et al. 2010) in addition to the actual information. We construct a null model by random shuffling of the data set for each system. Numerically, resultant correlation matrices from the shuffled data set are equivalent to the Wishart matrices. We use the analytical results of the Wishart matrices, which have well-defined spectral properties (Bowick and Brézin 1991) where the eigenvalue density is given by

$$P_W(\lambda) = \frac{Q}{2\pi\sigma^2}\frac{\sqrt{(\lambda_+ - \lambda)(\lambda - \lambda_-)}}{\lambda} \tag{3}$$

with $Q = \frac{W}{L} \geq 1$ and $\sigma = 1$ the standard deviation.

The abovementioned distribution is known as Marcenko–Pastur distribution which gives the theoretical bounds for the eigenvalues as

$$\lambda_\pm = \sigma^2\left(1 + \frac{1}{Q} \pm 2\sqrt{\frac{1}{Q}}\right) \tag{4}$$

For the current analysis, $W = 250$ is the size of the window and $N = 31$ for global financial indices and $N = 21$ for foreign exchange rates. The details of the bounds for both the system are given in Table 3.

Table 3 Details of the RMT bound and eigenvalue statistics for the two systems (global financial indices and foreign exchange rates)

System	W	N	Q	λ_+	λ_-	$\lambda < \lambda_-$	$\lambda_+ < \lambda$
Global indices	250	31	8.0645	1.8282	0.4197	15	2
FX rates	250	21	11.9047	1.6636	0.5043	11	2

4.1 Eigenvalue Distribution

In this work, we first compare the eigenvalue distribution of the correlation matrix corresponding to each year with the correlation matrix for the complete data set and then with the randomly shuffled data. As the financial markets are dynamical and change continuously with time, comparing the yearly distribution with the distribution of the complete period can give insights how the system at a small time scale differs from the overall behavior.

Figure 6 shows the complete eigenvalue distribution of the global financial indices matrices for each year from 2006 to 2015, the complete data set of 10 years and random shuffled distribution. We also analyze the behavior of the lower side of the eigenvalue spectrum for the global financial indices shown in Fig. 7. The distribution on both sides of the spectra differs significantly from random distribution as well as from the complete period. The differences between the complete data are higher for the crisis period (2007, 2008, 2011) where there is an increase in the magnitude of the largest eigenvalue as shown in Table 4. However, there is not much change in the number of eigenvalues outside the RMT bounds for each year which is nearly 2 or 3, but there is a significant increase in the number of eigenvalues outside the RMT bound on the lower side of eigenvalue spectra during the crisis period. For 2007, there are 19; 2008, we have 18; 2010, there are 18; and for 2011, a total of 19 eigenvalues are outside the lower RMT bound $\lambda < lambda_-$. There the number of eigenvalues of the lower side of the spectra is an indication of the crisis period.

Figure 8 shows the complete eigenvalue distribution and the lower part of eigenvalue distribution of the foreign exchange rates on a yearly basis from 2006 to 2015, with the complete analysis period of 10 years with random shuffled data. The FX

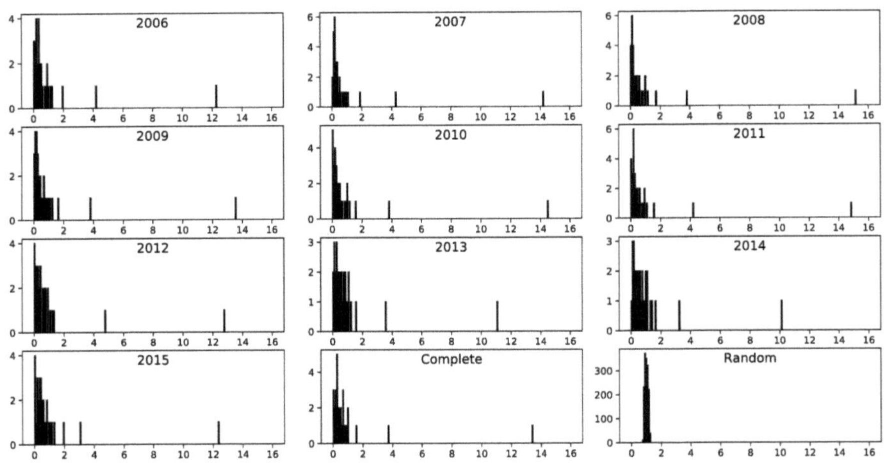

Fig. 6 Comparison of eigenvalue distribution of 31 global financial indices for each year from 2006 to 2015, for the complete period of 10 years and random shuffled system

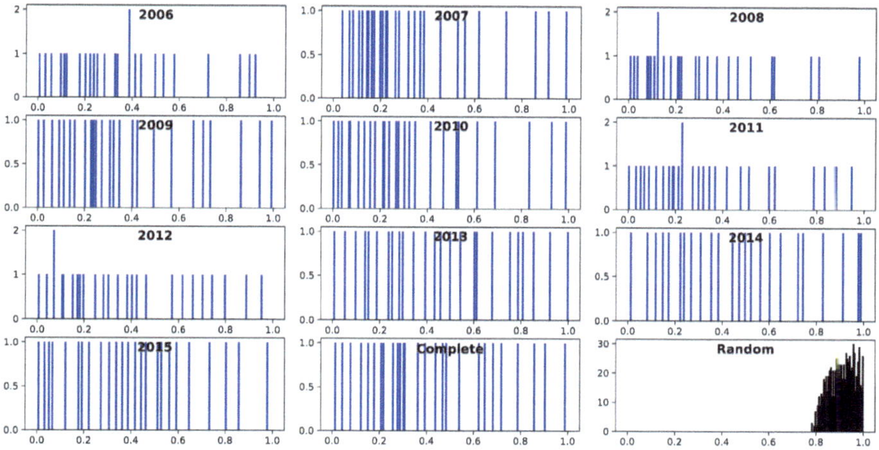

Fig. 7 Comparison of eigenvalue distribution on the lower side of the spectra of 31 global financial indices for each year from 2006 to 2015, for the complete period of 10 years and random shuffled system

Table 4 Comparison of the eigenvalue statistics of the global financial markets and foreign exchange rates

	Year	2006	2007	2008	2009	2010	2011	2012	2013	2014	2015
Global	$\lambda < \lambda_-$	17	19	18	17	18	19	16	12	11	14
	$\lambda_+ < \lambda$	3	3	2	2	2	2	2	2	2	3
	λ_{max}	12.31	14.23	15.19	13.52	14.53	14.81	12.75	11.11	10.09	12.39
	λ_{min}	0.007	0.038	0.008	0.005	0.005	0.005	0.005	0.007	0.011	0.007
FX	$\lambda < \lambda_-$	10	9	12	12	13	11	12	10	9	10
	$\lambda_+ < \lambda$	3	2	2	2	2	1	1	2	2	2
	λ_{max}	7.77	7.48	8.91	9.82	10.06	10.55	10.44	8.03	7.09	8.08
	λ_{min}	0.0013	0.0019	0.0013	0.0007	0.0006	0.0004	0.0005	0.0019	0.0011	0.0009

rates also show a sudden increase in the largest eigenvalue during the crisis period 2007–2012, there is also an increase in the number of eigenvalues outside the lower bound. However, on the higher side of the spectra, there is not much change in the number of eigenvalues outside RMT bounds which is on average is nearly 2.

Analyzing both the systems indicates that most of the number of eigenvalues which are outside the RMT bound are located on the lower side of the spectrum. Thus, the essence of the interaction among agents is indicated mostly by the eigenvalues which are outside the lower RMT bound, most of the information is concentrated there (Figs. 7 and 9).

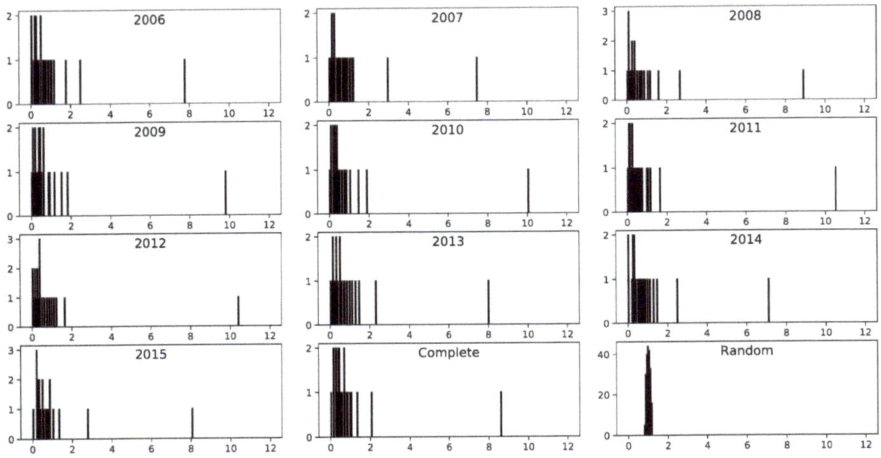

Fig. 8 Comparison of eigenvalue distribution of foreign exchange rates each year from 2006 to 2015, for the complete period of 10 years and random shuffled system

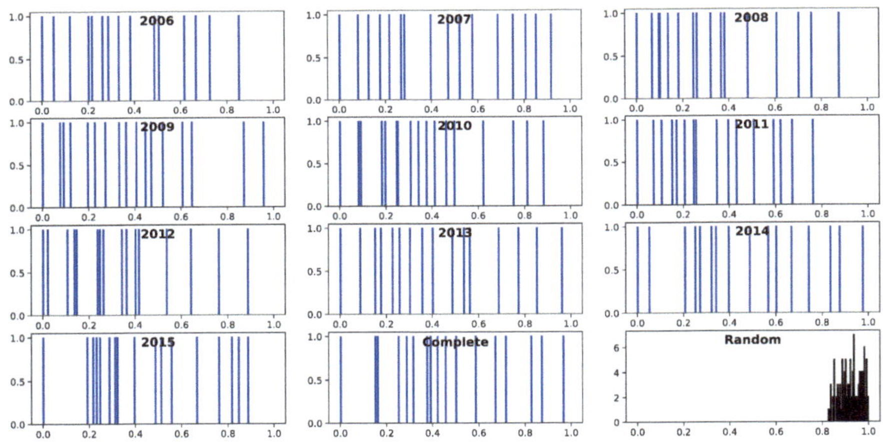

Fig. 9 Comparison of eigenvalue distribution on the lower side of the spectra of foreign exchange rates each year from 2006 to 2015, for the complete period of 10 years and random shuffled system

4.2 *Localization of Eigenvector: Eigenvector Entropy*

Localization of an eigenvector is estimated by the information content of the eigenvector. Higher the information content, more localized the eigenvector is. We use Shannon's entropy to assess the information contained in an eigenvector. Shannon's entropy of an eigenvector v_i is defined as

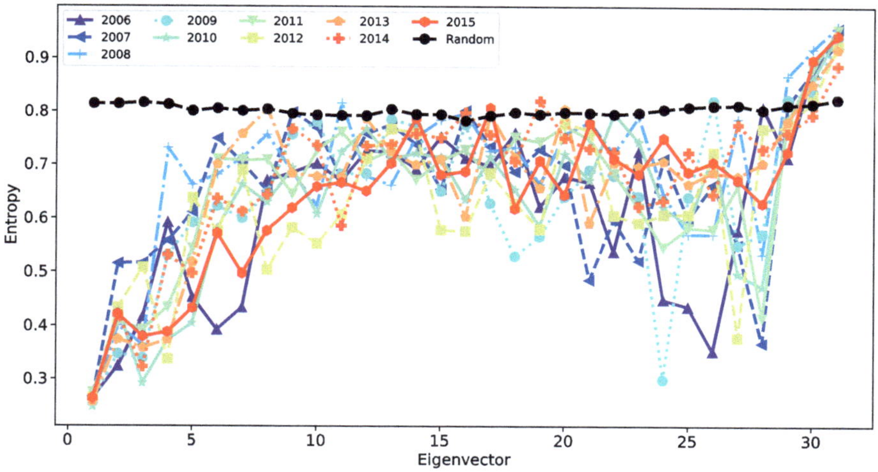

Fig. 10 Comparison of eigenvector entropy for the global financial market from 2006 to 2015

$$H_i = -\sum_{j=1}^{N} u_i(j) log_L(u_i(j)) \tag{5}$$

where N is the number of eigenvector components and $u_i(j) = (v_i(j))^2$ is the square of the jth component of the ith normalized vector v_i.

Eigenvector which has low entropy is highly localized with significant contribution from only some components. We find that in both the systems Figs. 10 and 11, we find that the eigenvector corresponding to the lower side of the spectrum has low entropy thus high localization as compared to the eigenvector on the higher side of spectra. The components of these eigenvectors have significant information about the different aspects of the system. The highly localized eigenvector gives the highly interacting group of commodities. It is established in many systems, for example, in protein families, the small eigenvectors of the correlation matrix have high information content can predict the functionally and structurally important groups of amino acids. The components of these eigenvector form highly interacting pairs that can be grouped into sectors or motif having an important functional role (Bhadola and Deo 2016). Also in finance, in Markowitz's theory of portfolio optimization (Markowitz 1971), low eigenvectors correspond to the least risky portfolios.

4.3 *Eigenvector*

As eigenvalues that diverge significantly from RMT predictions, the corresponding eigenvectors contains significant information. We perform a detailed analysis on the

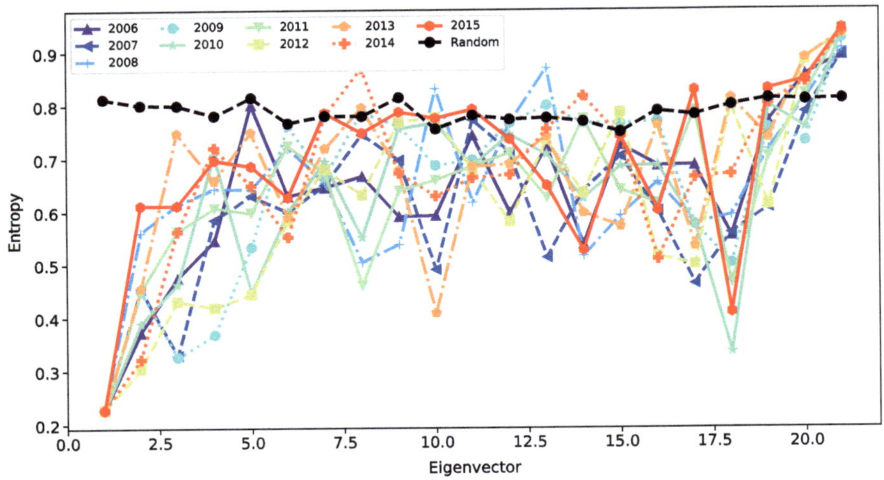

Fig. 11 Comparison of the eigenvector entropy for each year from 2006 to 2015 for the foreign exchange rate

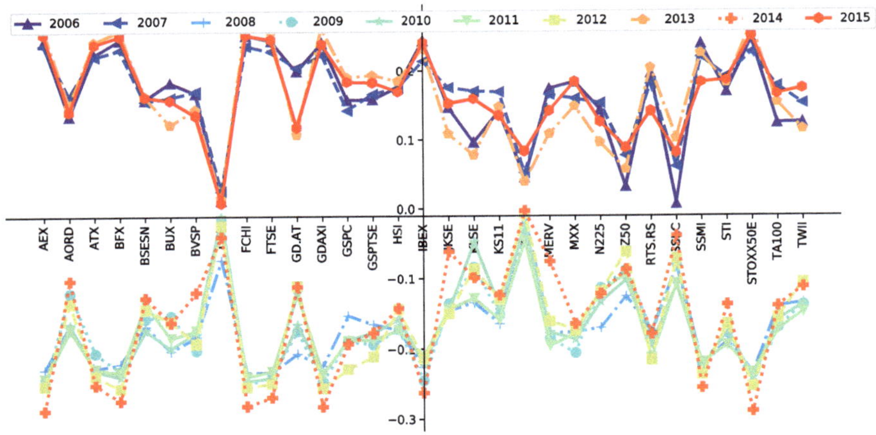

Fig. 12 Eigenvector components for the largest eigenvectors for the global financial indices for years 2006 to 2015

eigenvector components of the eigenvalues that are significantly far from the RMT bounds. The largest eigenvector shows the state of the system; in our analysis, it represents the global financial conditions.

Figures 12 and 13 show the components of the largest and the second largest eigenvector for each year from 2006 to 2015. As the largest eigenvector, gives the overall state of the system, it can be seen that the most of the Asia-Pacific countries contribute very little towards the state of the world financial condition. The global economic conditions are mainly decided by the performance of the European

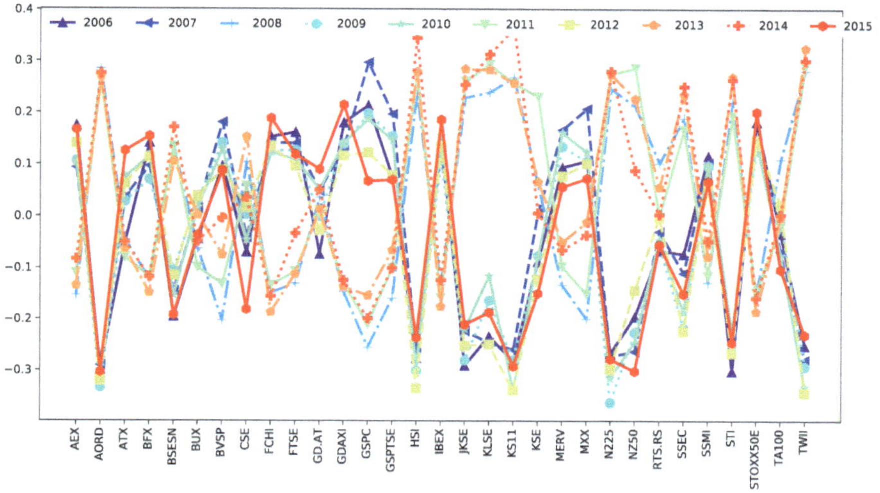

Fig. 13 Eigenvector components for the second largest eigenvectors for the global financial indices for years 2006 to 2015

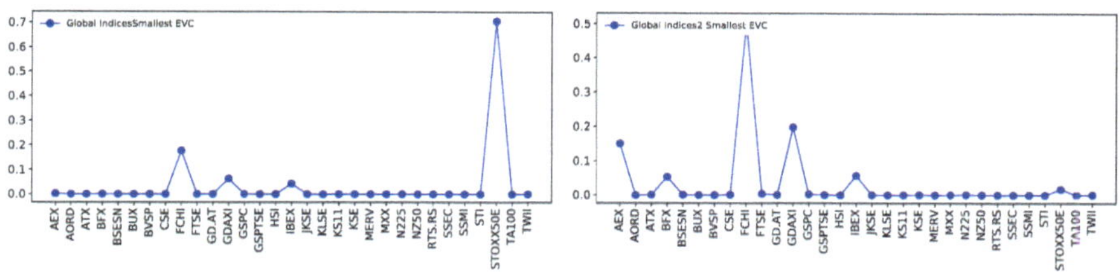

Fig. 14 Square of the eigenvector components for the smallest (left) and second smallest (right) eigenvectors for the global financial indices for years 2006 to 2015

markets. The second largest eigenvector divides the system into two categories: first those markets for which the components are positive, and second, the markets for which the components are negative. The components change signs in the groups with the change in time; the overall groups remain the same. Upon closer look, these two categories are based on the geographical location of the financial market. The two categories are first is the European and American market and the other group with opposite sign are the countries from Asia-Pacific region. The groups remain intact with time; all the countries change sign simultaneously. The groups are given by group 1 (European and American) that consists of 16 countries including Argentina, Austria, Belgium, Brazil, Canada, Euro Zone, France, Germany, Greece, Switzerland, Hungary, Spain, Netherland, Mexico, UK, and US). Group 2 consists of Australia, China, India, Indonesia, Israel, Japan, Malaysia, South Korea, Hong Kong, Singapore, New Zealand, Pakistan, Russia, Sri Lanka, and Taiwan from the Asia-Pacific region. Next, we investigate the eigenvector component corresponding to the smallest eigenvalues and the second smallest eigenvalues which are shown

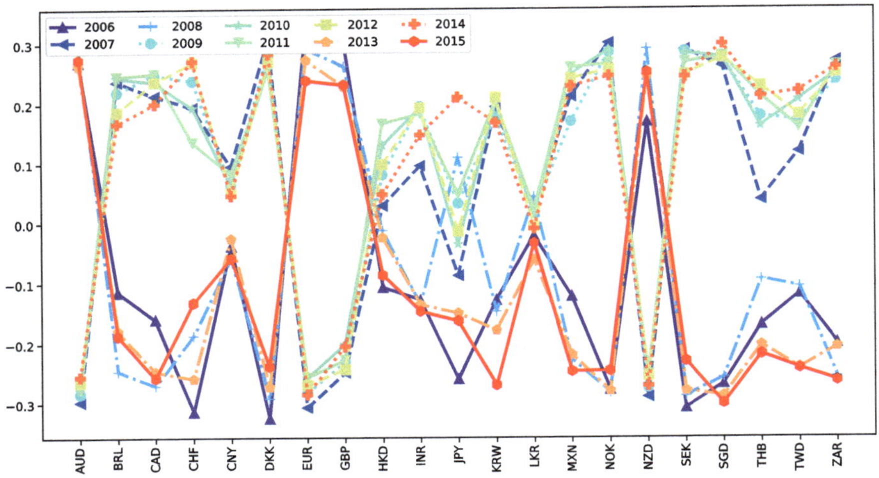

Fig. 15 Eigenvector components for the largest eigenvectors for the foreign exchange rates each year from 2006 to 2015

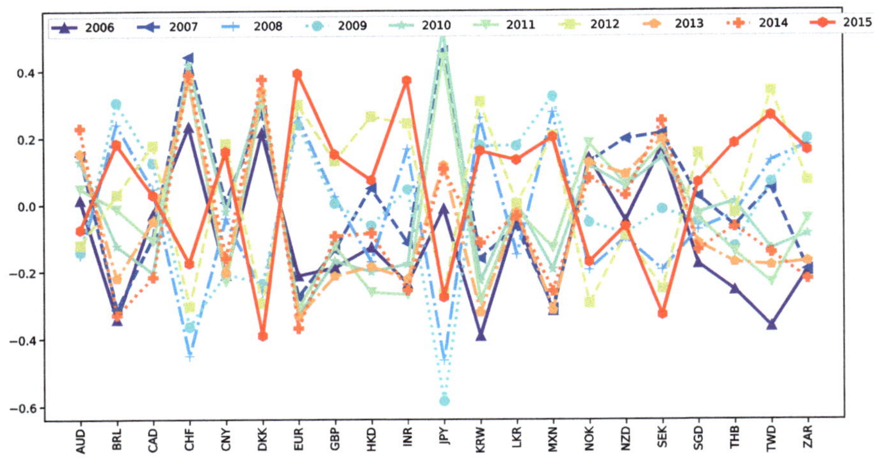

Fig. 16 Eigenvector components for the second largest eigenvectors for the foreign exchange rates each year from 2006 to 2015

in Fig. 14. These eigenvectors are highly localized with contribution from only a few components. These components form a highly interacting pair of markets that can be clustered together into sectors. For the smallest eigenvector Fig. 14, the main contribution is from France, Germany, Spain, and European Union; this forms a very highly interacting pair of markets, with high interdependence. European Union is most influential in the cluster. The second smallest eigenvector gives a cluster comprising Netherland, Belgium, France, Germany, Spain, and the European Union. In this cluster France is the most influential market.

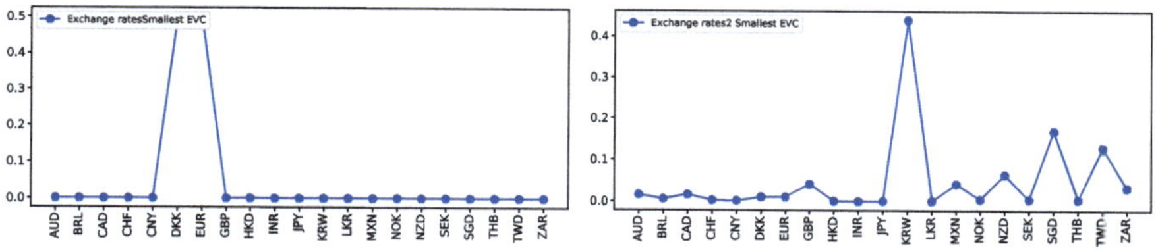

Fig. 17 Square of the eigenvector components for the smallest (left) and second smallest (right) eigenvectors for the foreign exchange rates each year from 2006 to 2015

Figures 15 and 16 show the variation of the largest and second largest eigenvectors for each currency with time. Figure 17 shows that the lowest eigenvectors are highly localized, and there are only a few contributing components. The largest eigenvector is not in one direction has both positive and negative component which is different from the largest eigenvector for the global financial markets where the components are completely positive or completely negative. The smallest eigenvector components show a very high interaction between the Denmark and Euro. The second smallest eigenvector however includes a large group of currencies including UK, South Korea, Mexico, New Zealand, Singapore, Taiwan, and South Africa.

5 Network Analysis

In this section, we use the threshold method to create a network from the correlation matrix. The correlation matrix can be converted into a graph based on the threshold θ. The resulting network $G^{\alpha}(N, E)$ with nodes (N) and edges $E^{\alpha}_{i,j}$ between two nodes i and j given by

$$E^{\alpha}_{i,j} = \begin{cases} 1 & if \quad i \neq j, C^{\alpha}_{i,j} \geq \theta \\ 0 & otherwise \end{cases} \tag{6}$$

where θ is the threshold value. The network is constructed for different threshold values ranging from [0, 1] and is shown in Fig. 18 for global financial indices and Fig. 19 for foreign exchange rates at different threshold values (0.1, 0.3, 0.5, 0.7).

Different threshold values generate different networks with the same set of nodes and different sets of edges. Using different properties results in a network with the same nodes but different links.

5.1 Network Properties

In this section, we discuss some of the basic properties of the correlation threshold network.

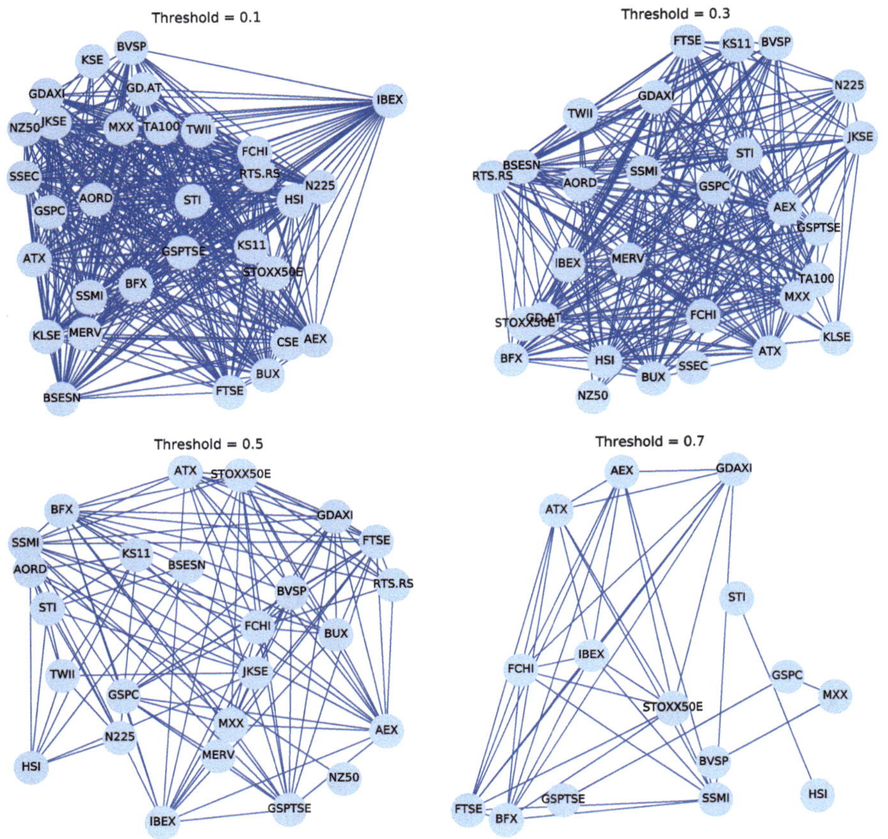

Fig. 18 Global financial network at different threshold values

5.1.1 Adjacency Matrix

Adjacency matrix or sometimes called as connection matrix is another representation of the connections in a network. The element of the adjacency matrix A_{ij} is 1 if there is a direct connection between node i and node j and A_{ij} is zero otherwise. In the case of correlation threshold network which is undirected network, the adjacency matrix is symmetric matrix which says that if there is a link from i to j, then there exists a reverse link j to i.

5.1.2 Network Density

Network density is defined as the ratio of the actual number of links in the networks to the maximum possible link in the network defined as $\rho = M/(N(N-1))$, where N is the number of nodes in the network and M is the connection in the network. We

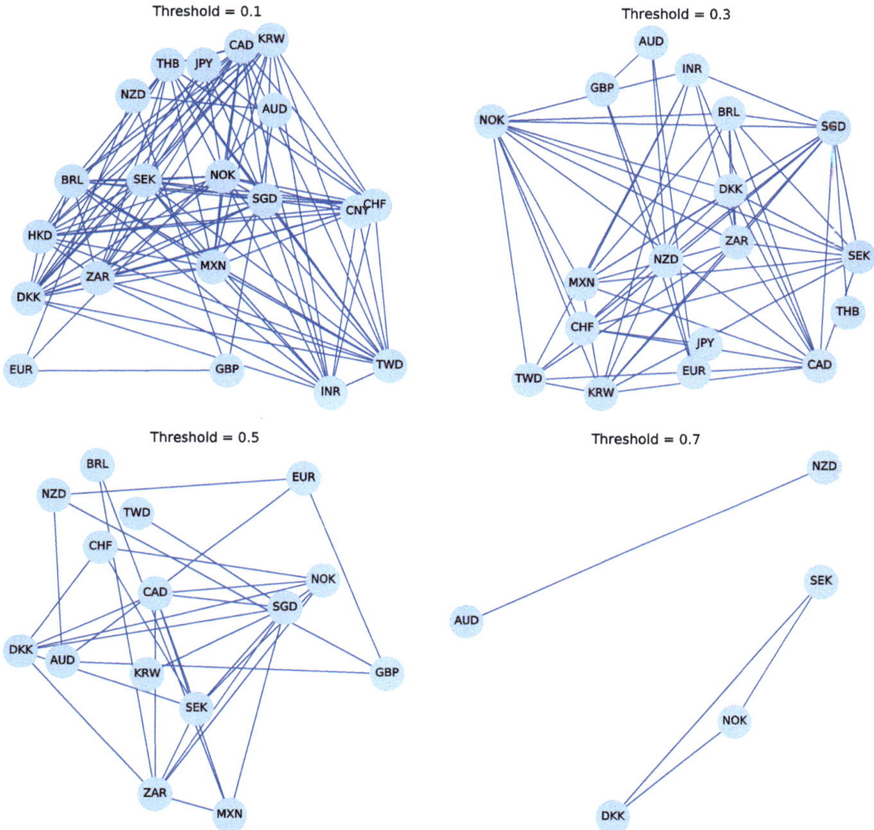

Fig. 19 Network of foreign exchange rates at different threshold values

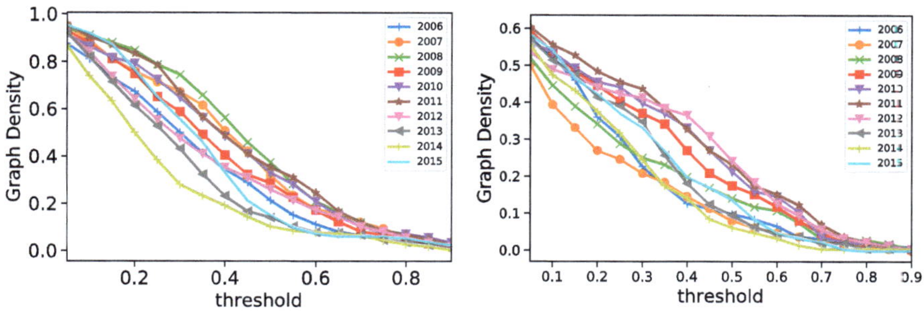

Fig. 20 Variation of network density with threshold for different windows for global financial markets (left) and .foreign exchange rates (right)

study change in network density at different threshold values with each year and i shown in Fig. 20 for global financial markets (left) and .foreign exchange rates (right). For the global financial markets, we observe an increase in the network density for the crisis period (subprime mortgage crisis (2007), global financial crisis (2008) and European sovereign debt crisis (2011)).

The foreign exchange network does not show any increase in network density for 2007 and 2008 and only shows a sudden increase in the density during the European sovereign debt crisis (2011).

5.2 *Average Degree*

Degree of a node gives the total number of interactions of that node with all other nodes. Mathematically the degree K_i of a node i is defined as

$$K_i = \sum_{j=1}^{N} A_{ij} \tag{7}$$

where N is the number of nodes in the network and A is the adjacency matrix of the network.

First, for each threshold, the degree of every node in the network is calculated for all windows. Then estimate the average degree of the network as $K_{avg} = \frac{1}{N}\sum_{i=1}^{N} K_i$, which gives the on an average what is the connectivity of each node. The variation of average degree with the threshold for both network is shown in Fig. 21. Figure 21 shows that there is an increase in the average degree in the network during the crisis period. This implies the increase in the number of connection within the network hence higher interaction.

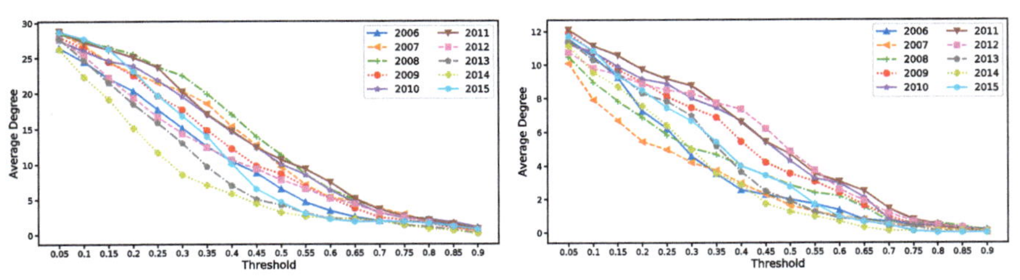

Fig. 21 Average degree with threshold for different windows for global financial markets (left) and foreign exchange rates (right)

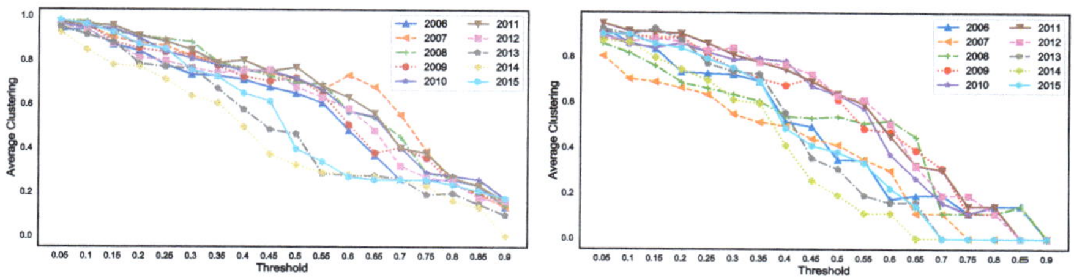

Fig. 22 Average clustering coefficient with threshold for different windows for global financial markets (left) and .foreign exchange rates (right)

6 Average Clustering Coefficient

Clustering coefficient is the measure of local connectivity in the network. It quantifies the interconnectivity between the neighbors of a node. The clustering coefficient of a node i is given as

$$C_i = \frac{2m_i}{n_i(n_i - 1)} \tag{8}$$

where m_i are the actual connection among the neighbors of i, and n_i is the number of neighbors of i. Clustering coefficient C_i is zero if $n_i \leq 2$. The average clustering coefficient is the average of clustering coefficient over all nodes in the network $C_{avg} = \frac{1}{N}\sum_{i=1}^{N} C_i$. We study the variation of the clustering coefficient with threshold and is shown in Fig. 22, which shows that there are significant changes in network with time. The clustering is high during the crisis for both the networks. The global financial indices 2007, 2008, and 2009 show high clustering. During the 2007 crisis, the foreign exchange network shows a higher value for the average clustering coefficient at large thresholds (0.5–0.8). Whereas at an intermediate threshold only sovereign debt crisis 2011 shows maximum clustering.

7 Conclusion

In this work, we discuss the spectral and network method based on correlation matrix to analyze the financial time series. Two independent systems, namely, global financial indices and foreign exchange rates, are studied in detail for a span of 10 years from 2006 to 2015. A correlation matrix is created from the time series data for each year from 2006 to 2015 and compared with the complete correlation matrix and the random shuffled matrix. Simple statistics on the correlation matrix shows that the financial markets are only positively correlated whereas the currency market shows both high positive correlation and negative anticorrelation. The eigenvalues and eigenvectors of the correlation matrices for each year and eigenvalue distribution are created. The eigenvalue distribution is compared with the random matrix results.

The distribution of eigenvalues reveals the distribution is different for the calm and the crisis period. An information theoretic measure on the eigenvector defined as eigenvector is calculated which quantifies the information content as well as the localization of the eigenvector. Eigenvector entropy shows that for both the systems, global financial indices and foreign exchange rates, the smallest eigenvectors are highly localized as compared to the eigenvalues on the upper side of the spectra. The eigenvector outside the random matrix bounds contains significant information for instance, for the global financial indices second largest eigenvector components can cluster the markets based on their geographical location. We study the variation of a number of eigenvalues out the RMT bounds with time indicate that during the crisis, there is a significant increase in the number of eigenvalues on the lower side of the spectra. The increase in the number of eigenvalues may be used as a measure to identify the crisis period. Analysis of these eigenvector gives the clusters of highly interacting pairs of commodities. At last, we created the threshold-based network form the correlation matrix. The evolution of some of the network properties with time is studied which shows that the network is more connected at the time of crisis hence indicating an increase in the interaction between the commodities during the crisis.

Acknowledgements ND thanks Department of Science and Technology (DST), India (SERB-DST No- EMR/2016/006536) for financial support, and PB acknowledges NU research grant, Naresuan University, Thailand for financial support.

References

Batushansky A, Toubiana D, Fait A (2016) Correlation-based network generation, visualization, and analysis as a powerful tool in biological studies: a case study in cancer cell metabolism. BioMed Res Int

Bhadola P, Deo N (2016) Targeting functional motifs of a protein family. Phys Rev E 94(4):042409

Bhadola P, Deo N (2017) Extreme eigenvector analysis of global financial correlation matrices. In: Econophysics and sociophysics: recent progress and future directions. Springer, pp 59–69

Bhadola P, Deo N (2019) Evolution and dynamics of the currency network. In:New perspectives and challenges in econophysics and sociophysics. Springer, pp 133–145

Billio M, Getmansky M, Lo AW, Pelizzon L (2012) Econometric measures of connectedness and systemic risk in the finance and insurance sectors. J Financ Econ 104(3):535–559

Birch J, Pantelous AA, Soramäki K (2016) Analysis of correlation based networks representing dax 30 stock price returns. Comput Econ 47(4):501–525

Cocco S, Monasson R, Weigt M (2013) From principal component to direct coupling analysis of coevolution in proteins: low-eigenvalue modes are needed for structure prediction. PLoS Comput Biol 9(8):e1003176

Federal reserve bank of St. Louis. https://fred.stlouisfed.org/

Junior LS, Franca IDP (2012) Correlation of financial markets in times of crisis. Phys A Stat Mech Appl 391(1–2):187–208

Kenett DY , Huang X, Vodenska I, Havlin S, Stanley HE (2015) Partial correlation analysis: applications for financial markets. Quant Financ 15(4):569–578

Kumar S, Deo N (2012) Correlation and network analysis of global financial indices. Phys Rev E 86(2):026101

Mantegna RN, Stanley HE (1999) Introduction to econophysics: correlations and complexity in finance. Cambridge University Press
Mark J Bowick and Édouard Brézin. Universal scaling of the tail of the density of eigenvalues in random matrix models. Phys Lett B 268(1):21–28,
Markowitz H (1971) Portfolio selection diversification of investments. By Harry M. Markowitz. YALE University Press
Nobi A, Lee S, Kim DH, Lee JW (2014) Correlation and network topologies in global and local stock indices. Phys Lett A 378(34):2482–2489
Pandey A et al (2010) Correlated wishart ensembles and chaotic time series. Phys Rev E 81(3):036202
Ren F, Zhou W-X (2014) Dynamic evolution of cross-correlations in the chinese stock market PloS one 9(5):e97711
Tulino AM, Verdú S, et al (2004) Random matrix theory and wireless communications. Found Trends® Commun Inf Theory 1(1):1–182
Wang G-J, Xie C, Stanley HE (2018) Correlation structure and evolution of world stock markets: evidence from pearson and partial correlation-based networks. Comput Econ 51(3):607–635
Yahoo finance database. https://finance.yahoo.com/
Zhu B, Han D, Wang P, Zhanchi W, Zhang T, Wei Y-M (2017) Forecasting carbon price using empirical mode decomposition and evolutionary least squares support vector regression. Appl Energy 191:521–530

Research on Loss Absorption of Financial Group

Morito Hashimoto and Setsuya Kurahashi

Abstract The mechanism of fund transaction conducted between banks is comprised of a network based on multiple banks. The suspension of bank funding directly causes a loss of depositor settlement methods. This subject attracts much attention from the nation because of the possibility of an increase in the national burden if public funds are injected to banks in the event of a financial crisis. Although many researchers have attempted to understand the propagation mechanisms of the crises in Systemic risk, the mechanism is not yet clearly understood. This research proposes a new agent-based modeling method for systemic risk based on a model that uses simple balance sheets and regulations of the financial institution. The model considers liquidity effects on interbank networks and endogenous risks. Using this model, this research validates the effect of bankruptcy in the case of the structure of the financial group in regional banks. In Japan, cases of financial groups, which is a kind of merger, are increasing recently in regional banks in particular, and it is significant to clarify the effect of the resilience of bankruptcy. In this research, our aim is to evaluate the effect of loss-absorbing mechanism that can be taken when banks constituting the financial group falls into a crisis of bankruptcy. And we made it clear that the selection of the banks that make up the financial group has an impact on the loss propagation.

Keywords Agent-based modeling · Systemic risk · Financial group · Network theory

---

M. Hashimoto (✉) · S. Kurahashi
Business science, Tsukuba University, 3-29-1 Otsuka, Bunkyo-ku, Tokyo, Japan
e-mail: Morito.Hashimoto@gmail.com

S. Kurahashi
e-mail: kurahashi.setsuya.gf@u.tsukuba.ac.jp
URL: http://www.gssm.otsuka.tsukuba.ac.jp/

A. S. Chakrabarti et al. (eds.), *Network Theory and Agent-Based Modeling in Economics and Finance*, https://doi.org/10.1007/978-981-13-8319-9_18

1 Introduction

Fund transactions performed between financial institutions through various securities could cause a chain reaction of financial institutions' failures that would destabilize financial markets. Financial institutions are generally under regulations that use net worth ratios that are stipulated by Basel III, which is the uniform international standard for international banks that are published by the Basel Committee on Banking Supervision. This standard serves as one of the measures of financial market stability. However, financial institutions compete intensely for interest rates and various charges because of financial liberalization. They, hence, can have problems with their operating foundations. Amid such circumstances, we must take necessary measures, that assume the failures of financial institutions could happen unexpectedly.

The mechanism of fund transactions conducted between financial institutions comprises of a network of multiple financial institutions. This financial institution network consists of individual interbank transactions. However, an interbank transaction is actually a market that is limited to financial institutions, with banks as market participants. These markets include the call market in Japan, which is a financial market for borrowing and lending short-term funds. This call market serves to adjust financial surpluses or deficits on a daily basis. Other markets, which exist all over the world, include the foreign spot and futures exchange markets. An interbank transaction market consists of financial institutions, money market brokers, and financial authorities. In such a market, transactions are conducted mainly by direct dealings or through money market brokers.

Taking a top-down view of the overall systemic risk, we can find some undiscovered points that are related to the risk of chain reaction failures in fund transaction networks. These have occurred in financial crises such as the savings and loan crisis and the Lehman collapse. In Fig. 1, the vertical axis indicates the number of financial institution failures, and the horizontal axis indicates the year. When comparing Japan and the United States, we see that there is no relationship between the peaks of financial institution failures.

The business suspension of one financial institution risks halt fund transactions and cause a chain reaction. The suspension of bank fund transactions directly causes the collapse of their means of support. The government establishes safety net measures such as the injection of public funds, formulation of deposit insurance systems, and lending by central banks. This subject attracts much attention from the nation because of the possibility of an increase in the national burden if public funds are injected into financial institutions in the event of a financial crisis. To rescue failed financial institutions, the country's Prime Minister must recognize the need for a well-ordered treatment of financial institutions through discussion at the Financial Crisis Response Council. However, it is necessary to carry out careful discussions to determine financial institutions that need rescue and the amount of public fund to inject in them. To do so, the mechanism of interbank networks' chain reaction failure needs to be clarified.

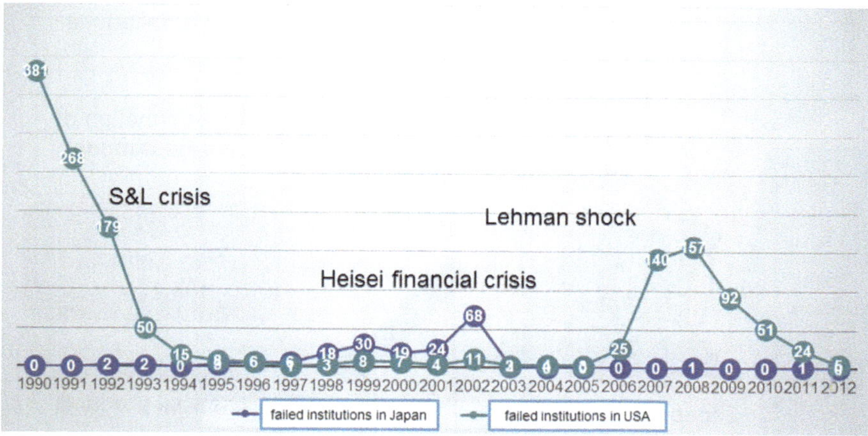

Fig. 1 Comparison of the number of failed financial institutions in Japan and the United States. The vertical axis is the number of failed financial institutions and the horizontal axis is the year. There is a difference in the peaks of the number of failed financial institutions in Japan and the United States

The financial system report of the International Monetary Fund (IMF) issued in June 2016 announced a danger of systemic risk for Deutsche Bank in Europe. The fact that the report named 28 banks with global systemic risks along with their respective degrees of risk as calculated by systemic risk simulation generated substantial debate (Diebold and Yilmaz 2014; Espinosa-Vega and Sole 2010). Research on systemic risk has been increasing in recent years (Prasanna and Sujit 2010; May and Arinaminpathy 2010; Nier et al. 2007; Mattia and Thomas 2015; Chen et al. 2016). Mathematical models, empirical research, and even agent-based models similar to the one presented here have become increasingly popular as research themes.

On the other hand, considering the situations of financial institutions and the direction of regulatory compliance with Basel III, a new Total Loss-absorbing Capacity (TLAC) framework will need to be examined (Shin 2015), and responding to further regulation will be required in Megabanks. In addition, the environments surrounding financial institutions are changing owing to 24 h availability resulting from the evolution of information technologies such as virtual currency (Bitcoins) and the emergence of new businesses in areas such as Fintech. If systemic risk increases, we need to take preemptive measures, such as strengthening regulations on financial institutions, accumulating reserves for deposit insurance, and promoting the consolidation of financial institutions.

As shown in Fig. 2, Japan's financial institutions have been forced to face tough conditions due to negative interest rates, declining birth rates and aging of the population, financial regulations, etc. In particular, regional banks and small- and medium-sized banks, whose scale is not large, are in a difficult situation. Among the new challenges, as shown in the comment of the Bank of Japan President Kuroda on February

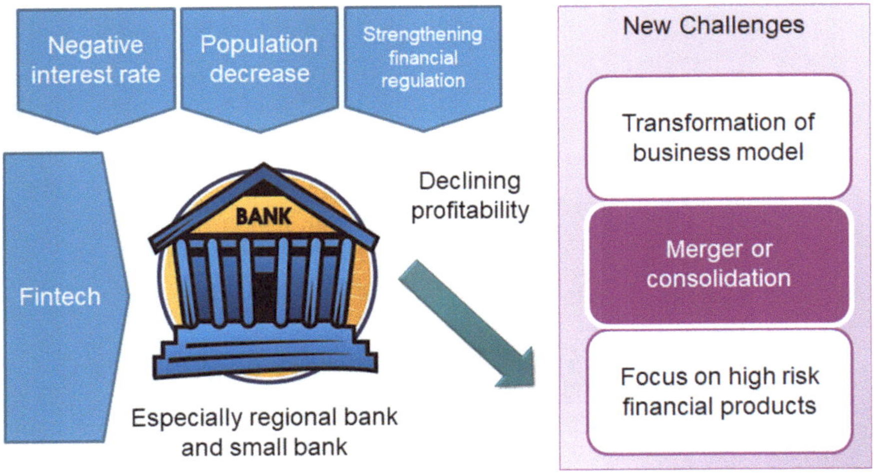

Fig. 2 Japan's financial institutions have been forced to face tough conditions due to negative interest rates, declining birth rates and aging of the population, financial regulations, etc. In particular, regional banks and small- and medium-sized banks, whose scale is not large, are in a difficult situation

16, 2017 (Haruhiko 2017), there are moves to strengthen profitability through mergers or consolidations in local and small banks in Japan.

In Japan's financial institutions, the number of financial institutions has gradually decreased as shown in Fig. 3. There were failures of financial institutions as a factor, but most were due to the mergers or consolidations of financial institutions. For example, in the case of MUFG which is one of the Japanese megabanks, many banks are merged, such as Mitsubishi Bank, Tokyo Bank, Sanwa Bank, and Tokai Bank. However, in recent years, there has been an increase in the number of cases where not only mergers which become one financial institution but individual financial institutions form financial groups that form new holding companies while remaining independent.

In this research, we will focus on the difficult regional banks or small- and medium-sized banks, and examine the impact of the recent increase of this financial group on the financial system. The financial group takes the same form in the megabank, but in these, it is managed in a framework called TLAC (Total Loss-Absorbing Capacity), and the influence on the scale is different, so it is not considered in this research.

The mergers or consolidations of regional banks in Japan has three forms. Its form and features are described in Table 1 and Fig. 4.

In Japan, there are moves to shift from No. 3 to No. 1 in stages, but in recent years, the number of FGs of No. 2 is increasing in particular, and including the group to be established, 15 regional bank holding companies become. The financial group of the regional bank has its own company independently, but it exchanges people and improves management efficiency, and they establish relationships complementing each other.

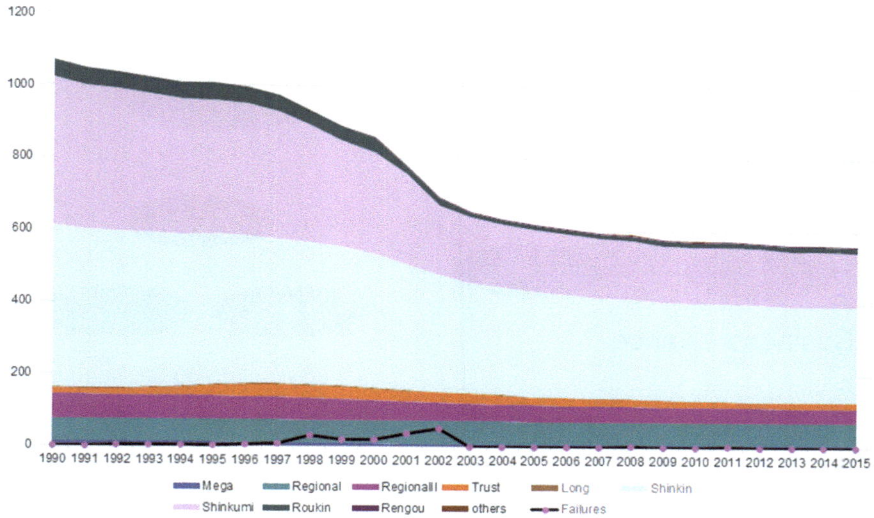

Fig. 3 The black line represents the number of bankruptcies. The vertical axis represents the number of banks and the horizontal axis represents the year. Trends in the number of financial institutions and a number of failures by fiscal year. From this figure, it is necessary to take into consideration the number decreasing due to the failures of financial institutions, but it is assumed that bank consolidation is progressing

Table 1 Merger form in Japanese banks. In recent cases, there are 15 cases of the financial group, and the number is increasing

Merger type	Description	Merit and Demerit	Sample	Case examples
Merger	More than one financial institution becomes one	It takes time until effects come out because different banks will be one	N/A	Tsukuba Bank (Kanto Bank, Tsukuba Bank, Ibaraki Bank 2010)
FG	Management integration by holding company	Postpone organizations and systems, strengthen management base and prioritize management efficiency	15	Hokuhoku FG (Hokuriku bank, Hokkaido bank 2003)
Business alliance	Alliance between independent banks	Establish a jointly funded company and promote common tasks efficiently	5	Chiba-Musashino Alliance (Chiba Bank, Musashino Bank 2016)

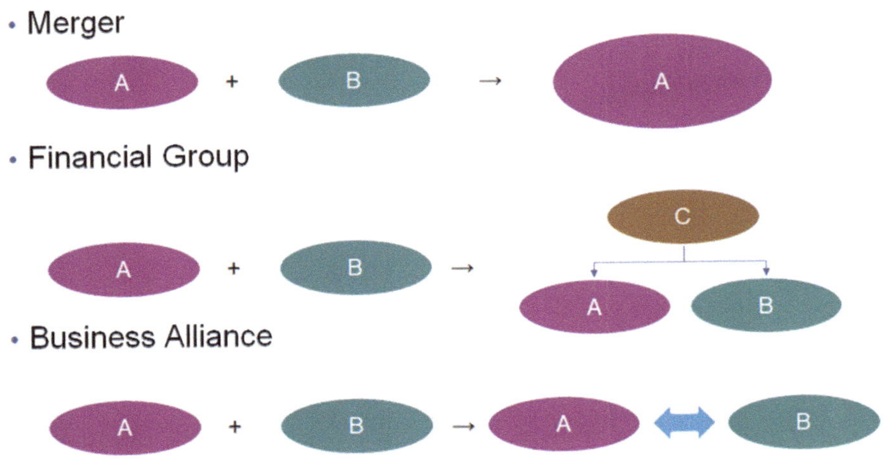

Fig. 4 The financial group of the regional bank has its own company independently, but it exchanges people and improves management efficiency, and they establish relationships complementing each other

In this research, we propose an agent-based model using a simple balance sheet that considers liquidity funds. In addition, as mentioned in a comment by Haruhiko Kuroda, President of the Bank of Japan on February 16, 2017 (Haruhiko 2017), particularly in small- and medium-sized banks in Japan, there is a movement to strengthen profitability through mergers and consolidations, so we examine the impact of the failure of financial institutions due to merger or integration. Consequently, we examine the influence of bank failures caused by mergers or consolidation. There are both positive and negative opinions regarding the effects of mergers or consolidations (Berger Allen et al. 1999), but there have been few evaluations of merger or consolidation effects using an agent-based model.

2 Proposed Method

In this section, we propose an agent-based model based on the balance sheet.

2.1 Modeling of the Balance Sheet

In this research, we extend the balance sheet model introduced by Eisenberg and Noe (2001) and adapt it to an agent-based model. The agent-based model is shown in Fig. 5, and the agent node represents a bank and the link represents interbank lending. When the arrow points from the B node to the A node, it indicates that the

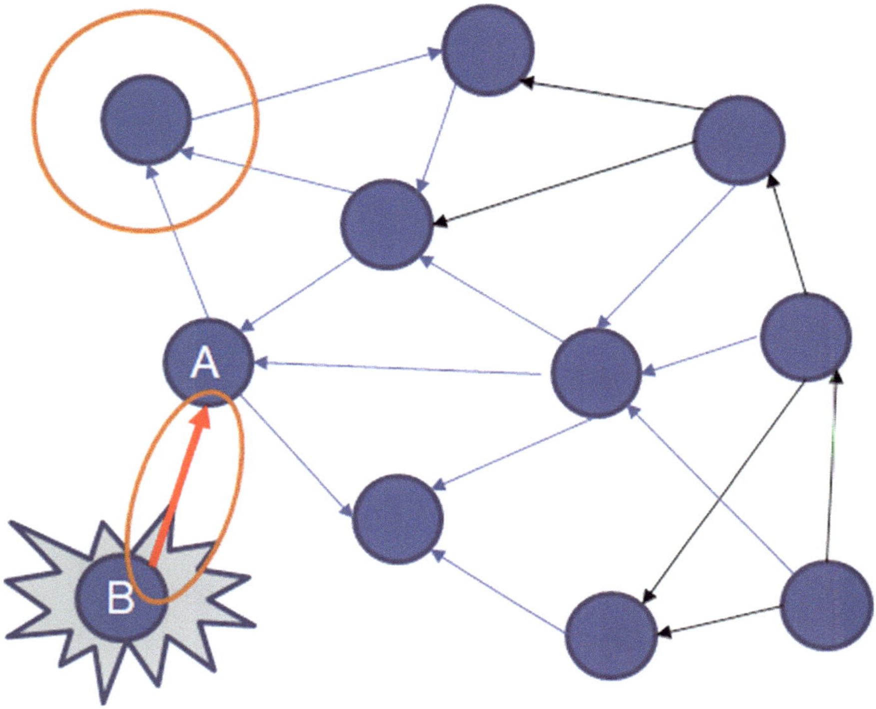

Fig. 5 The structure of Agent-based model

bank A is lending to the bank B. In this case, if Bank B becomes insolvent, Bank A loses an asset. Furthermore, the balance sheet adopted by this model and the account represented in it are shown in Fig. 6.

As the figure shows, the balance sheet contains the following items:

- Liabilities
  - capital(c)
  - deposits(d)
  - Interbank liabilities(Il)
- Assets
  - illiquidity asset(ia): Mortgage loans, individual or corporate loans
  - liquidity asset(la): cash
  - Interbank assets(Ia)

The balance of this balance sheet follows Eq. (1):

$$c + d + Il = ia + la + Ia \tag{1}$$

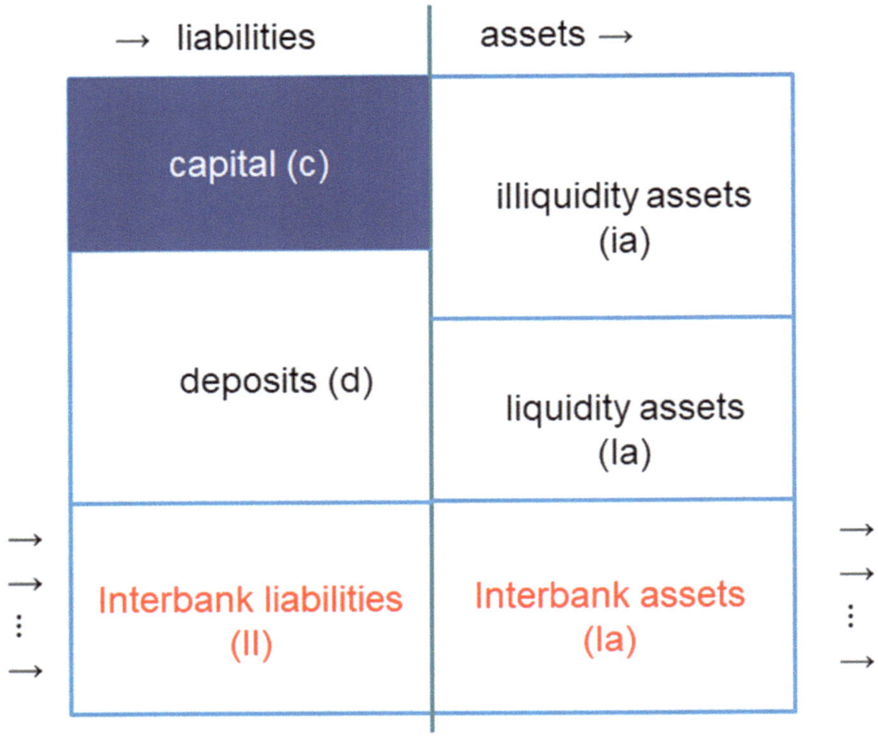

Fig. 6 Balance sheets

2.2 Consideration of the Reserve Ratio (RR) and Capital Adequacy Ratio (CAR)

In this study, we consider the RR and the CAR in the balance sheet. The RR is the proportion of liquid assets held by deposits in the central bank, which are required to be high in developing countries. The liquid assets in central bank is (la_c), and the RR is defined in Eq. (2).

$$ReserveRatio = \frac{la_c}{d} \tag{2}$$

The CAR is a ratio set according to the degree of influence on the systemic risk of financial institutions. An equation representing this ratio is shown in Eq. (3).

$$CapitalAdequacyRatio = \frac{c}{Risk - WeightedAssets} \tag{3}$$

Table 2 Weighting of risk-weighted assets

Account	Ratio (%)
W_{la}: $liquidityassets$	0
W_{iam}: $illiquidityassets : mortgage$	50
W_{ial}; $illiquidtyassets : Loans$	100
W_{ia}: $Interbankassets$	20

The CAR is defined as follows in Table 2, taking Risk-Weighted Assets into consideration. And, the illiquidity asset is defined that the ratio (q) is mortgage and the rest are Loans. In this research, (q) was defined as 50%.

In this research, we defined three types of balance sheets, megabanks, regional banks, and small- and medium-sized banks, and each node was classified according to the degree of each agent in the network model. The balance sheet of this research is shown in Table 3. Here, the degree is set to 25 or more for megabanks, 11–24 for medium banks, and 10 or less for small banks. In addition, the value of each account is the average value of the sum of bank accounts of each type, and a total of 330 banks of financial data were collected. The organized data is shown in Table 4.

2.3 Implementation of Agent-Based Modeling

In this research, we implement our agent-based modeling method. The model updates the balance sheet on a monthly basis in accordance with a time series. And, the soundness of the bank is judged by the capital adequacy ratio. The flow of agent-based modeling is as follows:

(1) "All banks" Check whether the capital adequacy ratio exceeds the specified value (CAR_r: 8%). If the capital exceeds the specified value (8%), the absorption of the shock is completed and go to (F). However, if it is below the specified value (8%), go to (2). The illiquidity asset defines that the ratio (q) is a mortgage and the rest are Loans.

- $CAR_r = \frac{c}{(ia \cdot q \cdot W_{iam} + ia \cdot (1-q) \cdot 1 + IBa \cdot W_{ia})} >= 0.08$

(2) "All banks" The impact of the market will reduce the value of illiquid assets, but here, we define the rate of decline (p) of non-liquid asset value with reference to Cifuentes et al. (2005) If even one bank's management deteriorates, the market deteriorates and the value of the non-liquidity asset is reduced. Capital of bank(i) absorbs shocks and define (c_a) as absorbed capital.

- $p = e^{(-\alpha) \cdot \Sigma il}$
- BS: $c + d + IBl > (1-p) \cdot ia + la + IBa$

Table 3 Balance sheet model of three types: megabanks, regional banks, small- and medium-sized banks

Bank type	Capital	Deposits	Interbank liabilities	Total liabilities	Liquidity assets	Illiquidity asset	Interbank assets	Total assets	CAR (%)	RR (%)
Mega bank	5318680	78007032	31366413	114692126	28244788	55080924	31366413	114692126	11.18	3.62
Regional bank	275786	3983502	604399	4863687	523768	3735519	604399	4863687	9.44	1.31
Small and medium-sized bank	30832	540570	19657	591059	147996	423406	19657	591059	9.59	2.74

Table 4 Source of bank accounts for each type. The accounts were combined for each type of bank, and a total of 330 bank accounts were used

Bank type	Number of banks used for average value calculation	Source of data
Mega bank	5	Japanese bankers association (Zenginkyo)
Regional bank	64	Regional banks association of Japan (Chiginkyo)
Small and medium-sized bank	261	Shinkin central bank research institute

- CAR: $\frac{c}{((1-p)\cdot ia\cdot q\cdot W_{iam} + (1-p)\cdot ia\cdot(1-q)\cdot W_{ial} + IBa\cdot W_{ia})}$

(3) "All banks" We absorb shocks by capital, check if the capital adequacy ratio exceeds the specified value (CAR_r: 8%), absorb the shock if the capital exceeds the prescribed (F), but if it is below, go to the next.

- BS: $(c - p\cdot ia) + d + IBl = (1-p)\cdot ia + la + IBa$
- CAR: $\frac{(c - p\cdot ia)}{((1-p)\cdot ia\cdot q\cdot W_{iam} + (1-p)\cdot ia\cdot(1-q)\cdot W_{ial} + IBa\cdot W_{ia})}$
- $CAR_r = 0.08$

(4) "All Banks" In order to maintain the capital adequacy ratio, we lend out the loan for the ratio (r) and supplement the liquidity assets with the non-liquid asset. Here, the exchange rate from non-liquid assets to liquid assets is defined as (s). In this research, (s) was defined as 100%.

- BS: $(c - p\cdot ia) + d + IBl = (1-r)\cdot(1-p)\cdot ia + (la + s\cdot r\cdot ia) + IBa$
- CAR: $\frac{(c - p\cdot ia)}{((1-r)\cdot(1-p)\cdot ia\cdot q\cdot W_{iam} + (1-p)\cdot ia\cdot(1-q)\cdot W_{ial} + IBa\cdot W_{ia})}$

(f) "All banks" When the capital adequacy ratio falls below the specified value (CAR_e: 0%) If the applicable bank fails and banks doing loans or purchasing shares between banks fail, bankruptcy bank-to-bank assets loaning to the bank that you did are lost. Let the number of banks doing loans be (n). In addition, in the event that the borrowing bank fails, all bank-to-bank assets will not be lost, but in the short term the asset cannot be used, so we will lose it in this study. After this failed process finishes, return to No. 1.

- BS: $c + d + IBl = ia + la + IBa\cdot((n-1)/n)$
- CAR: $\frac{c}{(ia\cdot q\cdot W_{iam} + ia\cdot(1-q)\cdot W_{ial} + IBa\cdot W_{ia}\cdot((n-1)/n))}$

2.4 Strengthen Management Foundation by Regional Bank Financial Group

As described above, we verify that the management base will be strengthened by forming a financial group at regional banks. In the financial group of regional banks, each bank in the financial group is maintained individually, but the merits of the financial group are as follows:

1. Compared with the merger, there is no need for the reorganization, share operation, integrate the system, etc., speed is fast.
2. Financial groups are often formed in neighboring areas, and it is possible to do good, such as making basic manuals common, while conducting marketing strategies that were in local areas individually.
3. In the situation of the management crisis, it is possible for banks in the financial group to transfer funds with low-interest rates.

Regarding No.3, there are concerns about Arms-length rules, and regulations are made so as not to impair the soundness of bank management through the conflict of interest transactions between banks and companies within the bank group. But, if it falls into a management crisis and in the situation where reconstruction assistance is necessary, it is possible to receive support as it is an inevitable situation. The CAR at that time shall be stipulated as 4%, from which the Financial Service Agency will trigger early corrective measures. Therefore, add No.5 after No.4 in the modeling flow as described above as follows:

(5) "All banks" If the capital adequacy ratio is below the specified value (CAR_r: 4%) and there are banks belonging to the same financial group, by having the purchase of the stock of the prescribed amount (z), Capital and liquidity assets increase to improve capital adequacy ratio. After that, we go to (F), including those not below 4%. In this research, (z) was defined as 0.3% of liquidity. The result of arranging the movement of loss absorption by the capital adequacy ratio is shown in Fig. 7.

- A bank that belongs to a financial group and needs reconstruction assistance
  - BS: $((c+z)-p\cdot ia)+d+IBl=(1-r)\cdot(1-p)\cdot ia+(la+s\cdot r\cdot ia+z)+IBa$
  - CAR: $\dfrac{((c+z)-p\cdot ia)}{((1-r)\cdot(1-p)\cdot ia\cdot q\cdot W_{iam}+(1-p)\cdot ia\cdot(1-q)\cdot W_{ial}+IBa\cdot W_{ia})}$
- The bank which belongs to the financial group and carries out reconstruction assistance
  - BS: $(c-p\cdot ia)+d+IBl=(1-r)\cdot(1-p)\cdot ia+(la+s\cdot r\cdot ia-z)+(IBa+z)$
  - CAR: $\dfrac{(c-p\cdot ia)}{((1-r)\cdot(1-p)\cdot ia\cdot q\cdot W_{iam}+(1-p)\cdot ia\cdot(1-q)\cdot W_{ial}+(IBa+z)\cdot W_{ia})}$

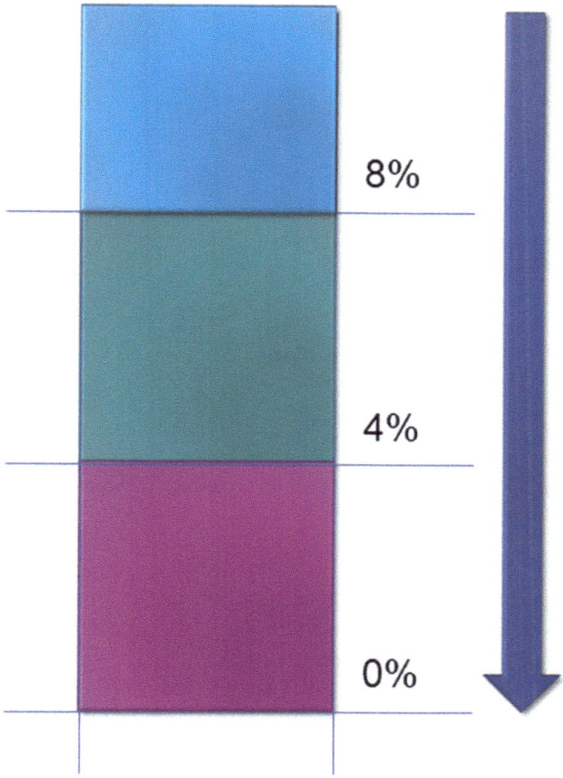

Fig. 7 More than 8%: safety areas, Less than 8%: convert illiquid assets into liquid assets by collecting loans with maturity, Below 4%: if the bank is one of the banks that make up the financial group, it will inject liquidity between the financial group

3 Verification Results

In this section, we explain the verification results obtained through the agent-based modeling conducted in this research.

3.1 Verification of Change in the Number of a Chain Reaction of Failures Resulting from Financial Groups of Regional Banks

In this study, we created 10 scale-free networks of 500 nodes with the number of edges (m) to be set to 3, and we defined them as a financial transaction network model of financial institutions. In that network, the 500 nodes represent financial institutions. Assuming a case in which one of the nodes failed, we measured the chain of bankruptcies of the financial institutions at that time. The procedure at that time is shown below. However, the directed link between the nodes in the network

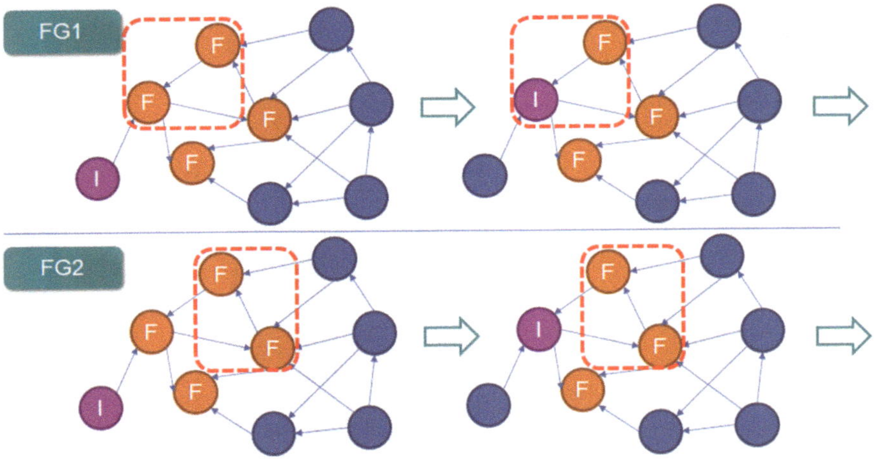

Fig. 8 The verification flow of this research. "I" represents an Initially failed node and "F" represents a Failed chains node

represents the financial transaction of the financial institution. In addition, in this model, a framework is built based on Hashimoto and Kurahashi (2017), and the basic behavior has been verified.

1. For one network, the node with the most chain reactions is selected as the initially failed node (specifically, loss of illiquid asset occurs).
2. Collect a number of failed chain reactions for each combination pattern of "FG".

In addition, the flow of agent-based simulation at that time is shown in Fig. 8, but since there are 500 financial institutions per network and 250 combination patterns of FG, 125,000 verifications are carried out and since there are 10 networks, a total of 1,250,000 verifications are carried out.

We investigate the change in the number of failed chains by constructing financial institutions as the main task of this research into the financial group. As described above, since nodes with the most failed chains are selected as initially failed nodes, The financial institution where the first loss occurs is fixed for each network. First, it shows how the chain reaction number of failures changed by constructing a financial group in Table 5. In all networks, a pattern in which the number of failure chain reactions decreased by one was observed, and in three networks, a pattern in which two chains of it were reduced was observed.

Here, the configuration of the financial group that reduces the chain reaction of failures is shown in the Table 6. In this table, some initially failed nodes are selected as one of the FG pairs.

Table 5 Verification of change in the number of a chain reaction of failures resulting from financial groups of regional banks. Patterns in which the number of a chain reaction of failures decreased by one in all networks were shown, and patterns in which the number of failures decreased by two in three networks were shown

Network no.	Number of failed chains before configuring FG (node number that failed first)	Difference of the number of failed chains using FG (number of FG patterns)
1	5(484)	−1(78), −2(2)
2	3(71)	−1(24)
3	4(492)	−1(74)
4	4(292)	−1(36)
5	5(375)	−1(62), −2(2)
6	6(454)	−1(28)
7	4(469)	−1(92), −2(2)
8	4(217)	−1(28)
9	4(399)	−1(48)
10	4(53)	−1(38)

Table 6 The configuration of the financial group that reduces the chain reaction of failures. Some initially failure nodes are selected as one of the FG pairs. Here, ∗ indicates the initial failed node

Network no.	Change in chain reaction by FG (number of nodes)	Node number when the failed chains decreased by two (degree)	Major FG nodes with reduced chain reactions (number of times configuring FG pairs)
1	−1(78), −2(2)	459(8), ∗484(27)	459(18), ∗484(17), 151(4), 172(3), 209(2), 272(2)
2	−1(24)		∗71(8), 280(5)
3	3(74)		∗492(22), 384(10), 375(7)
4	−1(36)		∗292(12), 136(5), 108(3), 483(2)
5	−1(62), −2(2)	∗375(14), 479(12)	∗375(14), 479(12), 429(7), 213(2), 233(2)
6	−1(28)		449(9), ∗454(6)
7	−1(92), −2(2)	∗469(18), 496(26)	496(26), ∗469(18), 426(5), 28(2), 165(2), 485(2)
8	−1(28)		379(7), ∗217(6), 92(3)
9	−1(48)		∗399(14), 275(6), 405(6)
10	−1(38)		377(10), ∗53(7), 347(4)

3.2 Evaluation of Verification Result

The configuration of the financial group that reduces the number of failed chains in the No.1 network is as shown in Fig. 9. In this network, choosing two nodes effectively reduces the number of failed chains, one of which is the initially failed node. If such a node is structured as a financial group, we can reduce the risk of chain reactions in the entire network, so we will consider indicators to identify the nodes to be included in the financial group.

In this research, in order to verify the network characteristics of the nodes of the financial group that reduces the number of failed chains, four network indices, betweenness centrality Fig. 10, eventness centrality Fig. 11, degreeness centrality Fig. 12, and pagerank Fig. 13, were collected in the No.1 network. But it was impossible to explain the nodes that sufficiently reduce the number of failed chains for every index. The correlation coefficient at this result is shown in Fig. 14. Among the four network indices, the evcentness centrality is the best value, but it is around 0.12 and can not be said to be a sufficiently good value.

Here, we examine Fig. 13 in detail, in the graph of pagerank, the node that reduces the number of failed chains was found to be located where the value of pagerank is close to zero. Since this tendency is possibly characteristic of this network No. 1 individually, for the 10 networks, the relationship between the node of the financial group that reduces the number of failed chains and pagerank is shown in Fig. 15.

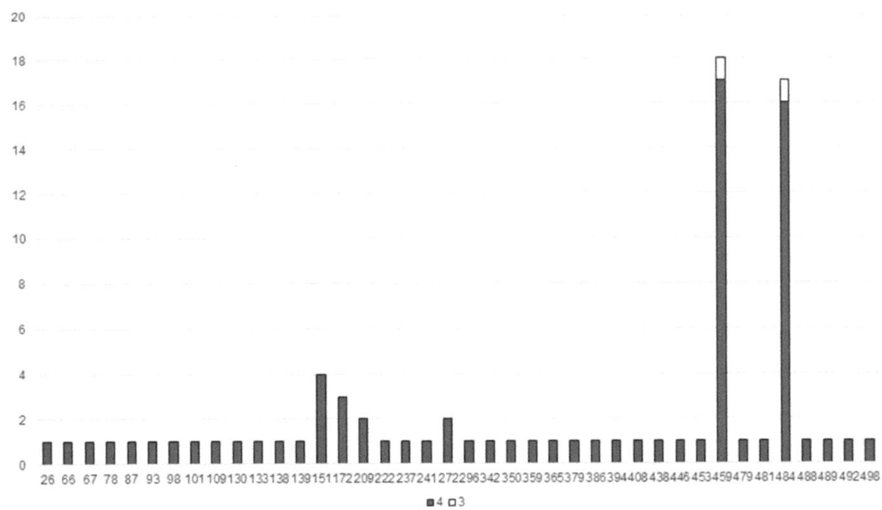

Fig. 9 The configuration of the financial group that reduces the number of failed chains in the No.1 network. The vertical axis represents the number of times configuring FG, and the horizontal axis represents the node number

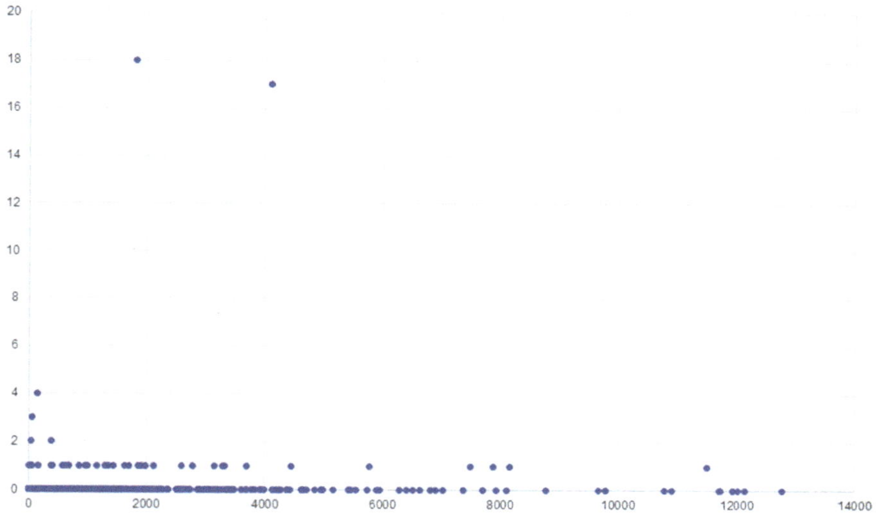

Fig. 10 Four network indices, betweenness centrality, degreeness centrality, evcentness centrality, and pagerank, were collected in the No.1 network. The vertical axis represents the number of times configuring in the financial group, the horizontal axis represents betweenness centrality

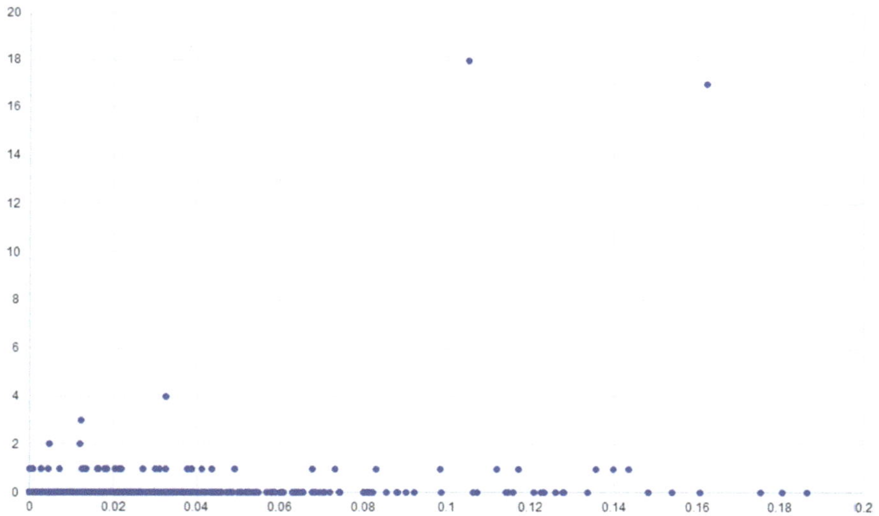

Fig. 11 The vertical axis represents the number of times configuring in the financial group and the horizontal axis represents evcentness centrality

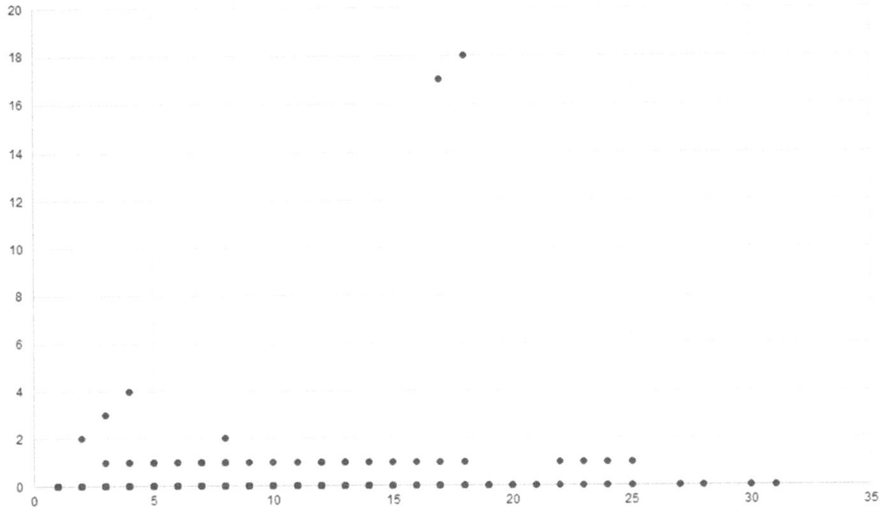

Fig. 12 The vertical axis represents the number of times configuring in the financial group and the horizontal axis represents degreeness centrality

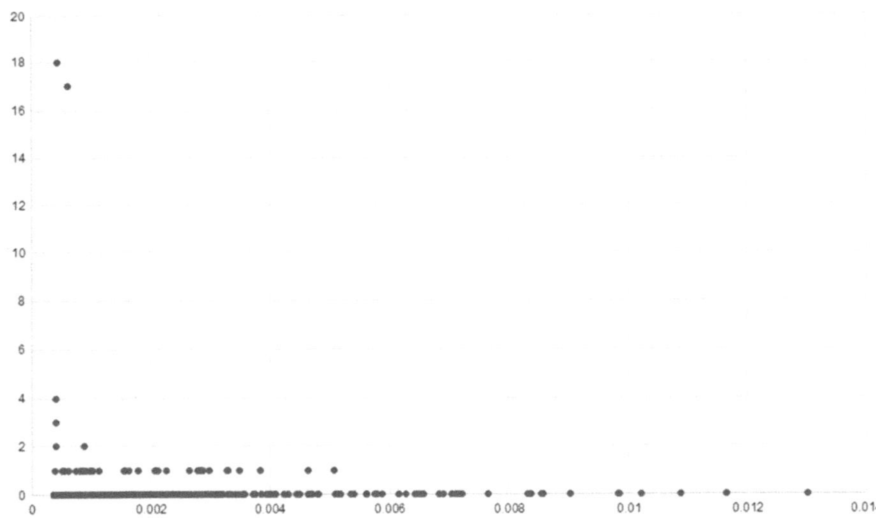

Fig. 13 The vertical axis represents the number of times configuring in the financial group and the horizontal axis represents pagerank. The node that reduces the number of the failed chain was found to be located where the value of pagerank is close to zero

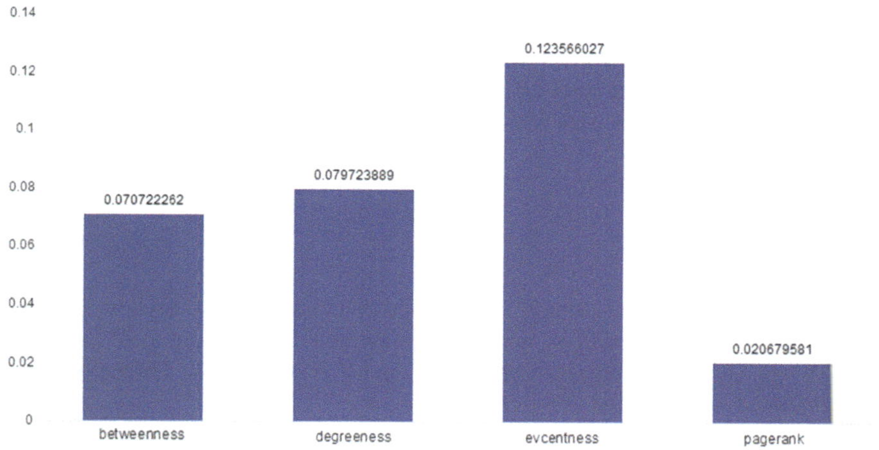

Fig. 14 The correlation coefficient at this result between four network indices and a node that reduces the number of failed chains in the No.1 network. The vertical axis represents the correlation coefficient and the horizontal axis represents four network indices

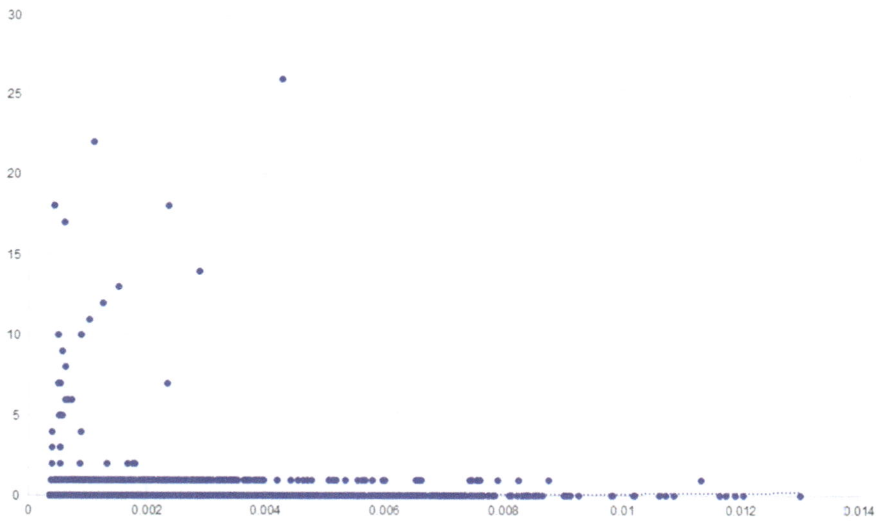

Fig. 15 For the 10 networks, the relationship between the node of the financial group that reduces the number of failed chains and pagerank. The vertical axis represents the number of times configuring in the financial group and the horizontal axis represents pagerank

From Fig. 15, it was found that in all 10 networks, the node that reduces the number of failed chain reactions tends to be located near the value of pagerank. From the above results, it was found that the number of failed chain reactions can be reduced by selecting a node with a page rank of 0.005 or less as a node of the financial group.

4 Conclusion

In this paper, we proposed a simple agent-based modeling method based on a balance sheet for interbank transactions, and we explained the chain reactions of failures in terms of an equilibrium constraint based on the CAR and the RR. Three points can be mentioned as the results of the research.

- We conducted agent-based modeling using a balance sheet model that takes equilibrium constraints based on the CAR and the RR into consideration.
- By configuring financial group at the initially failed node, the risk of a chain reaction of failures is reduced.
- For nodes configuring a financial group, nodes with pagerank 0.005 or higher should not be selected.

It can be assumed that reducing the risk of a chain reaction of failures by configuring financial group at the initially failed node can reduce the influence of diffusion by absorbing shock at the beginning of loss occurrence.

In addition, since nodes with large page ranks are expected to propagate easily if such nodes are included in the FG, it is assumed that the failure of chain risk will increase.

Financial assistance for injecting public funds into financial institutions is 140 billion yen per financial institution, and for the capital increase as well an average of 200 billion yen per financial institution is used. In practical terms, discrimination is required to provide financial assistance to financial institutions where the number of bankruptcies becomes large if bankruptcy occurs, but according to the results of this research, improvement of the discrimination accuracy of financial institutions that increase the number of a chain reaction of failures can be expected. This is a major economic effect in the sense that it proposes risk indicators to improve the certainty of injecting public funds, considering the characteristics of financial network analysis by Imakubo and Soejima 2008.

Finally, we will discuss future developments. In this research, we focus on maintaining the capital adequacy ratio as the base constraint, but by using this framework, we can estimate LCR (Liquidity Coverage Ratio), and consideration of the impact of increasing of the interest rate by central banks, etc. can be incorporated into the model. Increasing indices unnecessarily increases the complexity of the model and there is a risk that the characteristics will be diminished and consideration must be given to the construction of the model but we would like to expand and focus on the target index in the future.

References

Berger Allen N, Demsetz Rebecca S, Strahan PE (1999) The consolidation of the financial services industry: causes, consequences, and implications for the future. J Bank Financ 23(2–4):135–194
Chen N, Liu X, Yao DD (2016) An optimization view of financial systemic risk modeling: network effect and market liquidity effect. Oper Res
Cifuentes R, Ferrucci G, Shin HS (2005) Liquidity risk and contagion. J Eur Econ Assoc 3(2–3):556–566
Diebold FX, Yilmaz K (2014) On the network topology of variance decompositions: measuring the connectedness of financial firms. J Econom 182(1):119–134
Eisenberg L, Noe TH (2001) Systemic risk in financial systems. Manag Sci 47(2):236–249
Espinosa-Vega MA, Sole JA (2010) Cross-border financial surveillance: a network perspective. IMF Working Papers, p 1–27
Hashimoto M, Kurahashi S (2017) Management integration impact analysis of financial institutions based on a fund transaction network model. In: Social simulation conference
Imakubo K, Soejima Y (2008) Funds trading network of the call market. Stud Financ 27:47–99
Konno N, Masuda N (2010) Complex network. Kindai Kagaku-sya
Kuroda H (2017) Building a more robust financial system: where are we after the global financial crisis and where do we go from here? In: Speech at the DICJ-IADI international conference
Mattia M, Thomas L (2015) Hubs and resilience: towards more realistic models of the interbank markets. Some recent developments, Banking integration and financial crisis
May RM, Arinaminpathy N (2010) Systemic risk: the dynamics of model banking systems. J R Soc Interface R Soc 7(46):823–838
Nier E, Yang J, Yorulmazer T, Alentorn A (2007) Network models and financial stability. J Econ Dyn Control 31(6):2033–2060
Prasanna G, Sujit K (2010) Contagion in financial networks. Bank Engl Q Bull 50(2):124
Shin HS (2015) Risk and liquidity. Oxford university Press Inc

Knowledge-Driven Approaches for Financial News Analytics

Bikesh Raj Upreti, Philipp Martin Back, Pekka Malo, Oskar Ahlgren and Ankur Sinha

Abstract Computational finance is one of the fastest-growing application areas for natural language processing technologies. Already today, algorithmic trading funds are successfully using robo readers and sentiment analysis techniques to support adaptive algorithms that are capable of making automated decisions with little or no human intervention. However, these technologies are still in a nascent state and the competition to improve approaches within the industry is fierce. In this chapter, we discuss financial news analytics and learning strategies that help machines combine domain knowledge with other linguistic information that is extracted from text sources. We provide an overview of existing linguistic resources and methodological approaches that can be readily utilized to develop knowledge-driven solutions for financial news analysis.

1 Introduction

With the growth of digital media, news analytics has been gaining wide traction in modern-day trading for two important reasons. First, a large body of literature has documented the relationship between news in media and the sentiment of investors

B. R. Upreti (✉) · P. M. Back · P. Malo · O. Ahlgren
Department of Information and Service Economy, Aalto University School of Business, Helsinki, Finland
e-mail: bikesh.upreti@aalto.fi

P. M. Back
e-mail: philipp.back@aalto.fi

P. Malo
e-mail: pekka.malo@aalto.fi

O. Ahlgren
e-mail: oskar.ahlgren@aalto.fi

A. Sinha
Production and Quantitative Methods, Indian Institute of Management Ahmedabad, Ahmedabad 380015, India
e-mail: asinha@iima.ac.in

A. S. Chakrabarti et al. (eds.), *Network Theory and Agent-Based Modeling in Economics and Finance*, https://doi.org/10.1007/978-981-13-8319-9_19

and traders; with the sentiment ultimately reflecting on financial variables such as stock returns (Chen et al. 2013; Garcia 2013; Vu et al. 2012), risk (Yu et al. 2013; Groth and Muntermann 2011), takeover success (Buehlmaier 2013), and performance of IPOs (Ferris et al. 2013). Second, classical methods for studying and analyzing market sentiment, such as regular surveys, are restraining as their results become quickly outdated and because the entire survey setup is strongly controlled by the designers. In contrast, a more promising approach for retrieving market sentiment is based on financial news analysis (Barber and Odean 2008; Dzielinski et al. 2011; Mitra and Mitra 2011; Tetlock et al. 2008).

The anxiety resulting from the global financial downturn has further steered news analytics toward the study of market sentiment. It is a question of behavioral finance and mass psychology, where actual news information and pure speculation are entwined. Recently, behavioral market models (see, for instance, Kumar and Lee (2006), Kaniel et al. (2005), Shiller (2003) or Hirshleifer (2001)) have gained strong acceptance as an approach for explaining the dynamics of modern financial markets with various quantifiers. Since economies are driven by ordinary people who operate on a micro level, it is natural to expect their awareness and behavior to be eventually driven by the continuous flow of news and its sentiments. When uncertainty is overwhelming and fear takes over, people often overreact en masse. The media—from traditional newspapers to social media—plays a substantial role in driving such mass psychology. News sentiments often become opinions that cause agents to interact, integrate and act in a manner that would be difficult to explain without understanding the interplay of news and these network of agents. Especially social media allows agents to share their views and beliefs seamlessly in real time. The flow and diffusion of such information within a network often contain spatial and temporal footprints that provide rich opportunities to analyze patterns, to discover relationships, and to further understand the agents' behavior.

One of the key challenges in behavioral finance is to model market sentiments in a way that permits quantitative time series analysis as part of broader statistical models (see for example Schmeling (2007) and Baker and Stein (2004)). In recent years, the rapid development of natural language processing technologies has paved the way for semiautomated tools for quantitative text analysis. The field of natural language processing is thereby closely connected to computational linguistics, where the goal is to develop algorithms and software for the intelligent analysis of natural language. The use of machine learning and Bayesian statistics makes this a powerful approach that allows models to adapt to new information. The Bayesian approach to natural language processing is well motivated by the fact that sentences and individual words tend to be ambiguous, and that even the language itself is continuously developing.

Over the last decade, sentiment analysis has evolved into a highly multidisciplinary field. Most of the published work has, however, focused on the analysis of product reviews, informal text in social media, and opinionated blog writings. Surprisingly, only a few academic studies have investigated how sentiments should be modeled in the domain of economics and finance. Perhaps one of the most plausible explanations for this lack of scientific studies is the strong domain dependence of sentiment analysis, coupled with the difficulty of finding high-quality linguistic resources. An expression with a clear polarity (sentiment) in one domain may have a different interpretation when considered in another domain. In practice, this means that the background information (such as lexicons and ontologies) that is used to power sentiment models needs to be separately adjusted for different applications. As pointed out by Loughran and McDonald (2011), many concepts that are considered negative in general contexts (for instance "liability") are best viewed as neutral in finance, unless their polarity is modified by additional expressions. Another important reason for the relatively small amount of academic research on financial sentiment analysis may be the high commercial value of well-performing sentiment models. It does not come as a surprise that many of the operational sentiment models in the domain of finance are company-owned "black boxes" and very little is known about their actual implementations.

In this book chapter, we provide an overview of recent developments in the domain of financial news analytics, with a special focus on sentiment analysis and event detection. The chapter is organized into five sections: in Sect. 2, we discuss the methods used in financial news analytics. Section 3 highlights the financial and economic impacts demonstrated by research in financial news analytics. Next, in Sect. 4, we provide an overview of resources that are instrumental in constructing financial news analytics. Finally, in Sect. 5, we discuss the specific needs of the financial domain and further present one possible approach to address such needs.

2 Methods in Financial News Analytics

The main objective of any news analytics system is to extract knowledge, information, and insights from the news—especially textual news—in order to assist decision making. In finance, two streams of methods related to sentiment analysis and event detection are prominent. In this chapter, we provide a brief overview of news analytics methods in the financial domain.

2.1 Sentiment Analysis

Financial sentiments computed from the text have been used frequently to demonstrate the value of text-expressed sentiment in financial analysis. Table 1[1] lists some of the recent studies that have utilized financial sentiments derived from news articles for financial models. Researchers have used various news outlets such as *The Wall Street Journal* (Engelberg 2008; Tetlock et al. 2008), *The New York Times* (Garcia 2013), the *Dow Jones News Service* (Engelberg 2008; Tetlock et al. 2008), the *Thomson Reuters NewsScope Service* (Sinha 2010), 10-K (Li 2010), and analyst reports (Huang et al. 2014). Based on our analysis of the literature, we found that

[1] In the Table Sem/Syn refers to "Semantics Consideration/Syntactical Consideration".

Table 1 Applications of textual information in financial research (adapted from Nassirtoussi et al. (2015))

Literature	Algorithm	Sem/Syn	Tools
Wuthrich et al. (1998)	k-NN, ANNs, Naïve Bayes, rule-based	Yes/No	NA
Peramunetilleke and Wong (2002)	Rule classifier	Yes/No	NA
Das and Chen (2007)	Combination of different classifiers	Yes/Yes	General Inquirer
Soni et al. (2007)	SVM (standard linear kernel)	Yes/No	LibSVM package
Zhai et al. (2007)	SVM (Gaussian RBF kernel and polynomial kernel)	Yes/No	NA
Tetlock et al. (2008)	OLS Regression	Yes/No	Harvard-IV-4 psychosocial dictionary
Schumaker and Chen (2009)	SVM	Yes/Yes	Arizona Text Extractor (AzTeK) and AZFin Text
Huang et al. (2010)	Weighted association rules	Yes/Yes	NA
Lugmayr and Gossen (2012)	SVM	Yes/No	SentiWordNet
Mittal and Goel (2012)	Linear Regression, Logistic Regression, SVM, and Self Organizing Fuzzy Neural Network (SOFNN)	Yes/No	LibSVM package, POMS questionnaire and SentiWordNet
Schumaker et al. (2012)	SVR	Yes/No	OpinionFinder
Vu et al. (2012)	C4.5 Decision Tree	Yes/Yes	CRF++ toolkit, Firehose, TST, CMU POS Tagger, AltaVista
Hagenau et al. (2013)	SVM (linear kernel), SVR	Yes/Yes	NA
Yu et al. (2013)	Naïve Bayes	Yes/No	Open-source NLTK

the majority of financial studies seems to prefer dictionary-based approaches over machine learning-based methods. In this section, we present some notable results from studies that focused on the analysis of sentiment outputs from textual analyses. These studies can be broadly categorized as either lexicon-dependent or machine learning-dependent and we specifically highlight the computational approaches that were used to extract financial sentiments.

The lexicon-based sentiment extraction approach has garnered a profound application in the financial domain. To assign a sentiment to the news, the lexicon-based method relies on the use of sentiment dictionary or on the analysis of the text corpus (Medhat et al. 2014). In contrast, the dictionary-based method treats documents as a bag-of-words, i.e., as a collection of unordered words and relies on mapping the words into sentiment categories. The words and their respective sentiment categories or scores are predefined in the form of dictionaries (Li 2010). Similarly, corpus-based methods, like dictionary-based methods rely on lists of sentiment seed words. However, to assign the sentiment to the news text, the corpus-based method seeks to exploit the linguistic characteristics, co-occurrences, and semantic relationships among the words in the corpus and sentiment seed words. Compared to dictionary-based methods, this approach offers more flexibility in modeling domain-specific text. Due to the ease of use and efficiency in computation, dictionary-based methods are popular in business, finance, and economic applications (Kearney and Liu 2014). General Inquirer (GI) (2015) and Diction (Hart 2000) are some of the popular dictionaries used for sentiment analysis. Despite their popularity, dictionary-based approaches provide limited flexibility and lead sometimes to inaccurate estimations; they are therefore not enough to cover domain-specific semantic needs. In this regard, corpus-based methods provide an edge in modeling domain-specific text and can potentially lead to better learning, yet the effectiveness of this method is hugely influenced by the starting list of seed words.

In financial news analytics, researchers have applied various lexical resources; see, e.g., the financial sentiment dictionary by Loughran and McDonald (L&M) (Chen et al. 2013; Liu and McConnell 2013; Ferguson et al. 2014; Garcia 2013), a customized dictionary that combines a general-purpose dictionary like the Harvard-IV psychological dictionaries with L&M (Carretta et al. 2011; Ferris et al. 2013; Engelberg et al. 2012; Ozik and Sadka 2012), a customized dictionary for earning specific data (Price et al. 2012). These studies have covered a wide range of information sources from social media (Chen et al. 2013) and conventional media-based articles (Carretta et al. 2011; Engelberg et al. 2012; Garcia 2013; Liu and McConnell 2013; Ozik and Sadka 2012) to IPO prospectuses (Jegadeesh and Wu 2013) and transcriptions of earning conference calls (Price et al. 2012).

In addition to the lexicon-based method, financial sentiment analysis has also benefited from the advancements in machine learning and computational science. These methods are capable of learning classification rules even from complex context, given that appropriate feature sets and training data are provided. This suite of methods draws heavily on statistical techniques to infer sentiments from the text (Li 2010) and largely combines both sentiment lexicons and learning algorithms. In this setup, sentiment lexicons provide the necessary prerequisite for preparing the data for the learning algorithms. Although both supervised and unsupervised methods have their fair share in sentiment analysis, supervised methods remain more common in research and practice. However, supervised learning methods require human-annotated or -labeled sentiment data from which the algorithm can learn to generalize the sentiment classification rule. The range of supervised learning methods in financial news analytics is profound and spans from Naïve Bayesian (Li 2010; Huang et al. 2014; Yu et al. 2013; Buehlmaier 2013), to decision trees (C4.5) (Vu et al. 2012), SVMs (Groth and Muntermann 2011), SVR (Schumaker and Chen 2009), KNN (Groth and Muntermann 2011), and neural networks (Pagolu et al. 2016; Day and Lee 2016; Mittal and Goel 2012).

2.2 Event and Topic Recognition

Another stream of research aims at identifying important events that influence the market reaction. Financial economists hold the belief of information-efficient markets (Fama 1965), where stock prices fluctuate in response to unexpected events, reported in public news. Financial economists have long studied the dynamics of how markets react to new information. Early works in the form of event studies have measured the impact of a specific event on the value of a firm (see, e.g., Ball and Brown (1968), Fama et al. (1969), Mandelker (1974)). Measuring the market impact of specific news events is, however, a labor-intensive process that spans from the identification of a given event, over the estimation of abnormal stock returns (to isolate individual stock returns from the general market movement), to conducting statistical tests that verify an event's economic significance (Jacobs et al. 2018). It is thus of little surprise that researchers have studied ways for automatic event detection from text since the 90s (Allan et al. 1998). Early *topic detection and tracking* (TDT) methods, where news sources are analyzed to capture new events as well as recurring event coverage, were data-driven, based on clustering, and designed to identify any kind of event (Yang et al. 1999).

A related area of research, but with roots in the domain of information extraction, is *event recognition*. In contrast to TDT, event recognition aims at identifying and extracting more specific event information, including explicit semantic interpretation (Rönnqvist and Sarlin 2017). As the information of interest is often domain- or task-specific, event recognition methods tend to require extensive expert guidance for annotating text, handcrafting rule sets, encoding linguistic patterns, or creating ontology knowledge bases. Despite these requirements, the majority of automatic event recognition techniques remains knowledge- or pattern-based (Arendarenko and Kakkonen 2012; Hogenboom et al. 2013; Du et al. 2016). An example of such systems within the finance domain is the Stock Sonar project by Feldman et al. (2011), who provide a tool for rule-based stock sentiment analysis for which event rules are manually crafted by domain experts. Boudoukh et al. (2018) successfully used this technology to demonstrate the impact of firm-level news publications on stock market volatility. After the automatic classification of news articles into topics (e.g., products, acquisitions, analyst recommendations), only certain event types were considered relevant trading signals. Out of 1.9M stories from the Dow Jones Newswire between 2000 and 2009 for all S&P500 companies, only 50% were considered relevant events. This division into relevant and irrelevant news significantly changed the understanding of the connection between stock price changes and news. Days with relevant news were 31–34% more likely to yield extreme returns, whereas moderate returns were slightly more likely to occur on days with no or irrelevant news (Boudoukh et al. 2018).

Ben Ami and Feldman (2017) further built on the Stock Sonar technology (Feldman et al. 2011) to formulate event-based trading strategies. They found that event information produced noisy signals when used aggregately; for example, as done by Boudoukh et al. (2018) who assume all news within the relevant class to be equal.

Ben Ami and Feldman (2017) argue that event types should be used as isolated trading signals. They obtained superior returns when considering different event types for different investment periods. Other approaches for event-driven stock price prediction conceptualize financial event recognition as the extraction of event tuples (Ding et al. 2015) or as semantic frame parsing (Xie et al. 2013).

Despite a growing number of applications, knowledge-based event recognition does have its challenges. First, the manual creation of rules and ontologies is a difficult and time-consuming task. Second, a strict rule set often causes low recall scores, as rules tend to cover merely a subset of the manifold ways in which information can be lexicalized (Jacobs et al. 2018). Finally, rule-based systems that rely on expert guidance are less applicable to new domains that offer fewer resources to target specific tasks (Rönnqvist and Sarlin 2017). Consequently, researches have explored more flexible, data-driven technologies for automatic event recognition that do not require predefined rules or ontologies.

Rönnqvist and Sarlin (2017) provide a noticeable example of how data-driven, weakly supervised event detection can be successfully used to identify banks in financial distress. They utilize a deep learning approach for detecting relevant discussion in text and for extracting human-readable descriptions of events. After analyzing 6.6M news articles, they identified 243 events of bank distress and government interventions. Their indicators cannot only signal the level of bank distress but also provide explanations. A second example comes from Jacobs, Lefever and Hoste (2018) who utilize a supervised classification approach for company-specific economic event detection in English news articles. The authors demonstrate that data-driven approaches obtain good recall and can capture variation in lexicalizations of events to a satisfactory extent. Nevertheless, Jacobs et al. (2018) also state that more annotated data and augmentative resources are needed to offset ambiguous event expressions. Manual feature engineering appears to pay off for economic event extraction, for example by integrating additional linguistic information by including semantic knowledge from structured resources such as DBpedia and dedicated ontologies for economics (e.g., the NewsEvent ontology (Lösch and Nikitina 2009) and derived CoProE ontology (Kakkonen and Mufti 2011) as well as syntactic information extracted from dependency parses.

Data-driven technologies for automatic event recognition are, however, predominantly found in general domain event extraction (see, e.g., Ahn (2006), Mitamura et al. (2015)) and remain rare in the areas of economics and finance. One reason for the lack of high-quality publications on data-driven event recognition in finance is that scientific studies on "money machines" are generally unpublishable, thus offer little incentive to the scientific community. On the other hand, any bank or hedge fund with sufficient capital and expertise to successfully implement trading strategies based on data-driven event recognition will naturally protect the "secret sauce" of their proprietary models instead of publishing them in a scientific journal.

Companies that seek to utilize data-driven event recognition but lack the capital or know-how to implement them in-house, can purchase ready-made solutions from third-party providers such as RavenPack.[2] They report how customers are capitalizing on the RavenPack event detection solution to developing, for example, long/short strategies from news and social media. However, assuming market efficiency (and that customers know how to utilize these third-party datasets correctly), one has to wonder how many individual customers can obtain a competitive edge from the same, of-the-shelve event recognition solution; at least when formulating event-driven trading strategies.

3 News Analytics and Financial Variables

Researchers in financial news analytics have studied various information sources and methods, yet they have consistently found a relationship between financial sentiments and events, computed from news corpus and various financial and other economic variables. For instance, the tone of news media is featured in a number of financial news analytics-related researches with results showing a significant impact of media tone on various variable such as future earning (Huang et al. 2014), takeover success (Buehlmaier 2013), return from a spin-off deal (Carretta et al. 2011), stock returns (Garcia 2013), and future performance of fund (Ozik and Sadka 2012). Table 2 provides a short overview of findings from financial news analytics.

In addition to news media, financial news analytics have capitalized on various corporate disclosure and IPO reports in predicting important financial variables. Ferris et al. (2013) studied the tone of IPOs and concluded that a conservative tone in IPO prospectuses increased the underpricing; the effect was especially prevalent for technology firms. Similarly, Li (2010) found that the tone of forward-looking statements (FLS) in 10-k and 10-Q fillings contains explanatory power for predicting the future performance of a firm. Compared to firms with a negative tone in their FLS, those with a positive FLS tone achieved 5% point higher return on assets (ROA) in the subsequent quarter.

To study market tone and sentiment, news analytics systems in finance have also shown interest in unconventional information sources such as social media, blogs, and various conversational platforms. For instance, Chen et al. (2013) found that opinions expressed on investment-related social media sites such as *seekingAlpha*—a platform for messaging and advice-sharing related to financial markets—can strongly predict future stock returns and earnings surprises. Das and Chen (2007) analyzed the sentiment of 145,100 instances of small talk from the internet and found a significant relationship between the tone of the talk and stock performance. However, they also reported that the individual message tone did not bear any significant relationship with stock performance. In another instance, Price et al. (2012) found that the tone of a conference call is a strong predictor of abnormal returns and trading volume. Among the other digital platforms in the blogosphere, the volume of blogs mentioning a firm had a marginal negative effect on stock return but a significantly positive effect on

[2]https://www.ravenpack.com/research/nlp-machine-learning-investing/.

Table 2 Results from financial studies on lexicon-based methods

Literature	Findings
Carretta et al. (2011)	Media attention and sentiment influence investors' spin-off deal preferences
Chen et al. (2013)	Opinions expressed in *seekingAlpha* articles and commentaries predict stock returns over various horizons
Engelberg et al. (2012)	All short sales traders respond to news content, i.e., trade volume increases in news day. Relationship between short and future return is strongest close to the news events
Ferris et al. (2013)	Increase in conservative tone in the prospectus is related to increased underpricing of IPOS. This relationship is stronger for technology-related IPOs than non-technology related
Garcia (2013)	Media-expressed sentiment assists in predicting stock returns on a daily basis. Effect of media sentiment is significantly stronger during the recession than economic expansion
Jegadeesh et al. (2013)	During the filing period the markets do not react to the tone of 10-k filing. There exists a negative relationship between the tone of IPO filing and underpricing
Ozik and Sadka (2012)	Differences in media sources do not impact investors' behavior. Investors' fund flows do not react to media. Media information can predict future performances of funds
Price et al. (2012)	Earning specific dictionary is more powerful in gauging the tone of conference calls. The tone of the conference call is a significantly stronger predictor of abnormal returns and trading volumes
Buehlmaier et al. (2013)	Positive news are accompanied by takeover success, whereas negative news is followed by failure. Media is the most important explanatory variable in takeover prediction
Das and Chen (2007)	Naïve Bayesian performs better. At the aggregate level, sentiment expressed in the message has some explanatory power to explain the stock movement. At the individual stock level, this relationship is insignificant
Groth et al. (2011)	SVM gives the best results. Intraday text-based analysis can identify abnormal risk events
Huang et al. (2014)	Naïve Bayesian method outperforms a dictionary-based approach. In-sample and out-of-sample classification accuracy for Naïve are 80.9% and 76.9%, respectively. Financial word list by L&M and diction achieved 62.0% and 54.9% classification accuracy, respectively. Investors attach more than twice the weight to negative news than positive news and ignore neutral news. Negative news has 42.0% more informative about future earnings than positive news
Li et al. (2010)	Naïve Bayesian achieved tone classification with 67% accuracy with three classes and content accuracy of 63% in content classification with twelve classes. The tone of FLS is correlated with future earnings and liquidity. Compared to a firm with a negative tone in FLS, 5% point higher return on Assets (ROA) in the subsequent quarter was found for a firm with positive FLS
Schumaker and Chen (2009)	Financial model that combines textual news data and stock price information performs the best. It achieved 57.1% accuracy in predicting price direction. Bag-of-word approach performs poorly. Noun phrase presents better textual representation
Vu et al. (2012)	82.9%, 80.5%, 75.6% and 75.0% prediction accuracy in stock price directional changes for Apple, Google, Microsoft, and Amazon, respectively
Yu et al. (2013)	Social media metrics have a stronger impact than conventional media on a firm's stock performance. The number of blog mentions has a marginal negative effect on return but a significantly positive effect on risk. The number of forum mentions has a negative effect on return. The number of tweets has a significantly positive effect on risk

market risk (Yu et al. 2013). In the same vein, the authors also found a significantly positive effect between the number of tweets and risk whereas the volume of forum messages had a negative effect on returns.

Existing literature on news analytics has covered both sentiment analysis as well as event or topic detection. On one hand, the findings are encouraging as they establish the importance of news analytics for understanding finance and economic phenomena. On the other hand, the domain-specific needs of financial news analytics have not received enough attention yet. Sentiment in the financial domain differs from other areas, as the facts, modifiers, and long-distance dependencies of words are capable of altering the sentiment. Further, financial events and topics interact with the market via associated sentiment. In this context, we present suitable guidelines to design a financial news analytics system.

4 Tools and Resources for Financial News Analytics

If one wants to analyze a document, be it financial or of any other type, two different aspects need to come together. On the one hand, there is the language with its grammar and expressions, and on the other hand, there is the actual content or knowledge. These two aspects need their own approaches and tools. Grammar can be parsed and a sentence structure can be analyzed; the meaning of metaphors and expressions can be found out too, as well as any raw sentiment. But insofar it is only words and no content. A central challenge in building an intelligent news analytics system is in providing it with common world knowledge. As put by Hendler and Feigenbaum (Hendler and Feigenbaum 2001), in order to build any system with "significant levels of computational intelligence, we need significant bodies of knowledge in knowledge bases". Computational intelligence is achieved by creating a relationship between the words. For instance, "JPMorgan" is much closer related to "investment banking" than to "energy production". This relationship can be created with ontologies that basically represent the conceptualization of human knowledge. A good example is DBpedia, which is an effort to extract information from Wikipedia and to create a database on how things are connected. In this way, all the knowledge in Wikipedia can be harnessed (DBpedia 2015). By connecting the meaning of words with how they are used in a document, a much deeper understanding of the text content can be achieved. Works that have attempted to incorporate human knowledge into sentences are not limited to financial sentiment analysis, but have been considered in other areas as well (Daraselia et al. 2014; Malo et al. 2013).

4.1 Linguistic Resources in Finance

At first glance, there is no difference between the language that is used in financial texts and in general texts. There are, however, differences in how metaphors are used and in how certain sentiments should be understood. If these differences are taken into account, general text analysis tools can be applied. Thereby, texts can be analyzed in numerous ways. Questions like "Which words are frequently used?" are rather superficial and quite easy to answer. This particular question, for example, is easily answered with a bag-of-words approach (Harris 1954), which in essence just maps each word to a number, and this, in turn, enables simple indexing and frequency calculations. "What does the author really mean?" is a far more complex question, and thus to answer this question much more powerful tools are needed. Fortunately, computational language research goes back to the 1950s, so there is a plethora of both commercial and open-source software to choose from.

Here we will present various language resources for the financial domain and also briefly discuss related research. The main focus has been on publicly available resources that can be obtained from the resources' home page. Most of these are published under a Creative Commons (2015), GNU (2015) or Apache (2015) license.[3]

4.1.1 Preprocessing Tools

Natural language processing (NLP) uses techniques like artificial intelligence and linguistics to parse human languages. The first steps of NLP were taken by Turing (1950) with his article *Computing Machinery and Intelligence*. The first sentence of this milestone publication is "I propose to consider the question, 'Can machines think?'" He then continues to discuss the "Turing test", which became the criterion of an intelligence indistinguishable from human intelligence. From their humble beginnings and up to the 1980s, most NLP systems were nothing more than hugely complicated sets of hand-written rules. With the introduction of more affordable and powerful computers toward the end of that decade, machine learning algorithms revolutionized NLP. Early models often used decision trees, before the focus shifted toward various statistical approaches. Some of the most common preprocessing steps in these systems are the following.

Tokenization is the process of splitting a text into words, phrases, symbols, or any other meaningful elements called tokens. These tokens can then be used as input for other processes, for instance, parsing or text mining processes.

Stemming is the method of reducing inflected words to their word stem, base or root form, e.g., "read" or "reads" would both become "read". This is a common process today, as many searchable databases assume words with the same stem to be synonyms.

[3]These licenses tend to allow users to freely use and modify the resource in any way they see fit, sometimes even commercially. For instance, a Creative Commons Attribution 3.0 Unported License means that you are not only free to share, copy, and redistribute all the material in any medium or format, but also to adapt it for any purpose, even commercially. The only limitation is that you have to give appropriate credit, provide a link to the license, and indicate if changes were made.

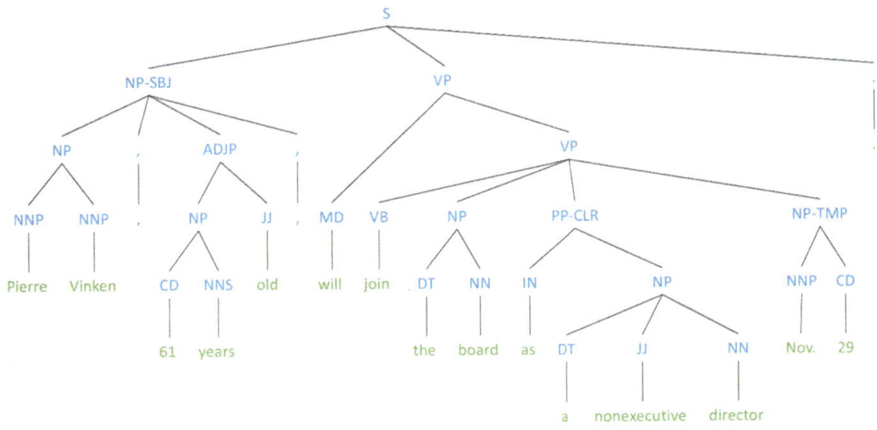

Fig. 1 The following sentence is parsed by NLTK; Pierre Vinken, 61 years old, will join the board as a nonexecutive director Nov. 29

Parsing is a method to extract the grammatical structure of a sentence; in other words, it is used to find subjects, objects, and adjectives. Figure 1 is an example of a parsed sentence.

Word sense disambiguation is used when a word has more than one meaning and the correct meaning must be determined. For this task resources such as WordNet or Wikipedia can be utilized.

For the inexperienced analyst, the Natural Language Toolkit (2015) provides a great introduction to this field. The included libraries contain among others tokenization, stemming, parsing, classification, and semantic reasoning. Figure 1 is an example of the parsing capabilities of NLTK. The toolkit also provides access to close to a hundred corpora and lexical resources. Bird, Loper, and Klein are the creators of the toolkit and they have published a very useful handbook for the toolkit: Natural Language Processing with Python (Bird et al. 2009). As NLTK is created using Python, it is platform-independent. The Apache OpenNLP—another open-source toolkit for processing natural language texts—also handles most common NLP tasks (OpenNLP 2015). One significant difference between NLTK and OpenNLP is that OpenNLP is based on Java instead of Python and hence the choice boils down to the application requirements and the programming language preferences.

The Stanford Natural Language Processing Group aims at creating algorithms that enable computers to understand human language. The Stanford CoreNLP is a collection of natural language processing tools that take raw texts as input and return, among other things, the base forms of the words, names of persons, times, numeric quantities, and even sentiments. The tools are developed for English but can be adapted for other languages as well. This framework was developed by Manning et al. (2014). One of the core technologies in the program suite is the natural language parser. Klein and Manning proposed a statistical approach for the parser. This approach requires a few hand-annotated sentences that serve as fundamental knowledge and based on those the parser tries to find the most likely structure of sentences (Klein and Manning 2003a, b). Apart from the more basic tools of the NLTK and OpenNLP, the Stanford NLP group also develops and maintains more advanced tools such as Named Entity Recognition (NER), Information Extraction (IE), and tools for matching patterns in linguistic trees (The Stanford Natural Language Processing Group 2015). NER is able to recognize and label the given names of persons, companies, or places. This technology was originally proposed by Lafferty et al. (2001) and Sutton and McCallum (2011). The Stanford NLP version was in turn developed by Finkel et al. (2005).

4.1.2 Metaphorical Resources

Financial language is full of metaphors and expressions that differ significantly in meaning from everyday language. A sailor "bails out", when he scoops water over the side of the boat. In finance, the State "bails out" a bank when it injects money into the financial institution to keep it from failing. The meaning of "credit", for example, has been reversed with time and now means "debt". If there are "synergies", people will be made redundant. All these are examples of the rich and evolving vocabulary in finance show that metaphors deserve special attention. A rich and progressive language is a pleasure to read, but new meanings can easily add confusion for an outsider. Metaphors and expressions are not only used for poetic and rhetoric reasons but according to Lakoff (1987) and Lakoff and Johnson (1980), they also affect our thinking and conceptualization. Kovacs (2007) showed that this general observation was also true for the domains of business, finance, and economics.

A common approach in linguistics research is to create a corpus containing a finite body of relevant text that is assumed to be representative of the language, time, and topic (McEnery and Wilson 2001). Cheng (2013) analyzed such a collection compiled from online financial analyst reports of BNP Paribas for the use of "emotions" together with metaphors. The study found that metaphors are helpful in portraying emotions in the reports of financial analysts. Cardini (2014) gathered texts on the financial crisis from *The Economist* and from *The International Economy*, published between Autumn 2008 and Winter 2012. His findings suggest that the most frequently used metaphors referred to something negative about a human being, an object, or a motion. "Damaged" or "destroyed" were among the most commonly used metaphors.

Despite the large interest in financial metaphors, there are still very few publicly available tools for financial metaphor research. The main reason for this is that metaphors rely heavily on context, which requires a considerable amount of text. This need is the main reason for the shortage of available tools, as news texts are copyrighted by news agencies that tend not to allow republication of their publications. One solution to this dilemma is to publish a list of news pieces so that the readers can generate the corpus themselves. Lopez and Llopis (2010) used this approach for their analysis of metaphors in English and Spanish financial news stories.

4.2 Resources for Financial Sentiment Analysis

Another important aspect of financial news analytics is sentiment analysis. As discussed in the introduction, sentiments in finance differ quite substantially from the sentiments expressed in other domains like movie reviews and product reviews. As a result, the common approach of sentiment analysis based on aggregating adjectives and adverbs is less likely to produce reliable results. The reason behind this is the way in which words hold sentiments in finance. For instance, contrary to everyday language, the word "profit" should be considered neutral, even if profit as such is very much desired. The underlying reason for this is the lack of information or reference points. If an investor reads that a company made a profit of "50 million", s/he cannot make any judgments solely on that statement. What was the profit last year? What was the revenue of the company? If the profit was 10 million last year or if the revenue was 80 million, the company is clearly doing well. If the profit is up, the sentiment will be positive, and if it is down, "profit" will have a negative sentiment. Most business-related words (such as profit, loss, sales, and assets) should be considered neutral by themselves, and any sentiments will be based on the entire phrase. In a financial news analysis, phrases, not individual words, form the fundamental unit of sentiment analysis. Generally speaking, opinions and sentiments in financial news are expressed subtly with lesser use of directly opinionating terms. Instead, domain-specific concepts that indirectly influence sentence-level polarity are common in financial text (Malo et al. 2013). In other words, financial news carries sentiments expressed with the use of more context-driven linguistic structure. Thus, extracting sentiments from financial news requires expertise in understanding the polarities that words carry (Loughran and McDonald 2011). Fortunately, there are both tools and corpora publicly available for financial news analysis.

SentiWordNet is a lexical resource based on WordNet.[4] In SentiWordNet each synset[5] is scored with three numbers—Obj(s), Pos(s), and Neg(s)—describing how objective, positive, and negative the terms are Baccianella et al. (2010).

SenticNet was published by Cambria et al. (2012) as a semantic resource for concept-level sentiment analysis. Its common-sense knowledge base utilizes sentic computing, which combines both AI and Semantic Web techniques in order to improve recognition, interpretation, and processing of natural language opinions.

MPQA Opinion Corpus contains annotated news articles from a variety of news sources. In the corpus, all news articles have been manually annotated for opinions and other private states, such as beliefs, emotions, sentiments, and speculations (Wiebe et al. 2005). In addition to the corpus, the authors have also published a subjectivity lexicon that can be used for general sentiment analysis. The group has also published the goodFor/badFor Corpus on events (Deng et al. 2013), the Subjectivity Lexicon (Wilson et al. 2005), and the +/−Effect Lexicon (Choi and Wiebe 2014).

---

[4] WordNet is a lexical database for the English language (Miller et al. 1990).

[5] A synset is set of synonyms in WordNet.

Financial Polarity Lexicon is a list of finance-specific words compiled explicitly for sentiment analysis. In addition to simply positive and negative words, the lexicon also includes new categories such as uncertain, litigious, strong modal, and weak modal (Loughran and McDonald 2011).

Financial Phrase Bank is a dataset that consists of almost 5000 phrases and sentences sampled from financial news texts and company press releases regarding Finnish companies listed on the NASDAQ OMX Nordic Exchange. All sentences have been annotated as positive/negative/neutral by a group of 16 annotators, both faculty and students, from Aalto University School of Business (Malo et al. 2014). The Financial Phrase Bank is publicly available.

Dual-Level News Corpus is an annotated dataset, that consists of 297 documents from Thomson Reuters newswire. Ten news topics with significant financial impact in 2010 were carefully defined, and relevant stories were manually selected from the newswire. The annotators were instructed to define up to three concepts per news story, which were important for the overall sentiment of the piece. After having found these concepts, they were asked to grade the sentiments on the document level, using a 7-point scale from "very positive" to "very negative". The same process was then to be repeated for all the individual sentences in the document. As the annotation was done at both the document and sentence level, the corpus is a dual-level news corpus (Takala et al. 2014). This resource is available from the authors by request only.

4.3 *Ontologies in Finance*

An ontology provides an answer to the question "What is a thing?" in a specific domain. As the answer depends heavily on the domain, ontologies are domain-dependent. As an example, you would get a completely different answer if you asked an investor and geographer what a "bank" is. Gruber (1993) gave the following definition of ontologies: *An ontology is a specification of a conceptualization.* In other words, an ontology is a model that defines the relationships between items and that contains logical information about these items. An ontology is a taxonomy[6] to which the logical information has been added. To fully benefit from a taxonomy or an ontology, it has to be machine-readable, which is achieved by using a well-defined ontology language.[7]

Ontologies tend to define the most general item as a "Thing". For example, in the Financial Report Ontology (see Table 3), "Thing" has a relationship to another thing called "Business Report", which in turn is related to a thing called "Financial Report". There is also a relationship between "Thing" and "Economic Entity". All

[6] A taxonomy is a simple hierarchical model, where the most general entity is at the top, and the most specific one is at the bottom.

[7] The most commonly used are OWL, RDF, RDF/S, and DAML+OIL. All these based on XML (Taye 2011).

these relationships are clearly defined in the ontology. Computers and machines are like babies; they need structure and they need to be led by the hand. Ontologies offer both structure and the parental guidance that the computer (software) needs in order to perform at its best. Proof of concept was provided by Malo and Siitari (2010), who proposed the Business Term Ontology (BTO). BTO provides a solid taxonomy of business domain concepts and thus serves as a backbone for analyses of financial news. This ontology has been integrated with Wikipedia-based concept-relatedness information and was used by Malo et al. (2010) to enhance content filtering accuracy.

As with the metaphorical resources, most of the ontologies currently used are available neither for commercial purposes nor for scientific research. However, there are exceptions and a few of them are described below:

Financial Report Ontology (FRO) is an ontology of a financial report. There is extensive documentation on the ontology at the resource's website (Hoffman 2015). The main purpose of FRO is to help accountants to create financial reports and to help auditors and financial analysts evaluate these reports. Even if the focus is on financial reporting, the ontology is still a tremendously useful tool for analyzing financial news and texts, as it provides extensive and solid relationships between large amounts of reporting terms.

Fadyart Finance Ontology was created by a consulting company specializing in knowledge management and in particular ontologies. This ontology contains information on different aspects of finance, for instance, markets, transactions, and instruments (Vanderlinden 2015). Unfortunately, the documentation on the ontology itself is rather sparse, but the documentation within it is both extensive and self-explanatory.

Financial Industry Business Ontology (FIBO) is developed by the Enterprise Data Management Council and is available from the Object Management Group's website. The ontology consists of several different repositories, Foundations, Business Entities, and Indices and Indicators. FIBO is a valuable source for knowledge about financial instruments, business entities, market data, and corporate actions.

OpenCyc is not primarily a financial ontology, but it still deserves being mentioned. OpenCyc (2015) is the open-source version of Cyc, which is often said to be the largest and most complete general knowledge base of today. The work on the ontology started in 1984 and the present version 4.0 contains 239,000 terms and 2,093,000 triples[8]; 28,000 of these terms are related to business. The ontology also contains 69,000 links to external resources such as DBpedia and WordNet.

[8]Triples define the relation of a thing to other things in an ontology.

Table 3 Part of the financial reporting ontology, adapted Hoffman (2015)

Line	Label	Object class	Report ElementName
1	Financial Reporting Ontology	[Network]	FRO
2	Thing	Abstract	fro:Thing
3	Business Report	Abstract	fro:BusinessReport
4	Financial Report	Abstract	fro:FinancialReport
5	Submission	Abstract	fro:Submission
6	Economic Entity	Abstract	fro:EconomicEntity
7	Reporting Unit	Abstract	fro:ReportingUnit
8	Business Segment	Abstract	fro:BusinessSegment
9	Geographic Area	Abstract	fro:GeographicArea
10	Asset Group	Abstract	fro:AssetGroup
11	Disposal Group	Abstract	fro:DisposalGroup
...	...	...	...

5 Addressing the Domain-Specific Needs in Financial News Analytics

Extracting sentiment from financial text is a challenging problem due to strong domain orientation, as mentioned earlier, and due to linguistic complexities. Despite the well-documented complexity, computing financial sentiments offer attractive benefits. Learning sentiments from financial text enhances understanding of financial markets, as the sentiments expressed through financial news and text are capable of influencing market behavior. One approach in computational intelligence is to build a knowledge-driven model for analyzing the sentiment of a financial text. Such a model combines a knowledge-base, financial ontologies, domain-specific knowledge, and machine-based learning strategies to optimally learn the sentiment from financial text (Malo et al. 2013, 2014).

5.1 Understanding Investor Perspective in Sentiment Analysis

Sentiment is commonly defined as emotions mediated to a human when reading a piece of text.[9] However, in financial applications, people are more interested in

[9]In financial literature, the term "sentiment" is sometimes also used when discussing "investor mood" or "market mood", which refers to the collective mood of the majority of market agents. These notions are, however, clearly different from the ones that are relevant in text analytics context, where the term "sentiment analysis" is used interchangeably with the task of polarity detection (i.e., categorization of news fragments into positive, negative or neutral classes depending on the tone of the text).

knowing whether a particular news item is good or bad from an investor's perspective, i.e., is the new information likely to influence asset prices and in what direction.

The strong focus on the investor's perspective leads to a somewhat different definition of sentiment than what is commonly considered in the text mining literature. In other domains, such as movie or product reviews, sentiments are closely connected to emotions and tones of opinions that are expressed through abundant combinations of adjectives and adverbs, whereas financial sentiments have more to do with expected favorable or unfavorable directions of events (such as "the sales are expected to increase this year"). As pointed out by Takala et al. (2014), the viewpoint also needs to be taken into account; although news about company lay-offs could be seen as negative by the general public, the same news may be received quite positively by profit-seeking investors.

It could even be argued that in a financial context, sentiments are surprisingly often expressed through factual statements that have a substantial impact on stock prices rather than through written individual opinions. Hence, instead of limiting sentiment to explicitly stated feelings, we may expand the definition to include also opinions/expectations regarding asset prices and their future development, which arise from facts featured in the text. Along with this viewpoint, the reliability of sources and timeliness of information play very important roles.

News and sentiment analysis can be powerful tools when combined with computer trading. There is a growing interest in research that connects news analytics and automated trading with a number of relevant applications, ranging from stock pricing and the discovery of stock alpha to risk control and fund management (Mitra and Mitra 2011). Commercial providers, such as *Thomson Reuters* and *RavenPack*, have already seized the opportunity. Below we provide a brief overview of the commonly encountered commercial use cases of financial sentiment analysis.

Media-driven high frequency trading: In finance, speed matters—the faster, the better. Information and sentiments age quickly, and old data offer very little value. Therefore, it is expected that news analytics are most effectively utilized in automated arbitrage trading strategies (event arbitrage), which are based on a market's response to events (e.g., earnings reports and macroeconomic releases). One of the recent developments is to create more granular sentiment metrics to replace traditional aggregate measures. It is reasonable to argue that not all news events are equally important from a financial perspective. Therefore, it is becoming a common practice to dissect the sentiment signals by themes or event types, so that traders can focus on the selected signals that are most relevant for their trading strategies (Hafez and Guerrero-Colón 2019).

Risk management: Along with the focus on improving returns, asset managers are also concerned with their ability to control portfolio risks and sensitivity to different types of news events. In this respect, news sentiment analysis tools play an important role in mitigating losses due to adverse events.

Compliance and surveillance: Investors, traders and asset managers are not the only groups who benefit from improved financial news analytics. Government officials and surveillance analysts also want to be alert about events that may signal market abuse.

5.2 *Computing Sentiment in Finance*

A knowledge base in financial sentiment analysis involves defining features and laying out rules for learning the sentiment from the text. Such a knowledge base needs to be built using financial ontologies and has to be designed in a way that captures domain-dependent polarities of financial text. The main objective when creating a financial knowledge base is to build a model that incorporates a human-like ability to recognize the financial sentiment. Building a model that emulates the human ability to recognize financial sentiments does not only depend on the underlying learning algorithms, but also on the quality of the sentiment databases that connect the algorithms with human- and society-level information. Thus, based on the work by Malo et al. (2014), we describe the process of creating such a financial knowledge base. The four aspects of financial polarity along with the rules for combining sentiments as proposed by Malo et al. (2014) provide the necessary groundwork for building an ontology-based knowledge base for financial sentiment analysis (for a detailed discussion see the works of Malo et al. (2013, 2014)).

5.2.1 General Polarity Bearing Expressions

The first and most abundant group of entries consists of commonly encountered opinionated or subjective expressions. A good starting point for compiling such resources can be found by using the MPQA corpus of opinion annotations (Wiebe et al. 2005; Wilson 2008; Wilson et al. 2009). Even though this is a very high-quality resource, it still needs to be treated with caution when used in a financial context. Due to the strong domain dependence, capturing sentiment expressions in finance requires the addition of human knowledge to the model. One can create finance-specific expressions by merging the MPQA-based list with the finance-specific lists compiled by Loughran and McDonald (2011). The lists are an outcome of a careful study for which the authors examined the quality of word classifications given by the Harvard dictionary and came up with word-sentiment classifications tailored for the financial domain.

5.2.2 Financial Entities

Financial entities are financial concepts with additional human knowledge about their semantic orientations. Basic financial concepts have neutral polarity by default and can be easily assessed from financial ontologies. However, at the phrase and sentence level, association with other entities can modify the polarity of the expression as a whole. For instance, "loss" in itself is neutral, whereas "loss has been slashed" or "loss has doubled" represent positive and negative polarities, respectively. Thus, extracting sentiments related to financial entities requires the addition of human knowledge to financial concepts. For this purpose, Malo et al. (2013), has listed in addition the following information: (i) description of the concept; (ii) anchor text, terms used to express financial concepts in text (e.g., EBIT,[10] ROI[11]); (iii) prior polarity, default polarities of financial concepts, usually neutral; and (iv) directional dependencies, possible semantic orientation after being modified by a modifier in a phrase. Directional dependencies for financial concepts can be roughly summed up as "positive-if-up" or "negative-if-up". For instance, "operating loss" is negative-if-up and "market value" is positive-if-up.

5.2.3 Directionalities

The direction of financial entities can be described as an event that is either going upward or downward. The terms that represent such changes in the direction of financial events are termed as "directionalities" (Malo et al. 2013). Words describing directionalities of financial events can be roughly categorized as an upward category that indicates increments (e.g., words like "boom" and "doubled") and a downward category that points toward decrements (e.g., words such as "reduced" and "fell"). Building financial sentiments requires listing out expressions that can describe the directionalities of financial events.

5.2.4 Polarity Influencers

As noted by Wilson et al. (2009), at the phrase level, sentiment analysis is complicated by the fact that the prior polarity of the financial concepts is often subjected to modifications by other expressions in the phrase. For example, "performance was good" indicates positive polarity whereas "performance was not good" expresses negative polarity. The class of words and operators that modify the semantic orientation at phrase level are labeled as "polarity influencers" (Wilson et al. 2009). According to Malo et al. (2013), polarity influencers can be broadly classified as follows: (i) negators, words that reverse the polarities (e.g., "not satisfactory"); (ii) boosters and diminishers, a category of words that influence the degree of positiveness or negativeness such as "rocket", "plummet", and "sink"; and (iii) modal operators, terms that determine the context in terms of possibility or necessity.

In addition to the abovementioned categories, one needs to consider the terms expressing uncertainties related to the financial and economic contexts (Malo et al. 2013). An illustration of sentiment formation is shown in Fig. 2. In the figure, the

[10]Earnings before interest and tax.

[11]Return on investment.

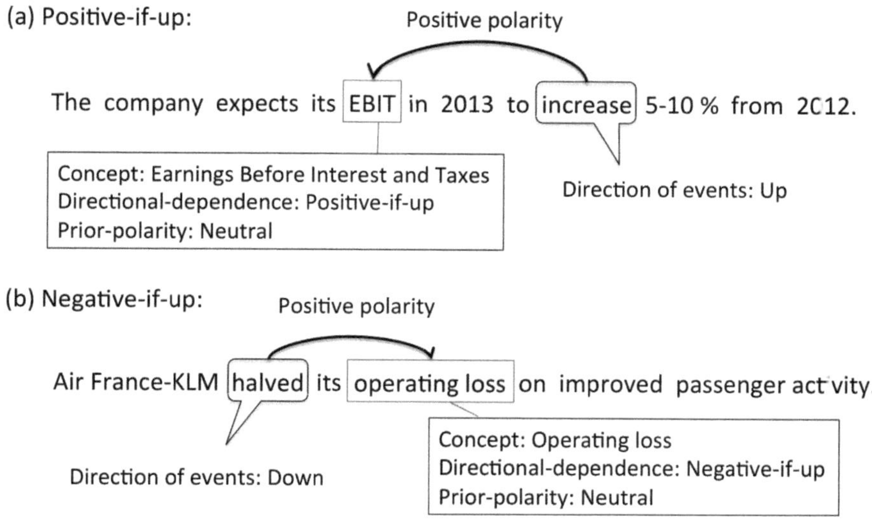

Fig. 2 Sentiment extraction based on four aspects of finical sentiment (Malo et al. 2013)

financial concept "operating profit" and "profits" are initially neutral but are modified by the polarity influencer.

5.2.5 Combining Sentiments

A sentence can be regarded as a sequence of words with the embedded syntactical relationship. Malo et al. (2014) have suggested an approach that takes a sentence as a sequence of words and transforms it into a polarity sequence without altering semantic and syntactic orientations. The approach can be described as a set of four rules for forming a sentence- or phrase-level sentiment, outlined as follows:

First, the syntactic and semantic structure of sentences or phrases should be preserved. Altering the structure can distort polarity of the expression as a whole. The semantic orientation of a sentence depends upon word order. When a sentence or phrase is constructed, financial concepts, negators, and other modifiers are arranged in an order to communicate intended polarity. The position of modifiers and influencers determines the polarity of financial concepts. Thus, preserving word order is a crucial consideration.

Second, a sentence can easily accommodate separated but still interacting concepts. In the phrase "While profits increased from last year, market share declined", profit is positive while market share has a negative polarity. Thus, at the sentence level, interactions between concepts residing even at a distance needs to be taken into account. A good model should be able to capture long-distance dependencies between the entities, directionalities, and their interactions. The ability to consider a complete sentence rather than just picking up a few adjacent words is a prerequisite for a good financial sentiment model. Representing sentences as a tree, using a sentence parser tree, can capture both the syntactical structure as well as the long-distance relationships between various entities within the sentences. The most notable choices in this regard are constituent trees and type-dependencies trees. A constituent tree is an ordered rooted tree, which depicts relationships between the words as "head" and "dependent". It also provides POS (parts of speech) tags for the words. Type-dependencies tree is an alternative approach in which relationships within the sentence are presented as uniform triplets of pairwise word relations. Both constituent and type-dependencies trees can be created using NLP software suites.

Third, the degree of polarity for entities that belong to the same class should be equivalent. For instance, all the negative entities, such as bad returns and worse returns are equally negative. In other words, the degree of positiveness and negativeness of a financial entity should be equal within the sentiment class.

Fourth, the polarity of a financial entity must be defined using domain-specific ontologies. Along with the method and data source, the choice of polarity definition used in sentiment extraction can potentially explain the performance differences of various financial sentiment models.

Using the sentiment knowledge base and sentiment combination rules as described earlier, a sentence tree can be converted to a polarized tree. Both constituents and type-dependencies trees can be modified into polarized trees by replacing terms with the corresponding sentiment class from a sentiment database. The polarized tree establishes relationships between the sentiment entities, directionalities, and influencers, which can be combined to compute the polarity of the main financial entity. Further, a sequence of neutral entities can be combined into one single neutral entity. The final modified polarized tree can be seen as a sequence of positive, negative, and neutral polarities in a tree structure. Using this polarized tree structure, the polarity of a sentence can be obtained using aggregation rules.

5.3 Neural Networks for Event Detection

The daily flood of business news undoubtedly helps investors to remain well-informed, yet it also makes it difficult to judge the actual information value of a single news item (Koudijs 2016; Yermack 2014; Boudoukh et al. 2013). Given the sheer mass of news produced every day, we might safely assume that only a tiny fraction will actually affect trading activities. Here, we showcase the architecture and use of a neural network for event recognition based on the work of Malo et al. (2018). The goal is to annotate news items with event tags that indicate whether a specific news event is found in a text. The network architecture is followed by change point analysis.

5.3.1 Network Architecture

Annotating news items with event tags is a fine-grained labeling task that may involve thousands of potential categories; a job that can be difficult to solve with a single classifier (Ahmed et al. 2016; Gao et al. 2017). A common remedy for this issue is to organize the output tags into semantically related subgroups (*verticals*) and separately train a specialist model for each subgroup. Such an approach is fairly straight forward in financial news analytics since there already exists a natural taxonomy for arranging events in a semantic hierarchy. For instance, all fine-grained *cash flow events* can be organized in one vertical, *legal events* go into another, and so on (see Fig. 3). The allocation of a news item to a specialized subgroup is guided by a specialist selector model on a sentence-by-sentence basis. After subgrouping the event tags, each sentence can be encoded to a semantic vector representation using word embedding algorithm such as word2vec. Word2Vec is a neural language model that captures the semantic similarities among the words through vector representation. This embedded sentence is combined with a bidirectional LSTM to capture the contextual information around the words in the sentence. Each specialist network and selector are modeled as bidirectional Long Short-Term Memory (LSTM) networks (Seo et al. 2016), followed by a word attention mechanism (Yang et al. 2016) that computes importance weighted encodings. Finally, these weighted encodings are passed to a fully connected layer with dropout and soft-max activation function that outputs the probabilities for different event labels. The final model structure of this neural architecture resembles a tree-shaped network, with the specialist models as branches.

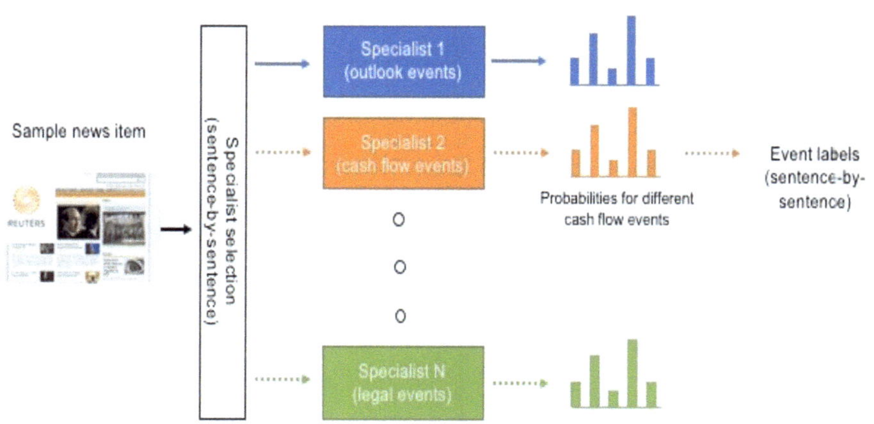

Fig. 3 Multi-specialist network for event tagging (Malo et al. 2018)

5.3.2 Change Point Analysis

Event recognition is an important task in financial news analytics. Yet, improved market efficiency has shortened the lifespan of news, and consequently, also the predictive relations between news and stock prices are shorter-lived. Statistical models trying to capture these dependencies will, therefore, experience structural changes, where both parameter estimates, as well as the set of contributing news variables, may change from one regime to another in a discontinuous manner. In particular, times of crisis are likely to exhibit nonstationary behavior (Münnix et al. 2012).

To address this problem, Malo et al. (2018) studied energy distance-based approaches for structural change point detection in linear regression models. The authors analyzed Thomson Reuters financial news data from years 2006 to 2009 to identify structural breakpoints (*change points*), as well as the subset of news variables that are relevant within each resulting regime for predicting banking sector returns. After annotating the news items with tags, using the deep neural network described above, the resulting event indicators were aggregated into time series that indicates how many times each event type is mentioned within a given time step.

Next, Malo et al. (2018) applied the nonparametric structural change detection algorithm NSA on the industry returns and found three distinct change points (four regimes). As shown from Fig. 4, the change points correspond well with macroeconomic events during the recent financial crisis: the first regime encapsulates the escalation of the subprime mortgage bubble into a recession; a structural change point in mid-May 2007 marks the beginning of the second regime when the Fed finally takes action to mitigate the crisis with a series of interest rate cuts; despite these

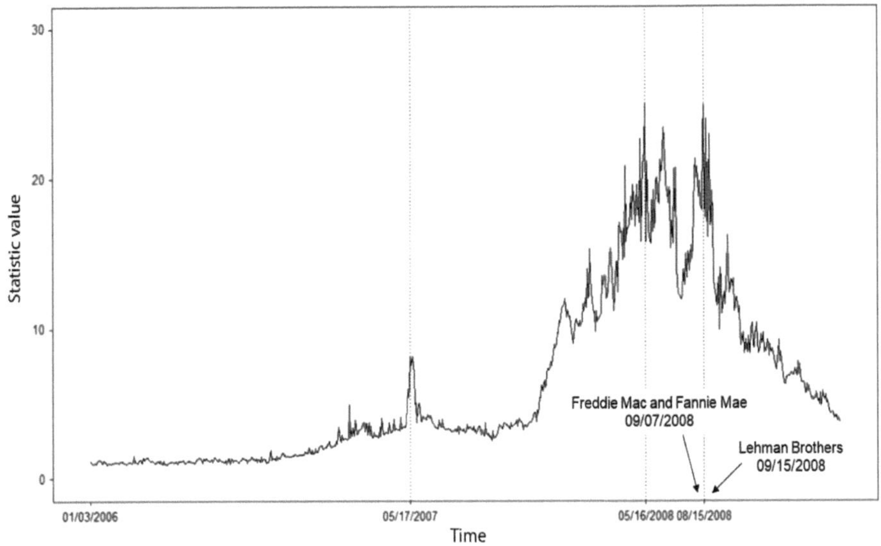

Fig. 4 Structural change points during the financial crisis. Malo et al. (2018)

measures, the entire economy fell into a recession during the third regime, which includes the bankruptcy of Fannie Mae and Freddie Mac in August 2008, followed by the default of Lehman Brothers and AIG; following another change point, the fourth regime describes the time of massive bailout programs.

Structural change point analysis as performed by Malo et al. (2018) can be a useful tool to identify early warning signals. The model settings can be adjusted to obtain a detection sensitivity that fits the specific use-case of analysts, traders, and policymakers.

6 Conclusions

The importance of financial news analytic in understanding and predicting financial and economic variables is ubiquitous. Traders gain a competitive advantage due to their superior capability to process news information (Engelberg et al. 2012). However, modern news analytics is complex as information is spread out across various sources such as news media, social media, blogs, forum discussion, and verbatim from the conversation. In addition, a general multipurpose text mining resources are likely to yield inferior performance as the finance differs from other domains in its specific need. Such domain-specific need demand addressing the investor's perspective and overall complexity embedded in financial sentiment analysis and event recognition. Failing to address the domain-specific needs can lead to inaccurate inferences that in turn can lead to unwarranted consequences, such as executing a trade on a false signal computed from the news (Von Beschwitz et al. 2015).

In this chapter, we highlighted the special ontological needs in financial text analysis and provided a brief introduction to the existing methods and key findings in relation to the financial domain. However, there is still a significant room for improvement within financial news analytics as the filed still have not received enough attention toward domain-specific need. With the use of more sophisticated knowledge-driven learning methods in text and content analysis, users can still achieve some degree of improved performance. To this end, we presented some of the linguistic resources that can potentially aid in building domain-specific news analytics systems. We also discussed one potential approach for sentiment analysis and event detection which addresses the specific needs of the financial domain.

References

Ahmed K, Baig MH, Torresani L (2016). Network of experts for large-scale image categorization. In: European conference on computer vision. Springer, Cham, pp 516–532
Ahn D (2006) The stages of event extraction. In: Proceedings of the workshop on annotating and reasoning about time and events. Association for Computer Linguistics, pp 1–8
Aggarwal CC, Zhai CX (eds) (2012) Mining text data. Springer Science and Business Media

Allan J, Carbonell JG, Doddington G, Yamron J, Yang Y (1998) Topic detection and tracking pilot study final report. In: Proceedings of the DARPA broadcast news transcription and understanding workshop
Apache License (2015) http://www.apache.org/licenses/LICENSE-2.0. Accessed 31 July 2015
Arendarenko E, Kakkonen T (2012) Ontology-based information and event extraction for business intelligence. In: International conference on artificial intelligence: methodology, systems, and applications. Springer, Berlin, pp 89–102
Baccianella S, Esuli A, Sebastiani F (2010) Sentiwordnet 3.0: an enhanced lexical resource for sentiment analysis and opinion mining. In: Proceedings of the seventh international conference on language resources and evaluation (LREC' 10), Valletta, Malta. European Language Resources Association (ELRA)
Baker M, Stein J (2004) Market liquidity as a sentiment indicator. J Financ 7:271–299
Ball R, Brown P (1968) An empirical evaluation of accounting income numbers. J Account Res, 159–178
Barber BM, Odean T (2008) All that glitters: the effect of attention and news on the buying behavior of individual and institutional investors. Rev Financ Stud 21:758–818
Ben Ami Z, Feldman R (2017) Event-based trading: building superior trading strategies with state-of-the-art information extraction tools
Bird S, Loper E, Klein E (2009) Natural language processing with python. O'Reilly Media Inc
Boudoukh J, Feldman R, Kogan S, Richardson M (2013) Which news moves stock prices? A textual analysis (No. w18725). National Bureau of Economic Research
Boudoukh J, Feldman R, Kogan S, Richardson M (2018) Information, trading, and volatility: evidence from firm-specific news. Rev Financ Stud 32(3):992–1033
Buehlmaier MM (2013) The role of the media in takeovers: theory and evidence
Cambria E, Havasi C, Hussain A (2012) SenticNet 2: a semantic and affective resource for opinion mining and sentiment analysis. AAAI FLAIRS, pp 202–207
Cardini F (2014) Analysing English metaphors of the economic crisis. Lingue Linguaggi 11:59–76
Carretta A, Farina V, Graziano EA, Reale M (2011) Does investor attention influence stock market activity? The case of spin-off deals, June 30, 2011
Chen H, De P, Hu YJ, Hwang BH (2013) Customers as advisors: the role of social media in financial markets. In: 3rd annual behavioural finance conference. Queen's University, Kingston, Canada
Cheng W (2013) Metaphors in financial analyst reports. In: CELC symposium 2013
Choi Y, Wiebe J (2014) +/−EffectWordNet: sense-level lexicon acquisition for opinion inference. In: Proceedings of the EMNLP 2014
Creative Commons (2015). http://creativecommons.org/licenses/by/3.0. Accessed 25 July 2015
Daraselia N, Yuryev A, Egorov S, Novichkova S, Nikitin A, Mazo I (2014) Extracting human protein interactions from MEDLINE using a full-sentence parser. Bioinformatics 20(5):604–611
Das SR, Chen MY (2007) Yahoo! for Amazon: sentiment extraction from small talk on the web. Manag Sci 53(9):1375–1388
Day MY, Lee CC (2016) Deep learning for financial sentiment analysis on finance news providers. In: 2016 IEEE/ACM international conference on advances in social networks analysis and mining (ASONAM). IEEE, pp 1127–1134
DBpedia (2015). http://wiki.dbpedia.org. Accessed 25 July 2015
Deng L, Choi Y, Wiebe J (2013) Benefactive/Malefactive event and writer attitude annotation. In: Annual meeting of the association for computational linguistics (ACL-2013, short paper)
Ding X, Zhang Y, Liu T, Duan J (2015) Deep learning for event-driven stock prediction. In: Twenty-fourth international joint conference on artificial intelligence
Du M, Pivovarova L, Yangarber R (2016) PULS: natural language processing for business intelligence. In: Proceedings of the 2016 workshop on human language technology, pp 1–8
Dzielinski M, Rieger MO, Talpsepp T (2011) Volatility asymmetry, news, and private investors. In: Mitra G, Mitra L (eds) The handbook of news analytics in finance. Wiley
Engelberg J (2008) Costly information processing: evidence from information announcements. In: AFA 2009 San Francisco meetings paper

Engelberg JE, Reed AV, Ringgenberg MC (2012) How are shorts informed?: short sellers, news, and information processing. J Financ Econ 105(2):260–278
Fama EF (1965) The behavior of stock-market prices. J Bus 38(1):34–105
Fama EF, Fisher L, Jensen MC, Roll R (1969) The adjustment of stock prices to new information. Int Econ Rev 10(1):1–21
Fan J, Lv J, Qi L (2011) Sparse high-dimensional models in economics. Annu Rev Econ 3(1):291–317
Feldman R, Rosenfeld B, Bar-Haim R, Fresko M (2011) The stock sonar-sentiment analysis of stocks based on a hybrid approach. In: Twenty-third IAAI conference
Ferguson NJ, Philip D, Lam HY, Guo JM (2014) Media content and stock returns: the predictive power of press. In: Midwest finance association 2013 annual meeting paper
Ferris SP, Hao Q, Liao MY (2013) The effect of issuer conservatism on IPO pricing and performance. Rev Financ 17(3):993–1027
Finkel J, Grenager T, Manning C (2005) Incorporating non-local information into information extraction systems by Gibbs sampling. In: Proceedings of the 43rd annual meeting of the association for computational linguistics (ACL 2005), pp 363–370
Gao J, Li Z, Nevatia R (2017) Knowledge concentration: learning 100 K object classifiers in a single CNN. arXiv:1711.07607
Garcia D (2013) Sentiment during recessions. J Financ 68(3):1267–1300
General Inquirer (2015). www.wjh.harvard.edu/~inquirer. Accessed 01 Aug 2015
GNU General Public License (2015). http://www.gnu.org/licenses/gpl.html. Accessed 31 July 2015
Groen JJ, Kapetanios G, Price S (2013) Multivariate methods for monitoring structural change. J Appl Econ 28(2):250–274
Groth SS, Muntermann J (2011) An intraday market risk management approach based on textual analysis. Decis Support Syst 50(4):680–691
Gruber TR (1993) A translation approach to portable ontologies. Knowl Acquis 5(2):199–220
Hafez P, Guerrero-Colón J (2019) White paper, Ravenpack tech
Hagenau M, Liebmann M, Neumann D (2013) Automated news reading: stock price prediction based on financial news using context-capturing features. Decis Support Syst 55:685–697
Harris Z (1954) Distributional structure. Word 10(2/3):146–62
Hart R (2000) DICTION 5.0. http://rhetorica.net/diction.htm. Accessed 01 Aug 2015
Hendler J, Feigenbaum E (2001) Knowledge is power: the semantic web vision. In: Web intelligence: research and development. Springer, Berlin
Hirshleifer D (2001) Investor psychology and asset pricing. J Financ 56:1533–1598
Hoffman C (2015) Digital Financial Reporting. http://xbrl.squa-respace.com/financialreport-ontology. Accessed 25 July 2015
Hogenboom A, Hogenboom F, Frasincar F, Schouten K, Van Der Meer O (2013) Semantics-based information extraction for detecting economic events. Multimed Tools Appl 64(1):27–52
Huang AH, Zang AY, Zheng R (2014) Evidence on the information content of text in analyst reports. Account Rev 89(6):2151–2180
Huang C-J, Liao J-J, Yang D-X, Chang T-Y, Luo Y-C (2010) Realization of a news dissemination agent based on weighted association rules and text mining techniques. Expert Syst Appl 37:6409–6413
Jacobs G, Lefever E, Hoste V (2018) Economic event detection in company-specific news text. In: The 56th annual meeting of the association for computational linguistics. Association for Computational Linguistics, pp 1–10
Jegadeesh N, Wu D (2013) Word power: a new approach for content analysis. J Financ Econ 110(3):712–729
Kakkonen T, Mufti T (2011) Developing and applying a company, product and business event ontology for text mining. In: Proceedings of the 11th international conference on knowledge management and knowledge technologies. ACM, p 24
Kaniel R, Saar G, Titman S (2005) Individual investor sentiment institution. Duke University

Kearney C, Liu S (2014) Textual sentiment in finance: a survey of methods and models. Int Rev Financ Anal 33:171–185

Klein D, Manning C (2003a) Accurate unlexicalized parsing. In: Proceedings of the 41st meeting of the association for computational linguistics, pp 423–430

Klein D, Manning C (2003b) Fast exact inference with a factored model for natural language parsing. In: Cambridge MA (ed) Advances in neural information processing systems 15 (NIPS 2002). MIT Press, pp 3–10

Koudijs P (2016) The boats that did not sail: asset price volatility in a natural experiment. J Financ 71(3):1185–1226

Kovacs E (2007) On the use of metaphors in the language of business, finance and economics. http://husse-esse.hu/conferences/husse-8-conference. Accessed 21 July 2015

Kumar A, Lee C (2006) Retail investor sentiment and return comovement. J Financ 61:2451–2486

Lafferty J, McCallum A, Pereira F (2001) Conditional random fields: probabilistic models for segmenting and labeling sequence data. Departmental Papers (CIS), Department of Computer & Information Science, University of Pennsylvania

Lakoff G (1987) Women, fire and dangerous things: what categories reveal about the mind?. University of Chicago Press, Chicago

Lakoff G, Johnson M (1980) Metaphors we live by. University of Chicago Press, Chicago

Li F (2010) The information content of forward-looking statements in corporate filings-a Naïve Bayesian machine learning approach. J Account Res 48(5):1049–1102

Liu B, McConnell JJ (2013) The role of the media in corporate governance: do the media influence managers' capital allocation decisions? J Financ Econ 110(1):1–17

Lopez A, Llopis M (2010) Metaphorical pattern analysis in financial texts: framing the crisis in positive or negative metaphorical terms. J Pragmat 42:3300–3313

Loughran T, McDonald B (2011) When is a liability not a liability? Textual analysis, dictionaries and 10-Ks. J Financ 66(1):35–66

Lösch U, Nikitina N (2009) The newsEvents ontology: an ontology for describing business events. In: Proceedings of the 2009 international conference on ontology patterns, vol 516, pp 187–193

Lugmayr A, Gossen G (2012) Evaluation of methods and techniques for language based sentiment analysis for DAX 30 stock exchange-a first concept of a "LUGO" sentiment indicator. In: Lugmayr A, Risse T, Stockleben B, Kaario J, Pogorelc B, Serral Asensio E (eds) SAME 2012-5th international workshop on semantic ambient media experience

Malo P, Siitari P (2010) A context-aware approach to user profiling with interactive preference learning. Aalto University working paper W-482

Malo P, Siitari P, Sinha A (2013) Automated query learning with wikipedia and genetic programming. Artif Intell 194:86–110

Malo P, Siitari P, Ahlgren O, Wallenius J, Korhonen P (2010) Semantic content filtering with wikipedia and ontologies. In: Proceedings of IEEE international conference on data mining workshops, pp 518–526

Malo P, Sinha A, Korhonen P, Wallenius J, Takala P (2014) Good debt or bad debt: detecting semantic orientations in economic texts. J Assoc Inf Sci Technol 65(4):782–796

Malo P, Sinha A, Takala P, Ahlgren O, Lappalainen I (2013) Learning the roles of directional expressions and domain concepts in financial news analysis. In: Proceedings of IEEE international conference of data mining workshops (ICDM SENTIRE). IEEE Press

Malo P, Viitasaari L, Gorskikh O, Ilmonen P (2018) Non-parametric structural change detection in multivariate systems. arXiv:1805.08512

Mandelker G (1974) Risk and return: the case of merging firms. J Financ Econ 1(4):303–335

Manning C, Raghavan P, Schütze H (2008) Introduction to information retrieval. Cambridge University Press

Manning C, Surdeanu M, Bauer J, Finkel J, Bethard SJ, McClosky D (2014) The stanford CoreNLP natural language processing toolkit. In: Proceedings of 52nd annual meeting of the association for computational linguistics: system demonstrations, pp 55–60

McEnery T, Wilson A (2001) Corpus linguistics, 2nd edn. Edinburgh University Press, Edinburgh

Medhat W, Hassan A, Korashy H (2014) Sentiment analysis algorithms and applications: a survey, pp 1093–1113
Miller GA, Beckwith R, Fellbaum CD, Gross D, Miller K (1990) WordNet: an online lexical database. Int J Lexicograph 3(4):235–244
Mittal A, Goel A (2012) Stock prediction using twitter sentiment analysis. Standford University, CS229
Mitamura T, Liu Z, Hovy EH (2015) Overview of TAC KBP 2015 event nugget track. In: Text analysis conference
Mitra L, Mitra G (2011) The handbook of news analysis in finance. Wiley
Münnix MC, Shimada T, Schäfer R, Leyvraz F, Seligman TH, Guhr T, Stanley HE (2012) Identifying states of a financial market. Sci Rep 2:644
Nassirtoussi AK, Aghabozorgi S, Wah TY, Ngo DCL (2015) Text mining of news-headlines for FOREX market prediction: a Multi-layer Dimension Reduction Algorithm with semantics and sentiment. Expert Syst Appl 42(1):306–324
Natural Language Toolkit (2015). http://www.nltk.org. Accessed 25 July 2015
OpenCyc (2015). http://opencyc.org. Accessed 25 July 2015
OpenNLP (2015). http://opennlp.apache.org/documentation/1.6.0/man-ual/opennlp.html. Accessed 25 July 2015
Ozik G, Sadka R (2012) Media and investment management
Pagolu VS, Reddy KN, Panda G, Majhi B (2016). Sentiment analysis of twitter data for predicting stock market movements. In: 2016 international conference on signal processing, communication, power and embedded system (SCOPES). IEEE, pp 1345–1350
Peramunetilleke D, Wong RK (2002) Currency exchange rate forecasting from news headlines. Aust Comput Sci Commun 24:131–139
Price SM, Doran JS, Peterson DR, Bliss BA (2012) Earnings conference calls and stock returns: the incremental informativeness of textual tone. J Bank Financ 36(4):992–1011
Rönnqvist S, Sarlin P (2017) Bank distress in the news: describing events through deep learning. Neurocomputing 264:57–70
Schmeling M (2007) Institutional and individual sentiment: smart money and noise trader risk? Int J Forecast 23:127–145
Schumaker RP, Chen H (2009) Textual analysis of stock market prediction using breaking financial news: the AZFin text system. ACM Trans Inf Syst (TOIS) 27(2):12
Schumaker RP, Zhang Y, Huang C-N, Chen H (2012) Evaluating sentiment in financial news articles. Decision support systems
Seo M, Kembhavi A, Farhadi A, Hajishirzi H (2016) Bidirectional attention flow for machine comprehension. arXiv:1611.01603
Shiller RJ (2003) From efficient markets theory to behavioral finance. J Econ Perspect 17:83–104
Sinha NR (2010) Underreaction to news in the US stock market
Soni A, van Eck NJ, Kaymak U (2007) Prediction of stock price movements based on concept map information. In: IEEE symposium on computational intelligence in multicriteria decision making, pp 205–211
Stock JH, Watson M (2009) Forecasting in dynamic factor models subject to structural instability. The methodology and practice of econometrics. A festschrift in Honour of David F Hendry, p 173, 205
Sutton C, McCallum A (2011) An introduction to conditional random fields. Found Trends R Mach Learn 4(4):267–373
Takala P, Malo P, Sinha A, Ahlgren O (2014) Gold-standard for topic-specific sentiment analysis of economic texts. In: Proceedings of The 9th edition of the language resources and evaluation conference (LREC), pp 2152–2157
Taye MM (2011) Web-based ontology languages and its based description logics. Res Bull Jordan ACM II(II). ISSN: 2078-7952
Tetlock PC, Saar-Tsechansky M, Macskassy S (2008) More than words: quantifying language to measure firms' fundamentals. J Financ 63:1437–1467

The Stanford Natural Language Processing Group (2015). http://nlp.stanford.edu/software/index.shtml. Aaccessed 25 July 2015
Turing AM (1950) Computing machinery and intelligence. Mind 59:433–460
Vanderlinden E (2015) Finance ontology. http://www.fadyart.com/onto-logies/documentation/finance/index.html. Accessed 27 July 2015
Von Beschwitz B, Keim DB, Massa M (2015). First to 'read' the news: news analytics and high frequency trading. In: Paris December 2015 finance meeting EUROFIDAI-AFFI
Vu TT, Chang S, Ha QT, Collier N (2012) An experiment in integrating sentiment features for tech stock prediction in twitter. In: Proceedings of the workshop on information extraction and entity analytics on social media data, Mumbai, India. The COLING 2012 Organizing Committee, pp 23–38
Wiebe J, Wilson T, Cardie C (2005) Annotating expressions of opinions and emotions in language. Lang Resour Eval 39:165–210
Wilson T, Wiebe J, Hoffmann P (2005) Recognizing contextual polarity in phrase-level sentiment analysis. In: Proceeding of the HLT-EMNLP-2005
Wilson T (2008) Fine-grained subjectivity analysis. School University of Pittsburgh
Wilson T, Wiebe J, Hoffman P (2009) Recognizing contextual polarity: an exploration of features for phrase-level sentiment analysis, computational linguistics, vol 35, no 3, pp 399–433
Wuthrich B, Cho V, Leung S, Permunetilleke D, Sankaran K, Zhang J (1998) Daily stock market forecast from textual web data. In: IEEE international conference on systems, man, and cybernetics, vol 3, pp 2720–2725
Xie B, Passonneau R, Wu L, Creamer GG (2013). Semantic frames to predict stock price movement. In: Proceedings of the 51st annual meeting of the association for computational linguistics, pp 873–883
Yang Y, Carbonell JG, Brown RD, Pierce T, Archibald BT, Liu X (1999) Learning approaches for detecting and tracking news events. IEEE Intell Syst Their Appl 14(4):32–43
Yang Z, Yang D, Dyer C, He X, Smola A, Hovy E (2016) Hierarchical attention networks for document classification. In: Proceedings of the 2016 conference of the North American chapter of the association for computational linguistics: human language technologies, pp 1480–1489
Yermack D (2014) Tailspotting: Identifying and profiting from CEO vacation trips. J Financ Econ 113(2):252–269
Yu Y, Duan W, Cao Q (2013) The impact of social and conventional media on firm equity value: a sentiment analysis approach. Decision support systems
Yuan G, Ho C, Lin C (2012) Recent advances of large-scale linear classification. Proc IEEE 100(9):2584–2603
Zhai Y, Hsu A, Halgamuge SK (2007) Combining news and technical indicators in daily stock price trends prediction. In Proceedings of the 4th international symposium on neural networks: advances in neural networks, Part III. Springer, Nanjing, China, pp 1087–1096

Network-Based Policies Versus Tax Evasion

Fernando Garcia Alvarado

Abstract This chapter explores the optimal income tax audit strategies from a social planner's perspective, whose objective is to minimize the aggregated tax evasion of a given society. Agents live in a social-cohesive network with homophilic linkages, meaning individuals connect only with people who are akin to them. Further, each period individuals share their memories about past audits and consequently update their subjective probability of being audited. The Tax Agency finds that network-based audit policies are inefficient, in the sense that they are just as good as random. Thus, the social planner credibly announces that, from now on, audit rates will be linearly proportionate to the agent's income, making richer people more prone of being audited. Audit rates are now endogenous and heterogeneous among agents, making it possible for the Tax Agency to find an optimal network-based policy following a local-average strategy where a "key sector" of society is predominantly audited every period. Following this strategy, under a dynamic framework, the Nash Equilibrium for the average subjective audit rate is swiftly raised after just a few fiscal years. The enhanced subjective audit rate, in turn, unfolds as a larger tax revenue collection and a significant deterrence of income tax evasion.

Keywords AB models · Networks · Quantitative policy · Tax evasion

JEL Classification: C54 · D85 · H26

This project has received funding from the ITN ExSIDE European Union's Horizon 2020 research and innovation programme under the Marie Sklodowska-Curie grant agreement No. 721846.

F. Garcia Alvarado (✉)
Ca' Foscari University of Venice, Venice, Italy
e-mail: fernando.garcia@unive.it

University of Paris 1 Pantheon-Sorbonne, Paris, France

A. S. Chakrabarti et al. (eds.), *Network Theory and Agent-Based Modeling in Economics and Finance*, https://doi.org/10.1007/978-981-13-8319-9_20

1 Introduction

Nowadays there exists a vast literature regarding the economics of crime, pioneered by the economic approach of crime and punishment by Becker (1968). Shortly after, Allingham and Sandmo (1972) studied the rational behavior behind the tax evasion phenomenon from an expected utility theory point of view. Despite the proposed model incurred in a few drawbacks, noted by the authors themselves, their study laid the foundations for plenty more research to come afterward. Mittone and Patelli (2000) pondered the decision process of tax evasion from an imitative behavior framework, where agents could fully comply, withhold a share of their income or free ride on their tax duties. Mittone (2006) enlarged the literature to account for attitude toward risk under dynamic choices, where agents faced a repetitive decision-making process integrated into a simulated environment.

Korobow et al. (2007) proposed a novel Agent-Based Model where agents held heterogeneous characteristics and accounted for an intrinsic perception regarding the enforcement policies practiced by the social planner, or Tax Agency, which audits a fraction of individuals each fiscal period. Hokamp and Pickhardt (2010) further delved the research of income tax evasion to encompass ABMs where individuals responded to different government policies. The model designed by the authors not only accounted for time effects, but it was also able to grasp how heterogeneous agents optimized an exponential utility model, which accounts for a more realistic setting than previous papers had proposed. Alm et al. (2017) conducted an economic experiment which uncovered how "neighbors", defined as all the agents with whom an individual shares information, have statistically significant peer effects and do "social pressure" on tax compliance. The authors identified how being surrounded by honest (cheating) taxpayers encouraged individuals to be honest (cheating) themselves.

Modeling peer effects and social norms in group behavior has recurrently resourced to network structures accounting for complementarity effects. Ballester et al. (2006) studied how under a noncooperative game with a quadratic utility function, a social planner could identify a "key player" based on a modified measure of Bonacich centrality. From a network-based point of view, whenever a social planner wants to minimize (maximize) the aggregate behavior, it must remove the node with highest (lowest) centrality measure. Whilst Ballester et al. (2006) consider complementarity effects in function of neighbors' *aggregate* actions, Ushchev and Zenou (2018) proposed an alternative mechanism where the effects are derived from the *average* value of neighbors' actions instead. Ushchev and Zenou proposed that whenever a game accounts for a social norm, network-based policies arising from local-aggregate models have little to no effect, as a local-average framework must be used instead. Specifically regarding networks of tax evasion, Andrei et al. (2014) studied how network effects played a non-negligible role in the aggregate tax behavior and concluded that network structures accounting for higher closeness centralities tend to enhance tax compliance. Garcia Alvarado (2019) further examined the "neighbor" peer effects on tax evasion under homophily, meaning agents with similar characteristics tend to link between each other, and proposed a social-cohesive network formation model preserving a high closeness centrality measure.

The rest of the current chapter develops as follows. Section 2 details an Agent-Based Model implementation which considers a social-cohesive network formation where agents share information about past audit schemes. Further, individuals update their own subjective probability of being audited and compute the optimal fraction of income they will disclose to the Tax Agency (or social planner). The social planner then performs audits on a fraction of individuals and the cycle repeats in a dynamic setup for several generations. Section 3 studies the model as a local-average (also called linear-in-means) game and finds the Nash Equilibrium and social norm of the game. Section 4 identifies the optimal audit policy from a network-based perspective and establishes the validity of such optimal audit strategy by performing robustness checks over different scenarios. Section 5 concludes the findings of this chapter and calls for further research regarding the use of Agent-Based Models and networks to study tax evasion and tax compliance.

2 Agent-Based Model Implementation

The current section explores the dynamic mechanism through which individual agents interact between each other, and with the social planner, in a tax compliance model. Figure 1 shows a flowchart depicting the five subroutines encompassed in the Agent-Based Model. First, agents are initialized with intrinsic, heterogeneous characteristics and assorted by means of a predefined network formation game. Later, agents derive the optimal fraction of income they wish to disclose $d^* \in [0, 1]$ to the Tax Agency. Afterward, the social planner singles out a group of agents to be audited by selecting one of the predefined audit strategies it might enforce. Immediately after, individuals communicate amongst themselves by sharing information with their set of "neighbors". Lastly, agents update their endogenous parameters, mainly their subjective probability of being audited and their loss-aversion coefficient. Once the fiscal cycle is closed, agents must optimally derive the fraction of income to be declared and repeat the game until they are retired from the labor market.

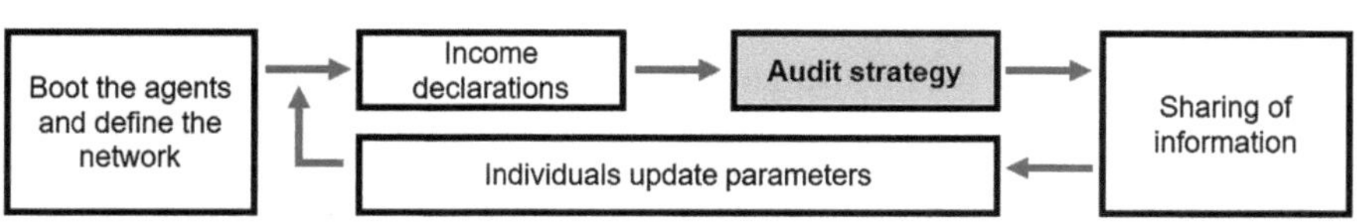

Fig. 1 A dynamic model of tax compliance

Network Formation

The launching of the model starts by generating a collection of agents and assigning to each one of them a set of exogenous characteristics: $\{\kappa$: tax-morale, ρ: loss-aversion, p: true audit rate, $\hat{p}$: subjective audit rate, I: income, and $age\}$. Individuals will face their first ever tax compliance decision using their exogenously given parameters and will convert their characteristics to endogenous with each decision cycle that takes place. Defining a society-level tax morale of $\kappa_S \in (0, 0.5)$ which remains exogenous and constant along time, each individual is initialized with a personal tax morale sampled from a uniform distribution $\kappa \sim U(0, 2\kappa_S)$. Age is distributed uniformly between 18 and 65 years; thereafter each player who reaches the age of 65 will be retired from the labor market and replaced by a new agent aged 18. The agents' initial loss-aversion coefficient is inputted from the distribution $\rho \sim U(0, 1)$, yet the parameter will become endogenous as soon as the agent coexists within its network.

Immediately upon the parameter booting lies an essential part of the ongoing Agent-Based model with peer effects and social norms: the network formation process. Understanding by *closeness*-centrality the efficiency to spread information inside a network, Andrei et al. (2014) advocate the implementation of highly closeness-central structures to model tax evasion. Garcia Alvarado (2019) tested nine different network formation process and concluded that fitness model compositions have a statistically significant advantage to enhance tax compliance metrics over any other proposed structure. A basic, yet convenient, fitness model network formation process which accounts for peer effects and social pressure may be mirrored from the one introduced by Billari et al. (2007).

The network construction commences by assigning to each individual exogenous and endogenous parameters: age and social angle, respectively. Further, agents are mapped in a two-dimensional spatial model where the x-axis represents age and the y-axis stands for social angle. The age criterion is non-mutable by the agents and increases by one unit each fiscal cycle, while the social angle may be understood as income, which stochastically increases for the individual with respect to the time axis. The tax evasion model will be further measured in fractions of income, therefore the absolute value of wealth is not relevant for the network formation, i.e., the income position mapped as the y-axis is sufficient. After all agents have been laid out, they tend to move North–East in the Cartesian plane until they reach the age of sixty years old. Once that occurs, the agent is replaced by an offspring which is mapped with 18 years of age and a fraction of the leaving agent's income. The model is ran for three generations until the final structure resembles the one depicted in Fig. 2.

Figure 2 represents each individual as a human-shaped node and each link by a direct connection between nodes. Moreover, agents are "homophilic", implying they tend to link exclusively with individuals which are relevant or akin to themselves. Therefore, a node's neighbors would be the nodes which are close to itself, both in x-axis and y-axis, bringing by consequence social-cohesion to the network. Such arrangement follows Ushchev and Zenou's (2018) call for interest in modeling local-average games with networks which take into consideration homophily and social-cohesion. The clearcut advantage of this network formation rule over any random

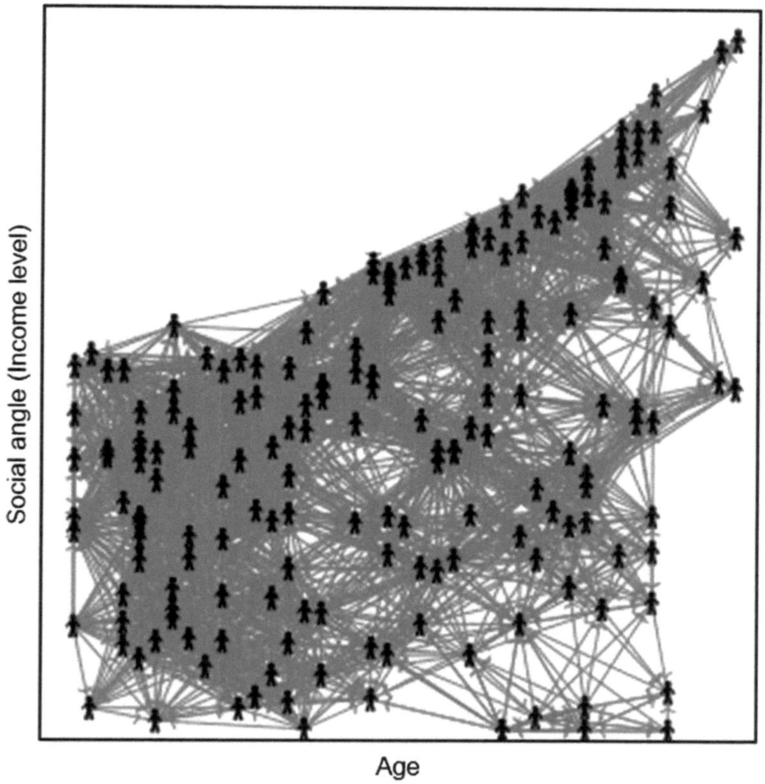

Fig. 2 A network with homophily and social-cohesion

process is its pragmatic resemblance to a real-life scenario where people discuss their fiscal issues and experiences exclusively with individuals who are socially akin to themselves and not with randomly chosen nodes in the network.

Tax Evasion from the Agent's Point of View

The seminal expected utility model of tax evasion considers two states of the world for any agent: to be audited or not. Considering the micro-founded optimization problem where there is a probability p of being audited, agents must derive the optimal fraction of disclosed income $d^* \in [0, 1]$ such that Eq. (1) is maximized:

$$EU[d] = p \cdot U(X) + (1 - p) \cdot U(Y), \quad (1)$$

where X is the net income after taxes and penalties in case an audit takes place, and Y is the net income in case no audit takes place. Expected utility models of this nature are inherently limited to study non-observable incomes, particularly the tax behavior of individuals whose true gross incomes are not known by the Tax Agency.

Likewise, even if a jurisdiction would be able to adopt an income-matching system for tax purposes, the model would still be valid for the non-observable fraction of the economy.

Expressing X and Y in terms of the gross earned income I, the applicable tax rate τ, and the penalty rate θ applied to the taxes due from the undisclosed fraction of income in case an audit occurs, Eq. (1) may be reformulated as

$$EU = p \cdot U[I - \tau(d \cdot I) - \theta\tau(I - d \cdot I)] + (1 - p) \cdot U[I - \tau(d \cdot I)] \quad (2)$$

Whenever the utility function to be maximized takes into consideration only the monetary values of the outcomes, Eq. (2) suffers from a commonly known drawback: the equation highly overestimates the level of evaded income tax. Solving for optimality conditions, a fully rational individual would understate its income whenever the composed enforcement parameter $p \cdot \theta$ is sufficiently low, i.e., whenever $p \cdot \theta < 1$. Given that the probability $p \in [0, 1]$ cannot be larger than one, the strict inequality implies that the fine rate θ must be strictly larger than one. This is a rather logical assumption, given that if the fine rate would be smaller than one, then it would be optimal for the taxpayer to be in the audit state of the world and face a tax payment of $\theta \cdot \tau < \tau$, thus the condition $\theta > 1$ is binding. Moreover, if the true audit rate p would be valued zero, i.e., audits for sure will never take place, then the agents would have no utilitarian incentive to fully disclose their income.

Nonetheless, the enforcement parameter $p \cdot \theta$ is hardly ever larger than one. In fact, audit rates tend to oscillate between one and two percent of the population and fine rates rarely exceed a value of 2, meaning that the enforcement condition is seldom met and $p \cdot \theta \ll 1$. In order to overcome this setback, a model with tax morale and loss-aversion parameters is proposed in a power utility function as follows:

$$U(d, W) = (1 + d)^{\kappa} W^{(1-\rho)}, \quad (3)$$

where $W = \{X, Y\}$ is understood as the end-of-period wealth, $\kappa \in (0, 1)$ is the tax-morale level and $\rho \in (0, 1)$ is the loss-aversion coefficient. Moreover, individuals are not assumed to be omni-conscious, thus agents do not know their true probability of being audited (p) but instead optimize under their subjective probability of being audited ($\hat{p}$). Considering for each agent i and each time t, yet discarding for simplicity the subindex, the expected utility to be maximized by each agent on every fiscal period is characterized as

$$EU = \hat{p}(1 + d)^{\kappa}[I - \tau(d \cdot I) - \theta\tau(I - d \cdot I)]^{(1-\rho)} + (1 - \hat{p})(1 + d)^{\kappa}[I - \tau(d \cdot I)]^{(1-\rho)} \quad (4)$$

The loss aversion coefficient employed in Eqs. (3) and (4) stands for the well-known wealth effect described by Tversky and Kahneman (1973). In this sense, a higher loss-aversion would concavely diminish the marginal utility of end-of-period wealth. On the other hand, a higher tax morale would yield a larger complying utility. Nonetheless, tax morale is a rather troublesome expression to be defined.

Tax morale has been a pivotal concern in tax compliance models ever since Myles and Naylor (1996) upheld a supplementary utility derived from conforming to societal rules. Accordingly, agents gain an additional benefit by following the established social norms. Although tax morale is ordinarily used as an umbrella term to accommodate a range of definitions, Luttmer and Singhal (2014) define tax morale as any intrinsic motivation, reciprocity, culture, biases and social influences which enhance tax compliance.

Audit Strategies

Taking advantage of the Agent-Based Model implementation, a wide set of audit strategies may be employed under diverse policy scenarios to computationally identify the optimal approach. Different audit schemes may be categorized into three main classifications: thumb-rules, simple strategies, and sophisticated measures. Table 1 shows the two thumb rules, four simple strategies based on the two-dimensional fitness model structure previously specified in this section, and four sophisticated schemes derived from their reciprocal network centrality measures.

Communicating Information

Besides the agents' predominant objective of maximizing their expected utility functions, individuals face a second noticeable ambition: to discover their true probability of being audited. Agents do not want to underestimate their faced audit rate given their loss-aversion of being audited, yet they do not wish to overestimate as they might disclose a fraction of income larger than optimal. Substituting p for the subjective audit rates $\hat{p}$ in Eq. (1) and deriving the First Order Conditions, it is trivial to see that the Expected Utility of an agent is reduced as the perceived audit rate increases:

$$\frac{\partial EU[d]}{\partial \hat{p}} = U(X) - U(Y) \leq 0 \tag{5}$$

Table 1 Set of possible audit strategies

Strategy	Method to choose the "n" agents to be audited
Random	Agents are randomly sampled with replacement
Cutoff	The agents who declared the lowest income values
Up-right	Randomly sampled from the North–East sector: $I_i \geq \overline{I}$ and $\text{age}_i \geq \overline{\text{age}}$
Up-left	Randomly sampled from the North–West sector: $I_i \geq \overline{I}$ and $\text{age}_i < \overline{\text{age}}$
Down-right	Randomly sampled from the South–East sector: $I_i < \overline{I}$ and $\text{age}_i \geq \overline{\text{age}}$
Down-left	Randomly sampled from the South–West sector: $I_i < \overline{I}$ and $\text{age}_i < \overline{\text{age}}$
Degree	Individuals with the highest number of direct connections
Betweenness	Agents who control the flow of information
Closeness	Individuals which make the flow of information more efficient
Intercentrality	"Key Players" as understood in Ballester et al. (2006)

Equation (5) is smaller or equal to zero given that the net income after an audit is performed, X, is smaller or equal to the net income when no audit is enforced, Y. The equality holds if and only if an individual decided to fully declare its income.

Thus, in order to discover their true audit rates, individuals mutually exchange information. At the end of each period, once the audit schemes have taken place, agents are limited to only communicate whether they have faced audits or not in the previous periods. Preceding Agent-Based Models have assumed that individuals are able to share their past income disclosures and their post-audit monetary payoffs, or even sharing their subjective probability of being audited during the next period. Notwithstanding, this chapter considers such assumptions to be too strong, as it would be hardly the case that in a real-life scenario people would be willing to freely share their tax evasion schemes along their neighborhood. Therefore, the assumption over their ability to communicate information is substantially relaxed to account for simple, observable, pieces of information. Moreover, Sect. 4 offers a robustness test to check how this limited information attains more realistic results than communicating more complex messages, and further considers the case where an individual's attention might be biased into listening to some of its neighbors with more consideration than to others.

Updating Parameters

Without a doubt, the paramount parameter encompassed in this Agent-Based network model is the agent's subjective probability of being audited. Individuals may update their perceived audit rate through temporal and geographical means. Temporal updating techniques encompass two mechanisms, the agent's prior subjective audit rate and the individual's own memory about the number of audits it may recall in the past years. Furthermore, geographical updating refers to the information received from its neighbors regarding their own memories about previous audit schemes. Equation (6) defines the subjective audit probability to be employed by agent i at time $t+1$ as a weighted average of the three updating channels: prior belief, personal memory and social influence.

$$\hat{p}_{i,t+1} = \lambda_1 \hat{p}_{i,t} + \lambda_2 \sum_{s=1}^{S_i} \frac{A_{i,t-s}}{S_i} + (1 - \lambda_1 - \lambda_2) \sum_{j \neq i}^{N_{i,t}} \frac{1}{N_{i,t}} \sum_{s=1}^{S_j} \frac{A_{j,t-s}}{S_j} \tag{6}$$

where λ_1 and λ_2 are convex averaging weights, $A_{i,t-s}$ is valued one if the agent i was audited in the period $(t-s)$ and zero otherwise, S_j is the memory or number of audit periods that agent j can recall in the past, and $N_{i,t-1}$ is the number of neighbors of agent i at time t (without counting itself).

Figures 3 and 4 show the long-run steady-state convergence of the average subjective audit rate. Remarkably, for practically all possible combinations of λ_1 and λ_2 the society reaches a steady-state convergence; with the exception of $\lambda_1 = 1$, in which case the agents never exchange information and keep no memory regarding audits in previous years. Particularly, Fig. 3 shows the weight on prior beliefs on the left axis and the importance of memory on the right axis. Figure 4 depicts the influence of

Fig. 3 Prior and memory

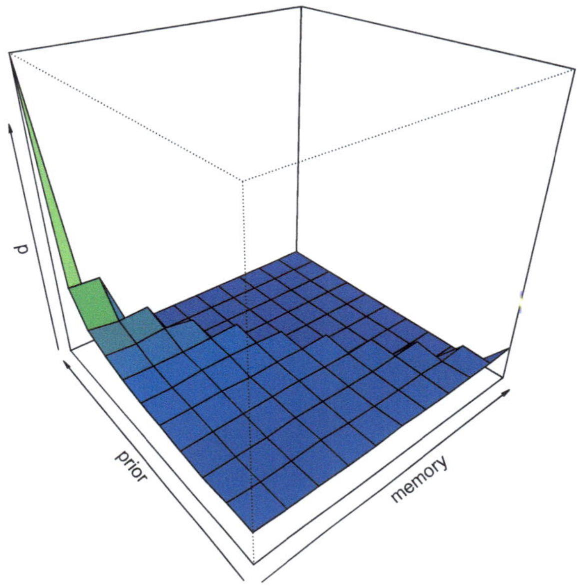

Fig. 4 Social influence and prior

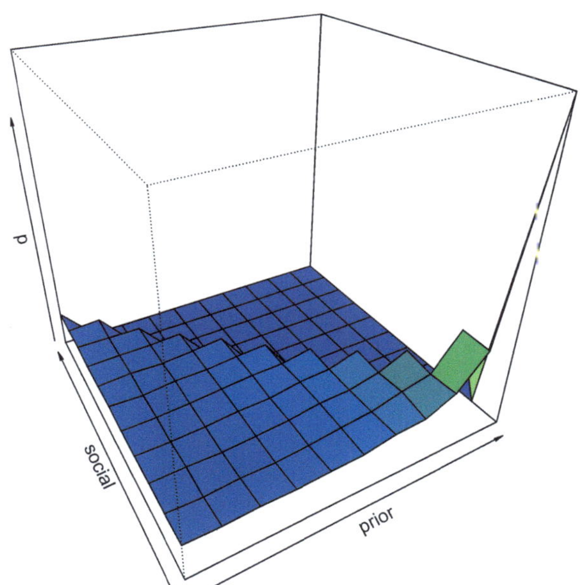

communication on the left axis and the importance of prior beliefs in the right axis. It may be observed how the steady state explodes whenever full attention is given to prior beliefs and zero emphasis is placed in the two remnant updating channels. A mathematical specification of this steady state is provided in Sect. 3.

Despite the weighted average expression in Eq. (6) allows for a different subjective audit rate per agent, the true audit rate is not endogenous itself. Hence, agents are still homogeneous amidst themselves with respect to the probability of being audited. Nonetheless, it would be safe to assume that any given Tax Agency would be more prone to audit individuals with soaring income levels and eye-catching wealth; whereas agents with low salaries might be less profitable to investigate.

In order to attain agent heterogeneity, an endogenous probability of being audited is fixed proportionally with respect to individuals' income level, thus agents with high endowments will be more susceptible to participate in audit schemes than their lower endowed counterparts. Following, the endogenous audit rate q for agent i is the homogeneous audit rate p multiplied by the ratio of the agent's income over the average income of the population.

$$q_i = \frac{I_i}{\sum_{j=1}^{N} I_j} \cdot Np \tag{7}$$

Furthermore, this convenient adjustment allows for consistent testing of parameters, as the expected value of being audited remains constant regardless of the modification. Taking Eq. (7) and summing over all agents i and dividing by N on both sides, we get $q = p$.

Analogously, other parameters are updated as well. The agents' income surges stochastically, their age increases by one unit and new agents enter and leave the labor market respectively. Another relevant benefit arising from the network formation process designed in this section is the possibility to transform the previously exogenous loss-aversion coefficient to an endogenous parameter. Recalling loss aversion as the agents' preference of avoiding a loss over attaining a win given its decreasing marginal utilities on wealth, it would be coherent to adjust such coefficient with respect to the individuals' income level. It follows from Eq. 8 that a local and endogenous loss-aversion parameter may be defined as the agent's relative income position with respect to its neighborhood N, divided by the total number of neighbors $|N|$ plus one.

$$\rho_i = \frac{\text{rank}(I_i)}{|N| + 1} \tag{8}$$

Accordingly, the highest (lowest) loss-aversion quotas relate to the most (least) affluent agents. It is interesting to notice how the proposed social-cohesive structure allows for the global distribution of individual loss-aversion of ρ to tend toward a bell shape; while maintaining its symmetry around 0.50 as in the exogenous scenario. Such bell-shaped distribution permits a more realistic characterization of the decision-making process.

3 Nash Equilibrium and Social Norms

The following section takes the social planner's point of view, whose primary objective is to find an audit policy such that the perceived audit rate is maximized given a limited budget of audits it may enforce in a single fiscal period. The standard *linear-in-means* model is suitable for games where peer effects and social norms play a role. Most of the results hereafter presented were first studied and derived by Ushchev and Zenou (2018). In this model, agents take into consideration the *average* value of actions enforced by their neighbors. Further, individuals exert an action defined by

$$x_{ig} = z_{ig}\beta + y_g\gamma + \frac{\psi}{N_g}\sum_{j=1, j\neq i}^{n} x_{jg} + \epsilon_{ig} \tag{9}$$

where x_{ig} is the action of agent i who lives in group g, z_{ig} are the individual-level covariates (tax morale and loss-aversion), y_g are the group-level variables (tax and fine rates), and ψ is defined as the agent's social interaction effect, which multiplies the average action exerted by its neighbors (not counting itself). It is important to notice how this particular game formalization is constitutionally different from the *local-aggregate* models studied in Ballester et al. (2006).

Consider an $n \times n$ adjacency matrix $\mathbf{H} = [h_{ij}]$ with entries $\{0, 1\}$, specifically, $[h_{ij}] = 1$ if there is a direct connection from i to j, and zero otherwise. Furthermore, $\mathbf{H}$ is an undirected matrix without self-loops. Consider $\mathbf{G}$ as the row-normalized version of $\mathbf{H}$ after dividing each entry by the degree of node i; $[g_{ij}] = [h_{ij}]/d_i$. It follows from this mathematical specification that $[g_{ij}]$ is the influence that player j exerts on agent i, which resembles the last term from Eq. (6).

Define a modified version of the well-known Bonacich centrality (1987) as specified by Ushchev and Zenou (2018):

$$\mathbf{M} := (1 - \lambda)(\mathbf{I} - \lambda\mathbf{G})^{-1}, \tag{10}$$

and rewrite the definition as

$$\mathbf{M} = (1 - \lambda)\sum_{k=0}^{\infty} \lambda^k \mathbf{G}^k. \tag{11}$$

The $n \times n$ matrix $\mathbf{M}$ is well-defined and row-normalized for any $\lambda \in [0, 1)$, given that the matrix $\mathbf{G}$ is row-normalized. The matrix $\mathbf{M}$ will come in very useful to compute the Nash equilibrium. Consequently, $\mathbf{M}$ may be understood as the matrix which keeps track of the number of walks of length k from node i to j ($g_{ij}^{[k]}$) discounted by a factor of λ^k, where each entry m_{ij} may be expressed as

$$m_{ij} = (1 - \lambda)\sum_{k=0}^{\infty} \lambda^k g_{ij}^{[k]}. \tag{12}$$

Analogous to Sect. 2, agents are interested in discovering the true audit rate they face on every fiscal period: underestimating the true value is risky while overestimating it would curtail their attained utility. Define x_{it} as the updated value of the subjective probability for agent i at time t. Moreover, $x_{it} = \hat{p}_{it}$ whenever true audit rates are homogeneous, whereas $x_{it} = \hat{q}_{it}$ for the case where rates are heterogeneous.

The concept of *social norm* for agent i in a local-average game, like this one, is defined as the mean value of the actions exercised by the neighbors of i. Following,

$$\bar{x}_i = \sum_{j=1}^{n} g_{ij} x_j, \tag{13}$$

and in matrix form notation:

$$\bar{\mathbf{x}} = \mathbf{G}\mathbf{x}. \tag{14}$$

Define an alternative utility function, v, which measures the agent's additional benefit from adhering to the *social norm*, $\bar{x}_i$, roughly interpretable as the "socially accepted" true audit probability. Agent's i utility function v_i is constructed as follows:

$$v_i(x_i, \mathbf{x}_{-i}, \mathbf{g}) = \alpha_i x_i - \frac{1}{2}x_i^2 - \frac{1}{2} \cdot \frac{\lambda}{1-\lambda}(x_i - \bar{x})^2 \tag{15}$$

The parameter $\alpha_i > 0$ in the linear-quadratic utility function specified in Eq. (15) corresponds to an individual heterogeneity factor. The coefficient α_i may be inferred as an adjustment rate to be multiplied by the updated subjective probability of being audited x_i. Roughly speaking, α_i may be derived as the relative income position that agent i has inside the network as previously proposed in Eq. (7). It is worth mentioning for the scenario with homogeneous audit rates that $\alpha_i = \alpha_j \ \forall\{i, j\}$.

The negative coefficient for the quadratic term of x_i makes it not optimal for agents to infinitively increase their action x_i. Finally, the last term subtracts the quadratic value of the spread between the exerted action and the social norm. Following, agents have a *taste for conforming* to the established social norm. The last term could also be interpreted as the *peer effects* motivated by *group pressure*.

Deriving the First Order Conditions from Eq. (15) with respect to x_i and equalizing to zero yields:

$$\frac{\partial v_i(x_i, \mathbf{x}_{-i}, \mathbf{g})}{\partial x_i} = \alpha_i - x_i - \frac{\lambda}{1-\lambda}(x_i - \bar{x}_i) = 0. \tag{16}$$

Computing x_i^* we obtain:

$$x_i^* = (1-\lambda)\alpha_i + \lambda \bar{x}_i, \tag{17}$$

reformulating the previous expression in matrix notation and substituting $\bar{\mathbf{x}} = \mathbf{G}\mathbf{x}$:

$$\mathbf{x} = (1-\lambda)\boldsymbol{\alpha} + \lambda \mathbf{G}\mathbf{x}, \tag{18}$$

applying linear algebra we compute the Nash Equilibrium $\mathbf{x}^*$ as

$$\mathbf{x}^* = (1-\lambda)[\mathbf{I} - \lambda\mathbf{G}]^{-1}\boldsymbol{\alpha}, \tag{19}$$

which, applying Eq. (10), may be simplified to:

$$\mathbf{x}^* = \mathbf{M}\boldsymbol{\alpha}. \tag{20}$$

Lastly, substituting Eq. (14) in Eq. (20) we obtain the equilibrium social norm $\bar{\mathbf{x}}^*$:

$$\bar{\mathbf{x}}^* = \mathbf{G}\mathbf{M}\boldsymbol{\alpha}. \tag{21}$$

Stochastic Nash Equilibrium

Consider now the matrix $\mathbf{G} = [g_{ij}]$ as a transition probability matrix in the Markovian sense, i.e., the probability of moving to the next state depends only on the current state. There are n states of the universe, characterized by the n positions of the nodes inside the network. Consider as well the random variable X_i as the action exerted by agent i, who has a heterogeneity factor α_i as previously defined. Seemingly, each agent has a true audit rate proportional to its relative income position α.

The stochastic approach to Nash Equilibrium contemplates the game as a sequence of mimicking behaviors, where each agent i at each step t must mimic the actions of another agent. We will understand "to mimic a behavior" as "believing my true audit rate to be the true audit rate of agent j". Therefore, agent i will choose its own α_i with a probability $(1-\lambda)$ and will choose a neighbor j with probability λ. Then, with probability $(1-\lambda)$, agent i will mimic α_j and with probability λ will choose a different neighbor k. The process will repeat until agent i has mimicked the audit rate of another player in the network.

The probability that agent i mimics the behavior of agent j is the total number of walks from i to j weighted by a decaying factor according to the length of the walk k. Mathematically speaking, the random variable may be well defined as

$$\mathbb{P}\{X_i = a_j\} = (1-\lambda)g_{ij}^{[0]}\alpha_j + (1-\lambda)\lambda g_{ij}^{[1]}\alpha_j + (1-\lambda)\lambda^2 g_{ij}^{[2]}\alpha_j + \dots \tag{22}$$

which simplifies to:

$$\mathbb{P}\{X_i = a_j\} = (1-\lambda)\sum_{k=0}^{\infty}\lambda^k g_{ij}^{[k]}, \tag{23}$$

where $g_{ij}^{[0]} = 1$ if $i = j$ and zero otherwise.

The last expression is equivalent to the one obtained in Eq. (12). That is, $\mathbb{P}\{X_i = a_j\}$ is the probability that agent i will mimic the behavior of agent j. Thus, substituting Eq. (12) in Eq. (23), the expected value of the random variable X_i is:

$$\mathbb{E}[X_i] = \sum_{j=1}^{n} m_{ij}\alpha_j, \tag{24}$$

which in matrix notation transforms to:

$$\mathbb{E}[\mathbf{X}] = (1-\lambda)\sum_{k=0}^{\infty} \lambda^k \mathbf{G}^k \boldsymbol{\alpha} = \mathbf{M}\boldsymbol{\alpha} = \mathbf{x}^*. \tag{25}$$

There are three reasons why the Nash Equilibrium computation is relevant for the current work. First, the equivalence between both deriving estimations suggest that the Nash Equilibrium is reached whether the agents are fully rational or merely stochastic decision makers. Therefore, bounded rationality or heuristic ruling would not limit the steady-state convergence. The second is that, using the row-normalized adjacency matrix, the Nash Equilibrium exists, is unique and it is interior as long as $\frac{\lambda}{1-\lambda} > 0$. Finally, the Nash Equilibrium is expressed in terms of the agents' heterogeneity parameter α. Patacchini and Zenou (2012) demonstrated that for local-average games where all the α coefficients are homogeneous among agents, the outcome does not depend on the network structure nor on the total number of agents involved; which is not true for local-aggregate games as studied by Ballester et al. (2006). This last remark has utter policy implications for the social planner pursuing to maximize the steady state $\mathbf{x}^*$ with respect to a network-based strategy, as the success of strategies is dependent on the heterogeneity levels inside the network.

4 Optimal Audit Policies

Ensuing the derivations computed in the previous section, it is germane to recall how the subjective audit rate, defined in Eq. (6), resembles a local-average game. The perceived audit rate is updated in function of the agent's prior, memory and the information received from its neighbors. The last term in Eq. (6) is analogous to the empirical mean audit rate experienced by i's neighborhood (N_i) over all the periods of time for which the agents keep memory. Redefining this last term as

$$\bar{A}_{i,t} = \sum_{j\neq i}^{N_{i,t}} \frac{1}{N_{i,t}} \sum_{s=1}^{S_j} \frac{A_{j,t-s}}{S_j}, \tag{26}$$

The updating mechanism may be reinterpreted as

$$\hat{p}_{i,t+1} = f(\text{prior}_t) + g(\text{memory}_t) + \psi \bar{A}_{i,t}, \tag{27}$$

where $\bar{A}_{i,t}$ is the sample estimate for the "social norm" of neighborhood N_i at time t, multiplied by an influence parameter ψ. Accordingly, the dynamics of subjective audit rates may be understood and studied as local-average repeated games.

Homogeneous Audit Rates

There is, however, an inconvenient in considering the subjective probability updating mechanism as a local-average game. Despite being able to compute the Nash Equilibrium in the previous section, there is no closed-form solution for a first-best maximization policy, given that the variable of interest is not observable, taxable, nor allows for subsidies. It is trivial, and perhaps humorous, to notice how it would be impossible for a social planner to "tax away" an agent's subjective probability of being audited and "redistribute" this belief to other individuals. Notwithstanding, the Agent-Based Model outlined in Sect. 2 was implemented to test ten different audit schemes, referring to the random audits method as the benchmark case, and assessed whether or not any of such strategies may be deemed as superior with respect to randomization.

Hereafter all simulations will be understood as dynamic games with fifty fiscal periods, where 300 agents face a 30% flat tax rate, a 2.0 fine rate if caught, a fiscal memory of four periods, a tax morale level of 0.25, endogenous loss-aversion, randomly sampled initial prior beliefs at time zero, balanced weighting averages $\lambda_1 = \lambda_2 = (1 - \lambda_1 - \lambda_2) = 1/3$, and, most importantly, an homogeneous true audit rate of 5% indistinctly applicable to all agents.

Figure 5 portrays the four sophisticated audit strategies contrasting against the random audit scheme (red). Optically, there seems to be little to no difference between all five strategies; the intercentrality measure has, however, a marginal and statistically significant higher level. In a similar way, Fig. 6 shows how the four two-dimensional audit instruments do not account for a statistically significant difference with respect to the random auditing process.

Table 2 summarizes the results obtained from the simulations for each audit strategy. The second column depicts the sample's average subjective probability of being audited during the final step of the simulation (which computationally approximates the Nash Equilibrium). Column $\Delta\hat{p}$ is a goodness-of-fit metric which shows the

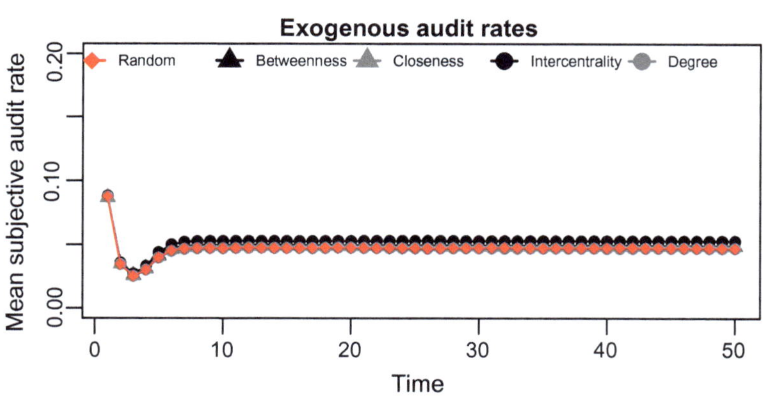

Fig. 5 Scenario where true audit rates are homogeneous

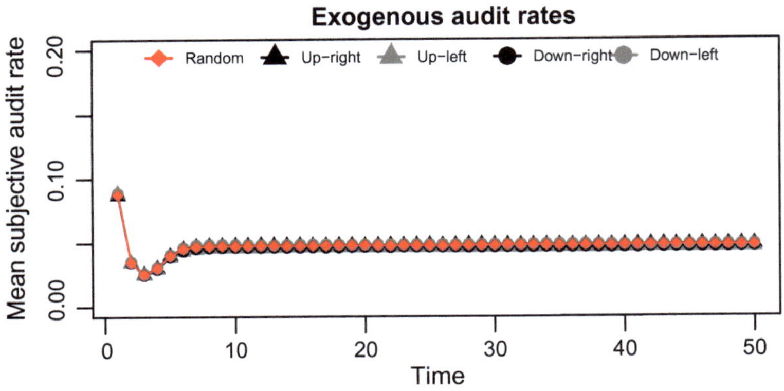

Fig. 6 Scenario where true audit rates are homogeneous

Table 2 Comparing strategies under homogeneous audit rates

Strategy	Average $\hat{p}$ (%)	$\Delta\hat{p}$ (%)	Δd^* (%)	Δ–Revenue (%)	*p*-Value (%)
Random	4.77	0	0	0	–
Cutoff	4.68	−2	−6	−6	0.626
Betweenness	4.93	3	3	1	0.243
Closeness	4.88	2	4	4	0.525
Intercentrality	5.31	11	3	1	0.001
Up-right	4.81	1	0	6	0.896
Up-left	4.61	−3	0	0	0.272
Down-right	4.67	−2	2	−4	0.414
Down-left	4.84	1	0	−6	0.629
Degree	4.76	0	−2	−3	0.827

Note Comparisons are made against the Random audit scenario

percentage change of the average $\hat{p}$ attained by each strategy with respect to the value attained by the benchmark (random audits).[1] The highest achiever according to this metric is the Intercentrality auditing scheme, which enlarged the sample's perceived audit rate by 11%. The next column, Δd^*, shows the increase gained for the average fraction of income declared by individuals. The Cutoff strategy seems to be counterproductive while three sophisticated strategies reaped an increase between three and four percent. Column Δ–Revenue shows the ancillary return on actual tax payments collected. Following this metric, the Intercentrality scheme seems to have lost its appeal versus the Closeness-centered audits. It turns out that an increase of just 11% in the mean perceived audit rate is not enough to significantly enhance tax compliance and escalate fiscal revenues.

[1] The score is computed by subtracting the benchmark value from the strategy's outcome and diving by the same benchmark value: $\Delta = \frac{\hat{p} - \hat{p}_{BM}}{\hat{p}_{BM}}$; analogous for all Δ–columns.

The last column of Table 2 shows the p-values scored by a Welch two-sample *t-test* between the observed steady state of $\hat{p}$ for each evaluated strategy versus the outcome $\hat{p}$ arising from randomized audits; where the null hypothesis supposes that the true difference in means is equal to zero. Following, with the exception of the Intercentrality scheme, no strategy encompasses a statistically significant effect on the long-run average value of $\hat{p}$ with respect to the benchmark scenario. As anticipated from Ushchev and Zenou (2018), sophisticated local-aggregate strategies will play little to no effect whenever in a local-average framework. Moreover, as foreseen from the previous section, simple network-based policies have no effect whenever all agents have a homogeneous α coefficient, i.e., there is no agent heterogeneity with respect to the true audit rate. Agents thus converge to a Nash Equilibrium which is roughly the true audit rate, homogeneously applicable for all individuals.

Heterogeneous Audit Rates

The previous subsection proved how, under homogeneous audit rates, network-based audit policies have little to no effect to enhance the audit rate perceived by agents. A social planner who is aware of this, might then attempt to add heterogeneity to the network. Assume a social planner, or Tax Agency, is capable of executing a credible message such that all individuals believe the news are true and binding. Aforesaid official communication would read something akin to *"From now on, the probability of being audited will be directly proportional to the income level of each individual."*

Subsequently, individuals would update their subjective probability of being audited in an endogenous fashion, as established in Eq. (7). Ensuing:

$$\hat{q}_{i,t+1} = \hat{p}_{i,t+1} \cdot \frac{I_i}{\sum_{j=1}^{N} I_j} \cdot N \tag{28}$$

Recurring to the same simulation parameters as previously defined, all strategies are re-examined under a new updating mechanism, where true audit rates are both endogenous and different for each individual. Figure 7 sketches a nonexistent improvement for the sophisticated audit strategies. The logic behind this ineffectiveness can be explained by understanding how, in a local-average game, targeting a few "key players" does not affect the "social norm" of the group. From a network-based point of view, an optimal policy should attempt to change the agents' perception; in other words, it should induce individuals to believe their likelihood to be audited is higher than it actually is. Therefore, the social planner should target a specific *group* of agents in lieu of key individuals.[2]

Figure 8 graphically depicts the main findings of this chapter: agent-based policies may be pertinent for local-aggregate models, however, group-based policies are more appropriate for local-average games. It is immediate to notice the remarkably high long-run Nash Equilibrium emerging from the simplistic Up-right audit policy.

Table 3 displays the performance of the audit strategies under heterogeneous audit rates and compares them versus the (homogeneous) random audit scheme presented

---

[2]For a more extensive discussion refer to Ushchev and Zenou (2018).

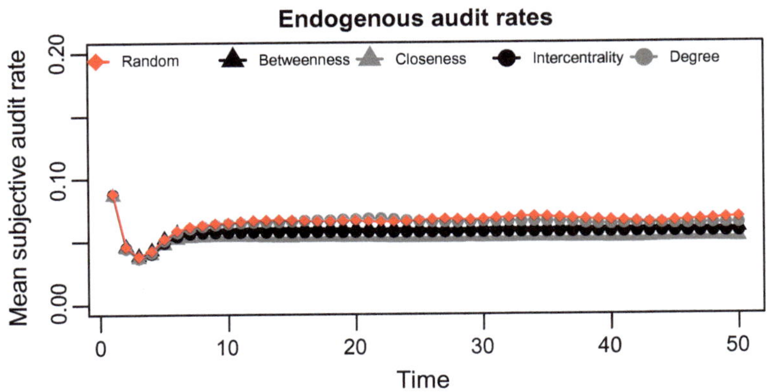

Fig. 7 Scenario where true audit rates are heterogeneous

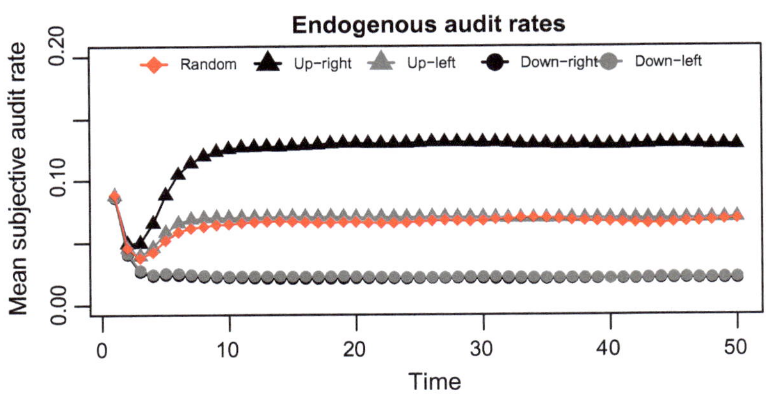

Fig. 8 Scenario where true audit rates are heterogeneous

in Table 2. Although several strategies earned a higher average $\hat{q}$ with respect to randomized audits, there is a striking steady-state value of 12.88% secured by the Up-right policy. The spiking $\Delta\hat{q}$ score for the Up-right strategy reaches 170%, meaning that it nearly doubled the benchmark's average perceived audit rate. Column Δd^* shows how the mean fraction of income declared rose an additional 13% for this same policy. Even more, metric Δ–Revenue announces how revenues augmented by about 16% whenever auditing the North–East sector of the population (bear in mind that this outcome pertains only to the simulation and cannot be straightforwardly employed to forecast how a country's shadow economy could be expeditiously curtailed). Lastly, the column with p-values displays near-zero values for almost all strategies; implicating that under endogenous true probabilities, the selection of one or other audit strategy will have real and statistically significant effects on the outcome.

Table 3 Comparing strategies under heterogeneous audit rates

Strategy	Average $\hat{q}$ (%)	$\Delta\hat{q}$ (%)	Δd^* (%)	Δ–Revenue (%)	p-Value (%)
Cutoff	4.24	−11	−5	−9	0.000
Betweenness	6.01	26	6	2	0.000
Closeness	5.25	10	5	−1	0.000
Intercentrality	5.71	20	−1	−7	0.000
Up-right	12.88	170	13	16	0.000
Up-left	6.95	46	9	5	0.052
Down-right	2.11	−56	−9	−15	0.000
Down-left	2.23	−53	−8	−14	0.000
Degree	6.40	34	3	0	0.169

Note Comparisons are made against the Random audit scenario presented in Table 2

Robustness Checks

As a consequence from the results presented in Table 2, which vastly favor the Up-right audit policy, this predominant strategy was further tested under a wide spectrum of diverse scenarios to allow for robustness checks.

Hereby and after, all *t-tests* are understood as two-sided Welch *t-tests* where the null hypothesis states that the difference between sample means is zero. Thirty simulations were ran per each scenario. The first robustness test, depicted in Fig. 9, studies the effect of modifying the initial prior distribution. The p-values attained from the *t-tests* proved that the null hypothesis cannot be rejected for any upper bound P of the initial prior distribution $\sim U(0, P)$: in other words, the initial prior belief of agents does not play any role in the long-run average subjective audit rate.

Figure 10 illustrates a positive effect of the steady state with respect to memory length (number of fiscal years an agent can recall into the past). In fact, the *t-tests* revealed that there is a positive elasticity effect of memory. However, there was no statistically significant difference for remembrance spans larger than seven fiscal periods; thus memory has a marginal effect up to a certain extent. The neighborhood size in Fig. 11, benchmarked as one, represents the spatial range of individuals that

Fig. 9 Initial prior beliefs

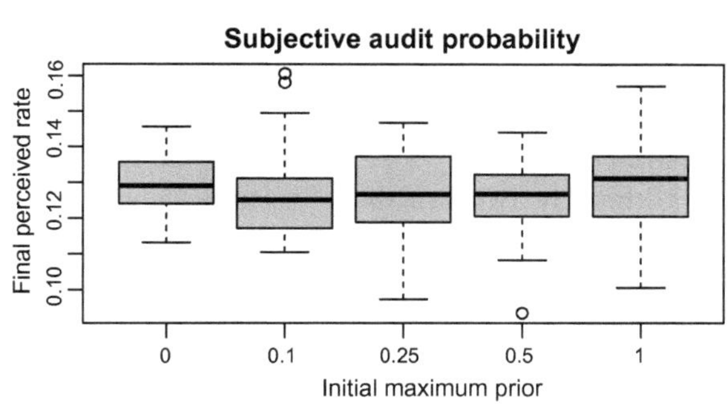

Fig. 10 Length of memory

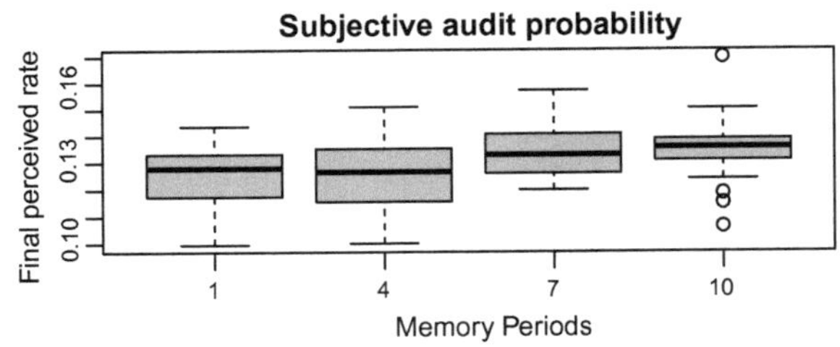

Fig. 11 Neighborhood size

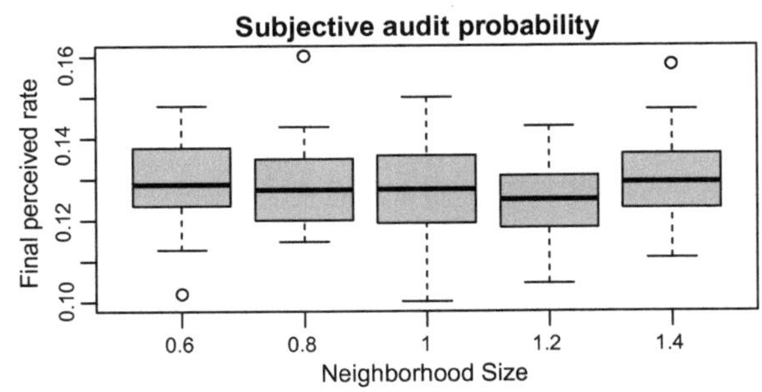

are available for each agent to form a link with. As long as the network remains ergodic, meaning that every node is reachable from any other node after a finite number of links, neighborhood size has no statistical significance in the perceived audit rate. The number of agents, as seen in Fig. 12 plays a nonlinear, statistically significant, positive effect on the steady state. Not only the long-run perceived audit rate marginally increases as the number of agents augments, but it also converges to the steady state with a smaller standard deviation. Figure 13, accompanied by its respective set of *t-tests*, supports the asseveration that tax rates do not significantly affect the perceived audit rate. Moreover, further analysis was done employing a stepped-tax schedule and no statistically significant difference was encountered with respect to modeling under flat tax rates.

A different approach was employed in Fig. 14. As it would be rather obvious that the perceived audit rate would be significantly affected by the true audit rate, the y-axis of Fig. 14 measures the relative overestimation of the average true audit rate in the long-run. Following, for a mean audit rate of 0.05%, agents tend to overestimate it by nearly 80% of the true value. Ongoing, the social norm overestimates the expected probability of being audited of 1%, 2%, 5%, 10% and 20% by 150%, 170%, 160%, 130% and 60%, respectively. Thus, the Up-right group-based audit instrument would perform satisfactorily for any nonzero true audit rate.

Ultimately, two structural scenarios were tested. First, consider a more realistic environment where individuals are more prone to take fiscal advices from people who are either more affluent, older, or both. Defining an agent performing such an attention favoritism as to be "Up-right biased", Fig. 15 portrays how this preference for counseling would lead to a statistically higher social norm whenever the Tax

Fig. 12 Number of agents

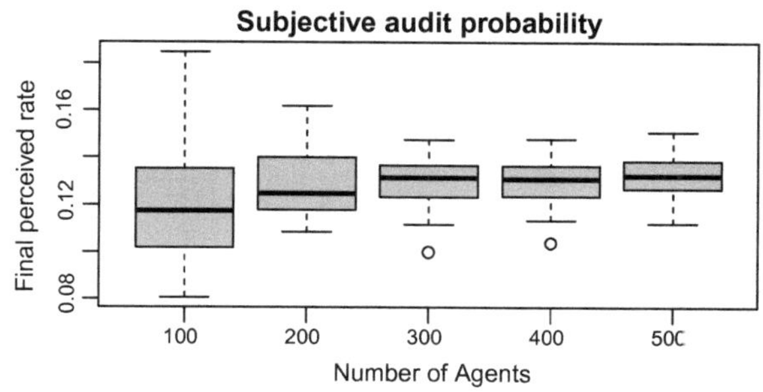

Fig. 13 Tax rate

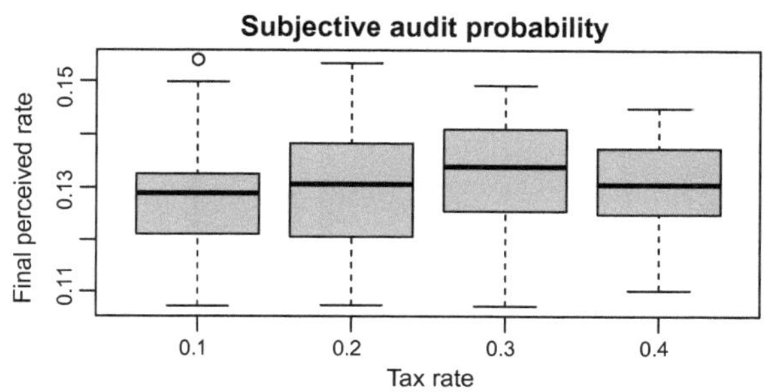

Fig. 14 Percentage increase in $\hat{q}$

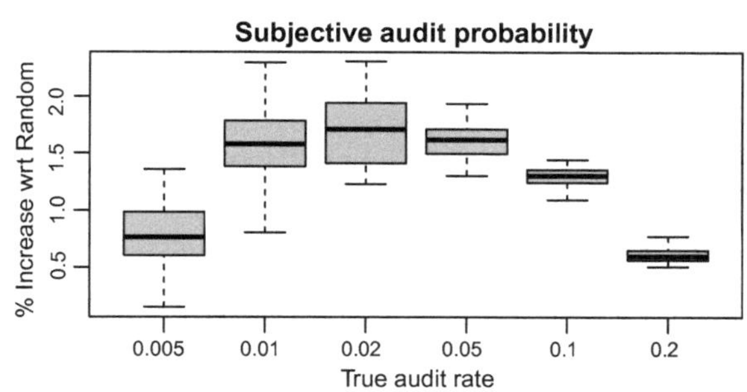

Agency enforces an Up-right policy scheme. The second structural setting modification allows agents to share their rationally updated subjective probability of being audited instead of being limited to communicate only their binomial memories about previous audit occurrences. Figure 16 demonstrates an abysmal disparity between the two signaling channels. Whenever agents share their rationally updated probability of being audited (which, to begin with, is in no way a pragmatic assumption about the way agents behave in the real-life) the long-run average perceived audit rate is ludicrously intensified. Figure 16, for example, shows how a true expected audit rate of 5% is largely overestimated and perceived as an audit rate of more than 30%. Ergo, the proposed audit strategy and signaling mechanisms not only resulted to be optimal in all simulated scenarios, but they also enjoy a certain anecdotal legitimacy.

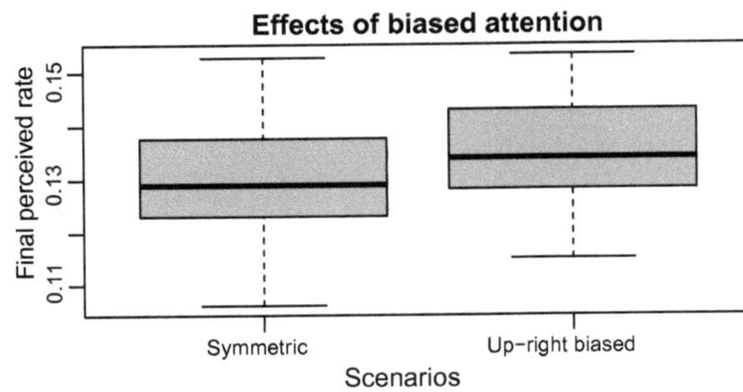

Fig. 15 Symmetric versus biased attention frames

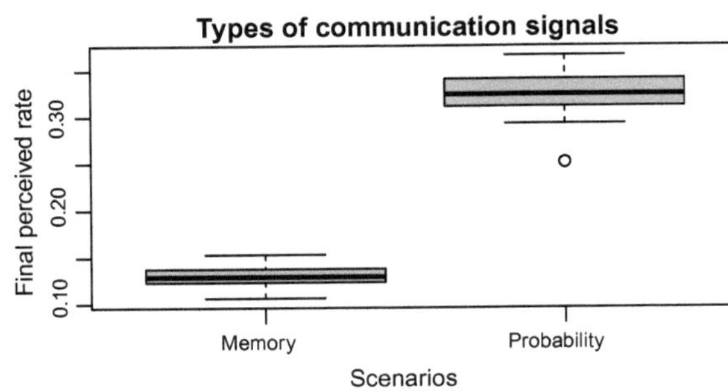

Fig. 16 Sharing $A_{i,t-s}$ versus sharing $\hat{p}_i$

It is worth mentioning that an extreme audit instrument, denoted "Corner" strategy, targeted the agents at the uppermost rightmost sector of the structure, yet was found to be less profitable than the simple "Up-right" scheme. Notwithstanding, more research is encouraged in the study of more complicated group-based audit policies.

Apparently, if a Tax Agency acts coherently respecting its announcement, the social norm of the group may be shifted to a higher level. Up to this time, targeting agents which are relatively more affluent, and older, in the North–East sector of the spatial network, yielded higher steady states of the subjective audit rates, enhanced the optimal fractions of income disclosed and augmented the revenue collections.

5 Conclusions

In this chapter a tax evasion model is studied from the social planner's (Tax Agency) point of view. The Tax Agency employs a two-step optimal solution, which consists in initially emitting a credible and binding message, stating: *"From now on, the probability of being audited will be directly proportional to the income level of each individual."* Assuming each agent faces a heterogeneous true audit rate, individuals attempt to discover their endogenous probability of being audited. Taxpayers are modeled as agents in a social-cohesive homophilic two-dimensional network, where individuals link only to "relevant others" which are akin to themselves. Moreover, each fiscal period, agents attempt to update their subjective probability of being audited by taking into consideration the signals received from their neighbors. Every individual interacts under limited communication, informing to its peers only the agent's own observed audit frequency faced in the previous years, up to where the individual is able to remember. Considering the updating mechanism as a local-average game, the average perceived audit rate of the whole society converges to a Nash Equilibrium steady state in the long run.

Attempting to optimize the revenues accrued from tax payment collections, the social planner resorts to maximize the subjective audit rate perceived by the individuals. An Agent-Based Model implementation allowed the model to be tested under a wide set of possible audit strategies, ranging from thumb-rules to simple group-based schemes, and even to audit instruments based on sophisticated centrality measures. The chapter found that, under homogeneous audit rates, practically no audit strategy is any better compared versus simple randomized income verifications. Under heterogeneous audit rates, however, a particular network-based audit policy deemed to be optimal by achieving an average perceived audit rate that almost doubled the benchmark and considerably increased the tax revenues in the simulated scenarios. Furthermore, the strategy proved to be robust under a collection of diverse settings, attaining a marginally stronger effect for larger number of agents and for longer memory spans. This chapter calls for further research about the implementation of network-based policies and Agent-Based Modeling, particularly for criminal behavior networks, analogous to the case of tax evasion and tax compliance.

References

Allingham MG, Sandmo A (1972) Income tax evasion: a theoretical analysis. J Public Econ 1(3–4):323–338

Alm J, Bloomquist KM, McKee M (2017) When you know your neigh- bour pays taxes: information, peer effects, and tax compliance (Working Paper Series No. 6775). Victoria University of Wellington, Chair in Public Finance

Andrei AL, Comer K, Koehler M (2014) An agent-based model of network effects on tax compliance and evasion. J Econ Psychol 40(C):119–133

Ballester C, Calvó-Armengol A, Zenou Y (2006) Who's who in networks. Wanted: The key player. Econometrica 74(5):1403–1417

Becker GS (1968) Crime and punishment: an economic approach. J Polit Econ 76:169

Billari F, Diaz BA, Fent T, Prskawetz A (2007) The "wedding-ring". Demogr Res 17(3):59–82

Bonacich P (1987) Power and centrality: a family of measures. Am J Sociol 92:1170–1182

Garcia Alvarado F (2019) Network effects in an agent-based model of tax evasion with social influence. In: Demazeau Y, Matson E, Corchado JM, de la Prieta F (eds) Advances in practical applications of survivable agents and multi- agent systems (chap. Forthcoming). Springer LNAI

Hokamp S, Pickhardt M (2010) Income tax evasion in a society of heterogeneous agents—evidence from an agent-based model. Int Econ J 24(4):541–553

Korobow A, Johnson C, Axtell R (2007) An agent-based model of tax compliance with social networks. Natl Tax J 60(3):589–610

Luttmer EFP, Singhal M (2014) Tax morale. J Econ Perspect 28(4):149–168

Mittone L (2006) Dynamic behaviour in tax evasion: an experimental approach. J Behav Exp Econ (formerly The Journal of Socio-Economics) 35(5):813–835
Mittone L, Patelli P (2000) Imitative behaviour in tax evasion (Economic simulations in swarm: agent-based modelling and object oriented programming)
Myles G, Naylor R (1996) A model of tax evasion with group conformity and social customs. Eur J Polit Econ 12(1):49–66
Patacchini E, Zenou Y (2012) Juvenile delinquency and conformism. J Law Econ Organ 28(1):1–31
Tversky A, Kahneman D (1973) Availability: a heuristic for judging frequency and probability. Cogn Psychol 5:207–232
Ushchev P, Zenou Y (2018) Social norms in networks. SSRN Electron J

Network Games: The Cooperative Approach

Surajit Borkotokey, Loyimee Gogoi and Rajnish Kumar

Abstract We survey the recent developments in network games under cooperative framework. Network structures are important in describing and analysing social and economic phenomena through which people interact with each other and are influenced by their peers. Game theory has been successfully applied as a tool for such descriptions and analysis. The corresponding games are called network games. Situations, where people cooperate under binding agreements, are best modelled by cooperative games. One of the primary concerns in the field of social and economic network theory is to design networks on the basis of such binding agreements among the interacting agents. Bi-cooperative games are a special class of cooperative games where people can choose to support or oppose an issue or remain indifferent while making binding agreements. Bi-cooperative network games are network games where agents can make positive or negative links or remain isolated based on their choices of supporting or opposing the issue. We primarily focus on both standard and non-trivial solution concepts of network games under cooperative and bi-cooperative set ups and highlight the recent applications of the theory in dealing with socio-economic issues.

Keywords TU cooperative games · Bi-cooperative games · Network games · Bi-cooperative network games

JEL Classifications: C71 · C72

S. Borkotokey (✉)
Department of Mathematics, Dibrugarh University, Assam, India
e-mail: surajitbor@yahoo.com

L. Gogoi
Department of Mathematical Sciences, Tezpur University, Assam, India

R. Kumar
Economics Group, Queen's Management School, Queen's University, Belfast, UK

A. S. Chakrabarti et al. (eds.), *Network Theory and Agent-Based Modeling in Economics and Finance*, https://doi.org/10.1007/978-981-13-8319-9_21

1 Introduction

Networks (social and economic networks, communication networks, etc.) are in the heart of everyday life. In the last two decades, the study of networks has received considerable interest in the scientific community, especially in Economics and other disciplines of Social Sciences, Finance and Managements, Computer Science and Biological Sciences, etc., to name a few. Our decisions and beliefs are often influenced by the people with whom we interact and share our ideas and opinions. Examples of such interactions are plenty, viz. in adopting new technologies and products, building political opinions, transmitting information on jobs, establishing formal and informal insurance, developing micro-finance that further influences trade patterns, policy framing in the macro level, etc., to name a few, see Jackson (2005b, 2011), Jackson et al. (2017). Jackson (2008) identifies two aspects of the study of networks from an Economist's point of view. The first aspect is to explore how the network activities influence the economic activities and second to determine the tools that can be used to analyse and model such network activities.

In this chapter, we confine ourselves to a much narrower objective stemming out from the second aspect, where the strategic interactions among agents in a socio-economic situation are considered to be the cause of network formation. We assume that the networks are not random. They are formed as a result of agents' incentives to maximize their own welfare. We also restrict ourselves to only those models where agents make binding agreements and can freely transfer their utilities among themselves. Formally we call them network games in cooperative framework or simply network games. Examples include informal risk sharing and favour exchange, provision of public goods and economic exchange, etc., see Jackson et al. (2017) for details.

Cooperative games with transferable utilities or TU games in short, deal with situations where players make binding agreements to form coalitions. A TU game is the pair (N, v) where N is the set of players or agents and $v : 2^N \to \mathbb{R}$ is a characteristic function such that $v(\emptyset) = 0$. For each subset S of N, which we call a coalition, the value $v(S)$ represents the worth of S. If there is no ambiguity on the player set N, we represent a cooperative game simply by its characteristic function v. The basic assumption of cooperative games is that the grand coalition forms. A solution is an n-vector assigning payoffs to each player in N. The standard requirement of a solution is that it satisfies efficiency, i.e. it should exhaust the total worth of the grand coalition among the players. Different solution concepts for TU games have been proposed, see Shapley (1953), Béal et al. (2014), Banzhaf (1965), Casajus and Huettner (2014), Joosten (1996), Nowak and Radzik (1994), Radzik and Driessen (2013), Sprumont (1990), etc., for example. Among the many valued solutions, the core, the weber set, the bargaining set, etc., are few of the significant solutions (for a detailed description of these values we refer to Peters (2008)). On the other hand, the Shapley value introduced by Shapley (1953) is perhaps the most popular one point solution for TU games. The one point solutions are also called values. Intuitively, we can represent a value by a function defined on the class of all TU games that assigns each game an n-vector in $\mathbb{R}^n$. Besides, the solidarity values given by Nowak and Radzik (1994), Béal et al. (2014), etc., the egalitarian rule, the egalitarian Shapley value given by Joosten (1996), etc., are few other prominent values.

Myerson (1977) introduces the Graph restricted TU games. The basic premise of his model is that due to various economic and social reasons, not all coalitions are feasible among players. The only admissible coalitions are those in which players make direct or indirect links among themselves. Therefore, the characteristic function is restricted to the subclass of graph restricted coalitions. Caulier et al. (2017) gives an example concerning the impossibility of coalition formations in a political situation. It is argued that, even though a coalitional government can be formed with many parties under some binding agreements, it is highly unlikely that the far left and the far-right parties can match up. Based on this idea, values for graph restricted games which were latter termed as communication situations by Jackson (2005a) are put forward. Myerson (1977) proposes a value for communication situations which was latter accepted as the Myerson value. Meessen (1998) proposes the Position value for communication situations. The Position value is also studied and characterized by Borm et al. (1992), Slikker (2005), Slikker (2007), van den Nouweland and Slikker (2012), etc.

Alternative to the communication situations, a model of network games under cooperative set up is proposed by Jackson and Wolinsky (1996). They argue that if the formation of links is costly, then how players make links: directly or indirectly, also matters. The resulting network game is a classical cooperative game on the set of possible (sub)networks. In such a model, a link between two agents of the network indicates a collaboration (cooperation) between these agents. In network games, one primarily concerns about the worth of the network, through which the players/agents are connected and supposed to communicate/collaborate/exchange. 'Worth' is taken in its broadest sense and depends on the peculiar context (e.g. this could be the total monetary value produced or handled, or the power of influence or control (e.g. for a mafia or terrorist network)). The worth of a network is produced by the network, in terms of, e.g. its monetary or economic value, its power of influence or control, etc. There has been a number of surveys on communication situations and their respective solution concepts, see Gogoi et al. (2018), Caulier et al. (2017). Therefore, here our primary concern is to the study the network games.

A crucial issue concerning network games is how to share between the agents the total worth produced by the network in a rational way. (or, in case worth reflects power: how to compute a power index of each agent? Who is most powerful/influential? This question is crucial in case of terrorist networks.) A general rule, called 'allocation rule', has to be defined, and is similar to what is called 'solution' of a cooperative game. Indeed, most allocation rules for network games hitertho given by Borkotokey et al. (2015), Jackson (2005a), van den Nouweland and Slikker (2012), Belau (2013, 2016), etc., are close to the Shapley value of some underlying cooperative game. Allocation rules are of two types: player based and link based, based on their affinities towards players and links, respectively. Following the same procedure of the computation of the Shapley value in TU games, the corresponding player-based and link-based, values are computed. They are, respectively, called the network Myerson value and the network Position value to distinguish them from their counterparts in communication situations, see Borkotokey et al. (2014). The network Position value is called link-based allocation rule because it allocates the payoff to links first and then to players. It is the average of the marginal contributions of each link at a time with the average being computed over all orders of possible link formations. The network Myerson value, on the other hand evaluates for each player the marginal contributions of all his links at a time. These two values, are therefore, the two extreme cases of the possible interactions of a player with her peers. Borkotokey et al. (2015) propose an allocation rule, called the interactive allocation rule, that accounts for multilateral interactions in a network. Each player can marginally contribute with some (and at least one) but not all his links. The authors characterize the interactive allocation rule using standard Shapley like axioms. Belau (2013, 2016) proposes and characterizes two new allocation rules for network games, namely the χ-value and the κ-value that generalize the network Myerson value and the network Position value, see Caulier et al. (2017).

As a generalization of the cooperative games, bi-cooperative games have been introduced by Bilbao (2000) and subsequently studied by Bilbao et al. (2008b, 2010), etc., in order to distinguish between positive collaborations (supposing to go in the direction of a common aim to achieve) and negative or harmful collaborations, which tends to prevent from realizing this common aim. Unlike cooperative game theory, bi-cooperative games are able to distinguish between positive (creative) collaboration/relation between agents, and negative (harmful) collaborations, the latter aspect being overlooked in most studies of networks so far and therefore we fail to take advantage of bi-cooperative games in the modelling of networks, which are particularly suited to the representation of positive and negative synergy/collaboration between agents. Solution models of bi-cooperative games can be found in Bilbao et al. (2007, 2008a, b), Labreuche and Grabisch (2008). Network games under bi-cooperative setups are studied by Borkotokey et al. (2018). As a motivating example, Borkotokey et al. (2018) consider a geopolitical issue on which countries with different motives agree to negotiate. The negotiations are facilitated/influenced by all existing pairwise trade relationships among these countries. A positive link in the trade network represents a complementary trade relation while a negative link indicates a competitive relationship. The motivational differences will influence which trade is suitable or subject to amendments. They also give an alternative situation citing (Manea 2011) where the players with opposing stance do not have any information about each other and a bargaining between randomly matched players leads to possible link formation and create components in a network, called mutually estranged sets. Note that as a third possibility, players can have positive and negative links simultaneously. This situation is best described by the example of double agents working for various spying agencies, see Isby (2004) or the duality of involvement of players in different networks according to their affinity to different groups of people. Another example is of a strict physical instructor who may be a very jovial person among friends. It is interesting to note that neither a network game under simple cooperative framework nor a bi-cooperative game without a network can model the above- mentioned situations.

The purpose of this survey is to make a brief discourse on various network inflicted games of cooperative and bi-cooperative nature and identify their possible research prospects. We present a historical perspective of the emergence of network games under cooperative set up. We survey the recent developments in bi-cooperative games and also in their subsequent network counterparts. The rest of the paper proceeds as follows. Section 2 presents the preliminary definitions and results of cooperative games and bi-cooperative games. In Sect. 3, we discuss the notion of network games. In Sect. 4 we discuss the notion of bi-cooperative network games and their respective solution concepts. Section 5 explores the application possibilities and the future prospects of the particular area of research.

2 Preliminaries

In this section, we present the preliminary notions of cooperative and bi-cooperative games relevant to the present study. Throughout this paper, we consider a fixed and finite set of players $N = \{1, 2, ..., n\}$. A coalition S is a subset of N and the set of possible coalitions in N is the power set of N, denoted by $2^N = \{S | S \subseteq N\}$. We call N the grand coalition. To avoid heavy notations we write the singleton coalition $\{i\}$ as i and the corresponding subtraction $S \setminus \{i\}$ as $S \setminus i$. Similarly we use $S \cup i$ instead of $S \cup \{i\}$ to describe coalition S with player i. Introducing the positive and negative contributions in coalitions, the notion of bi-coalitions is defined in the following manner. Let the player set N be partitioned into three subsets $\{S, T, N \setminus (S \cup T)\}$ such that $S \cap T = \emptyset$. The underlying assumption is that players in S contribute positively to the game, players in T contribute negatively and those in $N \setminus (S \cup T)$ are absentees. Each such pair (S, T) that partitions N into three subsets is called a bi-coalition. Let $\mathcal{Q}(N) = \{(S, T) \mid S, T \subseteq N, S \cap T = \emptyset\}$ be the set of all bi-coalitions in N. Define the order relation $\sqsubseteq_M$ in $\mathcal{Q}(N)$ which is implied by monotonicity, viz. for $(S, T), (S', T') \in \mathcal{Q}(N)$, $(S, T) \sqsubseteq_M (S', T')$ iff $S \subseteq S'$ and $T' \subseteq T$. Under this operation, $\mathcal{Q}(N)$ is a distributive lattice with the bottom element $(\emptyset, N)$ and the top element $(N, \emptyset)$. Labreuche and Grabisch (2008) argue that the game structure defined on $\mathcal{Q}(N)$ with the order implied by monotonicity is isomorphic to the multichoice game introduced by Hsiao and Raghavan (1993) with three levels of participation. Therefore, to make this structure non-trivial, Labreuche and Grabisch (2008) adopt the product order, viz. for $(S, T), (S', T') \in \mathcal{Q}(N)$, $(S, T) \sqsubseteq_P (S', T')$ iff $S \subseteq S'$ and $T \subseteq T'$. Under this order $\mathcal{Q}(N)$ is an inf-semilattice with $(\emptyset, \emptyset)$ as the least element and all $(S, N \setminus S)$, $S \subseteq N$, the maximal elements. Let us denote by the small letters the size of the coalitions, viz. s, t, etc., as the size of coalitions S, T, etc.

2.1 Cooperative Games

Recall from Sect. 1 that a cooperative game with transferable utility or simply a TU game (N, v) can be represented by its characteristic function v if the player set N is fixed. The set of all cooperative games with the player set N is denoted by $\mathcal{G}(N)$. The basic premise of TU games is that players cooperate to make the highest profit by forming the grand coalition N. Under the standard additions and scalar multiplications of functions over the field of reals, $\mathcal{G}(N)$ is a linear space. $v \in \mathcal{G}(N)$ is monotonic if $v(S_1) \leq v(S_2)$ for all $S_1, S_2 \in 2^N$ with $S_1 \subset S_2$. It is convex if $v(S_1 \cup S_2) + v(S_1 \cap S_2) \geq v(S_1) + v(S_2)$ for all $S_1, S_2 \in 2^N$ and superadditive if $v(S_1 \cup S_2) \geq v(S_1) + v(S_2)$ for all $S_1, S_2 \in 2^N$ with $S_1 \cap S_2 = \emptyset$. For each coalition $S \in 2^N$, the unanimity game (N, u_S) makes an important subclass of $\mathcal{G}(N)$ which is defined as follows:

$$u_S(T) = \begin{cases} 1 & \text{if } S \subseteq T \\ 0 & \text{otherwise.} \end{cases}$$

It can be shown without much efforts that the set of unanimity games $\{u_S \in \mathcal{G}(N) | S \subseteq N\}$ forms a basis for $\mathcal{G}(N)$, see Shapley (1953). Thus every cooperative game v can be uniquely represented by a linear combination of the unanimity games as follows:

$$v = \sum_{\emptyset \neq S \subseteq N} \lambda_S v_S \tag{2.1}$$

where $\lambda_\emptyset = 0$ and $\lambda_S = \sum_{T \subseteq S} (-1)^{s-t} v(T)$ for all $\emptyset \neq S \subseteq N$ are called Harsanyi dividends. Recall from Sect. 1 that a value is a function $\Phi : \mathcal{G}(N) \to \mathbb{R}^n$. In the latter sections, we mostly focus on the Shapley like solutions in networks, therefore, in the following we give a formal definition of the Shapley value. The Shapley value due to Shapley (1953) is given by

$$\Phi_i^{Sh}(v) = \sum_{S \subseteq N \setminus i} \frac{s!\,(n-s-1)!}{n!} v(S \cup i) - v(S) \tag{2.2}$$

The characterization of the Shapley value requires the Axioms $S1 - S4$ below. Prior to that, we define the following. A player $i \in N$ is a null player in v if $v(S \cup i) = v(S)$ for every coalition $S \subseteq N$. The productivity of the null player in any coalition is zero. Two players $i, j \in N$ are called symmetric with respect to the game v if for each $S \subseteq N \setminus \{i, j\}$, $v(S \cup i) = v(S \cup j)$.

Axiom S1. Efficiency (Eff): A value $\Phi : \mathcal{G}(N) \to \mathbb{R}^n$ satisfies efficiency (Eff) if for the game $v \in \mathcal{G}(N)$:

$$\sum_{i \in N} \Phi_i(v) = v(N)$$

Axiom S2. Null Player Property (NP): The value $\Phi : \mathcal{G}(N) \to \mathbb{R}^n$ satisfies the null player property if for all $v \in \mathcal{G}(N)$ it holds that $\Phi_i(v) = 0$ for each null player $i \in N$ in v.

Axiom S3. Symmetry (Sym): A value $\Phi : \mathcal{G}(N) \to \mathbb{R}^n$ is symmetric if for every pair of symmetric players $i, j \in N$ with respect to the game $v \in \mathcal{G}(N)$, $\Phi_i(v) = \Phi_j(v)$.

Axiom S4. Linearity (Lin): The value $\Phi : \mathcal{G}(N) \to \mathbb{R}^n$ satisfies linearity (Lin) if $u, w \in \mathcal{G}(N)$, each pair of $\gamma, \eta \in \mathbb{R}$ and each player $i \in N$:

$$\Phi_i(\gamma u + \eta w) = \gamma \Phi_i(u) + \eta \Phi_i(w) \tag{2.3}$$

Φ satisfies additivity if in particular, Eq. (2.3) holds only for $\gamma = \eta = 1$.

2.2 *Bi-cooperative Games*

Given the player set $N = \{1, 2, ..., n\}$ of players, and $\mathcal{Q}(N)$, the set of all pairs (S, T) with $S, T \subseteq N$, and $S \cap T = \emptyset$, a bi-cooperative game is defined as a function $b : \mathcal{Q}(N) \to \mathbb{R}$ satisfying $b(\emptyset, \emptyset) = 0$. Following the same terminologies as in TU games, we call $b(S, T)$ the worth of bi-coalition $(S, T) \in \mathcal{Q}(N)$. Note that here, worths may be both positive and negative. If no ambiguity arises on the player set N, we denote a bi-cooperative game (N, b) simply by b. Let $\mathcal{B}(N)$ denote the class of bi-cooperative games on $\mathcal{Q}(N)$. A single point solution for bi-cooperative games is a function $\Phi : \mathcal{B}(N) \to \mathbb{R}^n$, which assigns to every game (N, b) an n-dimensional real vector $\Phi(N, b)$ that represents a payoff distribution over the players. If no ambiguity arises on the player set N, with an abuse of notations, we use $\Phi(b)$ instead of $\Phi(N, b)$. Labreuche and Grabisch (2008) define the LG value[1] as a generalization of the Shapley value in the bi-cooperative framework. Recall that there are different proposals of values for bi-cooperative games, see Bilbao (2000), Bilbao et al. (2008a, b), Felsenthal and Machover (1997), Grabisch and Labreuche (2005), Labreuche and Grabisch (2008), etc. Among them the first proposal was put forward by Felsenthal and Machover (1997) especially for ternary voting games. In proposals made by Bilbao (2000) and similarly by Grabisch and Labreuche (2005, 2008), only the largest and least elements of the lattice $(\mathcal{Q}(N), \sqsubseteq)$ are important. The underlying assumption is that all players agree to a certain proposal or oppose it to form the bi-coalitions $(N, \emptyset)$ and $(\emptyset, N)$, respectively. However this requirement is too restrictive to appreciate and as already mentioned, in particular, fails to adhere to the notion of bipolarity present in a bi-cooperative game. Following Labreuche and Grabisch (2008), we have the next definition.

[1]The value proposed by Labreuche and Grabisch (2008) is named as the LG value by Borkotokey and Gogoi (2014), Borkotokey et al. (2018) to distinguish it from the original Shapley value of TU games.

Definition 1 A player $i \in N$ is called *positively monotone* with respect to $b \in \mathcal{B}(N)$ if $b(S \cup i, T) \geq b(S, T)$ for each $(S, T) \in \mathcal{Q}(N \setminus i)$. Player i is *negatively monotone* with respect to b if $b(S, T \cup i) \leq b(S, T)$ for each $(S, T) \in \mathcal{Q}(N \setminus i)$. The bi-cooperative game b is monotone if all players are both positively and negatively monotone with respect to b.

The expression $b(S \cup i, T) - b(S, T)$ is called the marginal contribution of player i with respect to $(S, T) \in \mathcal{Q}(N \setminus i)$ when she is a positive contributor and the expression $b(S, T) - b(S, T \cup i))$ is called the marginal contribution of player i with respect to $(S, T) \in \mathcal{Q}(N \setminus i)$ when she is a negative contributor. The definition of the LG value due to Labreuche and Grabisch (2008) for bi-cooperative games requires the following:

Definition 2 Player $i \in N$ is a null player for the bi-cooperative game b, if for every $(S, T) \in \mathcal{Q}(N \setminus i)$, we have $b(S \cup i, T) = b(S, T) = b(S, T \cup i)$.

Definition 3 A function $\Phi : \mathcal{B}(N) \to (\mathbb{R}^n)^{\mathcal{Q}(N)}$ defines the LG value, for every $(S, T) \in \mathcal{Q}(N)$, if it satisfies the following axioms:

Axiom B1 (Efficiency): For $b \in \mathcal{B}(N)$, we have, $\sum_{i \in N} \Phi_i(b)(S, T) = b(S, T)$.

Axiom B2 (Linearity): For $\alpha, \beta \in \mathbb{R}$, $b, v \in \mathcal{B}(N)$, $\Phi_i(\alpha b + \beta v) = \alpha \Phi_i(b) + \beta \Phi_i(v)$.

Axiom B3 (Null Player axiom): If player i is a null player for $b \in \mathcal{B}(N)$, then, $\Phi_i(b)(S, T) = 0$, for all $(S, T) \in \mathcal{Q}(N)$.

Axiom B4 (Intra-Coalition symmetry): If $b \in \mathcal{B}(N)$ and a permutation π is defined on N, such that $\pi S = S$ and $\pi T = T$, then it holds that, for all $i \in N$, $\Phi_{\pi i}(\pi b)(\pi S, \pi T) = \Phi_i(b)(S, T)$, where $\pi b(\pi S, \pi T) = b(S, T)$ and $\pi S = \{\pi i : i \in S\}$. This axiom implies that the players are anonymous. The payoff does not depend on the numbering of the players.

Axiom B5 (Inter-Coalition symmetry): Let $i \in S$ and $j \in T$, and $b_i, b_j \in \mathcal{B}(N)$ such that for all $(S', T') \in \mathcal{Q}(N \setminus \{i, j\})$,

$$b_i(S' \cup i, T') - b_i(S', T') = b_j(S', T') - b_j(S', T' \cup j)$$
$$b_i(S' \cup i, T' \cup j) - b_i(S', T' \cup j) = b_j(S' \cup i, T') - b_j(S' \cup i, T' \cup j)$$

Then,

$$\Phi_i(N, b_i)(S, T) = -\Phi_j(N, b_j)(S, T). \tag{2.4}$$

Axiom B5 implies that if a player $i \in S$ in a game b_i behaves exactly opposite to a player $j \in T$ in a game b_j, then the payoff of i in b_i is exactly opposite to j in b_j.

Axiom B6 (Monotonicity): Given $b, b' \in \mathcal{B}(N)$ such that $\exists i \in N$ with

$$b'(S', T') = b(S', T') \tag{2.5}$$
$$b'(S' \cup i, T') \geq b(S' \cup i, T') \tag{2.6}$$
$$b'(S', T' \cup i) \geq b(S', T' \cup i) \tag{2.7}$$

for all $(S', T') \in \mathcal{Q}(N \setminus i)$, then $\Phi_i(b')(S, T) \geq \Phi_i(b)(S, T);\ \forall\ (S, T) \in \mathcal{Q}(N)$.

Define the function $\Phi^{LG} : \mathcal{B}(N) \to (\mathbb{R}^n)^{\mathcal{Q}(N)}$ by

$$\Phi_i(b)(S, T) = \sum_{K \subseteq (S \cup T) \setminus i} \frac{k!(s + t - k - 1)!}{(s + t)!} [V(K \cup i) - V(K)] \tag{2.8}$$

where for $K \subseteq S \cup T$, $V(K) := b(S \cap K, T \cap K)$.

Labreuche and Grabisch (2008) prove that the function $\Phi^{LG} : \mathcal{B}(N) \to (\mathbb{R}^n)^{\mathcal{Q}(N)}$ given by Eq. (2.8) is the unique LG value characterized by Axioms B1-B6. The payoff given by the LG value in Eq. (2.8) for a positive (resp. negative) contributor depends only on her added value from being indifferent to become a positive contributor (resp. negative contributor). The information regarding how this player behaves when she switches from positive to negative is not relevant, see Labreuche and Grabisch (2008). In what follows next, the consistency of the LG value is discussed in the line of Hart and Mas-Colell (1989). This idea is necessary to obtain the Position value of bi-cooperative network games.

2.3 *Consistency of the LG Value*

The consistency of values for TU games is studied by Hart and Mas-Colell (1989). The idea is based on an intuitive notion of the potential function. Borkotokey et al. (2018) extends this idea to bi-cooperative games. In the following, we briefly mention the framework under which the LG value assumes a potential and is a consistent value in the sense of Hart and Mas-Colell (1989).

Definition 4 For $(S, T) \in \mathcal{Q}(N)$, the superior unanimity games $\{\overline{u}_{(S,T)} \mid (\emptyset, \emptyset) \neq (S, T) \in \mathcal{Q}(N)\}$ in $\mathcal{B}(N)$ are given by,

$$\overline{u}_{(S,T)}(A, B) = \begin{cases} 1, & \text{if } (S, T) \sqsubseteq_P (A, B) \\ 0, & \text{otherwise} \end{cases}$$

The above collection forms a basis for $\mathcal{B}(N)$ and so every $b \in \mathcal{B}(N)$ can be expressed as a linear combination of the superior unanimity games as follows:

$$b = \sum_{(K,L) \in \mathcal{Q}(N)} a_{K,L} \overline{u}_{(K,L)} \tag{2.9}$$

where $a_{K,L}$ are the real constants.

Following Shapley (1953), we can associate to every $b \in \mathcal{B}(N)$ and each $(S, T) \in \mathcal{Q}(N)$, a TU game v such that $v(M) = b(S \cap M, T \cap M),\ \forall\, M \subseteq S \cup T$. So, v has a corresponding representation in terms of the unanimity Cooperative games $\{U_M \mid M \subseteq N, M \neq \emptyset\}$ as follows, see Harsanyi (1963), Shapley (1953).

$$v = \sum_{\emptyset \neq M \subseteq S \cup T} a_M U_M, \quad \text{where } a_M = \sum_{L \subseteq M} (-1)^{m-l} v(L) \tag{2.10}$$

It follows that

$$\sum_{\emptyset \neq M \subseteq S \cup T} a_M = b(S, T)$$

Now after some simple computations, we find the expression of the LG value for b in terms of the Harsany's dividends for Φ^{sh}, the Shapley value of the associate cooperative game $(S \cup T, v)$ as follows, see Harsanyi (1963).

$$\Phi_i^{LG}(N, b)(S, T) = \sum_{i \ni M \subseteq (S \cup T)} \frac{a_M}{|M|} \tag{2.11}$$

Definition 5 For a given $b \in \mathcal{B}(N)$ and a function $P_{(N,b)} : \mathcal{Q}(N) \to \mathbb{R}$ associated with each (N, b), the positive marginal contribution of player i at some $(S, T) \in \mathcal{Q}(N)$ denoted by $D_i^+ P_{(N,b)}(S, T)$, is defined as

$$D_i^+ P_{(N,b)}(S, T) = \begin{cases} P_{(N,b)}(S, T) - P_{(N,b)}(S \setminus i, T), & \text{when } i \in S \\ 0, & \text{when } i \notin S \end{cases}$$

and the negative marginal contribution of player i at $(S, T) \in \mathcal{Q}(N)$ denoted by $D_i^- P_{(N,b)}(S, T)$, is defined as

$$D_i^- P_{(N,b)}(S, T) = \begin{cases} P_{(N,b)}(S, T) - P_{(N,b)}(S, T \setminus i), & \text{when } i \in T \\ 0, & \text{when } i \notin T \end{cases}$$

The expression $\Delta_i P_{(N,b)}(S, T) = D_i^+ P_{(N,b)}(S, T) + D_i^- P_{(N,b)}(S, T)$ denotes the marginal contribution of player i in (S, T) due to $P_{(N,b)}$. Note that a player can either be in S (a positive contributor) or T (a negative contributor) but not both, therefore exactly one of her marginal contributions in (S, T) due to $P_{(N,b)}$ is non-zero.

Definition 6 The Potential game $(P_{(N,b)}, N)$ or simply $P_{(N,b)}$ over the player set N is a function $P_{(N,b)} : \mathcal{Q}(N) \to \mathbb{R}$ associated with each $b \in \mathcal{B}(N)$ that satisfies $P_{(N,b)}(\emptyset, \emptyset) = 0$ and the following condition:

$$\sum_{i \in N} \Delta_i P_{(N,b)}(S, T) = b(S, T) \quad \forall\ (S, T) \in \mathcal{Q}(N) \tag{2.12}$$

Equation (2.12) can be simplified as follows:

$$\sum_{i \in S} D_i^+ P_{(N,b)}(S, T) + \sum_{i \in T} D_i^- P_{(N,b)}(S, T) = b(S, T) \quad \forall\ (S, T) \in \mathcal{Q}(N) \tag{2.13}$$

Theorem 1 *To any game $b \in \mathcal{B}(N)$, there associates a unique potential game $P_{(N,b)}$ such that the payoff to each player is given by her marginal contributions due to $P_{(N,b)}$. Moreover the resulting payoff vector, namely $\left\{\Delta_i P_{(N,b)}(S,T)\right\}_{i=1}^{n}$ coincides with the LG value of b for all $(S,T) \in \mathcal{Q}(N)$.*

For the proof of Theorem 1, we refer the reader to Borkotokey et al. (2018).

Take $\mathcal{B}(S \cup T)$ to be the class of all bi-cooperative games on $(S \cup T)$.

Definition 7 Consider the bi-coalition $(S,T) \in \mathcal{Q}(N)$, $\emptyset \neq A \subseteq (S \cup T)$ and $A \subseteq S$ or $A \subseteq T$, a bi-cooperative game $b \in \mathcal{B}(S \cup T)$ and a solution function $\Phi = (\Phi_i)_{i \in (S \cup T)}$. The upper reduced game $((S \cup T) \setminus A,\ U_b)$ and the lower reduced game $((S \cup T) \setminus A,\ L_b)$ are defined, respectively, as follows:

$$U_b(S',T') = b(S' \cup A, T') - \sum_{i \in A} \Phi_i(S \cup T, b)(S' \cup A, T') \tag{2.14}$$

$$L_b(S',T') = b(S', T' \cup A) - \sum_{i \in A} \Phi_i(S \cup T, b)(S', T' \cup A) \tag{2.15}$$

for all $(S',T') \in \mathcal{Q}((S \cup T) \setminus A)$ such that $S' \neq \emptyset$ and $T' \neq \emptyset$ and $U_b(\emptyset,\emptyset) = 0 = L_b(\emptyset,\emptyset)$.

Definition 8 Given a bi-network $(S,T) \in \mathcal{Q}(N)$. A solution Φ is said to be Consistent if for every Bi-cooperative game $b \in \mathcal{B}(S \cup T)$ and every $\emptyset \neq A \subseteq (S \cup T)$, one has

$$\Phi_j((S \cup T) \setminus A, U_b)(S',T') = \Phi_j(S \cup T, b)(S' \cup A, T') \tag{2.16}$$

$$\Phi_j((S \cup T) \setminus A, L_b)(S',T') = \Phi_j(S \cup T, b)(S', T' \cup A), \tag{2.17}$$

$\forall (S',T') \in \mathcal{Q}((S \cup T) \setminus A), \quad \forall j \in S' \cup T'$.

Lemma 1 *Let $b \in \mathcal{B}(N)$. Let the function $\mathcal{Q}_{(N,b)} : \mathcal{Q}(N) \to \mathbb{R}$ satisfy*

$$\sum_{i \in S} \left\{\mathcal{Q}_{(N,b)}(S,T) - \mathcal{Q}_{(N,b)}(S \setminus i, T)\right\} + \sum_{i \in T} \left\{\mathcal{Q}_{(N,b)}(S,T) - \mathcal{Q}_{(N,b)}(S, T \setminus i)\right\} = b(S,T) \tag{2.18}$$

for each $(S,T) \in \mathcal{Q}(N)$ with $S \neq \emptyset$ and $T \neq \emptyset$.
Then for each $(S,T) \in \mathcal{Q}(N)$,

$$\mathcal{Q}_{(N,b)}(S,T) = P_{(N,b)}(S,T) + \mathcal{Q}_{(N,b)}(\emptyset,\emptyset) \tag{2.19}$$

Theorem 2 *The LG value is Consistent.*

For the proof, we refer to Borkotokey et al. (2018).

3 Network Games

A network g builds on a finite set of nodes corresponding to players and a finite set of pairs of nodes called links which correspond to the bilateral interactions among players. Thus g is a list of unordered pairs of players $\{i, j\}$, where $\{i, j\} \in g$ represents a link in the network g. We write ij to represent the link $\{i, j\}$ to make notations simple. The set of all links in the network g is denoted by $L(g)$.

The degree of a node (player) in a network is the number of links it has in the network. The set of all possible links formed by the nodes of N is the complete network denoted by g^N. The set $G^N = \{g|g \subseteq g^N\}$ consists of all possible subnetworks on N.

$N(g)$ denotes the set of players having at least a link in g, i.e., $N(g) = \{i \mid \exists j \in N$ such that $ij \in g\}$. We assume $n(g) = \#N(g)$. For $g \in G^N$, $l(g)$ is the total number of links in g and $L_i(g)$ denotes the set of links of player i. Thus $L_i(g) = \{ij \mid \exists\, j \in N\ :\ ij \in g\}$. Let $l_i(g) = \#L_i(g)$. It follows that $l(g) = \frac{1}{2}\sum\limits_i l_i(g)$.

Given any $S \subseteq N$, g^S denotes the complete network formed by the players in S and $g|_S$ denotes the subnetwork of g formed by the players in S. Thus, $g|_S = \{ij \mid ij \in g$ and $i \in S, j \in S\}$.

For any $g_1, g_2 \in G^N$, the expression $g_1 + g_2$ denotes the network obtained through adding networks g_1 and g_2 and the expression $g_1 \setminus g_2$ denotes the network obtained from g_1 by subtracting its subnetwork g_2.

Nodes i and j are directly connected in g if $i = j$ or $ij \in L(g)$. i and j are connected in g even if i and j are not directly connected in g but there is a sequence $(i_1 = i, i_2, ..., i_k = j)$ such that $i_h i_{h+1} \in L(g)$ for $h = 1, 2, ..., k-1$. A network g is connected if any two nodes in it are connected. Induced by the notion of connectivity, a network g can be partitioned into components where the components are the maximally connected subnetworks of g. $C(g)$ denotes the set of all the components of g.

A value function is a function $v : G^N \rightarrow \mathbb{R}$ such that $v(\emptyset) = 0$, where $\emptyset$ represents the empty network, i.e., network without links. The value function specifies the total worth generated by a given network structure. It may involve both costs and benefits based on the information available.

A network game is a pair (N, v), consisting of a set N of players and a value function v. If N is fixed and no ambiguity arises due to N, the network game is denoted by only v. Let $\mathcal{NG}(N)$ denote the class of all network games defined on the player set N. A value function v is component additive if $v(g) = \sum\limits_{g' \in C(g)} v(g'),$ for any $g \in G^N$.

An allocation rule is a function $Y : G^N \times \mathcal{NG}(N) \rightarrow \mathbb{R}^n$ that allocates the worth $v(g)$ of a network g among the nodes, i.e., $\sum\limits_{i \in N} Y_i(g, v) = v(g),\ \forall g \in G^N,\ \ v \in \mathcal{NG}(N)$. Each component $Y_i(g, v)$ of $Y(g, v)$ represents the payoff to player i with respect to v and g. Recall from Sect. 1 that the Myerson value and the Position value are two important allocation rules. Allocation rules that consider multilateral interactions have also received attention recently, see Borkotokey et al. (2015). In the following we describe these rules along with their recent characterizations.

3.1 The Myerson Value and the Position Value

The definition and characterization of the Myerson value and the Position value for network games require the following definitions:

Definition 9 An allocation rule Y is component balanced (in short CB) if for any component additive v, and $g' \in C(g)$,

$$\sum_{i \in N(g')} Y_i(g', v) = v(g').$$

Component balanced allocation rules do not consider externalities across the components of a network when the corresponding network game does not show externalities across these components.

Definition 10 An allocation rule satisfies balanced contributions if for any component additive v, $g \in G^N$ and $i, j \in N$, it holds that,

$$Y_i(g, v) - Y_i(g \setminus L_j(g), v) = Y_j(g, v) - Y_j(g \setminus L_i(g), v).$$

Definition 11 An allocation rule Y satisfies balanced link contribution if

$$\sum_{l \in L_j(g)} [Y_i(g, v) - Y_i(g \backslash l, v)] = \sum_{l \in L_i(g)} [Y_j(g, v) - Y_j(g \backslash l, v)].$$

Balanced link contribution is another fairness axiom similar to the equal bargaining power, where players will have equal gain or loss on an average from removing one link at a time from the network.

Definition 12 An allocation rule Y on the domain of component additive value functions is additive if

$$Y(g, v_1 + v_2) = Y(g, v_1) + Y(g, v_2),$$

for each pair (N, v_1), (N, v_2) of network games with component additive value functions.

Additivity is a property that specifies that there are no externalities in the allocation rule when the players in a network are involved in several situations, each described by their own value function (van den Nouweland and Slikker 2012).

Definition 13 For $i, j \in N$, a link $ij \in g$ is superfluous in the network game (N, v) if

$$v(g') = v(g' \backslash ij),$$

for each network $g' \subseteq g$.

Definition 14 An allocation rule Y defined on the class of component additive value functions satisfies the superfluous link property if

$$Y(g, v) = Y(g \backslash ij, v)$$

for each network game (N, v) with a component additive value function and all links ij that are superfluous in (N, v).

The superfluous link property implies that if a link in the network is non-productive in terms of the value the network generates, the allocation rule is also indifferent to that link while computing the value for each player.

Definition 15 A value function v is link anonymous on g if $v(g') = v(g'')\ \forall\, g', g'' \subseteq g$ such that $l(g') = l(g'')$.

Definition 16 An allocation rule Y on the domain of component additive value functions is link anonymous if for each network $g \in G^N$ and link anonymous value function $v \in \mathcal{NG}(N)$ on g, there exists an $\alpha \in \mathbb{R}$ such that

$$Y_i(g, v) = \alpha l(g_i),\ \forall\, i \in N.$$

Link anonymity states that if a network after interchanging its links generates the same value, the allocations of the players in the network are dependent only on the relative number of links that each player is involved in. This idea is similar to that of the symmetry axiom of the Shapley value for cooperative games.

Two important allocation rules for network games, namely the network Myerson value $Y^{NMV} : G^N \times \mathcal{NG}(N) \to \mathbb{R}^n$ and the network Position value $Y^{NPV} : G^N \times \mathcal{NG}(N) \to \mathbb{R}^n$ are given below.

$$Y_i^{NMV}(g, v) = \sum_{S \subset N \backslash \{i\}} \frac{s!(n-s-1)!}{n!}(v(g|_{S \cup i}) - v(g|_S)) \tag{3.1}$$

$$Y_i^{NPV}(g, v) = \sum_{\substack{i \neq j, \\ ij \in g}} \left(\sum_{g' \subset N \backslash ij} \frac{1}{2}(v(g' + ij) - v(g')) \right) \frac{l(g')!(l(g) - l(g') - 1)!}{l(g)!} \tag{3.2}$$

The axioms of component balancedness and the balanced contributions are characterizing properties of the Myerson value for network games given by Slikker (2007) as follows:

Theorem 3 (Slikker (2007), pp. 497) *Y satisfies component balance and balanced contributions if and only if $Y(g, v) = Y^{NMV}(g, v)$ for all $g \in G^N$ and any component additive v.*

One of the recent characterizations of the Position value is due to van den Nouweland and Slikker (2012). In the following, we state their theorem.

Theorem 4 (van den Nouweland and Slikker (2012), pp. 13) *The Position value Y^{NPV} is the unique allocation rule on the domain of component additive value functions that is component balanced, additive and satisfies the superfluous link property and link anonymity.*

3.2 Allocation Rules Based on Multilateral Interactions

In order to obtain the multilateral interactive allocation rule introduced in Borkotokey et al. (2015) we slightly modify our notations as follows:

Definition 17 (Borkotokey et al. (2015), pp. 914) Fix a network g. Let G_g denote the set of all subnetworks of g. The class of all network games defined on G_g is given by $\mathcal{NG}_g(N)$. Now, $\mathcal{NG}_g(N)$ is a vector space over reals and of dimension $2^{l(g)} - 1$. An allocation rule on g is a function $Y_g : \mathcal{NG}_g(N) \to \mathbb{R}^n$ such that $Y_{g,i}(v)$ represents the payoff to player i with respect to $v \in \mathcal{NG}_g(N)$ and specific to the network g.

Note that so far, efficiency has been implicit to an allocation rule in network games. However, inefficient allocation rules are also important in cooperative set up, see Banzhaf (1965), van den Brink and van der Laan (1998) for example. The interactive allocation rule proposed in Borkotokey et al. (2015) has two variations, one efficient and the other inefficient. Thus we have the following definitions due to Borkotokey et al. (2015):

Definition 18 An efficient allocation rule Y_g on g is one in which $\sum_i Y_{g,i}(v) = v(g), \ \forall v \in \mathcal{NG}_g(N)$. It specifies how the value generated by the network is shared among the players forming the network g.

Definition 19 An allocation rule Y_g on g is said to satisfy the Linearity axiom (L) if for $v, \ v' \in \mathcal{NG}_g(N)$, and $a, b \in \mathbb{R}$,

$$Y_{g,i}(av + bv') = aY_{g,i}(v) + bY_{g,i}(v'), \ \forall i \in N.$$

A player i is *null* for the value function $v \in \mathcal{NG}_g(N)$ with respect to the network g, if for every $g' \subseteq g - L_i(g)$ and every $g_i \subseteq L_i(g)$, we have $v(g' + g_i) = v(g')$.

Definition 20 An allocation rule Y_g is said to satisfy the Null player axiom (NP) if for every null player $i \in N$ for $v \in \mathcal{NG}_g(N)$, we have $Y_{g,i}(v) = 0$.

Definition 21 Let π be a permutation on N. For a given network g, and for each $v \in \mathcal{NG}_g(N)$, define the game $\pi v \in V_{\pi g}$ by, $\pi v(\pi g') = v(g')\ \forall g' \in G_g$. The allocation rule Y_g on g is said to satisfy the Anonymity axiom (A) if for all $v \in \mathcal{NG}_g(N)$ and permutations π on N,

$$Y_{g,i}(v) = Y_{\pi g, \pi i}(\pi v), \quad \forall\, i \in N.$$

Definition 22 An allocation rule Y_g on g is said to satisfy the Monotonicity axiom (M) if for every monotonic value function v, $Y_{g,i}(v) \geq 0$ for every $i \in N$.

The next theorem characterizes an efficient allocation rule for network game different from the Myerson value and the Position value. The significance of this rule is that it is based on standard Shapley like axioms of TU cooperative games.

Theorem 5 *If Y_g satisfies axioms L, NP, A and M for a given network g, then for every $v \in \mathcal{NG}_g(N)$, there exists a family of non-negative real constants $p^{l(g)}_{l(g')+l(g_i)}$, where $g_i \subseteq L_i(g)$, $l(g_i) = 1, 2, ..., l_i(g)$ and $g' \subseteq g - g_i$, $l(g') = 0, ..., l(g) - l(g_i)$ such that,*

$$Y_{g,i}(v) = \sum_{\emptyset \neq g_i \subseteq L_i(g)} \left\{ \sum_{g' \subseteq g - g_i} p^{l(g)}_{l(g')+l(g_i)} \left\{ v(g' + g_i) - v(g') \right\} \right\} \tag{3.3}$$

Moreover Y_g satisfies E (efficiency) if and only if the coefficients $p^{l(g)}_{l(g')+l(g_i)}$ satisfy the following:

$$\begin{aligned} &\sum_{i \in N} \sum_{\emptyset \neq g_i \subseteq g} p^{l(g)}_{l(g)} = 1 \\ &\sum_{i \in N} \sum_{\emptyset \neq g_i \subseteq g'} p^{l(g)}_{l(g')} - \sum_{i \in N} \sum_{\emptyset \neq g_i \subseteq g - g'} p^{l(g)}_{l(g')+l(g_i)} = 0, \quad \forall g' \subsetneq g. \end{aligned} \tag{3.4}$$

The allocation rule proposed in Theorem 5 however applies to a very narrow class of network games where the networks are of special types. This is given in the next theorem as follows:

Theorem 6 (Borkotokey et al. (2015), pp. 916) *An allocation rule that satisfies L, NP, A, M and E on g for any $v \in \mathcal{NG}_g(N)$ if and only if either g is k-regular or is a star.*

The expression $v(g' + g_i) - v(g')$ in Eq. (3.3) is called the marginal multilateral contribution of player i due to her links $g_i \in L_i(g)$ with respect to the network game v over $g' \subseteq g - g_i$. In particular, if $l(g_i) = 1$ that is, g_i comprises of only a single link, the corresponding expression is called the marginal bilateral contribution. The marginal multilateral contribution measures the level of interactions made by player i through her links in the network game. The network Myerson value and the network Position value are actually based on the two extreme cases of interactions. The network Myerson value measures the marginal contribution of player i over each possible coalition of players connected through the links in the network. It follows that under this rule, i actually exhausts all her links in the network together to make her contributions felt. Contrary to this, in the network Position value, her contributions to the network is measured on the basis of her interactions through one link at a time (the marginal bilateral contribution). However, in addition to these two extreme cases of interactions, player i's marginal contributions due to her interactions through any number of links at a time between 1 and $l_i(g)$ can also be of some importance. A possible reason for such variable interactions shown by a player is the interdependence/interplay among her contributions through these links. Thus the expression $v(g' + g_i) - v(g')$ is termed as the marginal multilateral contribution of player i due to her links $g_i \in L_i(g)$ with respect to the network game v over $g' \subseteq g - g_i$ at level k, whenever $k = l(g_i)$.

The restriction imposed on the networks because of efficiency is avoided by introducing the axiom of *multilateral interactions* in Borkotokey et al. (2015) as follows. Let v be a game on the network g. Fix a player i and $g_i \subseteq L_i(g)$. The reduced game with respect to g_i is a game $v_{[g_i]}^{g-g_i+[g_i]}$ defined on the network $g - g_i + [g_i]$ of $l(g) - l(g_i) + 1$ links, where $[g_i]$ indicates a single hypothetical link such that as if i is connected to the network $g - g_i$ only through this link. This new network $g - g_i + [g_i]$ is called the reduced network of g for g_i. The reduced game is defined as for every $g' \subsetneq g - g_i$,

$$v_{[g_i]}^{g-g_i+[g_i]}(g') = v(g')$$
$$v_{[g_i]}^{g-g_i+[g_i]}(g' + [g_i]) = v(g' + g_i)$$

For simplicity, the reduced game is denoted by $v_{[g_i]}$ only. In a reduced game player i is connected to her peers through a single hypothetical link which is actually the subnetwork g_i. Thus, the multilateral interaction of a player at any level k can be realized through her bilateral interactions in the corresponding reduced game (through a single hypothetical link). Under this framework, the Position value is an aggregate of the marginal multilateral contributions of player i at level 1 (considering interaction due to one link at a time). Therefore, the allocation rule for a reduced game should yield the Position value. Thus, an allocation rule Y_g is said to admit the Position value Y^{NPV} for the game v at each level of interactions if for each $i \in N$, it can be expressed as,

$$Y_{g,i}(v) = \sum_{\emptyset \neq g_i \subseteq L_i(g)} Y_{g-g_i+[g_i],i}(v_{[g_i]}) \tag{3.5}$$

where for each $\emptyset \neq g_i \subseteq L_i(g)$,

$$Y_{g-g_i+[g_i],i}(v_{[g_i]}) = Y_i^{NPV}(g, v_{[g_i]}). \tag{3.6}$$

An allocation rule Y_g is said to satisfy the axiom of Multilateral Interactions (MI) if it admits the position value for any network game v at each level of interaction.

We identify each $Y_{g-g_i+[g_i]}$ with the Position value for network games of the corresponding reduced game. Theorem 7 below is the characterization of the interactive allocation rule which has the flavour of both player-based and link-based allocation rules.

Theorem 7 *An allocation rule Y_g satisfies the axioms L, NP, A, M and MI, if and only if for all $i \in N$, it can be expressed as follows:*

$$Y_{g,i}(v) = \sum_{\emptyset \neq g_i \subseteq L_i(g)} \sum_{g' \subseteq g - g_i} \frac{1}{2} \frac{(l(g) - l(g') - l(g_i))!\, l(g')!}{(l(g) - l(g_i) + 1)!} \left\{ v(g' + g_i) - v(g') \right\}, \tag{3.7}$$

The allocation rule given by Eq. (3.7) is an allocation rule based on multilateral interactions among players through their links.

4 Bi-cooperative Network Games

Recall the examples in Sect. 1 where players or nodes and/or the links in the network can be both positive and negative. Accordingly we obtain three distinct network structures. As already mentioned, cooperative games alone cannot describe the interactions nor the network as such. The description of such games is possible only through imposing bi-cooperative structure on networks. We first obtain the Position value of bi-cooperative network games (Borkotokey et al. 2018). The Myerson value for bi-cooperative games is introduced as an analogy to its counterpart in network games.

Let $g_1, g_2 \in G^N$ be such that $g_1 \cap g_2 = \emptyset$. Assume that g_1 consists of the positive links while g_2 builds on the negative links. The pair (g_1, g_2) so formed is called a bi-network. The players who have links only in g_1 are taken as positive contributors and those only in g_2 are taken as negative contributors. When a player has links in both g_1 and g_2, both her positive and negative contributions are realized through her positive and negative links, respectively. The links in the network $g^N \setminus (g_1 + g_2)$ are now called the absentee links. $\mathcal{Q}(G^N) = \{(g_1, g_2) | g_1, g_2 \in G^N \text{ with } g_1 \cap g_2 = \emptyset\}$ denotes the set of all bi-networks. For $g_1, g_2 \in G$, $l(g_1, g_2)$ denotes the total number of links in g_1 and g_2 together and similarly $L(g_1, g_2)$ denotes the set of all links in g_1 and g_2, i.e. $L(g_1, g_2) = \{ij \mid i, j \in N(g_1) \text{ or } N(g_2)\}$. The set $L_i(g_1, g_2)$ comprises of links that player i has in g_1 and g_2, i.e. $L_i(g_1, g_2) = \{ij \mid \exists j \; : \; ij \in g_1 \text{ or } ij \in g_2\}$. $l_i(g_1, g_2)$ is the number of links in player i's link set. Let $N(g_1, g_2)$ be the set of all players in g_1 and g_2, i.e. $N(g_1, g_2) = \{j \mid ij \in g_1 \text{ or } g_2 \text{ for some } i \in N : i \neq j\}$. A value function is a function $b : \mathcal{Q}(G^N) \to \mathbb{R}$, with $b(\emptyset, \emptyset) = 0$. The value $b(g_1, g_2)$ for every member (g_1, g_2) of $\mathcal{Q}(G^N)$ represents its worth under b when the links in g_1 contribute positively to the network, links in g_2 contribute negatively and the others remain indifferent.

A bi-cooperative network game is a pair (N, b), of a set N of players and a value function b. When the player set N is fixed, we simply denote the game by b. The class of all bi-cooperative network games with player set N is denoted by $\mathcal{NB}(N)$.

Definition 23 An allocation rule for $\mathcal{NB}(N)$ is a function $Y : \mathcal{Q}(G^N) \times \mathcal{NB}(N) \to \mathbb{R}^n$ such that

$$\sum_i Y_i((g_1, g_2), b) = b(g_1, g_2), \quad \forall\, (g_1, g_2) \in \mathcal{Q}(G^N), b \in \mathcal{NB}(N).$$

4.1 *The Position Value for Bi-cooperative Network Games*

In what follows next, the associated link game of a bi-cooperative network game is defined. Recall that the Position value for network games is the Shapley value of the associated link game divided equally between the players forming links where the links are assumed to be players. In a bi-cooperative network game also the Position value is defined in the same manner. For every $g \in G^N$, the symbol $[g]$ denotes the coalition when every link in g represents a hypothetical player. Thus given $(g_1, g_2) \in \mathcal{Q}(G^N)$ the pair $([g_1], [g_2])$ specifies a bi-coalition where $[g_1]$ is the set of all hypothetical players representing the positive links and $[g_2]$ is the set of all hypothetical players representing the negative links. Let N^* be the set of all such hypothetical players in G^N. Given a bi-cooperative network game $(N, b) \in \mathcal{NB}(N)$, the associated link game (N^*, b^*) of (N, b) is defined as follows. Given $(S, T) \in \mathcal{Q}(N^*)$, there exists $(g_1, g_2) \in \mathcal{Q}(G^N)$ such that $S = [g_1]$ and $T = [g_2]$. Define $b^* : \mathcal{Q}(N^*) \to \mathbb{R}$ by $b^*(S, T) = b(g_1, g_2)$. Given $(g_1, g_2) \in \mathcal{Q}(G^N)$, set $\mathcal{Q}([g_1], [g_2]) = \{(S, T) \mid S \subseteq [g_1], T \subseteq [g_2] : S \cap T = \emptyset\}$. It follows that for $(S, T) \in \mathcal{Q}([g_1], [g_2])$, there is a $(g_1', g_2') \in \mathcal{Q}(G^N)$ with $g_1' \subseteq g_1$ and $g_2' \subseteq g_2$ such that $S = [g_1'], T = [g_2']$. The class of the associated link games with the hypothetical player set N^* is denoted by $\mathcal{B}^*(N^*)$. It is worth to note that the members of $\mathcal{B}^*(N^*)$ are actually bi-cooperative games. Therefore the LG value Φ^{LG} of $\mathcal{B}^*(N^*)$ can also be obtained as in any other bi-cooperative games.

Definition 24 Given $(N, b) \in \mathcal{NB}(N)$, an allocation rule $Y : \mathcal{Q}(G^N) \times \mathcal{NB}(N) \to \mathbb{R}^n$ with respect to $(g_1, g_2) \in \mathcal{Q}(G^N)$ is a link-based allocation rule for the class $\mathcal{NB}(N)$ of bi-cooperative network games if there is a consistent value $\Psi : \mathcal{B}^*(N^*) \to (\mathbb{R}^{\frac{n(n-1)}{2}})^{\mathcal{Q}(N^*)}$ for the class of the associated link games $\mathcal{B}^*(N^*)$ such that,

$$\begin{aligned} \sum_{l \in (g_1 + g_2)} \Psi_l(N^*, b^*)([g_1], [g_2]) &= b^*([g_1], [g_2]) \\ Y_i((g_1, g_2), b) &= \sum_{l \in L_i(g_1, g_2)} \frac{\Psi_l(N^*, b^*)([g_1], [g_2])}{2} \end{aligned} \tag{4.1}$$

Definition 25 The Position value Y^{BNPV} is a link-based allocation rule where player $i \in N$ in (N, b) receives half of the LG values for each of her links in the associated node game (N^*, b^*), i.e.

$$Y_i^{BNPV}((g_1, g_2), b) = \sum_{l \in L_i(g_1, g_2)} \frac{\Phi_l^{LG}(N^*, b^*)([g_1], [g_2])}{2} \quad (4.2)$$

Remark 1 Given a bi-network $(g_1, g_2) \in \mathcal{Q}(G^N)$, $b \in \mathcal{NB}(N)$ and Y a link-based allocation rule for (N, b), then there exists a consistent value function $\mathcal{B}^*(N^*)$ such that Eq. (4.1) holds. Let $\emptyset \neq g^* \subseteq (g_1 + g_2)$. The upper reduced game $((N((g_1, g_2) \setminus g^*), U_b)$ and the lower reduced game $((N((g_1, g_2) \setminus g^*), L_b)$ with respect to the value function Ψ are defined as follows:

$$U_b(g_1', g_2') = b(g_1' + g^*, g_2') - \sum_{l \in L(g^*)} \Psi_l([g_1] \cup [g_2], b^*)([g_1'] \cup [g^*], [g_2'])$$

$$L_b(g_1', g_2') = b(g_1', g_2' + g^*) - \sum_{l \in L(g^*)} \Psi_l([g_1] \cup [g_2], b^*)([g_1'], [g_2'] \cup [g^*])$$

for all $(g_1', g_2') \in \mathcal{Q}((g_1, g_2) \setminus g^*)$, where $\mathcal{Q}((g_1, g_2) \setminus g^*) = \{(g_1', g_2') \in \mathcal{Q}(G^N) \mid g_1' \subseteq g_1 \setminus g^*, g_2' \subseteq g_2 \setminus g^*\}$.

The upper and lower reduced games can be interpreted along the line of their counterparts in TU games and bi-cooperative games. Let $b \in \mathcal{NB}(N)$ be fixed. Let us denote by $\mathcal{NB}_b(N)$ the subset of $\mathcal{NB}(N)$ that consists of only three elements, viz. b and the upper and the lower reduced games of b. The notion is important in characterizing the Position value for a bi-cooperative network game as shown below.

In what follows next, the Position value for a bi-cooperative network game is characterized using the following standard game-theoretic axioms.

Axiom B1 (Efficiency): For $\tilde{b} \in \mathcal{NB}_b(N)$ and $(g_1, g_2) \in \mathcal{Q}(G^N)$, an allocation rule Y satisfies Efficiency if

$$\sum_{i \in N(g_1 + g_2)} Y_i((g_1, g_2), \tilde{b}) = \tilde{b}(g_1, g_2).$$

For a given $(g_1, g_2) \in \mathcal{Q}(G^N)$, a player $i \in N$ is called a null player for all $\tilde{b} \in \mathcal{NB}_b(N)$ if for every $g_i \subseteq L_i(g_1, g_2)$ and every $(g_1', g_2') \in \mathcal{Q}(G^{N \setminus i})$ such that

$$\tilde{b}(g_1' + g_i, g_2') = \tilde{b}(g_1', g_2') = \tilde{b}(g_1', g_2' + g_i).$$

The null player property states that if the presence of a player's subnetworks in a bi-network has no influence on the value it generates then she gets nothing.

Axiom B2 (NP): An allocation rule Y satisfies the Null Player property (NP) if for every null player $i \in N$ and $\tilde{b} \in \mathcal{NB}_b(N)$ one has,

$$Y_i((g_1, g_2), \tilde{b}) = 0, \quad \forall (g_1, g_2) \in \mathcal{Q}(G^N).$$

Axiom B3 (Linearity): For all $\alpha, \beta \in \mathbb{R}$ and $\tilde{b}_1, \tilde{b}_2 \in \mathcal{NB}_b(N)$ and $(g_1, g_2) \in \mathcal{Q}(G^N)$,

$$Y((g_1, g_2), \alpha\tilde{b}_1 + \beta\tilde{b}_2) = \alpha Y((g_1, g_2), \tilde{b}_1) + \beta Y((g_1, g_2), \tilde{b}_2).$$

The Intra-network symmetry implies that the allocation rule is independent of players' identities and accounts only for their contributions in the game.

Axiom B4 (Intra-network symmetry): For each $\tilde{b} \in \mathcal{NB}_b(N)$ and $(g_1, g_2) \in \mathcal{Q}(G^N)$ let σ be a permutation on N such that $\sigma(g_1) = g_1$ and $\sigma(g_2) = g_2$, then it holds that for all $i \in N$

$$Y_{\sigma(i)}((g_1, g_2), \tilde{b} \circ \sigma^{-1}) = Y_i((g_1, g_2), \tilde{b}).$$

The following definition is needed for the next axiom 'Inter-network symmetry':

Definition 26 Let $(g_1, g_2) \in \mathcal{Q}(G^N)$ and $i, j \in N(g_1, g_2)$ and $b_i, b_j \in \mathcal{NB}(N)$ and $l_i(g_1) = l_j(g_2)$ and $l_j(g_1) = l_i(g_2)$. An upper (similarly lower) reduced game $\tilde{b}_i \in \mathcal{NB}_{b_i}(N)$ is said to correspond to a lower (similarly upper) reduced game $\tilde{b}_j \in \mathcal{NB}_{b_j}(N)$ if g_i is reduced from g_1 (similarly g_2) in $\tilde{b}_i$ by deleting the same number of links in g_j from g_2 (similarly g_1) in $\tilde{b}_j$, $\forall g_i \subseteq L_i(g_1, g_2)$ and $g_j \subseteq L_j(g_1, g_2)$.

The Inter-network symmetry ensures that when players have opposite links (i.e. positive and negative) providing identical contributions in the opposite directions (i.e. if one value is positive then the other one is negative) they will get identical but opposite (i.e. positive and negative) payoffs.

Axiom BN5 (Inter-network symmetry): For a given $(g_1, g_2) \in \mathcal{Q}(G^N)$, let $i, j \in N(g_1, g_2)$ and $b_i, b_j \in \mathcal{NB}(N)$ and $l_i(g_1) = l_j(g_2)$ and $l_j(g_1) = l_i(g_2)$ such that $\tilde{b}_i \in \mathcal{NB}_{b_i}(N)$ and corresponding $\tilde{b}_j \in \mathcal{NB}_{b_j}(N)$ and

$$\tilde{b}_i(g_1' + g_i, g_2') - \tilde{b}_i(g_1', g_2') = \tilde{b}_j(g_1', g_2') - \tilde{b}_j(g_1', g_2' + g_j)$$
$$\tilde{b}_i(g_1' + g_i, g_2' + g_j) - \tilde{b}_i(g_1', g_2' + g_j) = \tilde{b}_j(g_1' + g_i, g_2') - \tilde{b}_j(g_1' + g_i, g_2' + g_j),$$

$\forall (g_1', g_2') \in \mathcal{Q}((g_1 + g_2) \setminus \{[L_i(g_1, g_2)], [L_j(g_1, g_2)]\})$ and $g_i \subseteq L_i(g_1)$, $g_j \subseteq L_j(g_2)$ and

$$\tilde{b}_i(g_1', g_2') - \tilde{b}_i(g_1', g_2' + g_i) = \tilde{b}_j(g_1' + g_j, g_2') - \tilde{b}_j(g_1', g_2')$$
$$\tilde{b}_i(g_1' + g_j, g_2') - \tilde{b}_i(g_1' + g_j, g_2' + g_i) = \tilde{b}_j(g_1' + g_j, g_2' + g_i) - \tilde{b}_j(g_1', g_2' + g_i),$$

$\forall (g_1', g_2') \in \mathcal{Q}((g_1 + g_2) \setminus \{[L_i(g_1, g_2)], [L_j(g_1, g_2)]\})$ and $g_i \subseteq L_i(g_2)$, $g_j \subseteq L_j(g_1)$.

Then we must have,

$$Y_i((g_1, g_2), \tilde{b}_i) = -Y_j((g_1, g_2), \tilde{b}_j).$$

The Inter-network symmetry axiom precisely says that when the number of positive links (similarly negative links) of i equals the number of negative links (similarly positive links) of j in a bi-network (g_1, g_2) and their contributions in the respective reduced games (upper and lower) are numerically equal but opposite in signs, then the incentive payoff for i shall be exactly the opposite of the payoff for j. This axiom resembles the Inter-Coalition Symmetry axiom due to Labreuche and Grabisch (2008).

Axiom B6 (Monotonicity): For any $(g_1, g_2) \in \mathcal{Q}(G^N)$, let $i \in N(g_1, g_2)$ and b_1, b_2 be two bi-cooperative network games such that $\tilde{b}_1 \in \mathcal{NB}_{b_1}(N)$ and $\tilde{b}_2 \in \mathcal{NB}_{b_2}(N)$ and

$$\begin{aligned} \tilde{b}_1(g_1', g_2') &= \tilde{b}_2(g_1', g_2') \\ \tilde{b}_1(g_1' + g_i, g_2') &\geq \tilde{b}_2(g_1' + g_i, g_2') \\ \tilde{b}_1(g_1', g_2' + g_i) &\geq \tilde{b}_2(g_1', g_2' + g_i) \end{aligned} \tag{4.3}$$

for all $g_i \subseteq L_i(g_1, g_2)$ and $(g_1', g_2') \in \mathcal{Q}(G^{N\setminus i})$, then

$$Y((g_1, g_2), \tilde{b}_1) \geq Y((g_1, g_2), \tilde{b}_2).$$

Remark 2 The axiom of monotonicity suggests that when i adds some positive links to a bi-network if the added value for i is not less for game b_1 and its reduced games than for another game b_2 and its reduced games, respectively, and similarly when i adds some negative links to the bi-cooperative game, if the negative added value for i is less in absolute value for b_1 and its reduced game than for b_2 and its reduced games then her respective payoffs from b_1 is no less than that from b_2. This differs from the standard monotonicity axiom due to Labreuche and Grabisch (2008) in the sense that it is evokes at every reduced game in addition to the game itself.

Theorem 8 *The Position value Y^{BNPV} is the unique allocation rule for bi-cooperative network games that satisfies Axioms (B1)–(B6).*

The proof of Theorem 8 can be found in Borkotokey et al. (2018).

4.2 The Myerson Value for Bi-cooperative Network Games

In what follows we define the Myerson Value for $\mathcal{NB}(N)$ and provide a characterization using a new set of axioms: paired component balanced and bi-cooperative balanced contributions. These axioms are extensions of their counterparts namely component balanced and balanced contributions of classical network games. Let the bi-cooperative network game b be called paired component additive, if it satisfies

$$b(g_1, g_2) = \sum_{g' \in C(g_1),\ g'' \in C(g_2)} b(g', g'')$$

Definition 27 An allocation rule Y is paired component balanced if for any paired component additive $b \in \mathcal{NB}(N)$ and $(g_1, g_2) \in \mathcal{Q}(G^N)$ and $g' \in C(g_1)$, $g'' \in C(g_2)$

$$\sum_{i \in N(g', g'')} Y_i((g_1, g_2), b) = b(g', g'')$$

Definition 28 An allocation rule Y satisfies bi-cooperative balanced contributions if for any paired component additive $b \in \mathcal{NB}(N)$ and $(g_1, g_2) \in \mathcal{Q}(G^N)$,

(a) whenever $i, j \in N(g_1)$, $Y_i((g_1, g_2), b) - Y_i((g_1 \setminus L_j(g_1), g_2), b) = Y_j((g_1, g_2), b) - Y_j((g_1 \setminus L_i(g_1), g_2), b)$.
(b) whenever $i, j \in N(g_2)$, $Y_i((g_1, g_2), b) - Y_i((g_1, g_2 \setminus L_j(g_2)), b) = Y_j((g_1, g_2), b) - Y_j((g_1, g_2 \setminus L_i(g_2)), b)$.
(c) whenever $i \in N(g_1), j \in N(g_2)$,
$Y_i((g_1, g_2), b) - Y_i((g_1, g_2 \setminus L_j(g_2)), b) = Y_j((g_1, g_2), b) - Y_j((g_1 \setminus L_i(g_1), g_2), b)$.
(d) whenever $i \in N(g_2), j \in N(g_1)$,
$Y_i((g_1, g_2), b) - Y_i((g_1 \setminus L_j(g_1), g_2), b) = Y_j((g_1, g_2), b) - Y_j((g_1, g_2 \setminus L_i(g_2)), b)$.

Remark 3 The axiom of paired component balanced signifies that if the value does not have externalities across the networks, the allocation rule simply divides the worth of every bi-network among its members. The axiom of bi-cooperative balanced contributions is an equity axiom which deals with the contribution of a player to the payoff of another player.

Definition 29 If $b \in \mathcal{NB}(N)$ is paired component additive, then given $(g_1, g_2) \in \mathcal{Q}(G^N)$, the associated node game of b with respect to (g_1, g_2) denoted by $b_{(g_1, g_2)}$ is given by,

$$b_{(g_1, g_2)}(S, T) = \sum_{g' \in C(g_1|_S), g'' \in C(g_2|_T)} b(g', g'') \ \ \forall (S, T) \sqsubseteq_P (N(g_1), N(g_2))$$

Remark 4 For each $i \in N$, the LG value of the associated node game $b_{(g_1, g_2)}$, denoted by $\Phi_i^{LG}(b_{(g_1, g_2)})(N(g_1), N(g_2))$ is given after some simplification as:

$$
\begin{aligned}
&\Phi_i^{LG}(b_{(g_1,g_2),b})(N(g_1),N(g_2))\\
&= \sum_{K\subseteq (N(g_1)\cup N(g_2))\setminus\{i\}} \frac{k!(n(g_1)+n(g_2)-k-1)!}{(n(g_1)+n(g_2))!}[b_{(g_1,g_2)}(N(g_1)\cap(K\cup\{i\}),N(g_2)\cap(K\cup\{i\}))\\
&\quad - b_{(g_1,g_2)}(N(g_1)\cap K, N(g_2)\cap K)]\\
&= \sum_{K\subseteq N(g_1,g_2)\setminus\{i\}} \frac{k!(n(g_1)+n(g_2)-k-1)!}{(n(g_1)+n(g_2))!}\left[\sum_{\substack{g'\in C(g_1|_{N(g_1)\cap(K\cup\{i\})})\\ g''\in C(g_2|_{N(g_2)\cap(K\cup\{i\})})}} b(g',g'') - \sum_{\substack{g'\in C(g_1|_{N(g_1)\cap K})\\ g''\in C(g_2|_{N(g_2)\cap K})}} b(g',g'')\right]\\
&= \sum_{K\subseteq N(g_1,g_2)\setminus\{i\}} \frac{k!(n(g_1)+n(g_2)-k-1)!}{(n(g_1)+n(g_2))!}[b(g_1|_{K\cup\{i\}},g_2|_{K\cup\{i\}}) - b(g_1|_K,g_2|_K)] \qquad (4.4)
\end{aligned}
$$

Theorem 9 *If b is paired component additive, then the unique allocation rule Y which satisfies paired component balanced and bi-cooperative balanced contributions is the LG value of the associated node game $b_{(g_1,g_2)}$. We call Y the Myerson value for the bi-cooperative network game and denote by Y^{BNMV}. Formally we have, $Y^{BNMV}((g_1,g_2),b) = \Phi^{LG}(b_{(g_1,g_2)})(N(g_1),N(g_2))$.*

Proof Given that the LG value satisfies axioms b1-b6, we need to show that the allocation rule satisfies the paired component balanced and the bi-cooperative balanced contributions.

To prove paired component balanced we have,

$$
\begin{aligned}
b(g',g'') &= \sum_{g_1\in C(g'),g_2\in C(g'')} b(g_1,g_2) = \sum_{\substack{g_1\in C(g'|_{N(g')})\\ g_2\in C(g''|_{N(g'')})}} b(g_1,g_2)\\
&= b_{(g',g''),b}(N(g'),N(g'')) = \sum_{i\in N(g')\cup N(g'')} \Phi_i^{LG}(b_{(g',g''),b})(N(g'),N(g'')), \text{ using axiom b1}\\
&= \sum_{i\in N(g',g'')} Y_i((g',g''),b^*)
\end{aligned}
$$

To prove bi-cooperative balanced contributions, let $i,j\in N(g_1)$, $w_1 = b_{(g_1,g_2)} - b_{(g_1\setminus L_j(g_1),g_2),b}$ and $w_2 = b_{(g_1,g_2)} - b_{(g_1\setminus L_i(g_1),g_2),b}$.

Let π be a permutation on N such that $\pi i = j$ and $\pi k = k$, for all $k\in N\setminus\{i,j\}$. Then $\pi w_1(\pi S,\pi T) = w_2(S,T), \forall (S,T)\sqsubseteq_P (N(g_1),N(g_2))$.

By axiom b4 and axiom b2, we have

$$
\Phi_i^{LG}(w_1)(N(g_1),N(g_2)) = \Phi_j^{LG}(w_2)(N(g_1),N(g_2)).
$$

Thus

$$
\begin{aligned}
&\Phi_i^{LG}(b_{(g_1,g_2)})(N(g_1),N(g_2)) - \Phi_i^{LG}(b_{(g_1\setminus L_j(g_1),g_2),b})(N(g_1),N(g_2))\\
=&\Phi_j^{LG}(b_{(g_1,g_2)})(N(g_1),N(g_2)) - \Phi_j^{LG}(b_{(g_1\setminus L_i(g_1),g_2),b})(N(g_1),N(g_2))
\end{aligned}
$$

This further yields that $Y_i((g_1,g_2),b) - Y_i((g_1\setminus L_j(g_1),g_2),b) = Y_j((g_1,g_2),b) - Y_j((g_1\setminus L_i(g_1),g_2),b)$. The proof of the following cases are symmetric:

(a) $Y_i((g_1, g_2), b) - Y_i((g_1, g_2 \setminus L_j(g_2)), b) = Y_j((g_1, g_2), b) - Y_j((g_1, g_2 \setminus L_i(g_2)), b)$ for all $i, j \in N(g_2)$.
(b) $Y_i((g_1, g_2), b) - Y_i((g_1, g_2 \setminus L_j(g_2)), b) = Y_j((g_1, g_2), b) - Y_j((g_1 \setminus L_i(g_1), g_2), b)$ for all $i \in N(g_1), j \in N(g_2)$.
(c) $Y_i((g_1, g_2), b) - Y_i((g_1 \setminus L_j(g_1), g_2), b) = Y_j((g_1, g_2), b) - Y_j((g_1, g_2 \setminus L_i(g_2)), b)$ for all $i \in N(g_2), j \in N(g_1)$,

For the uniqueness, if possible let $Y^1 : \mathcal{Q}(G^N) \times \mathcal{NB}(N) \to \mathbb{R}^n$ and $Y^2 : \mathcal{Q}(G^N) \times \mathcal{NB}(N) \to \mathbb{R}^n$ be two different fair allocation rules that satisfy paired component balanced and bi-cooperative balanced contributions. We will show that $Y^1 = Y^2$. The proof is by induction on $l(g_1, g_2)$.

For $l(g_1, g_2) = 1$, then the link either in g_1 or in g_2. Thus paired component balanced of Y^1 and Y^2, we have $Y^1 = Y^2 = Y^{MV}$. Now, assume that for $l(g_1, g_2) = k$ where $k \in \mathbb{N}$, an allocation rule that satisfies component balance and the balanced contributions property is unique. Consider $l(g_1, g_2) = k + 1$. For $i, j \in N$, following exclusive cases arise. Either $i, j \in N(g_1)$ or $i, j \in N(g_2)$ or $i \in N(g_1), j \in N(g_2)$ or $i \in N(g_2), j \in N(g_1)$. We prove here the case $i, j \in N(g_1)$ as the proof of the other is analogous by symmetry.

By bi-cooperative balanced contributions and induction hypothesis, we have,

$$
\begin{aligned}
Y_i^1((g_1, g_2), b) - Y_i^1((g_1 \setminus L_j(g_1), g_2), b) &= Y_j^1((g_1, g_2), b) - Y_j^1((g_1 \setminus L_i(g_1), g_2), b) \\
\Rightarrow Y_i^1((g_1, g_2), b) - Y_j^1((g_1, g_2), b) &= Y_i^1((g_1 \setminus L_j(g_1), g_2), b) - Y_j^1((g_1 \setminus L_i(g_1), g_2), b) \\
&= Y_i^2((g_1 \setminus L_j(g_1), g_2), b) - Y_j^2((g_1 \setminus L_i(g_1), g_2), b) \\
&= Y_i^2((g_1, g_2), b) - Y_j^2((g_1, g_2), b)
\end{aligned}
$$

Since $i, j \in N(g_1)$, therefore they are in the same connected component g_1. Therefore,

$$Y_i^1((g_1, g_2), b) - Y_j^1((g_1, g_2), b) = d(g_1),$$

where $d(g_1)$ depends on $C(g_1)$ and $C(g_2)$ only, but not on i. By paired component balanced we have,

$$
b(g', g'') = \sum_{i \in N(g', g'')} Y_i^1((g_1, g_2), b) = \sum_{i \in N(g', g'')} Y_i^2((g_1, g_2), b)
$$

$$
\Rightarrow \sum_{i \in N(g', g'')} (Y_i^1((g_1, g_2), b) - Y_i^2((g_1, g_2), b)) = 0 \Rightarrow \; n(g' + g'')d(g_1) = 0 \Rightarrow \; d(g_1) = 0.
$$

It follows that $Y^1 = Y^2$,, i.e. there can be at most one fair allocation rule for b. This completes the proof.

5 Application Possibilities

Cooperative game theory has been used in describing social and economic networks since the work of Myerson (1977). The underlying objects in Myerson's model are however, graph restricted TU games. Jackson and Wolinsky (1996) extends this concept to introduce games on networks to include cost of network formation. In Jackson (2005a), Jackson introduces the flexible network allocation rules where players can explore outside possibilities of network formation and accordingly design their allocation rule. The player based, the link based and the other allocation rules have been based on different narratives useful for various applicable domains. In this section, we give an outline of the recent developments and application possibilities of the theory so far discussed in the earlier sections. The list is, however, not exhaustive but indicative of the emergence of a promising area of research.

Opportunistic Mobile Networks

In Singh and Borkotokey (2019), an algorithm is developed to control the selfish nodes in an opportunistic mobile communication network from overtaking the network. The opportunistic mobile communication network is a social communication network, where every node depends on the other in transferring data and makes a potential router. However, due to the lack of resource, nodes becomes selfish by dropping the messages relayed through them. Suitable allocation rules under cooperative set up can incentivise the sharing of the total resource among the nodes so that they make binding agreements and relay the messages without dropping. Cybersecurity has been seen to be another promising area of research for its applications in both industry and academia. Conventional security technology succumbs to the challenges from massive and complicated network traffic. Game theory techniques have emerged handy to tackle many of these challenges, see Amadi et al. (2017), Alpcan and Basar (2011), etc. However, most research works in this area are not network layer specific due to the complex nature of the cyber world. The efficiency and stability due to Jackson and Wolinsky (1996) of the cyber network can be studied employing various allocation rules under a cooperative set up. A model based on bi-cooperative network games can handle simultaneously both attackers and defenders of the cyber network considering them as positive and negative players.

Middlemen in Networks

In Gogoi et al. (2018), the notion of middlemen in network games is studied under cooperative framework. Referring to the situations where international trades among manufacturing and producing countries take place through a third country, political lobbying happens through negotiators or the common server networks are used by the telecom companies, etc., they suggest that the role of the middlemen and her strategic position in the network need to be explored explicitly. The middlemen are considered to be individually non-productive but they can increase the worth of the network when they form links with the productive players. It is shown in Gogoi et al. (2018), that after the payment of the intermediary fee to the middlemen, the allocation rule awards all the other players (non-middlemen) better payoffs than whatever they would get in absence of the middlemen. The literature of intermediary activities in supply chain network has been an established domain of studies, see Christopher (2016). Kalai et al. (1997) introduce the notion of a middleman under cooperative situations and compared the core allocations to the middlemen. Yavas (1994) uses this idea in Industrial Organization. Recently (Arya et al. 2015) also apply the middleman concept to analysing supply chains. Note that, the middleman is not necessarily always a facilitator to a group of agents but it may create concentration of bargaining power and reduce the gain possibilities of the other agents. Therefore, in a network game, both the positive and negative aspects of a middleman are necessary to be studied. This can be effectively pursued using the notion of bi-cooperative network games.

Microarray Network Games

Microarray tools generate large amount of information on gene expression. This information identifies the genes useful for the diagnosis and prognosis of the genetic diseases. Network games are seen to more accurately describe the interactions among genes. In Neog et al. (2019), a gene co-expression network is constructed and a special game on this network is defined. The Position value of this network game is used to select the salient genes which deserve further investigation. The justification of the use of the Position value in identifying the genes is made in terms of the genetic interpretations of the standard axioms. This provides an alternative statistical tool for analysing the data for the purpose of attributing causality. Other allocation rules can also be effectively applied to compare with the results obtained through non-game-theoretic methods.

The Probabilistic Network Game

The notion of probabilistic games on networks is discussed in Calvo et al. (1999), Gómez et al. (2008), Ghintran et al. (2012). All these models, however, focus on the graph restricted games proposed by Myerson (1977). The narrative in this area of research is that the players are endowed with a probability distribution to make links with the other players and the worths of the link restricted coalitions are essentially the expected worths computed with these probability distributions. This idea can be intuitively and very naturally extended to network games a lá (Jackson and Wolinsky 1996) by assigning a probability to each of the links in the complete network. Thus we define a probabilistic network, see Calvo et al. (1999) as a pair (N, p), where N is the player set and $p : L(g^N) \to [0, 1]$ is a function that assigns to each link ij its probability of realization. Therefore each probabilistic network p induces $2^{\binom{n}{2}}$ probabilistic subnetworks of the form $p^k = (p^k_{ij})_{ij \in L(g^N)}$), $k \in \{1, 2, ..., 2^{\binom{n}{2}}\}$ such that $p^k_{ij} = p_{ij}$ or $p^k_{ij} = 0$. Let $\mathcal{P}(p)$ denote the class of all such networks. A probabilistic network game is a pair (N, v, p) where (N, p) is a probabilistic network and $v : \mathcal{P}(p) \to \mathbb{R}$ be such that $v(0) = 0$. Let $G^P(N)$ be the class of all probabilistic network games. An allocation rule on the class $G^P(N)$ of probabilistic network games is a function $Y : G^P(N) \to \mathbb{R}^n$ such that $\sum_{i \in N} Y_i(N, v, p) = v(p)$. Allocation rules for probabilistic network games can be player-based, link-based or interactive similar to the ordinary network games. One can also think of non-expected value extensions for the probabilistic network games. Other possible extensions include the river sharing problem (Béal et al. 2014; Dong et al. 2012), the trade and financial network formation problem (Gale and Kariv 2007, 2009), the study of random geopolitical networks and cybersecurity networks, etc., to name just a few.

Acknowledgements The authors acknowledge the financial support from the UKIERI grant [184-15/2017(IC)].

References

Alpcan T, Basar T (2011) Network security: a decision and game-theoretic approach. Cambridge University Press, Cambridge

Amadi EC, Eheduru GE, Eze FU, Ikerionwu C (2017) Game theory application in cyber security; a review. In: IEEE 3rd international conference on electro-technology for national development (NIGERCON)

Arya A, Löffler C, Mittendorf B, Pfeiffer T (2015) The middleman as a panacea for supply chain coordination problem. Eur J Oper Res 240:393–400

Banzhaf JF (1965) Weighted voting doesn't work: a mathematical analysis. Rutgers Law Rev 19:317–343

Béal S, Rémila E, Solal P (2017) Axiomatization and implementation of a class of solidarity values for TU-games. Theory Decis 83:61–94

Belau J (2013) An outside-option-sensitive allocation rule for networks: the kappa-value. Econ Theory Bull 1(2):175–188

Belau J (2016) Outside option values for network games. Math Soc Sci 84:76–86

Bilbao JM (2000) Cooperative games on combinatorial structures. Kluwer Academic Publishers, Boston

Bilbao JM, Fernández JR, Jiménez N, López JJ (2007) The core and the weber set for bi-cooperative games. Int J Game Theory 36:209–222

Bilbao JM, Fernández JR, Jiménez N, López JJ (2008a) Biprobabilistic values for bi-cooperative games. Discret Appl Math 156:2698–2711

Bilbao JM, Fernández JR, Jiménez N, López JJ (2008b) The Shapley value for bi-cooperative games. Ann Oper Res 158:99–115

Bilbao JM, Fernández JR, Jiménez N, López JJ (2010) The selectope for bicooperative games. Eur J Oper Res 204:522–532

Borkotokey S, Kumar R, Sarangi S (2015) A solution concept for network games: the role of multilateral interactions. Eur J Oper Res 243(3):912–920

Borkotokey S, Gogoi L, Sarangi S (2014) A survey of player-based and link-based allocation rules for network games. Stud Microecon 2(1):5–26

Borkotokey S, Gogoi L (2014) Bi-cooperative network games: a solution concept. J Game Theory 3:35–40

Bicooperative network games: a linkbased allocation rule. Int J Econ Theory 14(2):103–128

Borm P, Owen G, Tijs S (1992) On the position value for communication situations. SIAM J Discret Math 5:305–320

Calvo E, Lasaga J, van den Nouweland A (1999) Values of games with probabilistic graphs. Math Soc Sci 37:79–95

Casajus A, Huettner F (2014) Weakly monotonic solutions for cooperative games. J Econ Theory 154:162–172

Caulier JF, Skoda A, Tanimura E (2017) Allocation rules for networks inspired by cooperative game-theory. Rev d'économie Polit 127(4):517–558
Christopher M (2016) Logistics & supply chain management, Pearson UK
Dong B, Ni D, Wang Y (2012) Sharing a polluted river network. Environ Resour Econ 53(3):367–387
Felsenthal D, Machover M (1997) Ternary voting games. Int J Game Theory 26:335–351
Gale D, Kariv S (2007) Financial networks. Am Econ Rev 97:99–103
Gale D, Kariv S (2009) Trading in networks: a normal form game experiment. Am Econ J Micro 1:114–132
Ghintran A, González-Arangüena E, Manuel C (2012) A probabilistic position value. Ann Oper Res 201:183–196
Gogoi L, Borkotokey S, Mukherjee D (2018) Multilateral interactions and isolation in middlemen-driven network games. In: Dey N, Borah S, Babo R, Ashour AS (eds) Social network analytics. Academic Press, pp 163–181
Gómez D, González-Arangüena E, Manuel C, Owen G (2008) A value for generalized probabilistic communication situations. Eur J Oper Res 190:539–556
Grabisch M, Labreuche Ch (2005) Bi-capacities I: definition, möbius transform and interaction. Fuzzy Sets Syst 151:211–236
Grabisch M, Labreuche Ch (2008) Bipolarization of posets and natural interpolation. J Math Anal Appl 343:1080–1097
Harsanyi JC (1963) A simplified bargaining model for the n-person cooperative game. Int Econ Rev 4:194–220
Hart S, Mas-Collel A (1989) Potential, value and consistency. Econometrica 57:589–614
Hsiao C, Raghavan T (1993) Shapley value for multichoice cooperative games(I). Games Econ Behav 5:240–256
Isby DC (2004) World War II: double agent's D-day victory, World War II. http://www.historynet.com/world-war-ii-double-agents-d-day-victory.htm. Accessed June 2004
Jackson MO, Wolinsky A (1996) A strategic model of social and economic networks. J Econ Theory 71:44–74
Jackson MO (2005a) Allocation rules for network games. Games Econ Behav 51:128–154
Jackson MO (2005b) The economics of social networks. In: Blundell R, Newey W, Persson T (eds) Proceedings of the 9th world congress of the econometric society. Cambridge University Press, Cambridge, UK
Jackson MO (2008) Social and economic networks. Princeton University Press, Princeton, NJ
Jackson MO (2011) An overview of social networks and economic applications. In: Benhabib J, Bisin A, Jackson MO (eds) Handbook of social economics, 1A. Elsevier Science, Amsterdam, pp 511–579
Jackson MO, Rogers B, Zenou Y (2017) The economic consequences of social-network structure. J Econ Lit 55(1):49–95. https://doi.org/10.1257/jel.20150694
Joosten R (1996) Dynamics, equilibria and values dissertation, Maastricht University
Kalai E, Postlewaite A, Roberts J (1997) Barriers to trade and disadvantageous middlemen: non-monotonicity of the core. J Econ Theory 19:200–209
Labreuche C, Grabisch M (2008) A value for bi-cooperative games. Int J Game Theory 37:409–438
Manea M (2011) Bargaining in stationary networks. Am Econ Rev 101:1042–1080
Meessen R (1988) Communication games, Masters thesis, Department of Mathematics, University of Nijmegen, The Netherlands
Myerson R (1977) Graph and cooperative in games. Math Oper Res 2:225–229
Nowak AS, Radzik T (1994) A solidarity value for n-person transferable utility games. Int J Game Theory 23(1):43–48
Neog P, Borkotokey S, Saikia K, Mahanta P, Sarma A, Kumar R (2019) Identifying the salient genes in microarray data: a game theoretic model of the co-expression networks (Submitted)
Park J-H (2000) International trade agreements between countries of asymmetric size. J Int Econ 50:473–495

Peters H (2008) Game theory: a multi-leveled approach. Springer, Berlin
Radzik T, Driessen T (2013) On a family of values for TU-games generalizing the Shapley value. Math Soc Sci 65:105–111
Shapley LS (1953) A value for n-person games. In: Kuhn H, Tucker AW (eds) Contribution to the theory of games II. Princeton University Press, Princeton, New Jersey, pp 307–317
Slikker M (2005) Link monotonic allocation schemes. Int Game Theory Rev 7:419–429
Slikker M (2007) Bidding for surplus in network allocation problems. J Econ Theory 137:493–511
Sprumont Y (1990) Population monotonic allocation schemes for cooperative games with transferable utility. Games Econ Behav 2:378–394
Singh MT, Borkotokey S (2019) Selfish controlled scheme in opportunistic mobile communication network. In: Kalita J, Balas VE, Borah S, Pradhan R (eds) Recent developments in machine learning and data analytics. Springer Singapore. https://doi.org/10.1007/978-981-13-1280-9
van den Brink R, van der Laan G (1998) Axiomatization of the normalized Banzhaf value and the Shapley value. Soc Choice Welf 15:567–582
van den Nouweland A, Slikker M (2012) An axiomatic characterization of the position value for network situations. Math Soc Sci 64:266–271
Yavas A (1994) Middlemen in bilateral search markets. J Labor Econ 12:406–429

MIX
Papier aus verantwortungsvollen Quellen
Paper from responsible sources
FSC® C105338

If you have any concerns about our products,
you can contact us on
ProductSafety@springernature.com

In case Publisher is established outside the EU,
the EU authorized representative is:
Springer Nature Customer Service Center GmbH
Europaplatz 3, 69115 Heidelberg, Germany

Printed by Libri Plureos GmbH
in Hamburg, Germany